QM Library
D1342836

Graduate Texts in Mathematics 254

Graduate Texts in Mathematics

1 Takeuti/Zaring. Introduction to Axiomatic Set Theory. 2nd ed.
2 Oxtoby. Measure and Category. 2nd ed.
3 Schaefer. Topological Vector Spaces. 2nd ed.
4 Hilton/Stammbach. A Course in Homological Algebra. 2nd ed.
5 Mac Lane. Categories for the Working Mathematician. 2nd ed.
6 Hughes/Piper. Projective Planes.
7 J.-P. Serre. A Course in Arithmetic.
8 Takeuti/Zaring. Axiomatic Set Theory.
9 Humphreys. Introduction to Lie Algebras and Representation Theory.
10 Cohen. A Course in Simple Homotopy Theory.
11 Conway. Functions of One Complex Variable I. 2nd ed.
12 Beals. Advanced Mathematical Analysis.
13 Anderson/Fuller. Rings and Categories of Modules. 2nd ed.
14 Golubitsky/Guillemin. Stable Mappings and Their Singularities.
15 Berberian. Lectures in Functional Analysis and Operator Theory.
16 Winter. The Structure of Fields.
17 Rosenblatt. Random Processes. 2nd ed.
18 Halmos. Measure Theory.
19 Halmos. A Hilbert Space Problem Book. 2nd ed.
20 Husemoller. Fibre Bundles. 3rd ed.
21 Humphreys. Linear Algebraic Groups.
22 Barnes/Mack. An Algebraic Introduction to Mathematical Logic.
23 Greub. Linear Algebra. 4th ed.
24 Holmes. Geometric Functional Analysis and Its Applications.
25 Hewitt/Stromberg. Real and Abstract Analysis.
26 Manes. Algebraic Theories.
27 Kelley. General Topology.
28 Zariski/Samuel. Commutative Algebra. Vol.I.
29 Zariski/Samuel. Commutative Algebra. Vol.II.
30 Jacobson. Lectures in Abstract Algebra I. Basic Concepts.
31 Jacobson. Lectures in Abstract Algebra II. Linear Algebra.
32 Jacobson. Lectures in Abstract Algebra III. Theory of Fields and Galois Theory.
33 Hirsch. Differential Topology.
34 Spitzer. Principles of Random Walk. 2nd ed.
35 Alexander/Wermer. Several Complex Variables and Banach Algebras. 3rd ed.
36 Kelley/Namioka et al. Linear Topological Spaces.
37 Monk. Mathematical Logic.
38 Grauert/Fritzsche. Several Complex Variables.
39 Arveson. An Invitation to C^*-Algebras.
40 Kemeny/Snell/Knapp. Denumerable Markov Chains. 2nd ed.
41 Apostol. Modular Functions and Dirichlet Series in Number Theory. 2nd ed.
42 J.-P. Serre. Linear Representations of Finite Groups.
43 Gillman/Jerison. Rings of Continuous Functions.
44 Kendig. Elementary Algebraic Geometry.
45 Loève. Probability Theory I. 4th ed.
46 Loève. Probability Theory II. 4th ed.
47 Moise. Geometric Topology in Dimensions 2 and 3.
48 Sachs/Wu. General Relativity for Mathematicians.
49 Gruenberg/Weir. Linear Geometry. 2nd ed.
50 Edwards. Fermat's Last Theorem.
51 Klingenberg. A Course in Differential Geometry.
52 Hartshorne. Algebraic Geometry.
53 Manin. A Course in Mathematical Logic.
54 Graver/Watkins. Combinatorics with Emphasis on the Theory of Graphs.
55 Brown/Pearcy. Introduction to Operator Theory I: Elements of Functional Analysis.
56 Massey. Algebraic Topology: An Introduction.
57 Crowell/Fox. Introduction to Knot Theory.
58 Koblitz. p-adic Numbers, p-adic Analysis, and Zeta-Functions. 2nd ed.
59 Lang. Cyclotomic Fields.
60 Arnold. Mathematical Methods in Classical Mechanics. 2nd ed.
61 Whitehead. Elements of Homotopy Theory.
62 Kargapolov/Merlzjakov. Fundamentals of the Theory of Groups.
63 Bollobas. Graph Theory.
64 Edwards. Fourier Series. Vol. I. 2nd ed.
65 Wells. Differential Analysis on Complex Manifolds. 2nd ed.

(continued after index)

Henning Stichtenoth

Algebraic Function Fields and Codes

Second Edition

With 14 Figures

Henning Stichtenoth
Sabanci University
Faculty of Engineering & Natural Sciences
34956 Istanbul
Orhanli, Tuzla
Turkey
henning@sabanciuniv.edu

ISSN: 0072-5285
ISBN: 978-3-540-76877-7 e-ISBN: 978-3-540-76878-4

Library of Congress Control Number: 2008938193

Mathematics Subject Classification (2000): 12xx, 94xx, 14xx, 11xx

Cover design: deblik, Berlin

Printed on acid-free paper

9 8 7 6 5 4 3 2 1

springer.com

Preface to the Second Edition

15 years after the first printing of *Algebraic Function Fields and Codes*, the mathematics editors of Springer Verlag encouraged me to revise and extend the book.

Besides numerous minor corrections and amendments, the second edition differs from the first one in two respects. Firstly I have included a series of exercises at the end of each chapter. Some of these exercises are fairly easy and should help the reader to understand the basic concepts, others are more advanced and cover additional material. Secondly a new chapter titled "Asymptotic Bounds for the Number of Rational Places" has been added. This chapter contains a detailed presentation of the asymptotic theory of function fields over finite fields, including the explicit construction of some asymptotically good and optimal towers. Based on these towers, a complete and self-contained proof of the Tsfasman-Vladut-Zink Theorem is given. This theorem is perhaps the most beautiful application of function fields to coding theory.

The codes which are constructed from algebraic function fields were first introduced by V. D. Goppa. Accordingly I referred to them in the first edition as *geometric Goppa codes*. Since this terminology has not generally been accepted in the literature, I now use the more common term *algebraic geometry codes* or AG *codes*.

I would like to thank Alp Bassa, Arnaldo Garcia, Cem Güneri, Sevan Harput and Alev Topuzoğlu for their help in preparing the second edition. Moreover I thank all those whose results I have used in the exercises without giving references to their work.

İstanbul, September 2008 Henning Stichtenoth

Preface to the First Edition

An algebraic function field over K is an algebraic extension of finite degree over the rational function field $K(x)$ (the ground field K may be an arbitrary field). This type of field extension occurs naturally in various branches of mathematics such as algebraic geometry, number theory and the theory of compact Riemann surfaces. Hence one can study algebraic function fields from very different points of view.

In algebraic geometry one is interested in the geometric properties of an algebraic curve $C = \{(\alpha, \beta) \in K \times K \mid f(\alpha, \beta) = 0\}$, where $f(X, Y)$ is an irreducible polynomial in two variables over an algebraically closed field K. It turns out that the field $K(C)$ of rational functions on C (which is an algebraic function field over K) contains a great deal of information regarding the geometry of the curve C. This aspect of the theory of algebraic function fields is presented in several books on algebraic geometry, for instance [11], [18], [37] and [38].

One can also approach function fields from the direction of complex analysis. The meromorphic functions on a compact Riemann surface S form an algebraic function field $\mathcal{M}(S)$ over the field $\mathbb{C}$ of complex numbers. Here again, the function field is a strong tool for studying the corresponding Riemann surface, see [10] or [20].

In this book a self-contained, purely algebraic exposition of the theory of algebraic functions is given. This approach was initiated by R. Dedekind, L. Kronecker and H. M. Weber in the nineteenth century (over the field $\mathbb{C}$), cf. [20]; it was further developed by E. Artin, H. Hasse, F. K. Schmidt and A. Weil in the first half of the twentieth century. Standard references are Chevalley's book 'Introduction to the Theory of Algebraic Functions of One Variable' [6], which appeared in 1951, and [7]. The close relationship with algebraic number theory is emphasized in [1] and [9].

The algebraic approach to algebraic functions is more elementary than the approach via algebraic geometry: only some basic knowledge of algebraic field extensions, including Galois theory, is assumed. A second advantage is that some principal results of the theory (such as the Riemann-Roch Theorem) can be derived very quickly for function fields over an arbitrary constant field K. This facilitates the presentation of some applications of algebraic functions to coding theory, which is the second objective of the book.

An error-correcting code is a subspace of $\mathbb{F}_q^n$, the n-dimensional standard vector space over a finite field $\mathbb{F}_q$. Such codes are in widespread use for the reliable transmission of information. As observed by V. D. Goppa in 1975, one can use algebraic function fields over $\mathbb{F}_q$ to construct a large class of interesting codes. Properties of these codes are closely related to properties of the corresponding function field, and the Riemann-Roch Theorem provides estimates, sharp in many cases, for their main parameters (dimension, minimum distance).

While Goppa's construction is the most important, it is not the only link between codes and algebraic functions. For instance, the Hasse-Weil Theorem (which is fundamental to the theory of function fields over a finite constant field) yields results on the weights of codewords in certain trace codes.

A brief summary of the book follows.

The general theory of algebraic function fields is presented in Chapters 1, 3 and 4. In the first chapter the basic concepts are introduced, and A. Weil's proof of the Riemann-Roch Theorem is given. Chapter 3 is perhaps the most important. It provides the tools necessary for working with concrete function fields: the decomposition of places in a finite extension, ramification and different, the Hurwitz Genus Formula, and the theory of constant field extensions. P-adic completions as well as the relation between differentials and Weil differentials are treated in Chapter 4.

Chapter 5 deals with function fields over a finite constant field. This chapter contains a version of Bombieri's proof of the Hasse-Weil Theorem as well as some improvements of the Hasse-Weil Bound. As an illustration of the general theory, several explicit examples of function fields are discussed in Chapter 6, namely elliptic and hyperelliptic function fields, Kummer and Artin-Schreier extensions of the rational function field.

The Chapters 2, 8 and 9 are devoted to applications of algebraic functions to coding theory. Following a brief introduction to coding theory, Goppa's construction of codes by means of an algebraic function field is described in Chapter 2. Also included in this chapter is the relation these codes have with the important classes of BCH and classical Goppa codes. Chapter 8 contains some supplements: the residue representation of geometric Goppa codes, automorphisms of codes, asymptotic questions and the decoding of geometric Goppa codes. A detailed exposition of codes associated to the Hermitian function field is given. In the literature these codes often serve as a test for the usefulness of geometric Goppa codes.

Chapter 9 contains some results on subfield subcodes and trace codes. Estimates for their dimension are given, and the Hasse-Weil Bound is used to obtain results on the weights, dimension and minimum distance of these codes.

For the convenience of the reader, two appendices are enclosed. Appendix A is a summary of results from field theory that are frequently used in the text. As many papers on geometric Goppa codes are written in the language of algebraic geometry, Appendix B provides a kind of dictionary between the theory of algebraic functions and the theory of algebraic curves.

Acknowledgements

First of all I am indebted to P. Roquette from whom I learnt the theory of algebraic functions. His lectures, given 20 years ago at the University of Heidelberg, substantially influenced my exposition of this theory.

I thank several colleagues who carefully read the manuscript: D. Ehrhard, P. V. Kumar, J. P. Pedersen, H.-G. Rück, C. Voss and K. Yang. They suggested many improvements and helped to eliminate numerous misprints and minor mistakes in earlier versions.

Essen, March 1993 Henning Stichtenoth

Contents

1

Foundations of the Theory of Algebraic Function Fields

In this chapter we introduce the basic definitions and results of the theory of algebraic function fields: valuations, places, divisors, the genus of a function field, adeles, Weil differentials and the Riemann-Roch Theorem.

Throughout Chapter 1 *we denote by* K *an arbitrary field.*

It is only in later chapters that we will assume that K has specific properties (for example, that K is a finite field – the case which is of particular interest to coding theory).

1.1 Places

Definition 1.1.1. *An algebraic function field* F/K *of one variable over* K *is an extension field* $F \supseteq K$ *such that* F *is a finite algebraic extension of* $K(x)$ *for some element* $x \in F$ *which is transcendental over* K.

For brevity we shall simply refer to F/K as a *function field.* Obviously the set $\tilde{K} := \{z \in F \mid z \text{ is algebraic over } K\}$ is a subfield of F, since sums, products and inverses of algebraic elements are also algebraic. $\tilde{K}$ is called the *field of constants* of F/K. We have $K \subseteq \tilde{K} \subsetneqq F$, and it is clear that $F/\tilde{K}$ is a function field over $\tilde{K}$. We say that K is *algebraically closed in* F (or K is the *full constant field* of F) if $\tilde{K} = K$.

Remark 1.1.2. The elements of F which are transcendental over K can be characterized as follows: $z \in F$ is transcendental over K if and only if the extension $F/K(z)$ is of finite degree. The proof is trivial.

Example 1.1.3. The simplest example of an algebraic function field is the *rational function field*; F/K is called *rational* if $F = K(x)$ for some $x \in F$ which is transcendental over K. Each element $0 \neq z \in K(x)$ has a unique representation

H. Stichtenoth, *Algebraic Function Fields and Codes*,
Graduate Texts in Mathematics 254,

$$z = a \cdot \prod_i p_i(x)^{n_i}\,, \tag{1.1}$$

in which $0 \neq a \in K$, the polynomials $p_i(x) \in K[x]$ are monic, pairwise distinct and irreducible and $n_i \in \mathbb{Z}$.

A function field F/K is often represented as a simple algebraic field extension of a rational function field $K(x)$; i.e., $F = K(x, y)$ where $\varphi(y) = 0$ for some irreducible polynomial $\varphi(T) \in K(x)[T]$. If F/K is a non-rational function field, it is not so clear, whether every element $0 \neq z \in F$ admits a decomposition into irreducibles analogous to (1.1); indeed, it is not even clear what we mean by an irreducible element of F. Another problem which is closely related to the representation (1.1) is the following: given elements $\alpha_1, \ldots, \alpha_n \in K$, find all rational functions $f(x) \in K(x)$ with a prescribed order of zero (or pole order) at $\alpha_1, \ldots, \alpha_n$. In order to formulate these problems for arbitrary function fields properly, we introduce the notions of valuation rings and places.

Definition 1.1.4. *A valuation ring of the function field F/K is a ring $\mathcal{O} \subseteq F$ with the following properties:*

(1) $K \subsetneqq \mathcal{O} \subsetneqq F$, and

(2) for every $z \in F$ we have that $z \in \mathcal{O}$ or $z^{-1} \in \mathcal{O}$.

This definition is motivated by the following observation in the case of a rational function field $K(x)$: given an irreducible monic polynomial $p(x) \in K[x]$, we consider the set

$$\mathcal{O}_{p(x)} := \left\{ \frac{f(x)}{g(x)} \,\middle|\, f(x), g(x) \in K[x],\ p(x) \nmid g(x) \right\}.$$

It is easily verified that $\mathcal{O}_{p(x)}$ is a valuation ring of $K(x)/K$. Note that if $q(x)$ is another irreducible monic polynomial, then $\mathcal{O}_{p(x)} \neq \mathcal{O}_{q(x)}$.

Proposition 1.1.5. *Let $\mathcal{O}$ be a valuation ring of the function field F/K. Then the following hold:*

(a) $\mathcal{O}$ is a local ring; i.e., $\mathcal{O}$ has a unique maximal ideal $P = \mathcal{O} \setminus \mathcal{O}^\times$, where $\mathcal{O}^\times = \{z \in \mathcal{O} \mid$ there is an element $w \in \mathcal{O}$ with $zw = 1\}$ is the group of units of $\mathcal{O}$.

(b) Let $0 \neq x \in F$. Then $x \in P \iff x^{-1} \notin \mathcal{O}$.

(c) For the field $\tilde{K}$ of constants of F/K we have $\tilde{K} \subseteq \mathcal{O}$ and $\tilde{K} \cap P = \{0\}$.

Proof. (a) We claim that $P := \mathcal{O} \setminus \mathcal{O}^\times$ is an ideal of $\mathcal{O}$ (from this it follows at once that P is the unique maximal ideal since a proper ideal of $\mathcal{O}$ cannot contain a unit).

(1) Let $x \in P$, $z \in \mathcal{O}$. Then $xz \notin \mathcal{O}^\times$ (otherwise x would be a unit), consequently $xz \in P$.

(2) Let $x, y \in P$. W.l.o.g. we can assume that $x/y \in \mathcal{O}$. Then $1 + x/y \in \mathcal{O}$ and $x + y = y(1 + x/y) \in P$ by (1). Hence P is an ideal of $\mathcal{O}$.

(b) is obvious.

(c) Let $z \in \tilde{K}$. Assume that $z \notin \mathcal{O}$. Then $z^{-1} \in \mathcal{O}$ as $\mathcal{O}$ is a valuation ring. Since z^{-1} is algebraic over K, there are elements $a_1, \dots, a_r \in K$ with $a_r(z^{-1})^r + \dots + a_1 z^{-1} + 1 = 0$, hence $z^{-1}(a_r(z^{-1})^{r-1} + \dots + a_1) = -1$. Therefore $z = -(a_r(z^{-1})^{r-1} + \dots + a_1) \in K[z^{-1}] \subseteq \mathcal{O}$, so $z \in \mathcal{O}$. This is a contradiction to the assumption $z \notin \mathcal{O}$. Hence we have shown that $\tilde{K} \subseteq \mathcal{O}$. The assertion $\tilde{K} \cap P = \{0\}$ is trivial. □

Theorem 1.1.6. *Let $\mathcal{O}$ be a valuation ring of the function field F/K and let P be its unique maximal ideal. Then the following hold:*

(a) P is a principal ideal.

(b) If $P = t\mathcal{O}$ then each $0 \neq z \in F$ has a unique representation of the form $z = t^n u$ for some $n \in \mathbb{Z}$ and $u \in \mathcal{O}^\times$.

(c) $\mathcal{O}$ is a principal ideal domain. More precisely, if $P = t\mathcal{O}$ and $\{0\} \neq I \subseteq \mathcal{O}$ is an ideal then $I = t^n\mathcal{O}$ for some $n \in \mathbb{N}$.

A ring having the above properties is called a *discrete valuation ring*. The proof of Theorem 1.1.6 depends essentially on the following lemma.

Lemma 1.1.7. *Let $\mathcal{O}$ be a valuation ring of the algebraic function field F/K, let P be its maximal ideal and $0 \neq x \in P$. Let $x_1, \dots, x_n \in P$ be such that $x_1 = x$ and $x_i \in x_{i+1}P$ for $i = 1, \dots, n-1$. Then we have*

$$n \leq [F : K(x)] < \infty \, .$$

Proof. From Remark 1.1.2 and Proposition 1.1.5(c) follows that $F/K(x)$ is a finite extension, so it is sufficient to prove that $x_1, \dots, x_n$ are linearly independent over $K(x)$. Suppose there is a non-trivial linear combination $\sum_{i=1}^n \varphi_i(x)x_i = 0$ with $\varphi_i(x) \in K(x)$. We can assume that all $\varphi_i(x)$ are polynomials in x and that x does not divide all of them. Put $a_i := \varphi_i(0)$, the constant term of $\varphi_i(x)$, and define $j \in \{1, \dots, n\}$ by the condition $a_j \neq 0$ but $a_i = 0$ for all $i > j$. We obtain

$$-\varphi_j(x)x_j = \sum_{i \neq j} \varphi_i(x)x_i \tag{1.2}$$

with $\varphi_i(x) \in \mathcal{O}$ for $i = 1, \dots, n$ (since $x = x_1 \in P$), $x_i \in x_j P$ for $i < j$ and $\varphi_i(x) = xg_i(x)$ for $i > j$, where $g_i(x)$ is a polynomial in x. Dividing (1.2) by x_j yields

$$-\varphi_j(x) = \sum_{i<j} \varphi_i(x)\frac{x_i}{x_j} + \sum_{i>j} \frac{x}{x_j} g_i(x)x_i \, .$$

All summands on the right hand side belong to P, therefore $\varphi_j(x) \in P$. On the other hand, $\varphi_j(x) = a_j + xg_j(x)$ with $g_j(x) \in K[x] \subseteq \mathcal{O}$ and $x \in P$, so that $a_j = \varphi_j(x) - xg_j(x) \in P \cap K$. Since $a_j \neq 0$, this contradicts Proposition 1.1.5(c). □

Proof of Theorem 1.1.6. (a) Assume that P is not principal, and choose an element $0 \neq x_1 \in P$. As $P \neq x_1\mathcal{O}$ there is $x_2 \in P \setminus x_1\mathcal{O}$. Then $x_2x_1^{-1} \notin \mathcal{O}$, thereby $x_2^{-1}x_1 \in P$ by Proposition 1.1.5(b), so $x_1 \in x_2P$. By induction one obtains an infinite sequence $x_1, x_2, x_3, \ldots$ in P such that $x_i \in x_{i+1}P$ for all $i \geq 1$, a contradiction to Lemma 1.1.7.

(b) The uniqueness of the representation $z = t^n u$ with $u \in \mathcal{O}^\times$ is trivial, so we only need to show the existence. As z or z^{-1} is in $\mathcal{O}$ we can assume that $z \in \mathcal{O}$. If $z \in \mathcal{O}^\times$ then $z = t^0 z$. It remains to consider the case $z \in P$. There is a maximal $m \geq 1$ with $z \in t^m\mathcal{O}$, since the length of a sequence

$$x_1 = z, \quad x_2 = t^{m-1}, \quad x_3 = t^{m-2}, \quad \ldots, \quad x_m = t$$

is bounded by Lemma 1.1.7. Write $z = t^m u$ with $u \in \mathcal{O}$. Then u must be a unit of $\mathcal{O}$ (otherwise $u \in P = t\mathcal{O}$, so $u = tw$ with $w \in \mathcal{O}$ and $z = t^{m+1}w \in t^{m+1}\mathcal{O}$, a contradiction to the maximality of m).

(c) Let $\{0\} \neq I \subseteq \mathcal{O}$ be an ideal. The set $A := \{r \in \mathbb{N} \mid t^r \in I\}$ is non-empty (in fact, if $0 \neq x \in I$ then $x = t^r u$ with $u \in \mathcal{O}^\times$ and therefore $t^r = xu^{-1} \in I$). Put $n := \min(A)$. We claim that $I = t^n\mathcal{O}$. The inclusion $I \supseteq t^n\mathcal{O}$ is trivial since $t^n \in I$. Conversely let $0 \neq y \in I$. We have $y = t^s w$ with $w \in \mathcal{O}^\times$ and $s \geq 0$, so $t^s \in I$ and $s \geq n$. It follows that $y = t^n \cdot t^{s-n}w \in t^n\mathcal{O}$. □

Definition 1.1.8. *(a) A place P of the function field F/K is the maximal ideal of some valuation ring $\mathcal{O}$ of F/K. Every element $t \in P$ such that $P = t\mathcal{O}$ is called a prime element for P (other notations are local parameter or uniformizing variable).*
(b) $\mathbb{P}_F := \{P \mid P$ is a place of $F/K\}$.

If $\mathcal{O}$ is a valuation ring of F/K and P is its maximal ideal, then $\mathcal{O}$ is uniquely determined by P, namely $\mathcal{O} = \{z \in F \mid z^{-1} \notin P\}$, cf. Proposition 1.1.5(b). Hence $\mathcal{O}_P := \mathcal{O}$ is called *the valuation ring of the place P*.

A second useful description of places is given in terms of valuations.

Definition 1.1.9. *A discrete valuation of F/K is a function $v : F \to \mathbb{Z} \cup \{\infty\}$ with the following properties :*

(1) $v(x) = \infty \iff x = 0$.

(2) $v(xy) = v(x) + v(y)$ for all $x, y \in F$.

(3) $v(x+y) \geq \min\{v(x), v(y)\}$ for all $x, y \in F$.

(4) There exists an element $z \in F$ with $v(z) = 1$.

(5) $v(a) = 0$ for all $0 \neq a \in K$.

In this context the symbol ∞ means some element not in $\mathbb{Z}$ such that $\infty + \infty = \infty + n = n + \infty = \infty$ and $\infty > m$ for all $m, n \in \mathbb{Z}$. From (2) and (4) it follows immediately that $v : F \to \mathbb{Z} \cup \{\infty\}$ is surjective. Property (3) is called the *Triangle Inequality*. The notions *valuation* and *triangle inequality* are justified by the following remark:

Remark 1.1.10. Let v be a discrete valuation of F/K in the sense of Definition 1.1.9. Fix a real number $0 < c < 1$ and define the function $|\ |_v : F \to \mathbb{R}$ by

$$|z|_v := \begin{cases} c^{v(z)} & \text{if } z \neq 0\,, \\ 0 & \text{if } z = 0\,. \end{cases}$$

It is easily verified that this function has the properties of an ordinary absolute value; the ordinary Triangle Inequality $|x+y|_v \leq |x|_v + |y|_v$ turns out to be an immediate consequence of condition (3) of Definition 1.1.9.

A stronger version of the Triangle Inequality can be derived from the axioms and is often very useful:

Lemma 1.1.11 (Strict Triangle Inequality). *Let v be a discrete valuation of F/K and let $x, y \in F$ with $v(x) \neq v(y)$. Then $v(x+y) = \min\{v(x), v(y)\}$.*

Proof. Observe that $v(ay) = v(y)$ for $0 \neq a \in K$ (by (2) and (5)), in particular $v(-y) = v(y)$. Since $v(x) \neq v(y)$ we can assume $v(x) < v(y)$. Suppose that $v(x+y) \neq \min\{v(x), v(y)\}$, so $v(x+y) > v(x)$ by (3). Then we obtain $v(x) = v((x+y) - y) \geq \min\{v(x+y), v(y)\} > v(x)$, a contradiction. □

Definition 1.1.12. *To a place $P \in \mathbb{P}_F$ we associate a function $v_P : F \to \mathbb{Z} \cup \{\infty\}$ (that will prove to be a discrete valuation of F/K) as follows: Choose a prime element t for P. Then every $0 \neq z \in F$ has a unique representation $z = t^n u$ with $u \in \mathcal{O}_P^\times$ and $n \in \mathbb{Z}$. Define $v_P(z) := n$ and $v_P(0) := \infty$.*

Observe that this definition depends only on P, not on the choice of t. In fact, if t' is another prime element for P then $P = t\mathcal{O} = t'\mathcal{O}$, so $t = t'w$ for some $w \in \mathcal{O}_P^\times$. Therefore $t^n u = (t'^n w^n) u = t'^n (w^n u)$ with $w^n u \in \mathcal{O}_P^\times$.

Theorem 1.1.13. *Let F/K be a function field.*

(a) For a place $P \in \mathbb{P}_F$, the function v_P defined above is a discrete valuation of F/K. Moreover we have

$$\begin{aligned} \mathcal{O}_P &= \{z \in F \mid v_P(z) \geq 0\}\,, \\ \mathcal{O}_P^\times &= \{z \in F \mid v_P(z) = 0\}\,, \\ P &= \{z \in F \mid v_P(z) > 0\}\,. \end{aligned}$$

(b) An element $x \in F$ is a prime element for P if and only if $v_P(x) = 1$.

(c) Conversely, suppose that v is a discrete valuation of F/K. Then the set $P := \{z \in F \mid v(z) > 0\}$ is a place of F/K, and $\mathcal{O}_P = \{z \in F \mid v(z) \geq 0\}$ is the corresponding valuation ring.

(d) Every valuation ring $\mathcal{O}$ of F/K is a maximal proper subring of F.

Proof. (a) Obviously v_P has the properties (1), (2), (4) and (5) of Definition 1.1.9. In order to prove the Triangle Inequality (3) consider $x, y \in F$ with $v_P(x) = n, v_P(y) = m$. We can assume that $n \leq m < \infty$, thus $x = t^n u_1$ and $y = t^m u_2$ with $u_1, u_2 \in \mathcal{O}_P^\times$. Then $x + y = t^n(u_1 + t^{m-n}u_2) = t^n z$ with $z \in \mathcal{O}_P$. If $z = 0$ we have $v_P(x+y) = \infty > \min\{n, m\}$, otherwise $z = t^k u$ with $k \geq 0$ and $u \in \mathcal{O}_P^\times$. Therefore

$$v_P(x+y) = v_P(t^{n+k}u) = n + k \geq n = \min\{v_P(x), v_P(y)\}.$$

We have shown that v_P is a discrete valuation of F/K. The remaining assertions of (a) are trivial, likewise (b) and (c).

(d) Let $\mathcal{O}$ be a valuation ring of F/K, P its maximal ideal, v_P the discrete valuation associated to P and $z \in F \setminus \mathcal{O}$. We have to show that $F = \mathcal{O}[z]$. To this end consider an arbitrary element $y \in F$; then $v_P(yz^{-k}) \geq 0$ for sufficiently large $k \geq 0$ (note that $v_P(z^{-1}) > 0$ since $z \notin \mathcal{O}$). Consequently $w := yz^{-k} \in \mathcal{O}$ and $y = wz^k \in \mathcal{O}[z]$. □

According to Theorem 1.1.13 places, valuation rings and discrete valuations of a function field essentially amount to the same thing.

Let P be a place of F/K and let $\mathcal{O}_P$ be its valuation ring. Since P is a maximal ideal, the residue class ring $\mathcal{O}_P/P$ is a field. For $x \in \mathcal{O}_P$ we define $x(P) \in \mathcal{O}_P/P$ to be the residue class of x modulo P, for $x \in F \setminus \mathcal{O}_P$ we put $x(P) := \infty$ (note that the symbol ∞ is used here in a different sense as in Definition 1.1.9). By Proposition 1.1.5 we know that $K \subseteq \mathcal{O}_P$ and $K \cap P = \{0\}$, so the residue class map $\mathcal{O}_P \to \mathcal{O}_P/P$ induces a canonical embedding of K into $\mathcal{O}_P/P$. Henceforth we shall always consider K as a subfield of $\mathcal{O}_P/P$ via this embedding. Observe that this argument also applies to $\tilde{K}$ instead of K; so we can consider $\tilde{K}$ as a subfield of $\mathcal{O}_P/P$ as well.

Definition 1.1.14. *Let $P \in \mathbb{P}_F$.*

(a) $F_P := \mathcal{O}_P/P$ is the residue class field of P. The map $x \mapsto x(P)$ from F to $F_P \cup \{\infty\}$ is called the residue class map with respect to P. Sometimes we shall also use the notation $x + P := x(P)$ for $x \in \mathcal{O}_P$.

(b) $\deg P := [F_P : K]$ is called the degree of P. A place of degree one is also called a rational place of F/K.

The degree of a place is always finite; more precisely the following holds.

Proposition 1.1.15. *If P is a place of F/K and $0 \neq x \in P$ then*

$$\deg P \leq [F : K(x)] < \infty.$$

Proof. First we observe that $[F : K(x)] < \infty$ by Remark 1.1.2. Thus it suffices to show that any elements $z_1, \ldots, z_n \in \mathcal{O}_P$, whose residue classes $z_1(P), \ldots, z_n(P) \in F_P$ are linearly independent over K, are linearly independent over $K(x)$. Suppose there is a non-trivial linear combination

$$\sum_{i=1}^{n} \varphi_i(x) z_i = 0 \tag{1.3}$$

with $\varphi_i(x) \in K(x)$. W.l.o.g. we assume that the $\varphi_i(x)$ are polynomials in x and not all of them are divisible by x; i.e., $\varphi_i(x) = a_i + x g_i(x)$ with $a_i \in K$ and $g_i(x) \in K[x]$, not all $a_i = 0$. Since $x \in P$ and $g_i(x) \in \mathcal{O}_P$, $\varphi_i(x)(P) = a_i(P) = a_i$. Applying the residue class map to (1.3) we obtain

$$0 = 0(P) = \sum_{i=1}^{n} \varphi_i(x)(P) z_i(P) = \sum_{i=1}^{n} a_i z_i(P)\,.$$

This contradicts the linear independence of $z_1(P), \ldots, z_n(P)$ over K. □

Corollary 1.1.16. *The field $\tilde{K}$ of constants of F/K is a finite field extension of K.*

Proof. We use the fact that $\mathbb{P}_F \neq \emptyset$ (which will be proved only in Corollary 1.1.20). Choose some $P \in \mathbb{P}_F$. Since $\tilde{K}$ is embedded into F_P via the residue class map $\mathcal{O}_P \to F_P$, it follows that $[\tilde{K} : K] \leq [F_P : K] < \infty$. □

Remark 1.1.17. Let P be a rational place of F/K; i.e., $\deg P = 1$. Then we have $F_P = K$, and the residue class map maps F to $K \cup \{\infty\}$. In particular, if K is an algebraically closed field, then all places are rational and we can read an element $z \in F$ as a function

$$z : \begin{cases} \mathbb{P}_F & \longrightarrow & K \cup \{\infty\}\,, \\ P & \longmapsto & z(P)\,. \end{cases} \tag{1.4}$$

This is why F/K is called a *function field.* The elements of K, interpreted as functions in the sense of (1.4), are constant functions. For this reason K is called the *constant field* of F. Also the following terminology is justified by (1.4):

Definition 1.1.18. *Let $z \in F$ and $P \in \mathbb{P}_F$. We say that P is a zero of z if $v_P(z) > 0$; P is a pole of z if $v_P(z) < 0$. If $v_P(z) = m > 0$, P is a zero of z of order m; if $v_P(z) = -m < 0$, P is a pole of z of order m.*

Next we shall be concerned with the question as to whether there exist places of F/K.

Theorem 1.1.19. *Let F/K be a function field and let R be a subring of F with $K \subseteq R \subseteq F$. Suppose that $\{0\} \neq I \subsetneqq R$ is a proper ideal of R. Then there is a place $P \in \mathbb{P}_F$ such that $I \subseteq P$ and $R \subseteq \mathcal{O}_P$.*

Proof. Consider the set

$$\mathcal{F} := \{\, S \,|\, S \text{ is a subring of } F \text{ with } R \subseteq S \text{ and } IS \neq S \,\}\,.$$

(IS is by definition the set of all finite sums $\sum a_\nu s_\nu$ with $a_\nu \in I, s_\nu \in S$; it is an ideal of S). $\mathcal{F}$ is non-empty as $R \in \mathcal{F}$, and $\mathcal{F}$ is inductively ordered by inclusion. In fact, if $\mathcal{H} \subseteq \mathcal{F}$ is a totally ordered subset of $\mathcal{F}$ then $T := \bigcup\{S \mid S \in \mathcal{H}\}$ is a subring of F with $R \subseteq T$. We have to verify that $IT \neq T$. Suppose this is false, then $1 = \sum_{\nu=1}^{n} a_\nu s_\nu$ with $a_\nu \in I, s_\nu \in T$. Since $\mathcal{H}$ is totally ordered there is an $S_0 \in \mathcal{H}$ such that $s_1, \ldots, s_n \in S_0$, so $1 = \sum_{\nu=1}^{n} a_\nu s_\nu \in IS_0$, a contradiction.

By Zorn's lemma $\mathcal{F}$ contains a maximal element; i.e., there is a ring $\mathcal{O} \subseteq F$ such that $R \subseteq \mathcal{O} \subseteq F$, $I\mathcal{O} \neq \mathcal{O}$, and $\mathcal{O}$ is maximal with respect to these properties. We want to show that $\mathcal{O}$ is a valuation ring of F/K.

As $I \neq \{0\}$ and $I\mathcal{O} \neq \mathcal{O}$ we have $\mathcal{O} \subsetneqq F$ and $I \subseteq \mathcal{O} \setminus \mathcal{O}^\times$. Suppose there exists an element $z \in F$ with $z \notin \mathcal{O}$ and $z^{-1} \notin \mathcal{O}$. Then $I\mathcal{O}[z] = \mathcal{O}[z]$ and $I\mathcal{O}[z^{-1}] = \mathcal{O}[z^{-1}]$, and we can find $a_0, \ldots, a_n, b_0, \ldots, b_m \in I\mathcal{O}$ with

$$1 = a_0 + a_1 z + \cdots + a_n z^n \quad \textit{and} \tag{1.5}$$
$$1 = b_0 + b_1 z^{-1} + \cdots + b_m z^{-m}\,. \tag{1.6}$$

Clearly $n \geq 1$ and $m \geq 1$. We can assume that m, n in (1.5) and (1.6) are chosen minimally and $m \leq n$. We multiply (1.5) by $1 - b_0$ and (1.6) by $a_n z^n$ and obtain

$$1 - b_0 = (1 - b_0)a_0 + (1 - b_0)a_1 z + \cdots + (1 - b_0)a_n z^n \quad \textit{and}$$
$$0 = (b_0 - 1)a_n z^n + b_1 a_n z^{n-1} + \cdots + b_m a_n z^{n-m}\,.$$

Adding these equations yields $1 = c_0 + c_1 z + \cdots + c_{n-1} z^{n-1}$ with coefficients $c_i \in I\mathcal{O}$. This is a contradiction to the minimality of n in (1.5). Thus we have proved that $z \in \mathcal{O}$ or $z^{-1} \in \mathcal{O}$ for all $z \in F$, hence $\mathcal{O}$ is a valuation ring of F/K. □

Corollary 1.1.20. *Let F/K be a function field, $z \in F$ transcendental over K. Then z has at least one zero and one pole. In particular $\mathbb{P}_F \neq \emptyset$.*

Proof. Consider the ring $R = K[z]$ and the ideal $I = zK[z]$. Theorem 1.1.19 ensures that there is a place $P \in \mathbb{P}_F$ with $z \in P$, hence P is a zero of z. The same argument proves that z^{-1} has a zero $Q \in \mathbb{P}_F$. Then Q is a pole of z. □

Corollary 1.1.20 can be interpreted as follows: each $z \in F$, which is not in the constant field $\tilde{K}$ of F/K, yields a non-constant function in the sense of Remark 1.1.17.

1.2 The Rational Function Field

For a thorough understanding of valuations and places in arbitrary function fields, a precise idea of these notions in the simplest case is indispensable. For this reason we investigate what these concepts mean in the case of the rational function field $F = K(x)$, where x is transcendental over K. Given an irreducible monic polynomial $p(x) \in K[x]$ we consider the valuation ring

$$\mathcal{O}_{p(x)} := \left\{ \frac{f(x)}{g(x)} \;\middle|\; f(x), g(x) \in K[x],\ p(x) \nmid g(x) \right\} \tag{1.7}$$

of $K(x)/K$ with maximal ideal

$$P_{p(x)} = \left\{ \frac{f(x)}{g(x)} \;\middle|\; f(x), g(x) \in K[x],\ p(x) | f(x),\ p(x) \nmid g(x) \right\}. \tag{1.8}$$

In the particular case when $p(x)$ is linear, i.e. $p(x) = x - \alpha$ with $\alpha \in K$, we abbreviate and write

$$P_\alpha := P_{x-\alpha} \in \mathbb{P}_{K(x)}. \tag{1.9}$$

There is another valuation ring of $K(x)/K$, namely

$$\mathcal{O}_\infty := \left\{ \frac{f(x)}{g(x)} \;\middle|\; f(x), g(x) \in K[x],\ \deg f(x) \le \deg g(x) \right\} \tag{1.10}$$

with maximal ideal

$$P_\infty = \left\{ \frac{f(x)}{g(x)} \;\middle|\; f(x), g(x) \in K[x],\ \deg f(x) < \deg g(x) \right\}. \tag{1.11}$$

This place is called the *infinite* place of $K(x)$. Observe that these labels depend on the specific choice of the generating element x of $K(x)/K$ (for example $K(x) = K(1/x)$, and the infinite place with respect to $1/x$ is the place P_0 with respect to x).

Proposition 1.2.1. *Let $F = K(x)$ be the rational function field.*
(a) Let $P = P_{p(x)} \in \mathbb{P}_{K(x)}$ be the place defined by (1.8), where $p(x) \in K[x]$ is an irreducible polynomial. Then $p(x)$ is a prime element for P, and the corresponding valuation v_P can be described as follows: if $z \in K(x) \setminus \{0\}$ is written in the form $z = p(x)^n \cdot (f(x)/g(x))$ with $n \in \mathbb{Z}$, $f(x), g(x) \in K[x]$, $p(x) \nmid f(x)$ and $p(x) \nmid g(x)$, then $v_P(z) = n$. The residue class field $K(x)_P = \mathcal{O}_P/P$ is isomorphic to $K[x]/(p(x))$; an isomorphism is given by

$$\phi : \begin{cases} K[x]/(p(x)) & \longrightarrow & K(x)_P\,, \\ f(x) \bmod p(x) & \longmapsto & f(x)(P)\,. \end{cases}$$

Consequently $\deg P = \deg p(x)$.
(b) In the special case $p(x) = x - \alpha$ with $\alpha \in K$ the degree of $P = P_\alpha$ is one, and the residue class map is given by

$$z(P) = z(\alpha) \quad \textit{for } z \in K(x)\,,$$

where $z(\alpha)$ is defined as follows: write $z = f(x)/g(x)$ with relatively prime polynomials $f(x), g(x) \in K[x]$. Then

$$z(\alpha) = \begin{cases} f(\alpha)/g(\alpha) & \text{if } g(\alpha) \neq 0\,, \\ \infty & \text{if } g(\alpha) = 0\,. \end{cases}$$

(c) Finally, let $P = P_\infty$ be the infinite place of $K(x)/K$ defined by (1.11). Then $\deg P_\infty = 1$. A prime element for P_∞ is $t = 1/x$. The corresponding discrete valuation v_∞ is given by

$$v_\infty(f(x)/g(x)) = \deg g(x) - \deg f(x)\,,$$

where $f(x), g(x) \in K[x]$. The residue class map corresponding to P_∞ is determined by $z(P_\infty) = z(\infty)$ for $z \in K(x)$, where $z(\infty)$ is defined as usual: if

$$z = \frac{a_n x^n + \cdots + a_0}{b_m x^m + \cdots + b_0} \quad \text{with } a_n, b_m \neq 0\,,$$

then

$$z(\infty) = \begin{cases} a_n/b_m & \text{if } n = m\,, \\ 0 & \text{if } n < m\,, \\ \infty & \text{if } n > m\,. \end{cases}$$

(d) K is the full constant field of $K(x)/K$.

Proof. We prove only some essentials of this proposition; the remaining parts of the proof are straightforward.

(a) Let $P = P_{p(x)}$, $p(x) \in K[x]$ irreducible. The ideal $P_{p(x)} \subseteq \mathcal{O}_{p(x)}$ is obviously generated by $p(x)$, hence $p(x)$ is a prime element for P. In order to prove the assertion about the residue class field we consider the ring homomorphism

$$\varphi : \begin{cases} K[x] & \longrightarrow & K(x)_P\,, \\ f(x) & \longmapsto & f(x)(P)\,. \end{cases}$$

Clearly the kernel of φ is the ideal generated by $p(x)$. Moreover φ is surjective: if $z \in \mathcal{O}_{p(x)}$, we can write $z = u(x)/v(x)$ with $u(x), v(x) \in K[x]$ such that $p(x) \nmid v(x)$. Thus there are $a(x), b(x) \in K[x]$ with $a(x)p(x) + b(x)v(x) = 1$, therefore

$$z = 1 \cdot z = \frac{a(x)u(x)}{v(x)} p(x) + b(x)u(x)\,,$$

and $z(P) = (b(x)u(x))(P)$ is in the image of φ. Thus φ induces an isomorphism ϕ of $K[x]/(p(x))$ onto $K(x)_P$.

(b) Now $P = P_\alpha$ with $\alpha \in K$. If $f(x) \in K[x]$ then $(x-\alpha)|(f(x)-f(\alpha))$, hence $f(x)(P) = (f(x)-f(\alpha))(P) + f(\alpha)(P) = f(\alpha)$. An arbitrary element $z \in \mathcal{O}_P$ can be written as $z = f(x)/g(x)$ with polynomials $f(x), g(x) \in K[x]$ and $(x-\alpha)\nmid g(x)$, therefore $g(x)(P) = g(\alpha) \neq 0$ and

$$z(P) = \frac{f(x)(P)}{g(x)(P)} = \frac{f(\alpha)}{g(\alpha)} = z(\alpha)\,.$$

(c) We will only show that $1/x$ is a prime element for P_∞. Clearly we have that $1/x \in P$. Consider some element $z = f(x)/g(x) \in P_\infty$; i.e., $\deg f < \deg g$. Then

$$z = \frac{1}{x} \cdot \frac{xf}{g}\ ,\ \textit{with}\ \deg(xf) \leq \deg g\,.$$

This proves that $z \in (1/x)\mathcal{O}_\infty$, hence $1/x$ generates the ideal P_∞ and it is therefore a P_∞-prime element.

(d) Choose a place P of $K(x)/K$ of degree one (e.g. $P = P_\alpha$ with $\alpha \in K$). The field $\tilde{K}$ of constants of $K(x)$ is embedded into the residue class field $K(x)_P$, hence $K \subseteq \tilde{K} \subseteq K(x)_P = K$. □

Theorem 1.2.2. *There are no places of the rational function field $K(x)/K$ other than the places $P_{p(x)}$ and P_∞, defined by* (1.8) *and* (1.11).

Corollary 1.2.3. *The places of $K(x)/K$ of degree one are in 1–1 – correspondence with $K \cup \{\infty\}$.*

The corollary is obvious by Proposition 1.2.1 and Theorem 1.2.2. In terms of algebraic geometry (cf. Appendix B) $K \cup \{\infty\}$ is usually interpreted as the *projective line* $\mathbf{P}^1(K)$ over K, hence the places of $K(x)/K$ of degree one correspond in a one-to-one way with the points of $\mathbf{P}^1(K)$.

Proof of Theorem 1.2.2. Let P be a place of $K(x)/K$. We distinguish two cases as follows:

Case 1. Assume that $x \in \mathcal{O}_P$. Then $K[x] \subseteq \mathcal{O}_P$. Set $I := K[x] \cap P$; this is an ideal of $K[x]$, in fact a prime ideal. The residue class map induces an embedding $K[x]/I \hookrightarrow K(x)_P$, consequently $I \neq \{0\}$ by Proposition 1.1.15. It follows that there is a (uniquely determined) irreducible monic polynomial $p(x) \in K[x]$ such that $I = K[x] \cap P = p(x) \cdot K[x]$. Every $g(x) \in K[x]$ with $p(x)\nmid g(x)$ is not in I , so $g(x) \notin P$ and $1/g(x) \in \mathcal{O}_P$ by Proposition 1.1.5. We conclude that

$$\mathcal{O}_{p(x)} = \left\{ \frac{f(x)}{g(x)} \,\Big|\, f(x), g(x) \in K[x],\ p(x)\nmid g(x) \right\} \subseteq \mathcal{O}_P\,.$$

As valuation rings are maximal proper subrings of $K(x)$, cf. Theorem 1.1.13, we see that $\mathcal{O}_P = \mathcal{O}_{p(x)}$.

Case 2. Now $x \notin \mathcal{O}_P$. We conclude that $K[x^{-1}] \subseteq \mathcal{O}_P$, $x^{-1} \in P \cap K[x^{-1}]$ and $P \cap K[x^{-1}] = x^{-1}K[x^{-1}]$. As in case 1,

$$\begin{aligned}
\mathcal{O}_P &\supseteq \left\{ \frac{f(x^{-1})}{g(x^{-1})} \;\middle|\; f(x^{-1}), g(x^{-1}) \in K[x^{-1}],\ x^{-1} \nmid g(x^{-1}) \right\} \\
&= \left\{ \frac{a_0 + a_1 x^{-1} + \cdots + a_n x^{-n}}{b_0 + b_1 x^{-1} + \cdots + b_m x^{-m}} \;\middle|\; b_0 \neq 0 \right\} \\
&= \left\{ \frac{a_0 x^{m+n} + \cdots + a_n x^m}{b_0 x^{m+n} + \cdots + b_m x^n} \;\middle|\; b_0 \neq 0 \right\} \\
&= \left\{ \frac{u(x)}{v(x)} \;\middle|\; u(x), v(x) \in K[x],\ \deg u(x) \leq \deg v(x) \right\} \\
&= \mathcal{O}_\infty .
\end{aligned}$$

Thus $\mathcal{O}_P = \mathcal{O}_\infty$ and $P = P_\infty$. □

1.3 Independence of Valuations

The main result of this section is the *Weak Approximation Theorem* 1.3.1 (which is also referred to as the *Theorem of Independence*). Essentially this says the following: If $v_1, \ldots, v_n$ are pairwise distinct discrete valuations of F/K and $z \in F$, and if we know the values $v_1(z), \ldots, v_{n-1}(z)$, then we cannot conclude anything about $v_n(z)$. A substantial improvement of Theorem 1.3.1 will be given later in Section 1.6.

Theorem 1.3.1 (Weak Approximation Theorem). *Let F/K be a function field, $P_1, \ldots, P_n \in \mathbb{P}_F$ pairwise distinct places of F/K, $x_1, \ldots, x_n \in F$ and $r_1, \ldots, r_n \in \mathbb{Z}$. Then there is some $x \in F$ such that*

$$v_{P_i}(x - x_i) = r_i \quad \textit{for } i = 1, \ldots, n .$$

Corollary 1.3.2. *Every function field has infinitely many places.*

Proof of Corollary 1.3.2. Suppose there are only finitely many places, say $P_1, \ldots, P_n$. By Theorem 1.3.1 we find a non-zero element $x \in F$ with $v_{P_i}(x) > 0$ for $i = 1, \ldots, n$. Then x is transcendental over K since it has zeros. But x has no pole; this is a contradiction to Corollary 1.1.20. □

Proof of Theorem 1.3.1. The proof is somewhat technical and therefore divided into several steps. For simplicity we write v_i instead of v_{P_i}.

Step 1. There is some $u \in F$ with $v_1(u) > 0$ and $v_i(u) < 0$ for $i = 2, \ldots, n$.
Proof of Step 1. By induction. For $n = 2$ we observe that $\mathcal{O}_{P_1} \not\subseteq \mathcal{O}_{P_2}$ and vice versa, since valuation rings are maximal proper subrings of F, cf. Theorem 1.1.13. Therefore we can find $y_1 \in \mathcal{O}_{P_1} \setminus \mathcal{O}_{P_2}$ and $y_2 \in \mathcal{O}_{P_2} \setminus \mathcal{O}_{P_1}$. Then $v_1(y_1) \geq 0$, $v_2(y_1) < 0$, $v_1(y_2) < 0$ and $v_2(y_2) \geq 0$. The element $u := y_1/y_2$ has the property $v_1(u) > 0$, $v_2(u) < 0$ as desired.

For $n > 2$ we have by induction hypothesis an element y with $v_1(y) > 0$, $v_2(y) < 0, \ldots, v_{n-1}(y) < 0$. If $v_n(y) < 0$ the proof is finished. In case $v_n(y) \geq 0$

we choose z with $v_1(z) > 0, v_n(z) < 0$ and put $u := y + z^r$. Here $r \geq 1$ is chosen in such a manner that $r \cdot v_i(z) \neq v_i(y)$ for $i = 1, \ldots, n-1$ (this is obviously possible). It follows that $v_1(u) \geq \min\{v_1(y), r \cdot v_1(z)\} > 0$ and $v_i(u) = \min\{v_i(y), r \cdot v_i(z)\} < 0$ for $i = 2, \ldots, n$ (observe that the Strict Triangle Inequality applies).

Step 2. There is some $w \in F$ such that $v_1(w-1) > r_1$ and $v_i(w) > r_i$ for $i = 2, \ldots, n$.

Proof of Step 2. Choose u as in Step 1 and put $w := (1+u^s)^{-1}$. We have, for sufficiently large $s \in \mathbb{N}$, $v_1(w-1) = v_1(-u^s(1+u^s)^{-1}) = s \cdot v_1(u) > r_1$, and $v_i(w) = -v_i(1+u^s) = -s \cdot v_i(u) > r_i$ for $i = 2, \ldots, n$.

Step 3. Given $y_1, \ldots, y_n \in F$, there is an element $z \in F$ with $v_i(z - y_i) > r_i$ for $i = 1, \ldots, n$.

Proof of Step 3. Choose $s \in \mathbb{Z}$ such that $v_i(y_j) \geq s$ for all $i, j \in \{1, \ldots, n\}$. By Step 2 there are $w_1, \ldots, w_n$ with

$$v_i(w_i - 1) > r_i - s \quad \textit{and} \quad v_i(w_j) > r_i - s \quad \text{for } j \neq i\,.$$

Then $z := \sum_{j=1}^{n} y_j w_j$ has the desired properties.

Now we are in a position to finish the proof of Theorem 1.3.1. By Step 3 we can find $z \in F$ with $v_i(z - x_i) > r_i$, $i = 1, \ldots, n$. Next we choose z_i with $v_i(z_i) = r_i$ (this is trivially done). Again by Step 3 there is z' with $v_i(z' - z_i) > r_i$ for $i = 1, \ldots, n$. It follows that

$$v_i(z') = v_i((z' - z_i) + z_i) = \min\{v_i(z' - z_i), v_i(z_i)\} = r_i\,.$$

Let $x := z + z'$. Then

$$v_i(x - x_i) = v_i((z - x_i) + z') = \min\{v_i(z - x_i), v_i(z')\} = r_i\,.$$

□

In Section 1.4 we shall show that an element $x \in F$ which is transcendental over K has as many zeros as poles (counted properly). An important step towards that result is our next proposition which sharpens both of Lemma 1.1.7 and Proposition 1.1.15. The Weak Approximation Theorem will play a significant role in the proof.

Proposition 1.3.3. *Let F/K be a function field and let $P_1, \ldots, P_r$ be zeros of the element $x \in F$. Then*

$$\sum_{i=1}^{r} v_{P_i}(x) \cdot \deg P_i \;\leq\; [F : K(x)]\,.$$

Proof. We set $v_i := v_{P_i}$, $f_i := \deg P_i$ and $e_i := v_i(x)$. For all i there is an element t_i with

$$v_i(t_i) = 1 \;\; \textit{and} \;\; v_k(t_i) = 0 \;\; \textit{for } k \neq i\,.$$

Next we choose $s_{i1},\ldots,s_{if_i} \in \mathcal{O}_{P_i}$ such that $s_{i1}(P_i),\ldots,s_{if_i}(P_i)$ form a basis of the residue class field F_{P_i} over K. By a weak application of Theorem 1.3.1 we can find $z_{ij} \in F$ such that the following holds for all i,j :

$$v_i(s_{ij}-z_{ij})>0 \quad and \quad v_k(z_{ij}) \geq e_k \quad for\ k \neq i\,. \tag{1.12}$$

We claim that the elements

$$t_i^a \cdot z_{ij}\ ,\ 1 \leq i \leq r,\ 1 \leq j \leq f_i,\ 0 \leq a < e_i$$

are linearly independent over $K(x)$. Their number is equal to $\sum_{i=1}^r f_i e_i = \sum_{i=1}^r v_{P_i}(x) \cdot \deg P_i$, so the proposition will follow from this claim.

Suppose there is a non-trivial linear combination

$$\sum_{i=1}^r \sum_{j=1}^{f_i} \sum_{a=0}^{e_i-1} \varphi_{ija}(x) t_i^a z_{ij} = 0 \tag{1.13}$$

over $K(x)$. W.l.o.g. we can assume that $\varphi_{ija}(x) \in K[x]$ and not all $\varphi_{ija}(x)$ are divisible by x. Then there are indices $k \in \{1,\ldots,r\}$ and $c \in \{0,\ldots,e_k-1\}$ such that

$$\begin{aligned} &x \mid \varphi_{kja}(x)\ for\ all\ a<c\ and\ all\ j \in \{1,\ldots,f_k\},\ and \\ &x \nmid \varphi_{kjc}(x)\ for\ some\ j \in \{1,\ldots,f_k\}\ . \end{aligned} \tag{1.14}$$

Multiplying (1.13) by t_k^{-c} we obtain

$$\sum_{i=1}^r \sum_{j=1}^{f_i} \sum_{a=0}^{e_i-1} \varphi_{ija}(x) t_i^a t_k^{-c} z_{ij} = 0\,. \tag{1.15}$$

For $i \neq k$ all summands of (1.15) are in P_k, since

$$\begin{aligned} v_k(\varphi_{ija}(x)t_i^a t_k^{-c} z_{ij}) &= v_k(\varphi_{ija}(x)) + a v_k(t_i) - c v_k(t_k) + v_k(z_{ij}) \\ &\geq 0+0-c+e_k > 0\,. \end{aligned}$$

For $i=k$ and $a<c$ we have

$$v_k(\varphi_{kja}(x)t_k^{a-c} z_{kj}) \geq e_k + a - c \geq e_k - c > 0\,.$$

(Note that $x | \varphi_{kja}(x)$ and therefore $v_k(\varphi_{kja}(x)) \geq e_k$.) For $i=k$ and $a>c$,

$$v_k(\varphi_{kja}(x)t_k^{a-c} z_{kj}) \geq a-c > 0\,.$$

Combining the above with (1.15) gives

$$\sum_{j=1}^{f_k} \varphi_{kjc}(x) z_{kj} \in P_k\,. \tag{1.16}$$

Observe that $\varphi_{kjc}(x)(P_k) \in K$, and not all $\varphi_{kjc}(x)(P_k) = 0$ (by (1.14)), so (1.16) yields a non-trivial linear combination

$$\sum_{j=1}^{f_k} \varphi_{kjc}(x)(P_k) \cdot z_{kj}(P_k) = 0$$

over K. This is a contradiction, as $z_{k1}(P_k), \ldots, z_{kf_k}(P_k)$ form a basis of F_{P_k}/K. □

Corollary 1.3.4. *In a function field F/K every element $0 \neq x \in F$ has only finitely many zeros and poles.*

Proof. If x is constant, x has neither zeros nor poles. If x is transcendental over K, the number of zeros is $\leq [F : K(x)]$ by Proposition 1.3.3. The same argument shows that x^{-1} has only a finite number of zeros. □

1.4 Divisors

The field $\tilde{K}$ of constants of an algebraic function field F/K is a finite extension field of K, cf. Corollary 1.1.16, and F can be regarded as a function field over $\tilde{K}$. Therefore the following assumption (which we maintain throughout the whole book) is not critical to the theory:

From here on, F/K will always denote an algebraic function field of one variable such that K is the full constant field of F/K.

Definition 1.4.1. *The divisor group of F/K is defined as the (additively written) free abelian group which is generated by the places of F/K; it is denoted by* Div(F). *The elements of* Div(F) *are called divisors of F/K. In other words, a divisor is a formal sum*

$$D = \sum_{P \in \mathbb{P}_F} n_P P \quad \textit{with } n_P \in \mathbb{Z}, \quad \textit{almost all } n_P = 0\,.$$

The support of D is defined as

$$\operatorname{supp} D := \{P \in \mathbb{P}_F \mid n_P \neq 0\}\,.$$

It will often be found convenient to write

$$D = \sum_{P \in S} n_P P\,,$$

where $S \subseteq \mathbb{P}_F$ is a finite set with $S \supseteq \operatorname{supp} D$.

A divisor of the form $D = P$ with $P \in \mathbb{P}_F$ is called a prime divisor. Two divisors $D = \sum n_P P$ and $D' = \sum n'_P P$ are added coefficientwise,

$$D + D' = \sum_{P \in \mathbb{P}_F} (n_P + n'_P)P .$$

The zero element of the divisor group $\mathrm{Div}(F)$ *is the divisor*

$$0 := \sum_{P \in \mathbb{P}_F} r_P P \ , \quad \text{all } r_P = 0 .$$

For $Q \in \mathbb{P}_F$ *and* $D = \sum n_P P \in \mathrm{Div}(F)$ *we define* $v_Q(D) := n_Q$, *therefore*

$$\mathrm{supp}\, D = \{P \in \mathbb{P}_F \mid v_P(D) \neq 0\} \quad \text{and} \quad D = \sum_{P \in \mathrm{supp}\, D} v_P(D) \cdot P .$$

A partial ordering on $\mathrm{Div}(F)$ *is defined by*

$$D_1 \leq D_2 :\iff v_P(D_1) \leq v_P(D_2) \quad \text{for all } P \in \mathbb{P}_F .$$

If $D_1 \leq D_2$ *and* $D_1 \neq D_2$ *we will also write* $D_1 < D_2$. *A divisor* $D \geq 0$ *is called positive (or effective). The degree of a divisor is defined as*

$$\deg D := \sum_{P \in \mathbb{P}_F} v_P(D) \cdot \deg P ,$$

and this yields a homomorphism $\deg : \mathrm{Div}(F) \to \mathbb{Z}$.

By Corollary 1.3.4 a nonzero element $x \in F$ has only finitely many zeros and poles in $\mathbb{P}_F$. Thus the following definition makes sense.

Definition 1.4.2. *Let* $0 \neq x \in F$ *and denote by* Z *(resp.* N*) the set of zeros (resp. poles) of* x *in* $\mathbb{P}_F$. *Then we define*

$$(x)_0 := \sum_{P \in Z} v_P(x)P \ , \quad \text{the zero divisor of } x ,$$

$$(x)_\infty := \sum_{P \in N} (-v_P(x))P \ , \quad \text{the pole divisor of } x ,$$

$$(x) := (x)_0 - (x)_\infty \ , \quad \text{the principal divisor of } x .$$

Clearly $(x)_0 \geq 0$, $(x)_\infty \geq 0$ and

$$(x) = \sum_{P \in \mathbb{P}_F} v_P(x)P . \tag{1.17}$$

The elements $0 \neq x \in F$ which are constant are characterized by

$$x \in K \iff (x) = 0 .$$

This follows immediately from Corollary 1.1.20 (note the general assumption made previously that K is algebraically closed in F).

Definition 1.4.3. *The set of divisors*

$$\mathrm{Princ}(F) := \{ \ (x) \mid 0 \neq x \in F \ \}$$

is called the group of principal divisors of F/K. This is a subgroup of $\mathrm{Div}(F)$, *since for $0 \neq x, y \in F$, $(xy) = (x) + (y)$ by (1.17). The factor group*

$$\mathrm{Cl}(F) := \mathrm{Div}(F)/\mathrm{Princ}(F)$$

is called the divisor class group of F/K. For a divisor $D \in \mathrm{Div}(F)$, the corresponding element in the factor group $\mathrm{Cl}(F)$ is denoted by $[D]$, the divisor class of D. Two divisors $D, D' \in \mathrm{Div}(F)$ are said to be equivalent, written

$$D \sim D' ,$$

if $[D] = [D']$; i.e., $D = D' + (x)$ for some $x \in F \setminus \{0\}$. This is easily verified to be an equivalence relation.

Our next definition plays a fundamental role in the theory of algebraic function fields.

Definition 1.4.4. *For a divisor $A \in \mathrm{Div}(F)$ we define the Riemann-Roch space associated to A by*

$$\mathscr{L}(A) := \{ \ x \in F \mid (x) \geq -A \ \} \cup \{0\} .$$

This definition has the following interpretation: if

$$A = \sum_{i=1}^{r} n_i P_i - \sum_{j=1}^{s} m_j Q_j$$

with $n_i > 0$, $m_j > 0$ then $\mathscr{L}(A)$ consists of all elements $x \in F$ such that

- x has zeros of order $\geq m_j$ at Q_j, for $j = 1, \ldots, s$, and
- x may have poles only at the places $P_1, \ldots, P_r$, with the pole order at P_i being bounded by n_i $(i = 1, \ldots, r)$.

Remark 1.4.5. Let $A \in \mathrm{Div}(F)$. Then

(a) $x \in \mathscr{L}(A)$ if and only if $v_P(x) \geq -v_P(A)$ for all $P \in \mathbb{P}_F$.

(b) $\mathscr{L}(A) \neq \{0\}$ if and only if there is a divisor $A' \sim A$ with $A' \geq 0$.

The proof of these remarks is trivial; nevertheless they are often very useful. In particular Remark 1.4.5(b) will be used frequently.

Lemma 1.4.6. *Let $A \in \mathrm{Div}(F)$. Then we have:*

(a) $\mathscr{L}(A)$ is a vector space over K.

(b) If A' is a divisor equivalent to A, then $\mathscr{L}(A) \simeq \mathscr{L}(A')$ (isomorphic as vector spaces over K).

Proof. (a) Let $x, y \in \mathscr{L}(A)$ and $a \in K$. Then for all $P \in \mathbb{P}_F$, $v_P(x+y) \geq \min\{v_P(x), v_P(y)\} \geq -v_P(A)$ and $v_P(ax) = v_P(a) + v_P(x) \geq -v_P(A)$. So $x+y$ and ax are in $\mathscr{L}(A)$ by Remark 1.4.5(a).

(b) By assumption, $A = A' + (z)$ with $0 \neq z \in F$. Consider the mapping

$$\varphi : \begin{cases} \mathscr{L}(A) & \longrightarrow & F\,, \\ x & \longmapsto & xz\,. \end{cases}$$

This is a K-linear mapping whose image is contained in $\mathscr{L}(A')$. In the same manner,

$$\varphi' : \begin{cases} \mathscr{L}(A') & \longrightarrow & F\,, \\ x & \longmapsto & xz^{-1} \end{cases}$$

is K-linear from $\mathscr{L}(A')$ to $\mathscr{L}(A)$. These mappings are inverse to each other, hence φ is an isomorphism between $\mathscr{L}(A)$ and $\mathscr{L}(A')$. □

Lemma 1.4.7. *(a) $\mathscr{L}(0) = K$.*
(b) If $A < 0$ then $\mathscr{L}(A) = \{0\}$.

Proof. (a) We have $(x) = 0$ for $0 \neq x \in K$, therefore $K \subseteq \mathscr{L}(0)$. Conversely, if $0 \neq x \in \mathscr{L}(0)$ then $(x) \geq 0$. This means that x has no pole, so $x \in K$ by Corollary 1.1.20.

(b) Assume there exists an element $0 \neq x \in \mathscr{L}(A)$. Then $(x) \geq -A > 0$, which implies that x has at least one zero but no pole. This is impossible. □

In the sequel we shall consider various K-vector spaces. The dimension of such a vector space V will be denoted by $\dim V$. Our next objective is to show that $\mathscr{L}(A)$ is finite-dimensional for each divisor $A \in \mathrm{Div}(F)$.

Lemma 1.4.8. *Let A, B be divisors of F/K with $A \leq B$. Then we have $\mathscr{L}(A) \subseteq \mathscr{L}(B)$ and*

$$\dim(\mathscr{L}(B)/\mathscr{L}(A)) \leq \deg B - \deg A\,.$$

Proof. $\mathscr{L}(A) \subseteq \mathscr{L}(B)$ is trivial. In order to prove the other assertion we can assume that $B = A + P$ for some $P \in \mathbb{P}_F$; the general case follows then by induction. Choose an element $t \in F$ with $v_P(t) = v_P(B) = v_P(A) + 1$. For $x \in \mathscr{L}(B)$ we have $v_P(x) \geq -v_P(B) = -v_P(t)$, so $xt \in \mathcal{O}_P$. Thus we obtain a K-linear map

$$\psi : \begin{cases} \mathscr{L}(B) & \longrightarrow & F_P\,, \\ x & \longmapsto & (xt)(P)\,. \end{cases}$$

An element x is in the kernel of ψ if and only if $v_P(xt) > 0$; i.e., $v_P(x) \geq -v_P(A)$. Consequently $\mathrm{Ker}(\psi) = \mathscr{L}(A)$, and ψ induces a K-linear injective mapping from $\mathscr{L}(B)/\mathscr{L}(A)$ to F_P. It follows that

$$\dim(\mathscr{L}(B)/\mathscr{L}(A)) \leq \dim F_P = \deg B - \deg A\,.$$

□

Proposition 1.4.9. *For each divisor $A \in \mathrm{Div}(F)$ the space $\mathscr{L}(A)$ is a finite-dimensional vector space over K. More precisely: if $A = A_+ - A_-$ with positive divisors A_+ and A_-, then*

$$\dim \mathscr{L}(A) \leq \deg A_+ + 1\,.$$

Proof. Since $\mathscr{L}(A) \subseteq \mathscr{L}(A_+)$, it is sufficient to show that

$$\dim \mathscr{L}(A_+) \leq \deg A_+ + 1\,.$$

We have $0 \leq A_+$, so Lemma 1.4.8 yields $\dim(\mathscr{L}(A_+)/\mathscr{L}(0)) \leq \deg A_+$. Since $\mathscr{L}(0) = K$ we conclude that $\dim \mathscr{L}(A_+) = \dim(\mathscr{L}(A_+)/\mathscr{L}(0)) + 1 \leq \deg A_+ + 1$. □

Definition 1.4.10. *For $A \in \mathrm{Div}(F)$ the integer $\ell(A) := \dim \mathscr{L}(A)$ is called the dimension of the divisor A.*

One of the most important problems in the theory of algebraic function fields is to calculate the dimension of a divisor. We shall be concerned with this question in the subsequent sections; the answer to the problem will be given by the Riemann-Roch Theorem 1.5.15.

We begin by proving a sharpening of Proposition 1.3.3. Roughly speaking, the next theorem states that an element $0 \neq x \in F$ has as many zeros as poles, provided the zeros and poles are counted properly.

Theorem 1.4.11. *All principal divisors have degree zero. More precisely: let $x \in F \setminus K$ and $(x)_0$ resp. $(x)_\infty$ denote the zero resp. pole divisor of x. Then*

$$\deg\,(x)_0 = \deg\,(x)_\infty = [F : K(x)]\,.$$

Corollary 1.4.12. *(a) Let A, A' be divisors with $A \sim A'$. Then we have $\ell(A) = \ell(A')$ and* $\deg A = \deg A'$.

(b) If $\deg A < 0$ *then $\ell(A) = 0$.*

(c) For a divisor A of degree zero the following assertions are equivalent:

(1) A is principal.

(2) $\ell(A) \geq 1$.

(3) $\ell(A) = 1$.

Proof of Corollary 1.4.12. (a) follows immediately from Lemma 1.4.6 and Theorem 1.4.11.

(b) Suppose that $\ell(A) > 0$. By Remark 1.4.5 there is some divisor $A' \sim A$ with $A' \geq 0$, hence $\deg A = \deg A' \geq 0$.

(c) (1) $\Rightarrow$ (2) : If $A = (x)$ is principal then $x^{-1} \in \mathscr{L}(A)$, so $\ell(A) \geq 1$.

(2) $\Rightarrow$ (3) : Assume now that $\ell(A) \geq 1$ and $\deg A = 0$. Then $A \sim A'$ for some $A' \geq 0$ (Remark 1.4.5(b)). The conditions $A' \geq 0$ and $\deg A' = 0$ imply that $A' = 0$, hence $\ell(A) = \ell(A') = \ell(0) = 1$, by Lemma 1.4.7.

(3) $\Rightarrow$ (1) : Suppose that $\ell(A) = 1$ and $\deg A = 0$. Choose $0 \neq z \in \mathscr{L}(A)$, then $(z) + A \geq 0$. Since $\deg((z) + A) = 0$, it follows that $(z) + A = 0$, therefore $A = -(z) = (z^{-1})$ is principal. □

Proof of Theorem 1.4.11. Set $n := [F : K(x)]$ and

$$B := (x)_\infty = \sum_{i=1}^{r} -v_{P_i}(x) P_i \,,$$

where $P_1, \ldots, P_r$ are all the poles of x. Then

$$\deg B = \sum_{i=1}^{r} v_{P_i}(x^{-1}) \cdot \deg P_i \leq [F : K(x)] = n$$

by Proposition 1.3.3, and thus it remains to show that $n \leq \deg B$ as well. Choose a basis $u_1, \ldots, u_n$ of $F/K(x)$ and a divisor $C \geq 0$ such that $(u_i) \geq -C$ for $i = 1, \ldots, n$. We have

$$\ell(lB + C) \geq n(l+1) \quad \textit{for all} \;\; l \geq 0 \,, \tag{1.18}$$

which follows immediately from the fact that $x^i u_j \in \mathscr{L}(lB + C)$ for $0 \leq i \leq l$, $1 \leq j \leq n$ (observe that these elements are linearly independent over K since $u_1, \ldots, u_n$ are linearly independent over $K(x)$). Setting $c := \deg C$ we obtain $n(l+1) \leq \ell(lB + C) \leq l \cdot \deg B + c + 1$ by Proposition 1.4.9. Thus

$$l(\deg B - n) \geq n - c - 1 \tag{1.19}$$

for all $l \in \mathbb{N}$. The right hand side of (1.19) is independent of l, therefore (1.19) is possible only when $\deg B \geq n$.

We have thus proved that $\deg (x)_\infty = [F : K(x)]$. Since $(x)_0 = (x^{-1})_\infty$, we conclude that $\deg (x)_0 = \deg (x^{-1})_\infty = [F : K(x^{-1})] = [F : K(x)]$. □

Example 1.4.13. Once again we consider the rational function field $F = K(x)$ as in Section 1.2. For $0 \neq z \in K(x)$ we have $z = a \cdot f(x)/g(x)$ with $a \in K \setminus \{0\}$, $f(x), g(x) \in K[x]$ monic and relatively prime. Let

$$f(x) = \prod_{i=1}^{r} p_i(x)^{n_i} \,, \quad g(x) = \prod_{i=1}^{s} q_j(x)^{m_j}$$

with pairwise distinct irreducible monic polynomials $p_i(x), q_j(x) \in K[x]$. Then the principal divisor of z in $\mathrm{Div}(K(x))$ appears thus :

$$(z) = \sum_{i=1}^{r} n_i P_i - \sum_{j=1}^{s} m_j Q_j + (\deg g(x) - \deg f(x)) P_\infty \,, \tag{1.20}$$

where P_i resp. Q_j are the places corresponding to $p_i(x)$ resp. $q_j(x)$, cf. Section 1.2. Therefore in arbitrary function fields, principal divisors can be considered as a substitute for the decomposition into irreducible polynomials that occurs in the rational function field.

Again we consider an arbitrary algebraic function field F/K. In Proposition 1.4.9 we have seen that the inequality

$$\ell(A) \leq 1 + \deg A \tag{1.21}$$

holds for all divisors $A \geq 0$. In fact (1.21) holds for every divisor of degree ≥ 0. In order to verify this, we can assume that $\ell(A) > 0$. Then $A \sim A'$ for some $A' \geq 0$ by Remark 1.4.5, so $\ell(A) = \ell(A') \leq 1 + \deg A' = 1 + \deg A$ by Corollary 1.4.12.

Next we want to prove the existence of a lower bound for $\ell(A)$, similar to the inequality in (1.21).

Proposition 1.4.14. *There is a constant $\gamma \in \mathbb{Z}$ such that for all divisors $A \in \mathrm{Div}(F)$ the following holds:*

$$\deg\, A - \ell(A) \leq \gamma \,.$$

The emphasis here lies on the fact that γ is independent of the divisor A; it depends only on the function field F/K.

Proof. To begin with, observe that

$$A_1 \leq A_2 \Rightarrow \deg A_1 - \ell(A_1) \leq \deg A_2 - \ell(A_2) \,, \tag{1.22}$$

by Lemma 1.4.8. We fix an element $x \in F \backslash K$ and consider the specific divisor $B := (x)_\infty$. As in the proof of Theorem 1.4.11 there exists a divisor $C \geq 0$ (depending on x) such that $\ell(lB+C) \geq (l+1) \cdot \deg B$ for all $l \geq 0$, see (1.18). On the other hand, $\ell(lB + C) \leq \ell(lB) + \deg C$ by Lemma 1.4.8. Combining these inequalities we find

$$\ell(lB) \geq (l+1)\deg B - \deg C = \deg(lB) + ([F : K(x)] - \deg C)\,.$$

Therefore

$$\deg(lB) - \ell(lB) \leq \gamma \quad \textit{for all } l > 0 \tag{1.23}$$

with some $\gamma \in \mathbb{Z}$. We want to show that (1.23) holds even when we substitute for lB any $A \in \mathrm{Div}(F)$ (with the above γ).

Claim. Given a divisor A, there exist divisors A_1, D and an integer $l \geq 0$ such that $A \leq A_1, A_1 \sim D$ and $D \leq lB$.

Using this claim, Proposition 1.4.14 will follow easily :

$$\begin{aligned}
\deg A - \ell(A) &\leq \deg A_1 - \ell(A_1) && (\textit{by } (1.22))\\
&= \deg D - \ell(D) && (\textit{by Corollary } 1.4.12)\\
&\leq \deg(lB) - \ell(lB) && (\textit{by } (1.22))\\
&\leq \gamma. && (\textit{by } (1.23))
\end{aligned}$$

Proof of the Claim. Choose $A_1 \geq A$ such that $A_1 \geq 0$. Then

$$\begin{aligned}
\ell(lB - A_1) &\geq \ell(lB) - \deg A_1 && (\textit{by Lemma } 1.4.8)\\
&\geq \deg(lB) - \gamma - \deg A_1 && (\textit{by } (1.23))\\
&> 0
\end{aligned}$$

for sufficiently large l. Thus there is some element $0 \neq z \in \mathscr{L}(lB - A_1)$. Setting $D := A_1 - (z)$ we obtain $A_1 \sim D$ and $D \leq A_1 - (A_1 - lB) = lB$ as desired. □

Definition 1.4.15. *The genus g of F/K is defined by*

$$g := \max\{\ \deg A - \ell(A) + 1 \mid A \in \mathrm{Div}(F)\ \}\,.$$

Observe that this definition makes sense by Proposition 1.4.14. It will turn out that the genus is the most important invariant of a function field.

Corollary 1.4.16. *The genus of F/K is a non-negative integer.*

Proof. In the definition of g, put $A = 0$. Then $\deg(0) - \ell(0) + 1 = 0$, hence $g \geq 0$. □

Theorem 1.4.17 (Riemann's Theorem). *Let F/K be a function field of genus g. Then we have:*
(a) For all divisors $A \in \mathrm{Div}(F)$,

$$\ell(A) \geq \deg A + 1 - g\,.$$

(b) There is an integer c, depending only on the function field F/K, such that

$$\ell(A) = \deg A + 1 - g\,,$$

whenever $\deg A \geq c$.

Proof. (a) This is just the definition of the genus.
(b) Choose a divisor A_0 with $g = \deg A_0 - \ell(A_0) + 1$ and set $c := \deg A_0 + g$. If $\deg A \geq c$ then

$$\ell(A - A_0) \geq \deg(A - A_0) + 1 - g \geq c - \deg A_0 + 1 - g = 1\,.$$

So there is an element $0 \neq z \in \mathscr{L}(A - A_0)$. Consider the divisor $A' := A + (z)$ which is $\geq A_0$. We have

$$\begin{aligned} \deg A - \ell(A) &= \deg A' - \ell(A') && (\textit{by Corollary } 1.4.12) \\ &\geq \deg A_0 - \ell(A_0) && (\textit{by Lemma } 1.4.8) \\ &= g - 1\,. \end{aligned}$$

Hence $\ell(A) \leq \deg A + 1 - g$. □

Example 1.4.18. We want to show that the rational function field $K(x)/K$ has genus $g = 0$. In order to prove this, let P_∞ denote the pole divisor of x (notation as in Proposition 1.2.1). Consider for $r \geq 0$ the vector space $\mathscr{L}(rP_\infty)$. Obviously the elements $1, x, \ldots, x^r$ are in $\mathscr{L}(rP_\infty)$, hence

$$r + 1 \leq \ell(rP_\infty) = \deg(rP_\infty) + 1 - g = r + 1 - g$$

for sufficiently large r. Thus $g \leq 0$. Since $g \geq 0$ holds for every function field, the assertion follows.

In general it is hard to determine the genus of a function field. Large parts of Chapter 3 will be devoted to this problem.

1.5 The Riemann-Roch Theorem

In this section F/K denotes an algebraic function field of genus g.

Definition 1.5.1. *For $A \in \mathrm{Div}(F)$ the integer*

$$i(A) := \ell(A) - \deg A + g - 1$$

is called the index of specialty of A.

Riemann's Theorem 1.4.17 states that $i(A)$ is a non-negative integer, and $i(A) = 0$ if $\deg A$ is sufficiently large. In the present section we will provide several interpretations for $i(A)$ as the dimension of certain vector spaces. To this end we introduce the notion of an *adele*.

Definition 1.5.2. *An adele of F/K is a mapping*

$$\alpha : \begin{cases} \mathbb{P}_F & \longrightarrow \quad F\,, \\ P & \longmapsto \quad \alpha_P\,, \end{cases}$$

such that $\alpha_P \in \mathcal{O}_P$ for almost all $P \in \mathbb{P}_F$. We regard an adele as an element of the direct product $\prod_{P \in \mathbb{P}_F} F$ and therefore use the notation $\alpha = (\alpha_P)_{P \in \mathbb{P}_F}$ or, even shorter, $\alpha = (\alpha_P)$. The set

$$\mathcal{A}_F := \{\, \alpha \,|\, \alpha \text{ is an adele of } F/K \,\}$$

is called the adele space of F/K. It is regarded as a vector space over K in the obvious manner (actually $\mathcal{A}_F$ can be regarded as a ring, but the ring structure will never be used).

The principal adele of an element $x \in F$ is the adele all of whose components are equal to x (note that this definition makes sense since x has only finitely many poles). This gives an embedding $F \hookrightarrow \mathcal{A}_F$. The valuations v_P of F/K extend naturally to $\mathcal{A}_F$ by setting $v_P(\alpha) := v_P(\alpha_P)$ (where α_P is the P-component of the adele α). By definition we have that $v_P(\alpha) \geq 0$ for almost all $P \in \mathbb{P}_F$.

We note that the notion of an adele is not consistent in the literature. Some authors use the name *repartition* for what we call an adele. Others mean by an adele (or a repartition) a mapping α such that $\alpha(P)$ is an element of the P-adic completion $\hat{F}_P$ for all $P \in \mathbb{P}_F$ (cf. Chapter 4).

Definition 1.5.3. *For $A \in \mathrm{Div}(F)$ we define*

$$\mathcal{A}_F(A) := \{\, \alpha \in \mathcal{A}_F \,|\, v_P(\alpha) \geq -v_P(A) \text{ for all } P \in \mathbb{P}_F \,\}\,.$$

Obviously this is a K-subspace of $\mathcal{A}_F$.

Theorem 1.5.4. *For every divisor A the index of specialty is*

$$i(A) = \dim(\mathcal{A}_F/(\mathcal{A}_F(A) + F))\,.$$

Here, as usual, dim means the dimension as a K-vector space. Note that although the vector spaces $\mathcal{A}_F$, $\mathcal{A}_F(A)$ and F are infinite-dimensional, the theorem states that the quotient space $\mathcal{A}_F/(\mathcal{A}_F(A)+F)$ has finite dimension over K. As a corollary, we obtain another characterization of the genus of F/K.

Corollary 1.5.5. $\quad g = \dim(\mathcal{A}_F/(\mathcal{A}_F(0) + F))\,.$

Proof of Corollary 1.5.5. $\quad i(0) = \ell(0) - \deg(0) + g - 1 = 1 - 0 + g - 1 = g.$ □

Proof of Theorem 1.5.4. We proceed in several steps.

Step 1. Let $A_1, A_2 \in \mathrm{Div}(F)$ and $A_1 \le A_2$. Then $\mathcal{A}_F(A_1) \subseteq \mathcal{A}_F(A_2)$ and

$$\dim(\mathcal{A}_F(A_2)/\mathcal{A}_F(A_1)) = \deg A_2 - \deg A_1\,. \tag{1.24}$$

Proof of Step 1. $\mathcal{A}_F(A_1) \subseteq \mathcal{A}_F(A_2)$ is trivial. It is sufficient to prove (1.24) in the case $A_2 = A_1 + P$ with $P \in \mathbb{P}_F$ (the general case follows by induction). Choose $t \in F$ with $v_P(t) = v_P(A_1) + 1$ and consider the K-linear map

$$\varphi : \begin{cases} \mathcal{A}_F(A_2) & \longrightarrow \quad F_P\,, \\ \quad \alpha & \longmapsto \quad (t\alpha_P)(P)\,. \end{cases}$$

One checks easily that φ is surjective and that the kernel of φ is $\mathcal{A}_F(A_1)$. Consequently

$$\deg A_2 - \deg A_1 = \deg P = [F_P : K] = \dim(\mathcal{A}_F(A_2)/\mathcal{A}_F(A_1))\,.$$

Step 2. Let $A_1, A_2 \in \mathrm{Div}(F)$ and $A_1 \le A_2$ as before. Then

$$\begin{aligned} &\dim((\mathcal{A}_F(A_2) + F)/(\mathcal{A}_F(A_1) + F) \\ &\qquad = (\deg A_2 - \ell(A_2)) - (\deg A_1 - \ell(A_1))\,. \end{aligned} \tag{1.25}$$

Proof of Step 2. We have an exact sequence of linear mappings

$$\begin{aligned} 0 \longrightarrow \mathscr{L}(A_2)/\mathscr{L}(A_1) &\xrightarrow{\sigma_1} \mathcal{A}_F(A_2)/\mathcal{A}_F(A_1) \\ &\xrightarrow{\sigma_2} (\mathcal{A}_F(A_2) + F)/(\mathcal{A}_F(A_1) + F) \longrightarrow 0 \end{aligned} \tag{1.26}$$

where σ_1 and σ_2 are defined in the obvious manner. In fact, the only non-trivial assertion is that the kernel of σ_2 is contained in the image of σ_1. In order to prove this, let $\alpha \in \mathcal{A}_F(A_2)$ with $\sigma_2(\alpha + \mathcal{A}_F(A_1)) = 0$. Then $\alpha \in \mathcal{A}_F(A_1) + F$, so there is some $x \in F$ with $\alpha - x \in \mathcal{A}_F(A_1)$. As $\mathcal{A}_F(A_1) \subseteq \mathcal{A}_F(A_2)$ we conclude that $x \in \mathcal{A}_F(A_2) \cap F = \mathscr{L}(A_2)$. Therefore $\alpha + \mathcal{A}_F(A_1) = x + \mathcal{A}_F(A_1) = \sigma_1(x + \mathscr{L}(A_1))$ lies in the image of σ_1.

From the exactness of (1.26) we obtain

$$\begin{aligned}\dim(\mathcal{A}_F(A_2)+F)/(\mathcal{A}_F(A_1)+F) \\ &= \dim(\mathcal{A}_F(A_2)/\mathcal{A}_F(A_1)) - \dim(\mathscr{L}(A_2)/\mathscr{L}(A_1)) \\ &= (\deg A_2 - \deg A_1) - (\ell(A_2) - \ell(A_1)),\end{aligned}$$

using (1.24).

Step 3. If B is a divisor with $\ell(B) = \deg B + 1 - g$, then

$$\mathcal{A}_F = \mathcal{A}_F(B) + F. \tag{1.27}$$

Proof of Step 3. To begin with, observe that for $B_1 \geq B$ we have (by Lemma 1.4.8)

$$\ell(B_1) \leq \deg B_1 + \ell(B) - \deg B = \deg B_1 + 1 - g.$$

On the other hand, $\ell(B_1) \geq \deg B_1 + 1 - g$ by Riemann's Theorem. Therefore

$$\ell(B_1) = \deg B_1 + 1 - g \quad \textit{for each} \quad B_1 \geq B. \tag{1.28}$$

Now we prove (1.27). Let $\alpha \in \mathcal{A}_F$. Obviously one can find a divisor $B_1 \geq B$ such that $\alpha \in \mathcal{A}_F(B_1)$. By (1.25) and (1.28),

$$\begin{aligned}\dim(\mathcal{A}_F(B_1)+F)/(\mathcal{A}_F(B)+F) &= (\deg B_1 - \ell(B_1)) - (\deg B - \ell(B)) \\ &= (g-1) - (g-1) = 0.\end{aligned}$$

This implies $\mathcal{A}_F(B) + F = \mathcal{A}_F(B_1) + F$. Since $\alpha \in \mathcal{A}_F(B_1)$ it follows that $\alpha \in \mathcal{A}_F(B) + F$, and (1.27) is proved.

End of the proof of Theorem 1.5.4. Now we consider an arbitrary divisor A. By Riemann's Theorem 1.4.17(b) there exists some divisor $A_1 \geq A$ such that $\ell(A_1) = \deg A_1 + 1 - g$. By (1.27), $\mathcal{A}_F = \mathcal{A}_F(A_1) + F$, and in view of (1.25) we obtain

$$\begin{aligned}\dim(\mathcal{A}_F/(\mathcal{A}_F(A)+F)) &= \dim(\mathcal{A}_F(A_1)+F)/(\mathcal{A}_F(A)+F) \\ &= (\deg A_1 - \ell(A_1)) - (\deg A - \ell(A)) \\ &= (g-1) + \ell(A) - \deg A = i(A).\end{aligned}$$

□

Theorem 1.5.4 can be restated as follows: for all $A \in \mathrm{Div}(F)$ holds

$$\ell(A) = \deg A + 1 - g + \dim(\mathcal{A}_F/(\mathcal{A}_F(A)+F)). \tag{1.29}$$

This is a preliminary version of the Riemann-Roch Theorem which we shall prove later in this section.

Next we introduce the concept of Weil differentials which will lead to a second interpretation for the index of specialty of a divisor.

Definition 1.5.6. *A Weil differential of F/K is a K-linear map $\omega : \mathcal{A}_F \to K$ vanishing on $\mathcal{A}_F(A) + F$ for some divisor $A \in \mathrm{Div}(F)$. We call*

$$\Omega_F := \{\omega \,|\, \omega \text{ is a Weil differential of } F/K\}$$

the module of Weil differentials of F/K. For $A \in \mathrm{Div}(F)$ let

$$\Omega_F(A) := \{\omega \in \Omega_F \,|\, \omega \text{ vanishes on } \mathcal{A}_F(A) + F\}.$$

We regard Ω_F as a K-vector space in the obvious manner (in fact, if ω_1 vanishes on $\mathcal{A}_F(A_1)+F$ and ω_2 vanishes on $\mathcal{A}_F(A_2)+F$ then $\omega_1+\omega_2$ vanishes on $\mathcal{A}_F(A_3) + F$ for every divisor A_3 with $A_3 \le A_1$ and $A_3 \le A_2$, and $a\omega_1$ vanishes on $\mathcal{A}_F(A_1) + F$ for $a \in K$). Clearly $\Omega_F(A)$ is a subspace of Ω_F.

Lemma 1.5.7. *For $A \in \mathrm{Div}(F)$ we have $\dim \Omega_F(A) = i(A)$.*

Proof. $\Omega_F(A)$ is in a natural way isomorphic to the space of linear forms on $\mathcal{A}_F/(\mathcal{A}_F(A) + F)$. Since $\mathcal{A}_F/(\mathcal{A}_F(A) + F)$ is finite-dimensional of dimension $i(A)$ by Theorem 1.5.4, our lemma follows immediately. □

A simple consequence of Lemma 1.5.7 is that $\Omega_F \neq 0$. To see this, choose a divisor A of degree ≤ -2. Then

$$\dim \Omega_F(A) = i(A) = \ell(A) - \deg A + g - 1 \ge 1,$$

hence $\Omega_F(A) \neq 0$.

Definition 1.5.8. *For $x \in F$ and $\omega \in \Omega_F$ we define $x\omega : \mathcal{A}_F \to K$ by*

$$(x\omega)(\alpha) := \omega(x\alpha).$$

It is easily checked that $x\omega$ is again a Weil differential of F/K. In fact, if ω vanishes on $\mathcal{A}_F(A) + F$ then $x\omega$ vanishes on $\mathcal{A}_F(A + (x)) + F$. Clearly our definition gives Ω_F the structure of a vector space over F.

Proposition 1.5.9. *Ω_F is a one-dimensional vector space over F.*

Proof. Choose $0 \neq \omega_1 \in \Omega_F$ (we already know that $\Omega_F \neq 0$). It has to be shown that for every $\omega_2 \in \Omega_F$ there is some $z \in F$ with $\omega_2 = z\omega_1$. We can assume that $\omega_2 \neq 0$. Choose $A_1, A_2 \in \mathrm{Div}(F)$ such that $\omega_1 \in \Omega_F(A_1)$ and $\omega_2 \in \Omega_F(A_2)$. For a divisor B (which will be specified later) we consider the K-linear injective maps

$$\varphi_i : \begin{cases} \mathscr{L}(A_i + B) & \longrightarrow & \Omega_F(-B)\,, \\ x & \longmapsto & x\omega_i\,. \end{cases} \qquad (i = 1, 2)$$

Claim. For an appropriate choice of the divisor B holds

$$\varphi_1(\mathscr{L}(A_1 + B)) \cap \varphi_2(\mathscr{L}(A_2 + B)) \neq \{0\}.$$

Using this claim, the proof of the proposition can be finished very quickly: we choose $x_1 \in \mathscr{L}(A_1 + B)$ and $x_2 \in \mathscr{L}(A_2 + B)$ such that $x_1\omega_1 = x_2\omega_2 \neq 0$. Then $\omega_2 = (x_1 x_2^{-1})\omega_1$ as desired.

Proof of the Claim. We start with a simple and well-known fact from linear algebra: if U_1, U_2 are subspaces of a finite-dimensional vector space V then

$$\dim(U_1 \cap U_2) \geq \dim U_1 + \dim U_2 - \dim V. \tag{1.30}$$

Now let $B > 0$ be a divisor of sufficiently large degree such that

$$\ell(A_i + B) = \deg(A_i + B) + 1 - g$$

for $i = 1, 2$ (this is possible by Riemann's Theorem 1.4.17). We set $U_i := \varphi_i(\mathscr{L}(A_i + B)) \subseteq \Omega_F(-B)$. Since

$$\begin{aligned} \dim \Omega_F(-B) = i(-B) &= \dim(-B) - \deg(-B) + g - 1 \\ &= \deg B - 1 + g, \end{aligned}$$

we obtain

$$\begin{aligned} &\dim U_1 + \dim U_2 - \dim \Omega_F(-B) \\ &= \deg(A_1 + B) + 1 - g + \deg(A_2 + B) + 1 - g - (\deg B + g - 1) \\ &= \deg B + (\deg A_1 + \deg A_2 + 3(1 - g)). \end{aligned}$$

The term in brackets is independent of B, so

$$\dim U_1 + \dim U_2 - \dim \Omega_F(-B) > 0$$

if $\deg B$ is sufficiently large. By (1.30) it follows that $U_1 \cap U_2 \neq \{0\}$ which proves our claim. □

We want to attach a divisor to each Weil differential $\omega \neq 0$. To this end we consider (for a fixed ω) the set of divisors

$$M(\omega) := \{ A \in \mathrm{Div}(F) \mid \omega \textit{ vanishes on } \mathcal{A}_F(A) + F \}. \tag{1.31}$$

Lemma 1.5.10. *Let $0 \neq \omega \in \Omega_F$. Then there is a uniquely determined divisor $W \in M(\omega)$ such that $A \leq W$ for all $A \in M(\omega)$.*

Proof. By Riemann's Theorem there exists a constant c, depending only on the function field F/K, with the property $i(A) = 0$ for all $A \in \mathrm{Div}(F)$ of degree $\geq c$. Since $\dim(\mathcal{A}_F/(\mathcal{A}_F(A) + F)) = i(A)$ by Theorem 1.5.4, we have that $\deg A < c$ for all $A \in M(\omega)$. So we can choose a divisor $W \in M(\omega)$ of maximal degree.

Suppose W does not have the property of our lemma. Then there exists a divisor $A_0 \in M(\omega)$ with $A_0 \not\leq W$, i.e. $v_Q(A_0) > v_Q(W)$ for some $Q \in \mathbb{P}_F$. We claim that

$$W + Q \in M(\omega)\,, \tag{1.32}$$

which is a contradiction to the maximality of W. In fact, consider an adele $\alpha = (\alpha_P) \in \mathcal{A}_F(W + Q)$. We can write $\alpha = \alpha' + \alpha''$ with

$$\alpha'_P := \begin{cases} \alpha_P & \text{for } P \neq Q, \\ 0 & \text{for } P = Q, \end{cases} \quad \text{and} \quad \alpha''_P := \begin{cases} 0 & \text{for } P \neq Q, \\ \alpha_Q & \text{for } P = Q. \end{cases}$$

Then $\alpha' \in \mathcal{A}_F(W)$ and $\alpha'' \in \mathcal{A}_F(A_0)$, therefore $\omega(\alpha) = \omega(\alpha') + \omega(\alpha'') = 0$. Hence ω vanishes on $\mathcal{A}_F(W + Q) + F$, and (1.32) is proved. The uniqueness of W is now obvious. □

The following definition makes sense by the preceding lemma.

Definition 1.5.11. *(a) The divisor (ω) of a Weil differential $\omega \neq 0$ is the uniquely determined divisor of F/K satisfying*

(1) ω vanishes on $\mathcal{A}_F((\omega)) + F$, and

(2) if ω vanishes on $\mathcal{A}_F(A) + F$ then $A \leq (\omega)$.

(b) For $0 \neq \omega \in \Omega_F$ and $P \in \mathbb{P}_F$ we define $v_P(\omega) := v_P((\omega))$.

(c) A place P is said to be a zero (resp. pole) of ω if $v_P(\omega) > 0$ (resp. $v_P(\omega) < 0$). The Weil differential ω is called regular at P if $v_P(\omega) \geq 0$, and ω is said to be regular (or holomorphic) if it is regular at all places $P \in \mathbb{P}_F$.

(d) A divisor W is called a canonical divisor of F/K if $W = (\omega)$ for some $\omega \in \Omega_F$.

Remark 1.5.12. It follows immediately from the definitions that

$$\Omega_F(A) = \{\omega \in \Omega_F \,|\, \omega = 0 \text{ or } (\omega) \geq A\}$$

and

$$\Omega_F(0) = \{\omega \in \Omega_F \,|\, \omega \text{ is regular}\}\,.$$

As a consequence of Lemma 1.5.7 and Definition 1.5.1 we obtain

$$\dim \Omega_F(0) = g\,.$$

Proposition 1.5.13. *(a) For $0 \neq x \in F$ and $0 \neq \omega \in \Omega_F$ we have $(x\omega) = (x) + (\omega)$.*

(b) Any two canonical divisors of F/K are equivalent.

It follows from this proposition that the canonical divisors of F/K form a whole class $[W]$ in the divisor class group $\mathrm{Cl}(F)$; this divisor class is called the *canonical class* of F/K.

Proof of Proposition 1.5.13. If ω vanishes on $\mathcal{A}_F(A)+F$ then $x\omega$ vanishes on $\mathcal{A}_F(A+(x))+F$, consequently

$$(\omega)+(x)\le(x\omega)\,.$$

Likewise $(x\omega)+(x^{-1})\le(x^{-1}x\omega)=(\omega)$. Combining these inequalities we obtain

$$(\omega)+(x)\le(x\omega)\le-(x^{-1})+(\omega)=(\omega)+(x)\,.$$

This proves (a). Item (b) follows from (a) and Proposition 1.5.9. □

Theorem 1.5.14 (Duality Theorem). *Let A be an arbitrary divisor and $W=(\omega)$ be a canonical divisor of F/K. Then the mapping*

$$\mu:\begin{cases}\mathscr{L}(W-A) & \longrightarrow & \Omega_F(A)\,,\\ x & \longmapsto & x\omega\end{cases}$$

is an isomorphism of K-vector spaces. In particular,

$$i(A)=\ell(W-A)\,.$$

Proof. For $x\in\mathscr{L}(W-A)$ we have

$$(x\omega)=(x)+(\omega)\ge-(W-A)+W=A\,,$$

hence $x\omega\in\Omega_F(A)$ by Remark 1.5.12. Therefore μ maps $\mathscr{L}(W-A)$ into $\Omega_F(A)$. Clearly μ is linear and injective. In order to show that μ is surjective, we consider a Weil differential $\omega_1\in\Omega_F(A)$. By Proposition 1.5.9 we can write $\omega_1=x\omega$ with some $x\in F$. Since

$$(x)+W=(x)+(\omega)=(x\omega)=(\omega_1)\ge A\,,$$

we obtain $(x)\ge-(W-A)$, so $x\in\mathscr{L}(W-A)$ and $\omega_1=\mu(x)$. We have thus proved that $\dim\Omega_F(A)=\ell(W-A)$. Since $\dim\Omega_F(A)=i(A)$ by Lemma 1.5.7, this implies $i(A)=\ell(W-A)$. □

Summing up the results of this section we obtain the Riemann-Roch Theorem; it is by far the most important theorem in the theory of algebraic function fields.

Theorem 1.5.15 (Riemann-Roch Theorem). *Let W be a canonical divisor of F/K. Then for each divisor $A\in\mathrm{Div}(F)$,*

$$\ell(A)=\deg A+1-g+\ell(W-A)\,.$$

Proof. This is an immediate consequence of Theorem 1.5.14 and the definition of $i(A)$. □

Corollary 1.5.16. *For a canonical divisor* W *we have*

$$\deg W = 2g - 2 \quad \textit{and} \quad \ell(W) = g\,.$$

Proof. For $A = 0$, the Riemann-Roch Theorem and Lemma 1.4.7 give

$$1 = \ell(0) = \deg 0 + 1 - g + \ell(W - 0)\,.$$

Thus $\ell(W) = g$. Setting $A = W$ we obtain

$$g = \ell(W) = \deg W + 1 - g + \ell(W - W) = \deg W + 2 - g\,.$$

Therefore $\deg W = 2g - 2$. □

From Riemann's Theorem we already know that there is some constant c such that $i(A) = 0$ whenever $\deg A \geq c$. We can now give a more precise description of how to choose this constant.

Theorem 1.5.17. *If* A *is a divisor of* F/K *of degree* $\deg A \geq 2g - 1$ *then*

$$\ell(A) = \deg A + 1 - g\,.$$

Proof. We have $\ell(A) = \deg A + 1 - g + \ell(W - A)$, where W is a canonical divisor. Since $\deg A \geq 2g - 1$ and $\deg W = 2g - 2$, we conclude that $\deg(W - A) < 0$. It follows from Corollary 1.4.12 that $\ell(W - A) = 0$. □

Observe that the bound $2g - 1$ in this theorem is the best possible, since for a canonical divisor W

$$\ell(W) > \deg W + 1 - g$$

by Corollary 1.5.16.

1.6 Some Consequences of the Riemann-Roch Theorem

As before, F/K denotes an algebraic function field of genus g. We want to discuss various consequences of the Riemann-Roch Theorem. Our first aim is to show that the Riemann-Roch Theorem characterizes the genus as well as the canonical class of F/K.

Proposition 1.6.1. *Suppose that* $g_0 \in \mathbb{Z}$ *and* $W_0 \in \mathrm{Div}(F)$ *satisfy*

$$\ell(A) = \deg A + 1 - g_0 + \ell(W_0 - A) \tag{1.33}$$

for all $A \in \mathrm{Div}(F)$. *Then* $g_0 = g$, *and* W_0 *is a canonical divisor.*

Proof. Setting $A = 0$ resp. $A = W_0$ in (1.33) we obtain $\ell(W_0) = g_0$ and $\deg W_0 = 2g_0 - 2$ (cf. the proof of Corollary 1.5.16). Let W be a canonical divisor of F/K. We choose a divisor A with $\deg A > \max\{2g-2, 2g_0-2\}$. Then $\ell(A) = \deg A + 1 - g$ by Theorem 1.5.17 and $\ell(A) = \deg A + 1 - g_0$ by (1.33). Therefore $g = g_0$. Finally we substitute $A = W$ in (1.33). This yields

$$g = (2g-2) + 1 - g + \ell(W_0 - W),$$

hence $\ell(W_0 - W) = 1$. Since $\deg(W_0 - W) = 0$, this implies that $W_0 - W$ is principal (cf. Corollary 1.4.12), so $W_0 \sim W$. □

Another useful characterization of canonical divisors is the following.

Proposition 1.6.2. *A divisor B is canonical if and only if $\deg B = 2g - 2$ and $\ell(B) \geq g$.*

Proof. Suppose that $\deg B = 2g - 2$ and $\ell(B) \geq g$. Choose a canonical divisor W. Then

$$g \leq \ell(B) = \deg B + 1 - g + \ell(W - B) = g - 1 + \ell(W - B),$$

therefore $\ell(W - B) \geq 1$. Since $\deg(W - B) = 0$, it follows from Corollary 1.4.12 that $W \sim B$. □

Next we come to a characterization of the rational function field.

Proposition 1.6.3. *For a function field F/K the following conditions are equivalent:*

(1) F/K is rational; i.e., $F = K(x)$ for some x which is transcendental over the field K.

(2) F/K has genus 0, and there is some divisor $A \in \mathrm{Div}(F)$ with $\deg A = 1$.

Proof. (1) $\Rightarrow$ (2): See Example 1.4.18.
(2) $\Rightarrow$ (1): Let $g = 0$ and $\deg A = 1$. As $\deg A \geq 2g - 1$ we have that $\ell(A) = \deg A + 1 - g = 2$ by Theorem 1.5.17. Thus $A \sim A'$ for some positive divisor A' (see Remark 1.4.5 (b)). Since $\ell(A') = 2$, there exists an element $x \in \mathscr{L}(A') \setminus K$, so $(x) \neq 0$ and $(x) + A' \geq 0$. As $A' \geq 0$ and $\deg A' = 1$, this is possible only if $A' = (x)_\infty$, the pole divisor of x. Now

$$[F : K(x)] = \deg (x)_\infty = \deg A' = 1$$

by Theorem 1.4.11, so $F = K(x)$. □

Remark 1.6.4. There exist non-rational function fields of genus 0 (these cannot have a divisor of degree 1 by Proposition 1.6.3). However, if K is an algebraically closed field or a finite field, there exists always a divisor of degree 1 (for an algebraically closed field this is trivial, for a finite constant field we shall prove it in Chapter 5), hence in these cases we have $g = 0$ if and only if F/K is rational.

It would seem appropriate at this point to give some examples of function fields of genus > 0. However, we defer such examples to Chapter 6 at which point we will have better methods at hand for calculating the genus.

Our next application of Section 1.5 is a strengthening of the Weak Approximation Theorem.

Theorem 1.6.5 (Strong Approximation Theorem). *Let $S \subsetneqq \mathbb{P}_F$ be a proper subset of $\mathbb{P}_F$ and $P_1, \ldots, P_r \in S$. Suppose there are given elements $x_1, \ldots, x_r \in F$ and integers $n_1, \ldots, n_r \in \mathbb{Z}$. Then there exists an element $x \in F$ such that*

$$v_{P_i}(x - x_i) = n_i \qquad (i = 1, \ldots, r), \qquad \textit{and}$$
$$v_P(x) \geq 0 \qquad \textit{for all } P \in S \setminus \{P_1, \ldots, P_r\}.$$

Proof. Consider the adele $\alpha = (\alpha_P)_{P \in \mathbb{P}_F}$ with

$$\alpha_P := \begin{cases} x_i & \text{for } P = P_i \,, \; i = 1, \ldots, r \,, \\ 0 & \text{otherwise}. \end{cases}$$

Choose a place $Q \in \mathbb{P}_F \setminus S$. For sufficiently large $m \in \mathbb{N}$ we have

$$\mathcal{A}_F = \mathcal{A}_F \left(mQ - \sum_{i=1}^{r} (n_i + 1) P_i \right) + F$$

by Theorem 1.5.4 and Theorem 1.5.17 (observe the definition of the index of specialty which is given in Definition 1.5.1). So there is an element $z \in F$ with $z - \alpha \in \mathcal{A}_F(mQ - \sum_{i=1}^{r}(n_i + 1)P_i)$. This means

$$v_{P_i}(z - x_i) > n_i \quad \text{for } i = 1, \ldots, r \,, \text{ and} \tag{1.34}$$

$$v_P(z) \geq 0 \quad \text{for } \; P \in S \setminus \{P_1, \ldots, P_r\}. \tag{1.35}$$

Now we choose $y_1, \ldots, y_r \in F$ with $v_{P_i}(y_i) = n_i$. In the same manner as above we construct $y \in F$ with

$$v_{P_i}(y - y_i) > n_i \quad \text{for } \; i = 1, \ldots, r \,, \text{ and} \tag{1.36}$$

$$v_P(y) \geq 0 \quad \text{for } \; P \in S \setminus \{P_1, \ldots, P_r\}. \tag{1.37}$$

Then we have for $i = 1, \ldots, r$,

$$v_{P_i}(y) = v_{P_i}((y - y_i) + y_i) = n_i \tag{1.38}$$

by (1.36) and the Strict Triangle Inequality. Putting $x := y + z$ we obtain

$$v_{P_i}(x - x_i) = v_{P_i}(y + (z - x_i)) = n_i \quad (i = 1, \ldots, r)$$

by (1.38). For $P \in S \setminus \{P_1, \ldots, P_r\}$, $v_P(x) = v_P(y + z) \geq 0$ holds by (1.35) and (1.37). □

Next we investigate elements in F which have only one pole.

Proposition 1.6.6. *Let $P \in \mathbb{P}_F$. Then for each $n \geq 2g$ there exists an element $x \in F$ with pole divisor $(x)_\infty = nP$.*

Proof. By Theorem 1.5.17 we know that $\ell((n-1)P) = (n-1)\deg P + 1 - g$ and $\ell(nP) = n \cdot \deg P + 1 - g$, hence $\mathscr{L}((n-1)P) \subsetneqq \mathscr{L}(nP)$. Every element $x \in \mathscr{L}(nP) \setminus \mathscr{L}((n-1)P)$ has pole divisor nP. □

Definition 1.6.7. *Let $P \in \mathbb{P}_F$. An integer $n \geq 0$ is called a pole number of P if there is an element $x \in F$ with $(x)_\infty = nP$. Otherwise n is called a gap number of P.*

Clearly n is a pole number of P if and only if $\ell(nP) > \ell((n-1)P)$. Moreover, the set of pole numbers of P is a sub-semigroup of the additive semigroup $\mathbb{N}$ (to see this note that, if $(x_1)_\infty = n_1 P$ and $(x_2)_\infty = n_2 P$ then $x_1 x_2$ has the pole divisor $(x_1 x_2)_\infty = (n_1 + n_2)P$).

Theorem 1.6.8 (Weierstrass Gap Theorem). *Suppose that F/K has genus $g > 0$ and P is a place of degree one. Then there are exactly g gap numbers $i_1 < \ldots < i_g$ of P. We have*

$$i_1 = 1 \quad \textit{and} \quad i_g \leq 2g - 1 \,.$$

Proof. Each gap number of P is $\leq 2g-1$ by Proposition 1.6.6, and 0 is a pole number. We have the following obvious characterization of gap numbers:

$$i \text{ is a gap number of } P \iff \mathscr{L}((i-1)P) = \mathscr{L}(iP) \,.$$

Consider now the sequence of vector spaces

$$K = \mathscr{L}(0) \subseteq \mathscr{L}(P) \subseteq \mathscr{L}(2P) \subseteq \cdots \subseteq \mathscr{L}((2g-1)P) \,, \tag{1.39}$$

where $\dim \mathscr{L}(0) = 1$ and $\dim \mathscr{L}((2g-1)P) = g$ by Theorem 1.5.17. Observe that

$$\dim \mathscr{L}(iP) \leq \dim \mathscr{L}((i-1)P) + 1$$

for all i, see Lemma 1.4.8. So we have in (1.39) exactly $g-1$ numbers $1 \leq i \leq 2g-1$ with $\mathscr{L}((i-1)P) \subsetneqq \mathscr{L}(iP)$. The remaining g numbers are gaps of P.

Finally we must show that 1 is a gap. Suppose the converse, so 1 is a pole number of P. Since the pole numbers form an additive semigroup, every $n \in \mathbb{N}$ is a pole number, and there are no gaps at all. This is a contradiction because $g > 0$. □

Remark 1.6.9. Suppose that K is algebraically closed. Then one can show that almost all places of F/K have the same sequence of gap numbers (which are therefore called the gap numbers of the function field F/K). Such places of F/K are said to be *ordinary* places. Every non-ordinary place is called a *Weierstrass point* of F/K. If the genus of F/K is ≥ 2, there exists at least one Weierstrass point, see [21] or [45].

For a divisor A of degree < 0 we have $\mathscr{L}(A) = \{0\}$ by Corollary 1.4.12. On the other hand, if $\deg A > 2g-2$ then $\ell(A) = \deg A + 1 - g$ by Theorem 1.5.17. So the dimension of A depends only on $\deg A$ (and the genus) in these cases. We want to consider more closely the case where $0 \leq \deg A \leq 2g-2$; here the situation is rather complicated, but there are still some general results.

Definition 1.6.10. *A divisor $A \in \mathrm{Div}(F)$ is called non-special if $i(A) = 0$; otherwise A is called special.*

We note some immediate consequences of this definition.

Remark 1.6.11. (a) A is non-special $\iff \ell(A) = \deg A + 1 - g$.

(b) $\deg A > 2g-2 \Rightarrow A$ is non-special.

(c) The property of a divisor A being special or non-special depends only on the class $[A]$ of A in the divisor class group.

(d) Canonical divisors are special.

(e) Every divisor A with $\ell(A) > 0$ and $\deg A < g$ is special.

(f) If A is non-special and $B \geq A$ then B is non-special.

Proof. (a) is clear from the definition of $i(A)$, (b) is just Theorem 1.5.17, and (c) follows from the fact that $\ell(A)$ and $\deg A$ depend only on the divisor class of A.

(d) For a canonical divisor W we have $i(W) = \ell(W - W) = 1$ by Theorem 1.5.14, hence W is special.

(e) $1 \leq \ell(A) = \deg A + 1 - g + i(A) \Rightarrow i(A) \geq g - \deg A > 0$ since $\deg A < g$. Thus A is special.

(f) A is non-special if and only if $\mathcal{A}_F = \mathcal{A}_F(A) + F$, see Theorem 1.5.4. If $B \geq A$ then $\mathcal{A}_F(A) \subseteq \mathcal{A}_F(B)$, so (f) follows. □

With regard to item (e) of the preceding remark, the following result is interesting.

Proposition 1.6.12. *Suppose that $T \subseteq \mathbb{P}_F$ is a set of places of degree one such that $|T| \geq g$. Then there exists a non-special divisor $B \geq 0$ with $\deg B = g$ and $\mathrm{supp}\, B \subseteq T$.*

Proof. The crucial step of the proof is the following claim:

Claim. Given g distinct places $P_1, \ldots, P_g \in T$ and a divisor $A \geq 0$ with $\ell(A) = 1$ and $\deg A \leq g-1$, there is an index $j \in \{1, \ldots, g\}$ such that $\ell(A + P_j) = 1$.

Suppose the claim is false; then $\ell(A + P_j) > 1$, and there are elements $z_j \in \mathscr{L}(A + P_j) \backslash \mathscr{L}(A)$ for $j = 1, \ldots, g$. Since

$$v_{P_j}(z_j) = -v_{P_j}(A) - 1 \quad \textit{and} \quad v_{P_i}(z_j) \geq -v_{P_i}(A) \quad \textit{for} \quad i \neq j \,,$$

the Strict Triangle Inequality implies that the $g+1$ elements $1, z_1, \ldots, z_g$ are linearly independent over K. Choose a divisor $D \geq A + P_1 + \ldots + P_g$ with $\deg D = 2g-1$. Then $1, z_1, \ldots, z_g \in \mathscr{L}(D)$, hence $\ell(D) \geq g+1$. On the other hand, $\ell(D) = \deg D + 1 - g = g$ by the Riemann-Roch Theorem. This contradiction proves our claim.

Now the proof of Proposition 1.6.12 is very simple. By the above claim we find divisors $0 < P_{i_1} < P_{i_1} + P_{i_2} < \ldots < P_{i_1} + P_{i_2} + \ldots + P_{i_g} =: B$ (with $i_\nu \in \{1, \ldots, g\}$, not necessarily distinct) such that $\ell(P_{i_1} + \ldots + P_{i_j}) = 1$ for $j = 1, \ldots, g$. In particular $\ell(B) = 1$. The divisor B is non-special because

$$\deg B + 1 - g = g + 1 - g = 1 = \ell(B)\,.$$

(cf. Remark 1.6.11(a).) □

We conclude this section with an inequality for the dimension of an arbitrary divisor of degree $\leq 2g-2$.

Theorem 1.6.13 (Clifford's Theorem). *For all divisors A with $0 \leq \deg A \leq 2g-2$ holds*

$$\ell(A) \leq 1 + \frac{1}{2} \cdot \deg A\,.$$

The main step in the proof of Clifford's Theorem is the following result.

Lemma 1.6.14. *Suppose that A and B are divisors such that $\ell(A) > 0$ and $\ell(B) > 0$. Then*

$$\ell(A) + \ell(B) \leq 1 + \ell(A+B)\,.$$

Proof of Lemma 1.6.14. Since $\ell(A) > 0$ and $\ell(B) > 0$ we can find $A_0, B_0 \geq 0$ with $A \sim A_0$ and $B \sim B_0$ (cf. Remark 1.4.5). The set

$$X := \{D \in \mathrm{Div}(F) \mid D \leq A_0 \text{ and } \mathscr{L}(D) = \mathscr{L}(A_0)\}$$

is non-empty because $A_0 \in X$. As $\deg D \geq 0$ for all $D \in X$, there is some divisor $D_0 \in X$ of minimal degree. It follows that

$$\ell(D_0 - P) < \ell(D_0) \quad \text{for all } P \in \mathbb{P}_F\,. \tag{1.40}$$

We want to show that

$$\ell(D_0) + \ell(B_0) \leq 1 + \ell(D_0 + B_0)\,. \tag{1.41}$$

From (1.41) the lemma will follow immediately:

$$\begin{aligned}\ell(A) + \ell(B) &= \ell(A_0) + \ell(B_0) = \ell(D_0) + \ell(B_0) \\ &\leq 1 + \ell(D_0 + B_0) \leq 1 + \ell(A_0 + B_0) = 1 + \ell(A+B)\,.\end{aligned}$$

In order to prove (1.41) we make the additional assumption that K is an infinite field (in fact, we shall show later that Lemma 1.6.14 also holds in the case of a finite constant field, see Theorem 3.6.3(d)). Let $\operatorname{supp} B_0 = \{P_1, \dots, P_r\}$. Then $\mathscr{L}(D_0 - P_i)$ is a proper subspace of $\mathscr{L}(D_0)$ for $i = 1, \dots, r$, and since a vector space over an infinite field is not the union of finitely many proper subspaces, we can find an element

$$z \in \mathscr{L}(D_0) \setminus \bigcup_{i=1}^{r} \mathscr{L}(D_0 - P_i). \tag{1.42}$$

Consider the K-linear map

$$\varphi : \begin{cases} \mathscr{L}(B_0) & \longrightarrow \quad \mathscr{L}(D_0 + B_0)/\mathscr{L}(A_0), \\ \quad x & \longmapsto \quad xz \mod \mathscr{L}(A_0). \end{cases}$$

From (1.42) follows easily that the kernel of φ is K, hence

$$\dim \mathscr{L}(B_0) - 1 \le \dim \mathscr{L}(D_0 + B_0) - \dim \mathscr{L}(A_0),$$

which proves (1.41). □

Proof of Theorem 1.6.13. The case $\ell(A) = 0$ is trivial. Likewise, if $\ell(W - A) = 0$ (where W is canonical), then

$$\ell(A) = \deg A + 1 - g = 1 + \frac{1}{2}\deg A + \frac{1}{2}(\deg A - 2g) < 1 + \frac{1}{2}\deg A,$$

since $\deg A \le 2g - 2$. It remains to consider the case where $\ell(A) > 0$ and $\ell(W - A) > 0$. We can apply Lemma 1.6.14 to obtain

$$\ell(A) + \ell(W - A) \le 1 + \ell(W) = 1 + g. \tag{1.43}$$

On the other hand,

$$\ell(A) - \ell(W - A) = \deg A + 1 - g \tag{1.44}$$

by the Riemann-Roch Theorem. Adding (1.43) and (1.44) yields the desired result. □

1.7 Local Components of Weil Differentials

In Section 1.5 we considered the *diagonal* embedding $F \hookrightarrow \mathcal{A}_F$ which maps $x \in F$ to the corresponding principal adele. Now we introduce for each place $P \in \mathbb{P}_F$ another *local* embedding $\iota_P : F \hookrightarrow \mathcal{A}_F$.

Definition 1.7.1. *Let $P \in \mathbb{P}_F$.*

(a) For $x \in F$ let $\iota_P(x) \in \mathcal{A}_F$ be the adele whose P-component is x, and all other components are 0.

(b) For a Weil differential $\omega \in \Omega_F$ we define its local component $\omega_P : F \to K$ by

$$\omega_P(x) := \omega(\iota_P(x)) .$$

Clearly ω_P is a K-linear mapping.

Proposition 1.7.2. *Let $\omega \in \Omega_F$ and $\alpha = (\alpha_P) \in \mathcal{A}_F$. Then $\omega_P(\alpha_P) \neq 0$ for at most finitely many places P, and*

$$\omega(\alpha) = \sum_{P \in \mathbb{P}_F} \omega_P(\alpha_P) .$$

In particular

$$\sum_{P \in \mathbb{P}_F} \omega_P(1) = 0 . \tag{1.45}$$

Proof. We can assume that $\omega \neq 0$ and we set $W := (\omega)$, the divisor of ω (see Definition 1.5.11). There is a finite set $S \subseteq \mathbb{P}_F$ such that

$$v_P(W) = 0 \quad \textit{and} \quad v_P(\alpha_P) \geq 0 \quad \textit{for all} \quad P \notin S .$$

Define $\beta = (\beta_P) \in \mathcal{A}_F$ by

$$\beta_P := \begin{cases} \alpha_P & \textit{for} \quad P \notin S , \\ 0 & \textit{for} \quad P \in S . \end{cases}$$

Then $\beta \in \mathcal{A}_F(W)$ and $\alpha = \beta + \sum_{P \in S} \iota_P(\alpha_P)$, hence $\omega(\beta) = 0$ and

$$\omega(\alpha) = \sum_{P \in S} \omega_P(\alpha_P) .$$

For $P \notin S$, $\iota_P(\alpha_P) \in \mathcal{A}_F(W)$ and therefore $\omega_P(\alpha_P) = 0$. □

We shall see in Chapter 4 that Equation (1.45) is nothing else but the *Residue Theorem* for differentials of F/K.

Next we show that a Weil differential is uniquely determined by each of its local components.

Proposition 1.7.3. *(a) Let $\omega \neq 0$ be a Weil differential of F/K and $P \in \mathbb{P}_F$. Then*

$$v_P(\omega) = \max\{ r \in \mathbb{Z} \mid \omega_P(x) = 0 \;\; \textit{for all} \;\; x \in F \;\; \textit{with} \;\; v_P(x) \geq -r \} .$$

In particular ω_P is not identically 0.

(b) If $\omega, \omega' \in \Omega_F$ and $\omega_P = \omega'_P$ for some $P \in \mathbb{P}_F$ then $\omega = \omega'$.

Proof. (a) Recall that, by definition, $v_P(\omega) = v_P(W)$ where $W = (\omega)$ denotes the divisor of ω. Let $s := v_P(\omega)$. For $x \in F$ with $v_P(x) \geq -s$ we have $\iota_P(x) \in \mathcal{A}_F(W)$, hence $\omega_P(x) = \omega(\iota_P(x)) = 0$. Suppose now that $\omega_P(x) = 0$ for all $x \in F$ with $v_P(x) \geq -s-1$. Let $\alpha = (\alpha_Q)_{Q \in \mathbb{P}_F} \in \mathcal{A}_F(W+P)$. Then

$$\alpha = (\alpha - \iota_P(\alpha_P)) + \iota_P(\alpha_P)$$

with $\alpha - \iota_P(\alpha_P) \in \mathcal{A}_F(W)$ and $v_P(\alpha_P) \geq -s-1$, hence

$$\omega(\alpha) = \omega(\alpha - \iota_P(\alpha_P)) + \omega_P(\alpha_P) = 0\,.$$

Therefore ω vanishes on $\mathcal{A}_F(W+P)$, a contradiction to the definition of W.
(b) If $\omega_P = \omega'_P$ then $(\omega - \omega')_P = 0$, hence $\omega - \omega' = 0$ by (a). □

Once again we consider the rational function field $K(x)$. We use the notation introduced in Section 1.2; i.e., P_∞ denotes the pole divisor of x and P_a denotes the zero divisor of $x-a$ (for $a \in K$). The following result will be important in Chapter 4.

Proposition 1.7.4. *For the rational function field $F = K(x)$ the following hold:*

(a) The divisor $-2P_\infty$ is canonical.

(b) There exists a unique Weil differential $\eta \in \Omega_{K(x)}$ with $(\eta) = -2P_\infty$ and $\eta_{P_\infty}(x^{-1}) = -1$.

(c) The local components η_{P_∞} resp. η_{P_a} of the above Weil differential η satisfy

$$\eta_{P_\infty}((x-a)^n) = \begin{cases} 0 & \textit{for} \quad n \neq -1\,, \\ -1 & \textit{for} \quad n = -1\,, \end{cases}$$

$$\eta_{P_a}((x-a)^n) = \begin{cases} 0 & \textit{for} \quad n \neq -1\,, \\ 1 & \textit{for} \quad n = -1\,. \end{cases}$$

Proof. (a) $\deg(-2P_\infty) = -2 = 2g-2$ and $\ell(-2P_\infty) = 0 = g$, hence $-2P_\infty$ is canonical by Proposition 1.6.2.
(b) Choose a Weil differential ω with divisor $(\omega) = -2P_\infty$. Then ω vanishes on the space $\mathcal{A}_{K(x)}(-2P_\infty)$, but it does not vanish identically on $\mathcal{A}_{K(x)}(-P_\infty)$. Since

$$\dim \mathcal{A}_{K(x)}(-P_\infty)/\mathcal{A}_{K(x)}(-2P_\infty) = 1$$

(see Equation (1.24) in the proof of Theorem 1.5.4) and

$$\iota_{P_\infty}(x^{-1}) \in \mathcal{A}_{K(x)}(-P_\infty) \setminus \mathcal{A}_{K(x)}(-2P_\infty)\,,$$

we conclude that

$$\omega_{P_\infty}(x^{-1}) = \omega(\iota_{P_\infty}(x^{-1})) =: c \neq 0\,.$$

Setting $\eta := -c^{-1}\omega$ we obtain $(\eta) = -2P_\infty$ and $\eta_{P_\infty}(x^{-1}) = -1$. The uniqueness of η is easily proved. If η^* has the same properties as η then $\eta - \eta^*$ vanishes on the space $\mathcal{A}_{K(x)}(-P_\infty)$, which implies $\eta - \eta^* = 0$.

(c) Since a Weil differential vanishes on principal adeles, we have by Proposition 1.7.2 that

$$0 = \eta((x-a)^n) = \sum_{P \in \mathbb{P}_{K(x)}} \eta_P((x-a)^n)\,. \tag{1.46}$$

For $P \neq P_\infty$ and $P \neq P_a$ holds $v_P((x-a)^n) = 0$, therefore $\eta_P((x-a)^n) = 0$ by Proposition 1.7.3, and then (1.46) yields

$$\eta_{P_\infty}((x-a)^n) + \eta_{P_a}((x-a)^n) = 0\,. \tag{1.47}$$

In the case $n \leq -2$ we have $v_{P_\infty}((x-a)^n) \geq 2$, hence $\eta_{P_\infty}((x-a)^n) = 0$ by Proposition 1.7.3. Now (1.47) implies that $\eta_{P_a}((x-a)^n) = 0$ as well. If $n \geq 0$, $\eta_{P_a}((x-a)^n) = 0$ by Proposition 1.7.3, and we obtain the result for $\eta_{P_\infty}((x-a)^n)$ again by (1.47).

Eventually we consider the case $n = -1$. Since

$$\frac{1}{x-a} = \frac{a}{x(x-a)} + \frac{1}{x} \quad \textit{and} \quad \iota_{P_\infty}\left(\frac{a}{x(x-a)}\right) \in \mathcal{A}_{K(x)}(-2P_\infty)\,,$$

we see that $\eta_{P_\infty}((x-a)^{-1}) = \eta_{P_\infty}(x^{-1}) = -1$ (by definition of η), and from (1.47) follows that $\eta_{P_a}((x-a)^{-1}) = 1$. □

1.8 Exercises

1.1. Consider the rational function field $K(x)/K$ and a non-constant element $z = f(x)/g(x) \in K(x) \setminus K$, where $f(x), g(x) \in K[x]$ are relatively prime. We call $\deg(z) := \max\{\deg f(x), \deg g(x)\}$ the degree of z.

(i) Show that $[K(x) : K(z)] = \deg(z)$, and write down the minimal polynomial of x over $K(z)$ (in order to avoid calculations, you may use Theorem 1.4.11 and Example 1.4.13).

(ii) Show that $K(x) = K(z)$ if and only if $z = (ax+b)/(cx+d)$ with $a, b, c, d \in K$ and $ad - bc \neq 0$.

1.2. For a field extension L/M we denote by $\mathrm{Aut}(L/M)$ the group of automorphisms of L/M (i.e., automorphisms of L which are the identity on M). Let $K(x)/K$ be the rational function field over K. Show:

(i) For every $\sigma \in \mathrm{Aut}(K(x)/K)$ there exist $a, b, c, d \in K$ such that $ad - bc \neq 0$ and $\sigma(x) = (ax+b)/(cx+d)$.

(ii) Given $a, b, c, d \in K$ with $ad - bc \neq 0$, there is a unique automorphism $\sigma \in \mathrm{Aut}(K(x)/K)$ with $\sigma(x) = (ax + b)/(cx + d)$.

(iii) Denote by $\mathrm{GL}_2(K)$ the group of invertible 2×2 - matrices over K. For $A = \begin{pmatrix} a & c \\ b & d \end{pmatrix} \in \mathrm{GL}_2(K)$ denote by σ_A the automorphism of $K(x)/K$ with $\sigma_A(x) = (ax + b)/(cx + d)$. Show that the map which sends A to σ_A, is a homomorphism from $\mathrm{GL}_2(K)$ onto $\mathrm{Aut}(K(x)/K)$. Its kernel is the set of diagonal matrices of the form $\begin{pmatrix} a & 0 \\ 0 & a \end{pmatrix}$ with $a \in K^\times$, hence

$$\mathrm{Aut}(K(x)/K) \simeq \mathrm{GL}_2(K)/K^\times .$$

(This group $\mathrm{GL}_2(K)/K^\times$ is called the projective linear group and is denoted by $\mathrm{PGL}_2(K)$.)

1.3. If L is a field and G is a group of automorphisms of L, we denote by

$$L^G := \{w \in L \mid \sigma(w) = w \ \ \textit{for all} \ \ \sigma \in G\}$$

the fixed field of G. It is well-known from algebra that if G is a finite group, then L/L^G is a finite extension of degree $[L : L^G] = \mathrm{ord}(G)$.
Now let $G \subseteq \mathrm{Aut}(K(x)/K)$ be a finite subgroup of the automorphism group of the rational function field $K(x)$ over K, and put

$$z := \sum_{\sigma \in G} \sigma(x) \ , \qquad u := \prod_{\sigma \in G} \sigma(x) \ .$$

Show:
(i) Either $z \in K$, or $K(z) = K(x)^G$.
(ii) Either $u \in K$, or $K(u) = K(x)^G$.
(iii) Find examples of finite subgroups $G \subseteq \mathrm{Aut}(K(x)/K)$ for both alternatives in (i) (and also in (ii)).

1.4. Let $K(x)$ be the rational function field over K. Find bases of the following Riemann-Roch spaces:

$$\mathscr{L}(rP_\infty) \ , \quad \mathscr{L}(rP_\alpha) \ , \quad \mathscr{L}(P_{p(x)}) \ ,$$

where $r \geq 0$, and the places P_∞, P_α and $P_{p(x)}$ are as in Section 1.2.

1.5. *(Representation of rational functions by partial fractions)*
(i) Show that every element $z \in K(x)$ can be written as

$$z = \sum_{i=1}^{r} \sum_{j=1}^{k_i} \frac{c_{ij}(x)}{p_i(x)^j} + h(x) \ ,$$

where

(a) $p_1(x), ..., p_r(x)$ are distinct monic irreducible polynomials in $K[x]$,

(b) $k_1, ..., k_r \geq 1$,

(c) $c_{ij}(x) \in K[x]$ and $\deg(c_{ij}(x)) < \deg(p_i(x))$,

(d) $c_{ik_i}(x) \neq 0$ for $1 \leq i \leq r$,

(e) $h(x) \in K[x]$.

Note that the case $r = 0$ is allowed; it just means that $z = h(x) \in K[x]$.

(ii) Show that the above representation of z is unique.

1.6. Let $F = K(x)$ be the rational function field over K. Show directly that $\mathcal{A}_F = \mathcal{A}_F(0) + F$ (by Corollary 1.5.5 this provides another proof that the rational function field has genus 0).

1.7. There are many analogies between algebraic function fields and algebraic number fields (i.e., finite extensions of the field $\mathbb{Q}$ of rational numbers). Here is a first example.

A valuation ring of a field L is a subring $\mathcal{O} \subsetneqq L$ such that for all $z \in L$ one has that $z \in \mathcal{O}$ or $z^{-1} \in \mathcal{O}$.

(i) Show that every valuation ring is a local ring (i.e., it has a unique maximal ideal).

(ii) Now we consider the field $L = \mathbb{Q}$. Show that for every prime number $p \in \mathbb{Z}$, the set $\mathbb{Z}_{(p)} := \{a/b \in \mathbb{Q} \mid a, b \in \mathbb{Z} \text{ and } p \nmid b\}$ is a valuation ring of $\mathbb{Q}$. What is the maximal ideal of $\mathbb{Z}_{(p)}$?

(iii) Let $\mathcal{O}$ be a valuation ring of $\mathbb{Q}$. Show that $\mathcal{O} = \mathbb{Z}_{(p)}$ for some prime number p.

In the following exercises, F/K always denotes a function field of genus g with full constant field K.

1.8. Assume that $g > 0$ and A is a divisor with $\ell(A) > 0$. Show that $\ell(A) = \deg(A) + 1$ if and only if A is a principal divisor.

1.9. Show that the following conditions are equivalent:

(a) $g = 0$.

(b) There is a divisor A with $\deg(A) = 2$ and $\ell(A) = 3$.

(c) There is a divisor A with $\deg(A) \geq 1$ and $\ell(A) > \deg(A)$.

(d) There is a divisor A with $\deg(A) \geq 1$ and $\ell(A) = \deg(A) + 1$.

In case of char$K \neq 2$, also the following condition is equivalent to the above:

(e) There are elements $x, y \in F$ such that $F = K(x, y)$ and $y^2 = ax^2 + b$, with $a, b \in K^\times$.

1.10. Let $\mathbb{R}(x)$ be the rational function field over the field of real numbers.

(i) Show that the polynomial $f(T) := T^2 + (x^2+1) \in \mathbb{R}(x)[T]$ is irreducible over $\mathbb{R}(x)$.

Let $F := \mathbb{R}(x,y)$, where $y^2 + x^2 + 1 = 0$. By (i), $[F : \mathbb{R}(x)] = 2$. Show:

(ii) $\mathbb{R}$ is the full constant field of F, and $F/\mathbb{R}$ has genus $g = 0$.

(iii) $F/\mathbb{R}$ is not a rational function field.

(iv) All places of $F/\mathbb{R}$ have degree 2.

1.11. Assume that $\mathrm{char}(K) \neq 2$. Let $F = K(x,y)$ with

$$y^2 = f(x) \in K[x]\ , \quad \deg f(x) = 2m+1 \geq 3\ .$$

Show:

(i) K is the full constant field of F.

(ii) There is exactly one place $P \in \mathbb{P}_F$ which is a pole of x, and this place is also the only pole of y.

(iii) For every $r \geq 0$, the elements $1, x, x^2, \ldots, x^r, y, xy, \ldots, x^s y$ with $0 \leq s < r-m$ are in $\mathscr{L}(2rP)$.

(iv) The genus of F/K satisfies $g \leq m$.

Remark. We will prove later that the genus is in fact $g = m$, if the polynomial $f(x)$ does not have multiple factors, see Example 3.7.6.

1.12. Let $K = \mathbb{F}_3$ be the field with 3 elements and $K(x)$ the rational function field over K. Show:

(i) The polynomial $f(T) = T^2 + x^4 - x^2 + 1$ is irreducible over $K(x)$.

(ii) Let $F = K(x,y)$ where y is a zero of the polynomial $f(T)$ as above, and let $\tilde{K}$ be the full constant field of F. Then $\tilde{K}$ has 9 elements, and $F = \tilde{K}(x)$.

1.13. Assume that F/K has a place $P \in \mathbb{P}_F$ of degree one. Show that there exist $x, y \in F$ such that $[F : K(x)] = [F : K(y)] = 2g+1$ and $F = K(x,y)$.

1.14. Let V, W be vector spaces over K. A non-degenerate pairing of V and W is a bilinear map $s : V \times W \to K$ such that the following hold: For every $v \in V$ with $v \neq 0$ there is some $w \in W$ with $s(v,w) \neq 0$, and for every $w \in W$ with $w \neq 0$ there is some $v \in V$ with $s(v,w) \neq 0$.

Now we consider a function field F/K, a divisor $A \in \mathrm{Div}(F)$ and a non-zero Weil differential $\omega \in \Omega_F$. Let $W := (\omega)$. Show that the map $s : \mathscr{L}(W-A) \times \mathcal{A}_F/(\mathcal{A}_F(A) + F) \to K$ given by $s(x,\alpha) := \omega(x\alpha)$ is well-defined, and it is a non-degenerate pairing.

1.15. Assume that the constant field K is algebraically closed. Show that for every integer $d \geq g$, there exists a divisor $A \in \mathrm{Div}(F)$ with $\deg(A) = d$ and $\ell(A) = \deg(A) + 1 - g$.

1.16. Let $i(A)$ denote the index of specialty of the divisor $A \in \mathrm{Div}(F)$. Show:
(i) $i(A) \le \max\{0, 2g-1-\deg(A)\}$.
(ii) Assume that $i(A) > 0$. Show that for every divisor B,

$$\ell(A-B) \le i(B) .$$

Hint. Find a monomorphism $\mu : \mathscr{L}(A-B) \to \Omega_F(B)$.
(iii) As a special case of (ii), show that

$$i(A) > 0 \quad \Rightarrow \quad \ell(A) \le g .$$

1.17. For a divisor $C \in \mathrm{Div}(F)$ with $\ell(C) > 0$ we define

$$|C| := \{A \in \mathrm{Div}(F) \mid A \sim C \textit{ and } A \ge 0\} .$$

This set is called the linear system corresponding to C. Obviously it depends only on the divisor class $[C] \in \mathrm{Cl}(F)$. The class $[C]$ is called primitive, if there is no divisor $B > 0$ such that $B \le A$ for all $A \in |C|$. Show:
(i) Every divisor class of degree $\ge 2g$ is primitive.
(ii) For $g \ge 1$, the canonical class is primitive.
(iii) Let $g \ge 1$, W a canonical divisor and P be a place of degree one. Then the class $[W+P]$ is not primitive.

1.18. The number $\gamma := \min\{[F : K(z)] | z \in F\}$ is called the gonality of F/K. We also define for all $r \ge 1$,

$$\gamma_r := \min\{\deg(A) \mid A \in \mathrm{Div}(F) \textit{ and } \ell(A) \ge r\} .$$

The sequence $(\gamma_1, \gamma_2, \gamma_3, \ldots)$ is called the gonality sequence of F/K.
(i) Show that $\gamma_1 = 0$ and $\gamma_2 = \gamma$.
In parts (ii) - (viii) we assume that there exists a rational place $P \in \mathbb{P}_F$.
Prove:
(ii) For all integers $r \ge 1$ there exists a divisor $A_r \ge 0$ with $\deg(A_r) = \gamma_r$ and $\ell(A_r) = r$.
(iii) $\gamma_r < \gamma_{r+1}$ for all $r \ge 1$.
(iv) $\gamma_r = r + g - 1$ for all $r > g$.
(v) If $g \ge 1$, then $\gamma_g = 2g - 2$.
(vi) $\gamma_r \ge 2(r-1)$ for all $r \in \{1, \ldots, g\}$.
(vii) If $g \ge 2$, then $\gamma \le g$.
(viii) Let $\Gamma := \{j \ge 0 \,|\, \text{there is no } r \text{ with } \gamma_r = j\}$. Then
  (1) $|\Gamma| = g$,
  (2) $1 \in \Gamma$ and $2g-1 \in \Gamma$, if $g \ge 1$.

2

Algebraic Geometry Codes

In this chapter we describe V.D.Goppa's construction of error-correcting codes using algebraic function fields. We start with a brief survey of the concepts of coding theory. Then we define algebraic geometry codes (AG codes) and develop their main properties. The codes constructed by means of a rational function field are discussed in detail in Section 2.3.

2.1 Codes

We are going to introduce some basic notions of coding theory. The reader who is not familiar with these concepts is referred to the introductory chapter of any book on error-correcting codes.

Let $\mathbb{F}_q$ denote the finite field with q elements. We consider the n-dimensional vector space $\mathbb{F}_q^n$ whose elements are n-tuples $a = (a_1, \ldots, a_n)$ with $a_i \in \mathbb{F}_q$.

Definition 2.1.1. *For $a = (a_1, \ldots, a_n)$ and $b = (b_1, \ldots, b_n) \in \mathbb{F}_q^n$ let*

$$d(a,b) := \left| \{ i \, ; \, a_i \neq b_i \} \right| .$$

This function d is called the Hamming distance on $\mathbb{F}_q^n$. The weight of an element $a \in \mathbb{F}_q^n$ is defined as

$$\mathrm{wt}(a) := d(a,0) = \left| \{ i \, ; \, a_i \neq 0 \} \right| .$$

The Hamming distance is a metric on $\mathbb{F}_q^n$ as one can verify immediately. In particular, the *Triangle Inequality* $d(a,c) \leq d(a,b) + d(b,c)$ holds for all $a, b, c \in \mathbb{F}_q^n$.

Definition 2.1.2. *A code C (over the alphabet $\mathbb{F}_q$) is a linear subspace of $\mathbb{F}_q^n$; the elements of C are called codewords. We call n the length of C and*

H. Stichtenoth, *Algebraic Function Fields and Codes*,
Graduate Texts in Mathematics 254,

$\dim C$ *(as* $\mathbb{F}_q$*-vector space) the dimension of* C. *An* $[n,k]$ *code is a code of length* n *and dimension* k.

The minimum distance $d(C)$ *of a code* $C \neq 0$ *is defined as*

$$d(C) := \min\{d(a,b) \mid a,b \in C \text{ and } a \neq b\} = \min\{\mathrm{wt}(c) \mid 0 \neq c \in C\}\,.$$

An $[n,k]$ *code with minimum distance* d *will be referred to as an* $[n,k,d]$ *code.*

Remark 2.1.3. More generally one can define a code to be an arbitrary non-empty subset $C \subseteq A^n$ where $A \neq \emptyset$ is a finite set. If $A = \mathbb{F}_q$ and $C \subseteq \mathbb{F}_q^n$ is a linear subspace, C is said to be a *linear* code. Most codes employed in practice belong to this class, therefore we will only consider linear codes in this book, without writing the attribute 'linear'.

For a code C with minimum distance $d = d(C)$ we set $t := [(d-1)/2]$ (where $[x]$ denotes the integer part of the real number x; i.e., $x = [x] + \varepsilon$ with $[x] \in \mathbb{Z}$ and $0 \leq \varepsilon < 1$). Then C is said to be *t-error correcting*. The following is obvious: if $u \in \mathbb{F}_q^n$ and $d(u,c) \leq t$ for some $c \in C$ then c is the only codeword with $d(u,c) \leq t$.

A simple way to describe a specific code C explicitly is to write down a basis of C (as a vector space over $\mathbb{F}_q$).

Definition 2.1.4. *Let* C *be an* $[n,k]$ *code over* $\mathbb{F}_q$. *A generator matrix of* C *is a* $k \times n$ *matrix whose rows are a basis of* C.

Definition 2.1.5. *The canonical inner product on* $\mathbb{F}_q^n$ *is defined by*

$$\langle a,b\rangle \;:=\; \sum_{i=1}^{n} a_i b_i\;,$$

for $a = (a_1,\ldots,a_n)$ *and* $b = (b_1,\ldots,b_n) \in \mathbb{F}_q^n$.

Obviously this is a non-degenerate symmetric bilinear form on $\mathbb{F}_q^n$.

Definition 2.1.6. *If* $C \subseteq \mathbb{F}_q^n$ *is a code then*

$$C^{\perp} := \{u \in \mathbb{F}_q^n \mid \langle u,c\rangle = 0 \text{ for all } c \in C\}$$

is called the dual of C. *The code* C *is called self-dual (resp. self-orthogonal) if* $C = C^{\perp}$ *(resp.* $C \subseteq C^{\perp}$*).*

It is well-known from linear algebra that the dual of an $[n,k]$ code is an $[n,n-k]$ code, and $(C^{\perp})^{\perp} = C$. In particular, the dimension of a self-dual code of length n is $n/2$.

Definition 2.1.7. *A generator matrix H of $C^\perp$ is said to be a parity check matrix for C.*

Clearly a parity check matrix H of an $[n,k]$ code C is an $(n-k)\times n$ matrix of rank $n-k$, and we have

$$C = \{\, u \in \mathbb{F}_q^n \mid H \cdot u^t = 0 \,\}$$

(where u^t denotes the transpose of u). Thus a parity check matrix 'checks' whether a vector $u \in \mathbb{F}_q^n$ is a codeword or not.

One of the basic problems in algebraic coding theory is to construct - over a fixed alphabet $\mathbb{F}_q$ - codes whose dimension and minimum distance are large in comparison with their length. However there are some restrictions. Roughly speaking, if the dimension of a code is large (with respect to its length), then its minimum distance is small. The simplest bound is the following.

Proposition 2.1.8 (Singleton Bound). *For an $[n,k,d]$ code C holds*

$$k + d \le n + 1\,.$$

Proof. Consider the linear subspace $E \subseteq \mathbb{F}_q^n$ given by

$$E := \{\, (a_1,\ldots,a_n) \in \mathbb{F}_q^n \mid a_i = 0 \quad \textit{for all} \quad i \ge d \,\}\,.$$

Every $a \in E$ has weight $\le d-1$, hence $E \cap C = 0$. As $\dim E = d-1$ we obtain

$$\begin{aligned} k + (d-1) &= \dim C + \dim E \\ &= \dim (C+E) + \dim (C \cap E) = \dim (C+E) \le n\,. \end{aligned}$$

□

Codes with $k + d = n + 1$ are in a sense optimal; such codes are called MDS *codes* (*maximum distance separable codes*). If $n \le q+1$, there exist MDS codes over $\mathbb{F}_q$ for all dimensions $k \le n$ (this will be shown in Section 2.3).

The Singleton Bound does not take into consideration the size of the alphabet. Several other *upper* bounds for the parameters k and d (involving the length n of the code and the size q of the alphabet) are known. They are stronger than the Singleton Bound if n is large with respect to q. We refer to [25],[28], see also Chapter 8, Section 8.4.

It is in general a much harder problem to obtain *lower* bounds for the minimum distance of a given code (or a given class of codes). Only few such classes are known, for instance BCH *codes*, *Goppa codes* or *quadratic residue codes* (cf. [25],[28]). One of the reasons for the interest in algebraic geometry codes (to be defined in the next section) is that for this large class of codes a good lower bound for the minimum distance is available.

2.2 AG Codes

Algebraic geometry codes (AG codes) were introduced by V.D. Goppa in [15]. Therefore they are sometimes also called *geometric Goppa codes.* As a motivation for the construction of these codes we first consider Reed-Solomon codes over $\mathbb{F}_q$. This important class of codes is well-known in coding theory for a long time. Algebraic geometry codes are a very natural generalization of Reed-Solomon codes.

Let $n = q - 1$ and let $\beta \in \mathbb{F}_q$ be a primitive element of the multiplicative group $\mathbb{F}_q^\times$; i.e., $\mathbb{F}_q^\times = \{\beta, \beta^2, \ldots, \beta^n = 1\}$. For an integer k with $1 \le k \le n$ we consider the k-dimensional vector space

$$\mathscr{L}_k := \{f \in \mathbb{F}_q[X] \,|\, \deg f \le k - 1\} \tag{2.1}$$

and the *evaluation map* $\mathrm{ev} : \mathscr{L}_k \to \mathbb{F}_q^n$ given by

$$\mathrm{ev}(f) := (f(\beta), f(\beta^2), \ldots, f(\beta^n)) \in \mathbb{F}_q^n\,. \tag{2.2}$$

Obviously this map is $\mathbb{F}_q$-linear, and it is injective because a non-zero polynomial $f \in \mathbb{F}_q[X]$ of degree $< n$ has less than n zeros. Therefore

$$C_k := \{(f(\beta), f(\beta^2), \ldots, f(\beta^n)) \,|\, f \in \mathscr{L}_k\} \tag{2.3}$$

is an $[n, k]$ code over $\mathbb{F}_q$; it is called an RS *code* (*Reed-Solomon code*). The weight of a codeword $0 \ne c = \mathrm{ev}(f) \in C_k$ is given by

$$\begin{aligned} \mathrm{wt}(c) &= n - \left|\{i \in \{1, \ldots, n\}\,;\, f(\beta^i) = 0\}\right| \\ &\ge n - \deg f \ge n - (k-1)\,. \end{aligned}$$

Hence the minimum distance d of C_k satisfies the inequality $d \ge n+1-k$. On the other hand, $d \le n + 1 - k$ by the Singleton Bound. Thus Reed-Solomon codes are MDS codes over $\mathbb{F}_q$. Observe however that RS codes are short in comparison with the size of the alphabet $\mathbb{F}_q$, since $n = q - 1$.

Now we introduce the notion of an algebraic geometry code. Let us fix some notation valid for the entire section.

> $F/\mathbb{F}_q$ *is an algebraic function field of genus* g.
>
> $P_1, \ldots, P_n$ *are pairwise distinct places of* $F/\mathbb{F}_q$ *of degree* 1.
>
> $D = P_1 + \ldots + P_n$.
>
> G *is a divisor of* $F/\mathbb{F}_q$ *such that* $\mathrm{supp}\, G \cap \mathrm{supp}\, D = \emptyset$.

Definition 2.2.1. *The algebraic geometry code (or* AG *code)* $C_{\mathscr{L}}(D, G)$ *associated with the divisors* D *and* G *is defined as*

$$C_{\mathscr{L}}(D, G) := \{(x(P_1), \ldots, x(P_n)) \,|\, x \in \mathscr{L}(G)\} \subseteq \mathbb{F}_q^n\,.$$

Note that this definition makes sense: for $x \in \mathscr{L}(G)$ we have $v_{P_i}(x) \geq 0$ $(i = 1, \ldots, n)$ because $\operatorname{supp} G \cap \operatorname{supp} D = \emptyset$. The residue class $x(P_i)$ of x modulo P_i is an element of the residue class field of P_i (see Definition 1.1.14). As $\deg P_i = 1$, this residue class field is $\mathbb{F}_q$, so $x(P_i) \in \mathbb{F}_q$.

As in (2.2) we can consider the *evaluation map* $\mathrm{ev}_D : \mathscr{L}(G) \to \mathbb{F}_q^n$ given by

$$\mathrm{ev}_D(x) := (x(P_1), \ldots, x(P_n)) \in \mathbb{F}_q^n \,. \tag{2.4}$$

The evaluation map is $\mathbb{F}_q$-linear, and $C_{\mathscr{L}}(D,G)$ is the image of $\mathscr{L}(G)$ under this map. The analogy with the definition of Reed-Solomon codes (2.3) is obvious. In fact, choosing the function field $F/\mathbb{F}_q$ and the divisors D and G in an appropriate manner, RS codes are easily seen to be a special case of AG codes, see Section 2.3.

Definition 2.2.1 looks like a very artificial way to define certain codes over $\mathbb{F}_q$. The next theorem will show why these codes are interesting: one can calculate (or at least estimate) their parameters n, k and d by means of the Riemann-Roch Theorem, and one obtains a non-trivial lower bound for their minimum distance in a very general setting.

Theorem 2.2.2. *$C_{\mathscr{L}}(D,G)$ is an $[n,k,d]$ code with parameters*

$$k = \ell(G) - \ell(G-D) \quad \textit{and} \quad d \geq n - \deg G \,.$$

Proof. The evaluation map (2.4) is a surjective linear map from $\mathscr{L}(G)$ to $C_{\mathscr{L}}(D,G)$ with kernel

$$\mathrm{Ker}(\mathrm{ev}_D) = \{x \in \mathscr{L}(G) \,|\, v_{P_i}(x) > 0 \textit{ for } i = 1, \ldots, n\} = \mathscr{L}(G-D) \,.$$

It follows that $k = \dim C_{\mathscr{L}}(D,G) = \dim \mathscr{L}(G) - \dim \mathscr{L}(G-D) = \ell(G) - \ell(G-D)$. The assertion regarding the minimum distance d makes sense only if $C_{\mathscr{L}}(D,G) \neq 0$, so we will assume this. Choose an element $x \in \mathscr{L}(G)$ with $\mathrm{wt}(ev_D(x)) = d$. Then exactly $n-d$ places $P_{i_1}, \ldots, P_{i_{n-d}}$ in the support of D are zeros of x, so

$$0 \neq x \in \mathscr{L}(G - (P_{i_1} + \ldots + P_{i_{n-d}})) \,.$$

We conclude by Corollary 1.4.12.(b) that

$$0 \leq \deg\left(G - (P_{i_1} + \ldots + P_{i_{n-d}})\right) = \deg G - n + d \,.$$

Hence $d \geq n - \deg G$. □

Corollary 2.2.3. *Suppose that the degree of G is strictly less than n. Then the evaluation map $\mathrm{ev}_D : \mathscr{L}(G) \to C_{\mathscr{L}}(D,G)$ is injective, and we have:*

(a) $C_{\mathscr{L}}(D,G)$ is an $[n,k,d]$ code with

$$d \geq n - \deg G \quad \textit{and} \quad k = \ell(G) \geq \deg G + 1 - g\,.$$

Hence

$$k + d \geq n + 1 - g\,. \tag{2.5}$$

(b) If in addition $2g - 2 < \deg G < n$, then $k = \deg G + 1 - g$.

(c) If $\{x_1, \ldots, x_k\}$ is a basis of $\mathscr{L}(G)$ then the matrix

$$M = \begin{pmatrix} x_1(P_1) & x_1(P_2) & \ldots & x_1(P_n) \\ \vdots & \vdots & & \vdots \\ x_k(P_1) & x_k(P_2) & \ldots & x_k(P_n) \end{pmatrix}$$

is a generator matrix for $C_{\mathscr{L}}(D,G)$.

Proof. By assumption we have $\deg(G-D) = \deg G - n < 0$, so $\mathscr{L}(G-D) = 0$. Since $\mathscr{L}(G-D)$ is the kernel of the evaluation map, this is an injective mapping. The remaining assertions are trivial consequences of Theorem 2.2.2 and the Riemann-Roch Theorem. □

We point out that the *lower* bound (2.5) for the minimum distance looks very similar to the *upper* Singleton Bound. Putting both bounds together we see that for $\deg G < n$,

$$n + 1 - g \leq k + d \leq n + 1\,. \tag{2.6}$$

Note that $k + d = n + 1$ if F is a function field of genus $g = 0$. Hence the AG codes constructed by means of a rational function field $\mathbb{F}_q(z)$ are always MDS codes. For more details see Section 2.3.

In order to obtain a meaningful bound for the minimum distance of $C_{\mathscr{L}}(D,G)$ by Theorem 2.2.2, we often assume that $\deg G < n$.

Definition 2.2.4. *The integer $d^* := n - \deg G$ is called the designed distance of the code $C_{\mathscr{L}}(D,G)$.*

Theorem 2.2.2 states that the minimum distance d of an AG code cannot be less than its designed distance. The question whether $d^* = d$ or $d^* < d$ is answered by the following remark.

Remark 2.2.5. Suppose that $\ell(G) > 0$ and $d^* = n - \deg G > 0$. Then $d^* = d$ if and only if there exists a divisor D' with $0 \leq D' \leq D$, $\deg D' = \deg G$ and $\ell(G - D') > 0$.

Proof. First we assume $d^* = d$. Then there is an element $0 \neq x \in \mathcal{L}(G)$ such that the codeword $(x(P_1), \ldots, x(P_n)) \in C_{\mathcal{L}}(D, G))$ has precisely $n - d = n - d^* = \deg G$ zero components, say $x(P_{i_j}) = 0$ for $j = 1, \ldots, \deg G$. Put

$$D' := \sum_{j=1}^{\deg G} P_{i_j} .$$

Then $0 \leq D' \leq D$, $\deg D' = \deg G$ and $\ell(G - D') > 0$ (as $x \in \mathcal{L}(G - D')$).

Conversely, if D' has the above properties then we choose an element $0 \neq y \in \mathcal{L}(G - D')$. The weight of the corresponding codeword $(y(P_1), \ldots, y(P_n))$ is $n - \deg G = d^*$, hence $d = d^*$. □

Another code can be associated with the divisors G and D, by using local components of Weil differentials. We recall some notation introduced in Chapter 1. For a divisor $A \in \mathrm{Div}(F)$, $\Omega_F(A)$ is the space of Weil differentials ω with $(\omega) \geq A$. This is a finite-dimensional vector space over $\mathbb{F}_q$ of dimension $i(A)$ (the index of specialty of A). For a Weil differential ω and a place $P \in \mathbb{P}_F$, the map $\omega_P : F \to \mathbb{F}_q$ denotes the local component of ω at P.

Definition 2.2.6. *Let G and $D = P_1 + \ldots + P_n$ be divisors as before (i.e., the P_i are pairwise distinct places of degree one, and $\mathrm{supp}\, G \cap \mathrm{supp}\, D = \emptyset$). Then we define the code $C_\Omega(D, G) \subseteq \mathbb{F}_q^n$ by*

$$C_\Omega(D, G) := \{(\omega_{P_1}(1), \ldots, \omega_{P_n}(1)) \mid \omega \in \Omega_F(G - D)\} .$$

Also the code $C_\Omega(D, G)$ is called an algebraic geometry code. The relation between the codes $C_{\mathcal{L}}(D, G)$ and $C_\Omega(D, G)$ will be explained in Theorem 2.2.8 and Proposition 2.2.10. Our first result about $C_\Omega(D, G)$ is an analogue to Theorem 2.2.2.

Theorem 2.2.7. *$C_\Omega(D, G)$ is an $[n, k', d']$ code with parameters*

$$k' = i(G - D) - i(G) \quad \textit{and} \quad d' \geq \deg G - (2g - 2) .$$

Under the additional hypothesis $\deg G > 2g - 2$, we have $k' = i(G - D) \geq n + g - 1 - \deg G$. If moreover $2g - 2 < \deg G < n$ then

$$k' = n + g - 1 - \deg G .$$

Proof. Let $P \in \mathbb{P}_F$ be a place of degree one and let ω be a Weil differential with $v_P(\omega) \geq -1$. We claim that

$$\omega_P(1) = 0 \quad \Longleftrightarrow \quad v_P(\omega) \geq 0 . \tag{2.7}$$

In order to prove this we use Proposition 1.7.3 which states that for an integer $r \in \mathbb{Z}$,

$$v_P(\omega) \geq r \iff \omega_P(x) = 0 \text{ for all } x \in F \text{ with } v_P(x) \geq -r\,. \tag{2.8}$$

The implication $\Leftarrow$ of (2.7) is an obvious consequence of (2.8). Conversely, suppose that $\omega_P(1) = 0$. Let $x \in F$ with $v_P(x) \geq 0$. Since $\deg P = 1$, we can write $x = a + y$ with $a \in \mathbb{F}_q$ and $v_P(y) \geq 1$. Then

$$\omega_P(x) = \omega_P(a) + \omega_P(y) = a \cdot \omega_P(1) + 0 = 0\,.$$

(Observe that $\omega_P(y) = 0$ because $v_P(\omega) \geq -1$ and $v_P(y) \geq 1$, cf. (2.8).) Hence (2.7) is proved.

Now we consider the $\mathbb{F}_q$-linear mapping

$$\varrho_D : \begin{cases} \Omega_F(G-D) & \longrightarrow \quad C_\Omega(D,G)\,, \\ \omega & \longmapsto (\omega_{P_1}(1), \ldots, \omega_{P_n}(1))\,. \end{cases}$$

ϱ_D is surjective, and its kernel is $\Omega_F(G)$ by (2.7). Therefore

$$k' = \dim \Omega_F(G-D) - \dim \Omega_F(G) = i(G-D) - i(G)\,. \tag{2.9}$$

Let $\varrho_D(\omega) \in C_\Omega(D,G)$ be a codeword of weight $m > 0$. Then $\omega_{P_i}(1) = 0$ for certain indices $i = i_1, \ldots, i_{n-m}$, so

$$\omega \in \Omega_F\big(G - \big(D - \sum_{j=1}^{n-m} P_{i_j}\big)\big)$$

by (2.7). Since $\Omega_F(A) \neq 0$ implies $\deg A \leq 2g-2$ (by Theorem 1.5.17), we obtain

$$2g-2 \geq \deg G - \big(n - (n-m)\big) = \deg G - m\,.$$

Hence the minimum distance d' of $C_\Omega(D,G)$ satisfies the inequality $d' \geq \deg G - (2g-2)$.

Assume now that $\deg G > 2g-2$. By Theorem 1.5.17 we obtain $i(G) = 0$. Now (2.9) and the Riemann-Roch Theorem yield

$$\begin{aligned} k' = i(G-D) &= \ell(G-D) - \deg(G-D) - 1 + g \\ &= \ell(G-D) + n + g - 1 - \deg G\,. \end{aligned}$$

The remaining assertions of Theorem 2.2.7 follow immediately. □

In analogy to Definition 2.2.4, the integer $\deg G - (2g-2)$ is called the *designed distance* of $C_\Omega(D,G)$.

There is a close relation between the codes $C_\mathscr{L}(D,G)$ and $C_\Omega(D,G)$:

Theorem 2.2.8. *The codes $C_\mathscr{L}(D,G)$ and $C_\Omega(D,G)$ are dual to each other; i.e.,*

$$C_\Omega(D,G) = C_\mathscr{L}(D,G)^\perp\,.$$

Proof. First we note the following fact: Consider a place $P \in \mathbb{P}_F$ of degree one, a Weil differential ω with $v_P(\omega) \geq -1$ and an element $x \in F$ with $v_P(x) \geq 0$. Then

$$\omega_P(x) = x(P) \cdot \omega_P(1)\,. \tag{2.10}$$

In order to prove (2.10) we write $x = a+y$ with $a = x(P) \in \mathbb{F}_q$ and $v_P(y) > 0$. Then $\omega_P(x) = \omega_P(a) + \omega_P(y) = a \cdot \omega_P(1) + 0 = x(P) \cdot \omega_P(1)$, by (2.8).

Next we show that $C_\Omega(D,G) \subseteq C_\mathscr{L}(D,G)^\perp$. So let $\omega \in \Omega_F(G-D)$ and $x \in \mathscr{L}(G)$. We obtain

$$\begin{aligned}
0 = \omega(x) &= \sum_{P \in \mathbb{P}_F} \omega_P(x) && (2.11)\\
&= \sum_{i=1}^{n} \omega_{P_i}(x) && (2.12)\\
&= \sum_{i=1}^{n} x(P_i) \cdot \omega_{P_i}(1) && (2.13)\\
&= \langle (\omega_{P_1}(1), \ldots, \omega_{P_n}(1)), (x(P_1), \ldots, x(P_n)) \rangle\,,
\end{aligned}$$

where $\langle\ ,\ \rangle$ denotes the canonical inner product on $\mathbb{F}_q^n$. We still have to justify the single steps in the above computation. (2.11) follows from Proposition 1.7.2 and the fact that Weil differentials vanish on principal adeles. For $P \in \mathbb{P}_F \backslash \{P_1, \ldots, P_n\}$ we have $v_P(x) \geq -v_P(\omega)$ (as $x \in \mathscr{L}(G)$ and $\omega \in \Omega(G-D)$), so $\omega_P(x) = 0$ by (2.8). This proves (2.12). Finally, (2.13) follows from (2.10). Hence $C_\Omega(D,G) \subseteq C_\mathscr{L}(D,G)^\perp$.

It is now sufficient to show that the codes $C_\Omega(D,G)$ and $C_\mathscr{L}(D,G)^\perp$ have the same dimension. Using Theorems 2.2.2, 2.2.7 and the Riemann-Roch Theorem we find:

$$\begin{aligned}
\dim C_\Omega(D,G) &= i(G-D) - i(G)\\
&= \ell(G-D) - \deg\,(G-D) - 1 + g - (\ell(G) - \deg G - 1 + g)\\
&= \deg D + \ell(G-D) - \ell(G)\\
&= n - (\ell(G) - \ell(G-D))\\
&= n - \dim C_\mathscr{L}(D,G) = \dim C_\mathscr{L}(D,G)^\perp\,.
\end{aligned}$$

□

Our next aim is to prove that $C_\Omega(D,G)$ can be represented as $C_\mathscr{L}(D,H)$ with an appropriate divisor H. For this purpose we need the following lemma.

Lemma 2.2.9. *There exists a Weil differential η such that*

$$v_{P_i}(\eta) = -1 \quad \textit{and} \quad \eta_{P_i}(1) = 1 \quad \textit{for } i = 1, \ldots, n.$$

Proof. Choose an arbitrary Weil differential $\omega_0 \neq 0$. By the Weak Approximation Theorem there is an element $z \in F$ with $v_{P_i}(z) = -v_{P_i}(\omega_0) - 1$ for $i = 1, \dots, n$. Setting $\omega := z\omega_0$ we obtain $v_{P_i}(\omega) = -1$. Therefore $a_i := \omega_{P_i}(1) \neq 0$ by (2.7). Again by the Approximation Theorem we find $y \in F$ such that $v_{P_i}(y - a_i) > 0$. It follows that $v_{P_i}(y) = 0$ and $y(P_i) = a_i$. We put $\eta := y^{-1}\omega$ and obtain $v_{P_i}(\eta) = v_{P_i}(\omega) = -1$, and

$$\eta_{P_i}(1) = \omega_{P_i}(y^{-1}) = y^{-1}(P_i) \cdot \omega_{P_i}(1) = a_i^{-1} \cdot a_i = 1\,.$$

□

Proposition 2.2.10. *Let η be a Weil differential such that $v_{P_i}(\eta) = -1$ and $\eta_{P_i}(1) = 1$ for $i = 1, \dots, n$. Then*

$$C_{\mathscr{L}}(D, G)^{\perp} = C_{\Omega}(D, G) = C_{\mathscr{L}}(D, H) \quad \textit{with} \quad H := D - G + (\eta)\,.$$

Proof. The equality $C_{\mathscr{L}}(D, G)^{\perp} = C_{\Omega}(D, G)$ was already shown in Theorem 2.2.8. Observe that $\operatorname{supp}(D - G + (\eta)) \cap \operatorname{supp} D = \emptyset$ since $v_{P_i}(\eta) = -1$ for $i = 1, \dots, n$. Hence the code $C_{\mathscr{L}}(D, D - G + (\eta))$ is defined. By Theorem 1.5.14 there is an isomorphism $\mu : \mathscr{L}(D - G + (\eta)) \to \Omega_F(G - D)$ given by $\mu(x) := x\eta$. For $x \in \mathscr{L}(D - G + (\eta))$ we have

$$(x\eta)_{P_i}(1) = \eta_{P_i}(x) = x(P_i) \cdot \eta_{P_i}(1) = x(P_i),$$

cf. (2.10). This implies $C_{\Omega}(D, G) = C_{\mathscr{L}}(D, D - G + (\eta))$. □

Corollary 2.2.11. *Suppose there is a Weil differential η such that*

$$2G - D \leq (\eta) \quad \textit{and} \quad \eta_{P_i}(1) = 1 \quad \textit{for} \quad i = 1, \dots, n\,.$$

Then the code $C_{\mathscr{L}}(D, G)$ is self-orthogonal; i.e., $C_{\mathscr{L}}(D, G) \subseteq C_{\mathscr{L}}(D, G)^{\perp}$. If

$$2G - D = (\eta) \quad \textit{and} \quad \eta_{P_i}(1) = 1 \quad \textit{for} \quad i = 1, \dots, n\,,$$

then $C_{\mathscr{L}}(D, G)$ is self-dual.

Proof. The assumption $2G - D \leq (\eta)$ is equivalent to $G \leq D - G + (\eta)$. Hence Proposition 2.2.10 implies

$$C_{\mathscr{L}}(D, G)^{\perp} = C_{\mathscr{L}}(D, D - G + (\eta)) \supseteq C_{\mathscr{L}}(D, G)\,.$$

This proves the first assertion. If we assume equality $2G - D = (\eta)$ then we have $G = D - G + (\eta)$ and therefore

$$C_{\mathscr{L}}(D, G)^{\perp} = C_{\mathscr{L}}(D, D - G + (\eta)) = C_{\mathscr{L}}(D, G)\,.$$

□

Remark 2.2.12. Using Proposition 2.2.10, one can reduce Theorem 2.2.7 to Theorem 2.2.2. This gives another proof of Theorem 2.2.7.

Definition 2.2.13. *Two codes $C_1, C_2 \subseteq \mathbb{F}_q^n$ are said to be equivalent if there is a vector $a = (a_1, \ldots, a_n) \in (\mathbb{F}_q^\times)^n$ such that $C_2 = a \cdot C_1$; i.e.,*

$$C_2 = \{(a_1c_1, \ldots, a_nc_n) \,|\, (c_1, \ldots, c_n) \in C_1\}\,.$$

Evidently equivalent codes have the same dimension and the same minimum distance. Note however that equivalence does not preserve all interesting properties of a code. For instance, equivalent codes may have non-isomorphic automorphism groups. We will consider automorphisms of codes in Chapter 8.

Proposition 2.2.14. *(a) Suppose G_1 and G_2 are divisors with $G_1 \sim G_2$ and $\operatorname{supp} G_1 \cap \operatorname{supp} D = \operatorname{supp} G_2 \cap \operatorname{supp} D = \emptyset$. Then the codes $C_{\mathscr{L}}(D, G_1)$ and $C_{\mathscr{L}}(D, G_2)$ are equivalent. The same holds for $C_\Omega(D, G_1)$ and $C_\Omega(D, G_2)$.*
(b) Conversely, if a code $C \subseteq \mathbb{F}_q^n$ is equivalent to $C_{\mathscr{L}}(D, G)$ (resp. $C_\Omega(D, G)$) then there exists a divisor $G' \sim G$ such that $\operatorname{supp} G' \cap \operatorname{supp} D = \emptyset$ and $C = C_{\mathscr{L}}(D, G')$ (resp. $C = C_\Omega(D, G')$).

Proof. (a) By assumption we have that $G_2 = G_1 - (z)$ with $v_{P_i}(z) = 0$ for $i = 1, \ldots, n$. Hence $a := (z(P_1), \ldots, z(P_n))$ is in $(\mathbb{F}_q^\times)^n$, and the mapping $x \mapsto xz$ from $\mathscr{L}(G_1)$ to $\mathscr{L}(G_2)$ is bijective (cf. Lemma 1.4.6). This implies that $C_{\mathscr{L}}(D, G_2) = a \cdot C_{\mathscr{L}}(D, G_1)$. The equivalence of $C_\Omega(D, G_1)$ and $C_\Omega(D, G_2)$ is proved similarly.
(b) Let $C = a \cdot C_{\mathscr{L}}(D, G)$ with $a = (a_1, \ldots, a_n) \in (\mathbb{F}_q^\times)^n$. Choose $z \in F$ with $z(P_i) = a_i$ $(i = 1, \ldots, n)$ and set $G' := G - (z)$. Then $C = C_{\mathscr{L}}(D, G')$. □

Remark 2.2.15. If G is a divisor whose support is not disjoint from $\operatorname{supp} D$, we can still define an algebraic geometry code $C_{\mathscr{L}}(D, G)$ associated with D and G as follows: Choose a divisor $G' \sim G$ with $\operatorname{supp} G' \cap \operatorname{supp} D = \emptyset$ (which is possible by the Approximation Theorem) and set $C_{\mathscr{L}}(D, G) := C_{\mathscr{L}}(D, G')$. The choice of G' is not unique and the code $C_{\mathscr{L}}(D, G)$ is well-defined only up to equivalence, by Proposition 2.2.14.

2.3 Rational AG Codes

In this section we investigate AG codes associated with divisors of a rational function field. We shall describe these codes very explicitly by means of generator and parity check matrices. In coding theory this class of codes is known by the name of *Generalized Reed-Solomon codes.* Some of the most important codes used in practice (such as BCH codes and Goppa codes; these codes will be defined later in this section) can be represented as subfield subcodes of Generalized Reed-Solomon codes in a natural manner.

Definition 2.3.1. *An algebraic geometry code $C_{\mathscr{L}}(D,G)$ associated with divisors G and D of a rational function field $\mathbb{F}_q(z)/\mathbb{F}_q$ is said to be rational (as in Section 2.2 it is assumed that $D = P_1 + \ldots + P_n$ with pairwise distinct places of degree one, and $\operatorname{supp} G \cap \operatorname{supp} D = \emptyset$).*

Observe that the length of a rational AG code is bounded by $q+1$ because $\mathbb{F}_q(z)$ has only $q+1$ places of degree one: the pole P_∞ of z and for each $\alpha \in \mathbb{F}_q$, the zero P_α of $z-\alpha$ (see Proposition 1.2.1). The following results follow immediately from Section 2.2.

Proposition 2.3.2. *Let $C = C_{\mathscr{L}}(D,G)$ be a rational AG code over $\mathbb{F}_q$, and let n,k,d be the parameters of C. Then we have:*
(a) $n \le q+1$.
(b) $k=0$ if and only if $\deg G < 0$, and $k=n$ if and only if $\deg G > n-2$.
(c) For $0 \le \deg G \le n-2$,

$$k = 1 + \deg G \quad \text{and} \quad d = n - \deg G\,.$$

In particular, C is an MDS code.
(d) $C^\perp$ is also a rational AG code.

Next we determine specific generator matrices for rational AG codes.

Proposition 2.3.3. *Let $C = C_{\mathscr{L}}(D,G)$ be a rational AG code over $\mathbb{F}_q$ with parameters n,k and d.*
(a) If $n \le q$ then there exist pairwise distinct elements $\alpha_1,\ldots,\alpha_n \in \mathbb{F}_q$ and $v_1,\ldots,v_n \in \mathbb{F}_q^\times$ (not necessarily distinct) such that

$$C = \{(v_1 f(\alpha_1), v_2 f(\alpha_2), \ldots, v_n f(\alpha_n)) \mid f \in \mathbb{F}_q[z] \text{ and } \deg f \le k-1\}\,.$$

The matrix

$$M = \begin{pmatrix} v_1 & v_2 & \ldots & v_n \\ \alpha_1 v_1 & \alpha_2 v_2 & \ldots & \alpha_n v_n \\ \alpha_1^2 v_1 & \alpha_2^2 v_2 & \ldots & \alpha_n^2 v_n \\ \vdots & \vdots & & \vdots \\ \alpha_1^{k-1} v_1 & \alpha_2^{k-1} v_2 & \ldots & \alpha_n^{k-1} v_n \end{pmatrix} \tag{2.14}$$

is a generator matrix for C.
(b) If $n = q+1$, C has a generator matrix

$$M = \begin{pmatrix} v_1 & v_2 & \ldots & v_{n-1} & 0 \\ \alpha_1 v_1 & \alpha_2 v_2 & \ldots & \alpha_{n-1} v_{n-1} & 0 \\ \alpha_1^2 v_1 & \alpha_2^2 v_2 & \ldots & \alpha_{n-1}^2 v_{n-1} & 0 \\ \vdots & \vdots & & \vdots & \vdots \\ \alpha_1^{k-1} v_1 & \alpha_2^{k-1} v_2 & \ldots & \alpha_{n-1}^{k-1} v_{n-1} & 1 \end{pmatrix} \tag{2.15}$$

where $\mathbb{F}_q = \{\alpha_1,\ldots,\alpha_{n-1}\}$ and $v_1,\ldots,v_{n-1} \in \mathbb{F}_q^\times$.

Proof. (a) Let $D = P_1 + \ldots + P_n$. As $n \le q$, there is a place P of degree one which is not in the support of D. Choose a place $Q \ne P$ of degree one (e.g., $Q = P_1$). By Riemann-Roch, $\ell(Q - P) = 1$, hence $Q - P$ is a principal divisor (Corollary 1.4.12). Let $Q - P = (z)$; then z is a generating element of the rational function field over $\mathbb{F}_q$ and P is the pole divisor of z. As usually we write $P = P_\infty$. By Proposition 2.3.2 we can assume that $\deg G = k - 1 \ge 0$ (the case $k = 0$ being trivial). The divisor $(k-1)P_\infty - G$ has degree zero, so it is principal (Riemann-Roch and Corollary 1.4.12), say $(k-1)P_\infty - G = (u)$ with $0 \ne u \in F$. The k elements $u, zu, \ldots, z^{k-1}u$ are in $\mathscr{L}(G)$ and they are linearly independent over $\mathbb{F}_q$. Since $\ell(G) = k$, they constitute a basis of $\mathscr{L}(G)$; i.e.,

$$\mathscr{L}(G) = \{uf(z) \mid f \in \mathbb{F}_q[z] \text{ and } \deg f \le k-1\}.$$

Setting $\alpha_i := z(P_i)$ and $v_i := u(P_i)$ we obtain

$$(uf(z))(P_i) = u(P_i)f(z(P_i)) = v_i f(\alpha_i)$$

for $i = 1, \ldots, n$. Therefore

$$C = C_{\mathscr{L}}(D, G) = \{(v_1 f(\alpha_1), \ldots, v_n f(\alpha_n)) \mid \deg f \le k-1\}.$$

The codeword in C corresponding to uz^j is $(v_1\alpha_1^j, v_2\alpha_2^j, \ldots, v_n\alpha_n^j)$, so the matrix (2.14) is a generator matrix of C.

(b) The proof is essentially the same as in the case $n \le q$. Now we have $n = q + 1$ and we can choose z in such a way that $P_n = P_\infty$ is the pole of z. As above, $(k-1)P_\infty - G = (u)$ with $0 \ne u \in F$, and $\{u, zu, \ldots, z^{k-1}u\}$ is a basis of $\mathscr{L}(G)$. For $1 \le i \le n - 1 = q$ the elements $\alpha_i := z(P_i) \in \mathbb{F}_q$ are pairwise distinct, so $\mathbb{F}_q = \{\alpha_1, \ldots, \alpha_{n-1}\}$. Moreover, $v_i := u(P_i) \in \mathbb{F}_q^\times$ for $i = 1, \ldots, n-1$. For $0 \le j \le k-2$ we obtain

$$((uz^j)(P_1), \ldots, (uz^j)(P_n)) = (\alpha_1^j v_1, \ldots, \alpha_{n-1}^j v_{n-1}, 0),$$

but for $j = k - 1$ holds

$$((uz^{k-1})(P_1), \ldots, (uz^{k-1})(P_n)) = (\alpha_1^{k-1} v_1, \ldots, \alpha_{n-1}^{k-1} v_{n-1}, \gamma)$$

with an element $0 \ne \gamma \in \mathbb{F}_q$. Substituting u by $\gamma^{-1}u$ yields the generator matrix (2.15). □

Definition 2.3.4. *Let $\alpha = (\alpha_1, \ldots, \alpha_n)$ where the α_i are distinct elements of $\mathbb{F}_q$, and let $v = (v_1, \ldots, v_n)$ where the v_i are nonzero (not necessarily distinct) elements of $\mathbb{F}_q$. Then the Generalized Reed-Solomon code, denoted by $\mathrm{GRS}_k(\alpha, v)$, consists of all vectors*

$$(v_1 f(\alpha_1), \ldots, v_n f(\alpha_n))$$

with $f(z) \in \mathbb{F}_q[z]$ and $\deg f \le k-1$ (for a fixed $k \le n$).

In the case $\alpha = (\beta, \beta^2, \ldots, \beta^n)$ (where $n = q-1$ and β is a primitive n-th root of unity) and $v = (1, 1, \ldots, 1)$, $\mathrm{GRS}_k(\alpha, v)$ is a Reed-Solomon code, cf. Section 2.2.

Obviously $\mathrm{GRS}_k(\alpha, v)$ is an $[n, k]$ code, and Proposition 2.3.3.(a) states that all rational AG codes over $\mathbb{F}_q$ of length $n \le q$ are Generalized Reed-Solomon codes. The converse is also true:

Proposition 2.3.5. *Every generalized Reed-Solomon code* $\mathrm{GRS}_k(\alpha, v)$ *can be represented as a rational* AG *code.*

Proof. Let $\alpha = (\alpha_1, \ldots, \alpha_n)$ with $\alpha_i \in \mathbb{F}_q$ and $v = (v_1, \ldots, v_n)$ with $v_i \in \mathbb{F}_q^\times$. Consider the rational function field $F = \mathbb{F}_q(z)$. Denote by P_i the zero of $z - \alpha_i$ $(i = 1, \ldots, n)$ and by P_∞ the pole of z. Choose $u \in F$ such that

$$u(P_i) = v_i \quad \textit{for} \quad i = 1, \ldots, n. \tag{2.16}$$

Such an element exists by the Approximation Theorem. (One can also determine a polynomial $u = u(z) \in \mathbb{F}_q[z]$ satisfying (2.16), by using Lagrange interpolation.) Now let

$$D := P_1 + \ldots + P_n \quad \textit{and} \quad G := (k-1)P_\infty - (u).$$

The proof of Proposition 2.3.3 shows that $\mathrm{GRS}_k(\alpha, v) = C_{\mathscr{L}}(D, G)$. □

The same arguments apply to a code of length $n = q + 1$ over $\mathbb{F}_q$ which has a generator matrix of the specific form (2.15). All such codes can be represented as rational AG codes.

In order to determine the dual of a rational AG code $C = C_{\mathscr{L}}(D, G)$, we need (by Theorem 2.2.8 and Proposition 2.2.10) a Weil differential ω of $\mathbb{F}_q(z)$ such that

$$v_{P_i}(\omega) = -1 \quad \textit{and} \quad \omega_{P_i}(1) = 1 \quad \textit{for} \quad i = 1, \ldots, n. \tag{2.17}$$

Lemma 2.3.6. *Consider the rational function field $F = \mathbb{F}_q(z)$ and n distinct elements $\alpha_1, \ldots, \alpha_n \in \mathbb{F}_q$. Let $P_i \in \mathbb{P}_F$ be the zero of $z - \alpha_i$ and $h(z) := \prod_{i=1}^n (z - \alpha_i)$. Suppose y is a element of F such that $y(P_i) = 1$ for $i = 1, \ldots, n$. Then there exists a Weil differential ω of $F/\mathbb{F}_q$ with the property* (2.17) *and the divisor*

$$(\omega) = (y) + (h'(z)) - (h(z)) - 2P_\infty \tag{2.18}$$

(where $h'(z) \in \mathbb{F}_q[z]$ is the derivative of the polynomial $h(z)$).

Proof. There is a Weil differential η of F with $(\eta) = -2P_\infty$ and $\eta_{P_\infty}(z^{-1}) = -1$ (see Proposition 1.7.4). We set

$$\omega := y \cdot (h'(z)/h(z)) \cdot \eta.$$

The divisor of ω is $(\omega) = (y) + (h'(z)) - (h(z)) - 2P_\infty$, in particular holds $v_{P_i}(\omega) = -1$ for $i = 1, \ldots, n$. We have to verify that $\omega_{P_i}(1) = 1$. Write $h(z) = (z - \alpha_i)g_i(z)$; then

$$y \cdot \frac{h'(z)}{h(z)} = \big(1 + (y-1)\big) \cdot \left(\frac{g_i'(z)}{g_i(z)} + \frac{1}{z - \alpha_i} \right) = \frac{1}{z - \alpha_i} + u$$

with $u \in F$ and $v_{P_i}(u) \geq 0$ (because $v_{P_i}(y-1) > 0$ and $v_{P_i}(g_i(z)) = 0$). Since $\eta_{P_i}((z-\alpha_i)^{-1}) = 1$ and $\eta_{P_i}(u) = 0$ (by Proposition 1.7.4(c) and Proposition 1.7.3(a)), we obtain

$$\omega_{P_i}(1) = \eta_{P_i}\left(y \cdot \frac{h'(z)}{h(z)} \right) = \eta_{P_i}\left(\frac{1}{z - \alpha_i} + u \right) = 1\,.$$

□

Note that Lemma 2.3.6 - combined with Theorem 2.2.8, Proposition 2.2.10 and Proposition 2.3.3 - enables us to specify a parity check matrix for $C_{\mathscr{L}}(D, G)$.

Next we would like to describe BCH codes and Goppa codes by means of rational AG codes. To this end we need the following concept:

Definition 2.3.7. *Consider an extension field $\mathbb{F}_{q^m}$ of $\mathbb{F}_q$ and a code C over $\mathbb{F}_{q^m}$ of length n. Then*

$$C|_{\mathbb{F}_q} := C \cap \mathbb{F}_q^n$$

is called the subfield subcode of C (or the restriction of C to $\mathbb{F}_q$).

$C|_{\mathbb{F}_q}$ is a code over $\mathbb{F}_q$. Its minimum distance cannot be less than the minimum distance of C, and for the dimension of $C|_{\mathbb{F}_q}$ we have the trivial estimate $\dim C|_{\mathbb{F}_q} \leq \dim C$. In general this can be a strict inequality.

Definition 2.3.8. *Assume that $n|(q^m - 1)$ and let $\beta \in \mathbb{F}_{q^m}$ be a primitive n-th root of unity. Let $l \in \mathbb{Z}$ and $\delta \geq 2$. Define a code $C(n, l, \delta)$ over $\mathbb{F}_{q^m}$ by the generator matrix*

$$H := \begin{pmatrix} 1 & \beta^l & \beta^{2l} & \ldots & \beta^{(n-1)l} \\ 1 & \beta^{l+1} & \beta^{2(l+1)} & \ldots & \beta^{(n-1)(l+1)} \\ \vdots & \vdots & \vdots & & \vdots \\ 1 & \beta^{l+\delta-2} & \beta^{2(l+\delta-2)} & \ldots & \beta^{(n-1)(l+\delta-2)} \end{pmatrix} \tag{2.19}$$

The code $C := C(n, l, \delta)^\perp|_{\mathbb{F}_q}$ is called a BCH *code with designed distance δ. In other words,*

$$C = \{ c \in \mathbb{F}_q^n \,|\, H \cdot c^t = 0 \}\,, \tag{2.20}$$

where the matrix H is given by (2.19).

We note that BCH codes are usually defined as special cyclic codes. One can easily show however, that our Definition 2.3.8 coincides with the usual definition, cf. [25],[28].

Proposition 2.3.9. *Let $n|(q^m-1)$ and let $\beta \in \mathbb{F}_{q^m}$ be a primitive n-th root of unity. Let $F = \mathbb{F}_{q^m}(z)$ be the rational function field over $\mathbb{F}_{q^m}$ and P_0 (resp. P_∞) be the zero (resp. pole) of z. For $i = 1, \ldots, n$ denote by P_i the zero of $z - \beta^{i-1}$, and set $D_\beta := P_1 + \ldots + P_n$. Suppose that $a, b \in \mathbb{Z}$ are integers with $0 \le a + b \le n-2$. Then we have*

(a) $C_{\mathscr{L}}(D_\beta, aP_0 + bP_\infty) = C(n, l, \delta)$ with $l = -a$ and $\delta = a + b + 2$ (where $C(n, l, \delta)$ is as in Definition 2.3.8).

(b) The dual of $C_{\mathscr{L}}(D_\beta, aP_0 + bP_\infty)$ is given by

$$C_{\mathscr{L}}(D_\beta, aP_0 + bP_\infty)^\perp = C_{\mathscr{L}}(D_\beta, rP_0 + sP_\infty)$$

with $r = -(a+1)$ and $s = n - b - 1$. Hence the BCH code $C(n, l, \delta)^\perp|_{\mathbb{F}_q}$ is the restriction to $\mathbb{F}_q$ of the code $C_{\mathscr{L}}(D_\beta, rP_0 + sP_\infty)$, with $r = l - 1$ and $s = n + 1 - \delta - l$.

Proof. (a) We consider the code $C_{\mathscr{L}}(D_\beta, aP_0 + bP_\infty)$ where $0 \le a + b \le n - 2$. The elements $z^{-a} \cdot z^j$ with $0 \le j \le a + b$ constitute a basis of $\mathscr{L}(aP_0 + bP_\infty)$. Hence the matrix

$$\begin{pmatrix} 1 & \beta^{-a} & \beta^{-2a} & \dots & (\beta^{n-1})^{-a} \\ 1 & \beta^{-a+1} & \beta^{-2a+2} & \dots & (\beta^{n-1})^{-a+1} \\ \vdots & \vdots & \vdots & & \vdots \\ 1 & \beta^{-a+(a+b)} & \beta^{-2a+2(a+b)} & \dots & (\beta^{n-1})^{-a+(a+b)} \end{pmatrix}$$

is a generator matrix of $C_{\mathscr{L}}(D_\beta, aP_0 + bP_\infty)$. Substituting $l := -a$ and $\delta := a + b + 2$ we obtain the matrix (2.19), so $C_{\mathscr{L}}(D_\beta, aP_0 + bP_\infty) = C(n, l, \delta)$.

(b) We use the notation of Lemma 2.3.6 and set

$$y := z^{-n} \quad \text{and} \quad h(z) := \prod_{i=1}^{n} (z - \beta^{i-1}) = z^n - 1\,.$$

Proposition 2.2.10 yields $C_{\mathscr{L}}(D_\beta, aP_0 + bP_\infty)^\perp = C_{\mathscr{L}}(D_\beta, B)$ with

$$\begin{aligned} B &= D_\beta - (aP_0 + bP_\infty) + (z^{-n}) + (h'(z)) - (h(z)) - 2P_\infty \\ &= D_\beta - (aP_0 + bP_\infty) + n(P_\infty - P_0) + (n-1)(P_0 - P_\infty) \\ &\qquad\qquad\qquad\qquad\qquad\qquad - (D_\beta - nP_\infty) - 2P_\infty \\ &= (-a-1)P_0 + (n-b-1)P_\infty\,. \end{aligned}$$

Since $l = -a$ and $\delta = a + b + 2$ (by (a)), we find $C_{\mathscr{L}}(D_\beta, aP_0 + bP_\infty)^\perp = C_{\mathscr{L}}(D_\beta, rP_0 + sP_\infty)$ with $s = n - b - 1 = n - (\delta - a - 2) - 1 = n + 1 - \delta - l$ and $r = -a - 1 = l - 1$. □

Next we introduce *Goppa codes*. As with BCH codes, our definition of Goppa codes differs from the usual definition given in most books on coding theory. However both definitions are equivalent.

Definition 2.3.10. *Let $L = \{\alpha_1, \ldots, \alpha_n\} \subseteq \mathbb{F}_{q^m}$ with $|L| = n$, and let $g(z) \in \mathbb{F}_{q^m}[z]$ be a polynomial of degree t such that $1 \le t \le n-1$ and $g(\alpha_i) \ne 0$ for all $\alpha_i \in L$.*
(a) We define a code $C(L, g(z)) \subseteq (\mathbb{F}_{q^m})^n$ by the generator matrix

$$H := \begin{pmatrix} g(\alpha_1)^{-1} & g(\alpha_2)^{-1} & \ldots & g(\alpha_n)^{-1} \\ \alpha_1 g(\alpha_1)^{-1} & \alpha_2 g(\alpha_2)^{-1} & \ldots & \alpha_n g(\alpha_n)^{-1} \\ \vdots & \vdots & & \vdots \\ \alpha_1^{t-1} g(\alpha_1)^{-1} & \alpha_2^{t-1} g(\alpha_2)^{-1} & \ldots & \alpha_n^{t-1} g(\alpha_n)^{-1} \end{pmatrix} \tag{2.21}$$

(b) The code $\Gamma(L, g(z)) := C(L, g(z))^\perp|_{\mathbb{F}_q}$ is called the Goppa code with Goppa polynomial $g(z)$. This means

$$\Gamma(L, g(z)) = \{c \in \mathbb{F}_q^n \mid H \cdot c^t = 0\}$$

with H as in (2.21).

Note that the matrix (2.21) is a special case of (2.14) (with $v_i = g(\alpha_i)^{-1}$), hence $C(L, g(z))$ and $C(L, g(z))^\perp$ are Generalized Reed-Solomon codes. Now we give an explicit description of these codes as rational AG codes.

Proposition 2.3.11. *In addition to the notation of Definition 2.3.10, let P_i denote the zero of $z - \alpha_i$ (for $\alpha_i \in L$), P_∞ the pole of z and $D_L := P_1 + \ldots + P_n$. Let G_0 be the zero divisor of $g(z)$ (in the divisor group of the rational function field $F = \mathbb{F}_{q^m}(z)$). Then we have*

$$C(L, g(z)) = C_{\mathscr{L}}(D_L, G_0 - P_\infty) = C_{\mathscr{L}}(D_L, A - G_0)^\perp \tag{2.22}$$

and

$$\Gamma(L, g(z)) = C_{\mathscr{L}}(D_L, G_0 - P_\infty)^\perp|_{\mathbb{F}_q} = C_{\mathscr{L}}(D_L, A - G_0)|_{\mathbb{F}_q},$$

where the divisor A is determined as follows: Set

$$h(z) := \prod_{\alpha_i \in L} (z - \alpha_i) \quad \text{and} \quad A := (h'(z)) + (n-1)P_\infty.$$

Proof. It is sufficient to prove (2.22). For $0 \le j \le t-1$, the element $z^j g(z)^{-1}$ is in $\mathscr{L}(G_0 - P_\infty)$ because

$$(z^j g(z)^{-1}) = j(P_0 - P_\infty) - (G_0 - tP_\infty) \ge -G_0 + P_\infty\,.$$

Since $\dim \mathscr{L}(G_0 - P_\infty) = t$, the elements $g(z)^{-1}$, $zg(z)^{-1}, \ldots, z^{t-1}g(z)^{-1}$ form a basis of $\mathscr{L}(G_0 - P_\infty)$. Thus (2.21) is a generator matrix of $C_{\mathscr{L}}(D_L, G_0 - P_\infty)$; i.e.,

$$C(L, g(z)) = C_{\mathscr{L}}(D_L, G_0 - P_\infty).$$

From Proposition 2.2.10 and Lemma 2.3.6 we obtain $C_{\mathscr{L}}(D_L, G_0 - P_\infty)^\perp = C_{\mathscr{L}}(D_L, B)$ with

$$\begin{aligned} B &= D_L - (G_0 - P_\infty) + (h'(z)) - (h(z)) - 2P_\infty \\ &= D_L - G_0 + P_\infty + A - (n-1)P_\infty - (D_L - nP_\infty) - 2P_\infty \\ &= A - G_0 . \end{aligned}$$

□

In coding theory the so-called BCH *Bound* (resp. the *Goppa Bound*) for the minimum distance of BCH codes (resp. Goppa codes) is well-known. Both bounds can easily be derived from the above results.

Corollary 2.3.12. *(a) (*BCH *Bound.) The minimum distance of a* BCH *code with designed distance δ is at least δ.*

(b) (Goppa Bound.) The minimum distance of a Goppa code $\Gamma(L, g(z))$ is at least $1 + \deg g(z)$.

Proof. (a) Using notation as in Proposition 2.3.9 we represent the BCH code in the form $C = C_{\mathscr{L}}(D_\beta, rP_0 + sP_\infty)|_{\mathbb{F}_q}$. The minimum distance of the code $C_{\mathscr{L}}(D_\beta, rP_0 + sP_\infty)$ is, by Propositions 2.3.2 and 2.3.9(b),

$$d = n - \deg(rP_0 + sP_\infty) = n - \big((l-1) + (n+1-\delta-l)\big) = \delta .$$

Since the minimum distance of a subfield subcode is not less than the minimum distance of the original code, the minimum distance of C is $\geq \delta$.

(b) In the same manner we represent $\Gamma(L, g(z))$ as $C_{\mathscr{L}}(D_L, A - G_0)|_{\mathbb{F}_q}$ (notation as in Proposition 2.3.11). As $C_{\mathscr{L}}(D_L, A - G_0)$ has the minimum distance

$$d = n - \deg(A - G_0) = n - \big((n-1) - \deg g(z)\big) = 1 + \deg g(z) ,$$

the assertion follows. □

Remark 2.3.13. The subfield subcode construction makes possible to construct codes over $\mathbb{F}_q$ of arbitrary length, by considering codes over an appropriate extension field $\mathbb{F}_{q^m}$ and restricting them to $\mathbb{F}_q$. Note however, that a code C over $\mathbb{F}_{q^m}$ may have good parameters (i.e., large dimension and minimum distance) whereas the restricted code $C|_{\mathbb{F}_q}$ can be very poor (since the dimension of $C|_{\mathbb{F}_q}$ may be much less than the dimension of C, cf. Chapter 9).

Remark 2.3.14. Subfield subcodes of Generalized Reed-Solomon codes are known as *alternant codes*, cf. [28]. Propositions 2.3.3 and 2.3.5 state that the class of alternant codes over $\mathbb{F}_q$ corresponds to the class of subfield subcodes of rational AG codes which are defined over extension fields $\mathbb{F}_{q^m} \supseteq \mathbb{F}_q$.

Remark 2.3.15. From the point of view of algebra, the rational function field $\mathbb{F}_q(z)$ is the most trivial example of an algebraic function field. Nevertheless the AG codes associated with divisors of $\mathbb{F}_q(z)$ are already interesting codes as we have seen in this section. So it looks promising to consider algebraic geometry codes associated with non-rational function fields $F/\mathbb{F}_q$.

Often a function field F is represented in the form

$$F = \mathbb{F}_q(x, y) \quad \textit{with} \quad \varphi(x, y) = 0\,,$$

where φ is a non-constant irreducible polynomial in two variables with coefficients in $\mathbb{F}_q$. Then F can be regarded as a finite algebraic extension of the rational function field $\mathbb{F}_q(x)$ (or $\mathbb{F}_q(y)$). Several problems arise:

(1) Is $\mathbb{F}_q$ the full constant field of F?

(2) Calculate the genus of F.

(3) Describe the places of F explicitly. In particular, which places are of degree one?

(4) Construct a basis for the spaces $\mathscr{L}(G)$, at least in specific cases.

(5) Give a convenient description of Weil differentials and of their local components.

Another interesting question is:

(6) How many places of degree one can a function field $F/\mathbb{F}_q$ of genus g have?

This question is important for coding theory since one often wants to construct long codes over $\mathbb{F}_q$, and the length of an AG code associated with a function field is bounded by the number of places of degree one.

In order to tackle these problems it is necessary to develop further the theory of algebraic function fields. This will be done in the subsequent chapters. We will continue the discussion of codes in Chapter 8.

2.4 Exercises

2.1. For a non-empty subset $M \subseteq \mathbb{F}_q^n$ we define the support of M as $\operatorname{supp} M = \{\, i \,|\, 1 \le i \le n, \text{ there is some } c = (c_1, \ldots, c_n) \in M \text{ with } c_i \neq 0\}$.

Now let C be an $[n, k]$ code over $\mathbb{F}_q$. For all r with $1 \le r \le k$, the r-th Hamming weight of C is defined as follows:

$$d_r(C) := \min\{\, |\operatorname{supp} W|\,;\ W \subseteq C \textit{ is an } r\textit{-dimensional subspace of } C\}\,.$$

The sequence $(d_1(C), d_2(C), \ldots, d_k(C))$ is called the weight hierarchy of the code C. Show:

(i) $d_1(C)$ is equal to the minimum distance of C.

(ii) $0 < d_1(C) < d_2(C) < \ldots < d_k(C) \leq n$.

(iii) (Singleton Bound) For all r with $1 \leq r \leq k$ one has $d_r(C) \leq n - k + r$.

2.2. In this exercise we study relations between the weight hierarchy of AG codes and the gonality sequence of function fields, cf. Exercise 1.18. Let $F/\mathbb{F}_q$ be a function field of genus g. As usual, we consider divisors $D = P_1 + \ldots + P_n$ with n distinct places P_i of degree one, and G with $\operatorname{supp} D \cap \operatorname{supp} G = \emptyset$. Let $C_{\mathscr{L}}(D, G)$ be the corresponding AG code and set $k := \dim C_{\mathscr{L}}(D, G)$. Show:

(i) For all r with $1 \leq r \leq k$ one has

$$d_r(C_{\mathscr{L}}(D, G)) \geq n - \deg G + \gamma_r \,.$$

Observe that this is a generalization of Goppa's Bound for the minimum distance of AG codes (Theorem 2.2.2).

(ii) Assume in addition that $\deg G < n$. Then

$$d_r(C_{\mathscr{L}}(D, G)) = n - k + r \quad \textit{for all } r \textit{ with } \quad g + 1 \leq r \leq k \,.$$

2.3. Let $C = C_{\mathscr{L}}(D, G)$ be the AG code associated to the divisors G and $D = P_1 + \ldots + P_n$, with n distinct places P_i of degree one and $\operatorname{supp} D \cap \operatorname{supp} G = \emptyset$. The integer $a := \ell(G - D)$ is called the abundance of C. Show that the r-th minimum distance $d_r = d_r(C_{\mathscr{L}}(D, G))$ of $C_{\mathscr{L}}(D, G)$ satisfies the estimate

$$d_r \geq n - \deg G + \gamma_{r+a} \,,$$

where γ_j denotes the j-th gonality of the function field $F/\mathbb{F}_q$. Conclude the estimate

$$d \geq n - \deg G + \gamma \,,$$

if the abundance of $C_{\mathscr{L}}(D, G)$ is ≥ 1 (and $\gamma = \gamma_2$ is the gonality of $F/\mathbb{F}_q$).

2.4. Let $C \subseteq \mathbb{F}_q^n$ be a rational AG code of dimension $k > 0$.

(i) Prove that for every m with $n + 1 - k \leq m \leq n$ there is a codeword $c \in C$ with $\operatorname{wt}(c) = m$.

(ii) Determine the weight hierarchy of C.

2.5. Let $F = \mathbb{F}_q(z)$ be the rational function field over $\mathbb{F}_q$. For $\alpha \in \mathbb{F}_q$ let P_α be the zero of $z - \alpha$, and denote by P_∞ the pole of z in F. Let $D := \sum_{\alpha \in \mathbb{F}_q} P_\alpha$ and $G = rP_\infty$ with $r \leq (q-2)/2$.

(i) Show that the code $C_{\mathscr{L}}(D, G)$ is self-orthogonal.

(ii) For $q = 2^s$ and $r = (q-2)/2$ show that $C_{\mathscr{L}}(D, G)$ is self-dual.

2.6. *(i)* Take the definition of a BCH code from any textbook on coding theory, and show that it is equivalent to our Definition 2.3.8.

(ii) The same problem for Goppa codes, see Definition 2.3.10.

2.7. Let C be an $[n,k]$ code over the field $\mathbb{F}_{q^m}$.

(i) Show that $\dim C|_{\mathbb{F}_q} \geq n - m(n-k)$.

(ii) Find non-trivial examples where equality holds.

(iii) Find examples where the minimum distance of $C|_{\mathbb{F}_q}$ is larger than the minimum distance of C.

2.8. (Generalized AG codes) Let $F/\mathbb{F}_q$ be a function field of genus g. Assume that $P_1,\ldots,P_s \in \mathbb{P}_F$ are distinct places and G is a divisor with $P_i \notin \operatorname{supp} G$. For $i = 1,\ldots,s$ let $\pi_i : F_{P_i} \to C_i$ be an $\mathbb{F}_q$-linear isomorphism from the residue class field $F_{P_i} = \mathcal{O}_{P_i}/P_i$ onto a linear code $C_i \subseteq \mathbb{F}_q^{n_i}$, whose parameters are $[n_i, k_i = \deg P_i, d_i]$. Let $n := \sum_{i=1}^{s} n_i$ and define the linear map $\pi : \mathscr{L}(G) \to \mathbb{F}_q^n$ by $\pi(f) := (\pi_1(f(P_1)),\ldots,\pi_s(f(P_s)))$. The image of π is then called a generalized AG code. Formulate and prove estimates for the dimension and the minimum distance of such codes, analogous to the Goppa Bound for AG codes (Theorem 2.2.2).

2.9. Let $F/\mathbb{F}_q$ be a function field of genus g, and let $D = P_1 + \ldots + P_n$ with distinct places P_i of degree 1. Consider a divisor G of the form $G = A + B$ with $A \geq 0$ and $B \geq 0$. Assume that $Z \geq 0$ is another divisor such that the following hold:

(1) $\operatorname{supp} G \cap \operatorname{supp} D = \operatorname{supp} Z \cap \operatorname{supp} D = \emptyset$,

(2) $\ell(A - Z) = \ell(A)$ and $\ell(B + Z) = \ell(B)$.

Show that the minimum distance d of the code $C_\Omega(D,G)$ satisfies

$$d \geq \deg G - (2g-2) + \deg Z\,.$$

Compare with Theorem 2.2.7.

3

Extensions of Algebraic Function Fields

Every function field over K can be regarded as a finite field extension of a rational function field $K(x)$. This is one of the reasons why it is of interest to investigate field extensions F'/F of algebraic function fields. In this chapter we shall study, among other things, the relationship between places, divisors, Weil differentials and the genera of F' and F. Let us first fix some notation to be maintained throughout the entire chapter.

F/K denotes an algebraic function field of one variable with full constant field K. The field K is assumed to be perfect; i.e., there does not exist a purely inseparable extension L/K with $1 < [L : K] < \infty$. We consider function fields F'/K' (where K' is the full constant field of F') such that $F' \supseteq F$ is an algebraic extension and $K' \supseteq K$. For convenience, we fix some algebraically closed field $\Phi \supseteq F$ and consider only extensions $F' \supseteq F$ with $F' \subseteq \Phi$.

Actually the perfectness of K will be essential only in a few places of Chapter 3, in particular in Section 3.6. The fact that we consider only extensions of F which are contained in Φ is no restriction at all, since Φ is algebraically closed and thereby every algebraic extension F'/F can be embedded into Φ.

Since this chapter is rather long, we first give a brief survey. In Section 3.1 we introduce the basic concepts: algebraic extensions of function fields, extensions of places, ramification index and residue class degree as well as the Fundamental Equality $\sum e_i f_i = n$.

Subrings of algebraic function fields, in particular holomorphy rings, are studied in Section 3.2.

In the next section we investigate the integral closure of a subring of F/K in a finite separable field extension F'/F, and we prove the existence of local integral bases. This section also contains Kummer's Theorem which is useful in determining the decomposition of a place in a finite extension of function fields.

H. Stichtenoth, *Algebraic Function Fields and Codes*,
Graduate Texts in Mathematics 254,

Let F'/F be a finite separable extension of algebraic function fields. The Hurwitz Genus Formula provides a relation between the genus of F, the genus of F' and the different of F'/F. Section 3.4 is concerned principally with this result.

In Section 3.5 we study the relation between ramified places and the different (Dedekind's Different Theorem), and we show how to calculate the different in specific cases.

Constant field extensions are considered in Section 3.6. The study of such extensions reduces many problems to the case where the constant field is algebraically closed (which is often simpler because all places have degree one).

Section 3.7 is concerned with Galois extensions of algebraic function fields. For some particular types of Galois extensions F'/F (Kummer and Artin-Schreier extensions), we determine the genus of F'.

The topic of Section 3.8 is Hilbert's theory of higher ramification groups, including Hilbert's Different Formula.

In Section 3.9 we discuss ramification and splitting of places in the compositum of two extensions; one of the main results here is Abhyankar's Lemma.

In Section 3.10 we consider purely inseparable extensions of an algebraic function field.

Finally we give in Section 3.11 some upper bounds for the genus of a function field: Castelnuovo's Inequality, Riemann's Inequality, and an estimate for the genus of the function field of a plane algebraic curve of degree n.

Many results of this chapter (in particular most of Sections 3.1, 3.3, 3.5, 3.7, 3.8 and 3.9) hold not only in the case of algebraic function fields but, more generally, for extensions of Dedekind domains. Hence the reader who is familiar with algebraic number theory can skip the appropriate sections.

3.1 Algebraic Extensions of Function Fields

We begin with some basic definitions.

Definition 3.1.1. *(a) An algebraic function field F'/K' is called an algebraic extension of F/K if $F' \supseteq F$ is an algebraic field extension and $K' \supseteq K$.*

(b) The algebraic extension F'/K' of F/K is called a constant field extension if $F' = FK'$, the composite field of F and K'.

(c) The algebraic extension F'/K' of F/K is called a finite extension if $[F' : F] < \infty$.

One can also consider arbitrary (not necessarily algebraic) extensions of function fields. However, we shall restrict ourselves to algebraic extensions since these are by far the most important ones. Arbitrary extensions of function fields are studied in detail in [7].

We note some simple consequences of the above definitions.

Lemma 3.1.2. *Let F'/K' be an algebraic extension of F/K. Then the following hold:*

(a) K'/K is algebraic and $F \cap K' = K$.

(b) F'/K' is a finite extension of F/K if and only if $[K' : K] < \infty$.

(c) Let $F_1 := FK'$. Then F_1/K' is a constant field extension of F/K, and F'/K' is a finite extension of F_1/K' (having the same constant field).

Proof. (a) and (c) are trivial. As to (b), we assume first that F'/K' is a finite extension of F/K. Then F' can be considered as an algebraic function field over K whose full constant field is K'. By Corollary 1.1.16 we conclude that $[K' : K] < \infty$.

Conversely, suppose that $[K' : K] < \infty$. Choose $x \in F \backslash K$; then $F'/K'(x)$ is a finite field extension (since x is transcendental over K'), and

$$[K'(x) : K(x)] \leq [K' : K] < \infty .$$

(Actually it holds that $[K'(x) : K(x)] = [K' : K]$, but we do not need this here.) Therefore

$$[F' : K(x)] = [F' : K'(x)] \cdot [K'(x) : K(x)] < \infty .$$

Since $K(x) \subseteq F \subseteq F'$, this implies $[F' : F] < \infty$. □

Now let us study the relation between the places of F and F'.

Definition 3.1.3. *Consider an algebraic extension F'/K' of F/K. A place $P' \in \mathbb{P}_{F'}$ is said to lie over $P \in \mathbb{P}_F$ if $P \subseteq P'$. We also say that P' is an extension of P or that P lies under P', and we write $P'|P$.*

Proposition 3.1.4. *Let F'/K' be an algebraic extension of F/K. Suppose that P (resp. P') is a place of F/K (resp. F'/K'), and let $\mathcal{O}_P \subseteq F$ (resp. $\mathcal{O}_{P'} \subseteq F'$) denote the corresponding valuation ring, v_P (resp. $v_{P'}$) the corresponding discrete valuation. Then the following assertions are equivalent:*

(1) $P'|P$.

(2) $\mathcal{O}_P \subseteq \mathcal{O}_{P'}$.

(3) There exists an integer $e \geq 1$ such that $v_{P'}(x) = e \cdot v_P(x)$ for all $x \in F$.

Moreover, if $P'|P$ then

$$P = P' \cap F \quad \text{and} \quad \mathcal{O}_P = \mathcal{O}_{P'} \cap F .$$

For this reason, P is also called the restriction of P' to F.

Proof. (1) $\Rightarrow$ (2): Suppose that $P'|P$ but $\mathcal{O}_P \not\subseteq \mathcal{O}_{P'}$. Then there is some $u \in F$ with $v_P(u) \geq 0$ and $v_{P'}(u) < 0$. As $P \subseteq P'$ we conclude $v_P(u) = 0$. Choose $t \in F$ with $v_P(t) = 1$, then $t \in P'$ and $r := v_{P'}(t) > 0$. Consequently

$$v_P(u^r t) = r \cdot v_P(u) + v_P(t) = 1\,,$$

$$v_{P'}(u^r t) = r \cdot v_{P'}(u) + v_{P'}(t) \leq -r + r = 0\,.$$

Thus $u^r t \in P \backslash P'$, a contradiction to $P \subseteq P'$.

Before proving (2) $\Rightarrow$ (1) we show the following:

$$\mathcal{O}_P \subseteq \mathcal{O}_{P'} \Rightarrow \mathcal{O}_P = F \cap \mathcal{O}_{P'}\,. \tag{3.1}$$

Clearly $F \cap \mathcal{O}_{P'}$ is a subring of F with $\mathcal{O}_P \subseteq F \cap \mathcal{O}_{P'}$, therefore $F \cap \mathcal{O}_{P'} = \mathcal{O}_P$ or $F \cap \mathcal{O}_{P'} = F$ by Theorem 1.1.13(c). Assume that $F \cap \mathcal{O}_{P'} = F$; i.e., $F \subseteq \mathcal{O}_{P'}$. Choose an element $z \in F' \backslash \mathcal{O}_{P'}$. Since F'/F is algebraic, there is an equation

$$z^n + c_{n-1} z^{n-1} + \cdots + c_1 z + c_0 = 0 \tag{3.2}$$

with $c_\nu \in F$. We have $v_{P'}(z^n) = n \cdot v_{P'}(z) < 0$ as $z \notin \mathcal{O}_{P'}$, therefore

$$v_{P'}(z^n) < v_{P'}(c_\nu z^\nu) \quad \textit{for} \quad \nu = 0, \cdots, n-1\,.$$

The Strict Triangle Inequality yields

$$v_{P'}(z^n + c_{n-1} z^{n-1} + \cdots + c_1 z + c_0) = n \cdot v_{P'}(z) \neq v_{P'}(0)\,.$$

This contradiction to (3.2) proves (3.1).

(2) $\Rightarrow$ (1): Now we suppose $\mathcal{O}_P \subseteq \mathcal{O}_{P'}$. Let $y \in P$; then $y^{-1} \notin \mathcal{O}_P$ by Proposition 1.1.5, therefore $y^{-1} \notin \mathcal{O}_{P'}$ by (3.1). Applying Proposition 1.1.5 once again, we obtain $y = (y^{-1})^{-1} \in P'$, hence $P \subseteq P'$.

(2) $\Rightarrow$ (3): Let $u \in F$ be an element with $v_P(u) = 0$. Then $u, u^{-1} \in \mathcal{O}_{P'}$ by (2), and so we have $v_{P'}(u) = 0$. Now choose $t \in F$ with $v_P(t) = 1$ and set $e := v_{P'}(t)$. Since $P \subseteq P'$ it follows that $e \geq 1$. Let $0 \neq x \in F$ and $v_P(x) =: r \in \mathbb{Z}$. Then $v_P(xt^{-r}) = 0$, and we obtain

$$v_{P'}(x) = v_{P'}(xt^{-r}) + v_{P'}(t^r) = 0 + r \cdot v_{P'}(t) = e \cdot v_P(x)\,.$$

(3) $\Rightarrow$ (2): $x \in \mathcal{O}_P \Rightarrow v_P(x) \geq 0 \Rightarrow v_{P'}(x) = e \cdot v_P(x) \geq 0 \Rightarrow x \in \mathcal{O}_{P'}$.

So we have proved the equivalences (1) $\iff$ (2) $\iff$ (3), and $\mathcal{O}_P = \mathcal{O}_{P'} \cap F$ if $P'|P$. The assertion $P = P' \cap F$ is now trivial (for example, from (3)). □

A consequence of the preceding proposition is that for $P'|P$ there is a canonical embedding of the residue class field $F_P = \mathcal{O}_P/P$ into the residue class field $F'_{P'} = \mathcal{O}_{P'}/P'$, given by

$$x(P) \mapsto x(P') \quad \textit{for} \quad x \in \mathcal{O}_P\,.$$

Therefore we can consider F_P as a subfield of $F'_{P'}$.

Definition 3.1.5. *Let F'/K' be an algebraic extension of F/K, and let $P' \in \mathbb{P}_{F'}$ be a place of F'/K' lying over $P \in \mathbb{P}_F$.*
(a) The integer $e(P'|P) := e$ with

$$v_{P'}(x) = e \cdot v_P(x) \quad \text{for all } x \in F$$

is called the ramification index of P' over P. We say that $P'|P$ is ramified if $e(P'|P) > 1$, and $P'|P$ is unramified if $e(P'|P) = 1$.
(b) $f(P'|P) := [F'_{P'} : F_P]$ is called the relative degree of P' over P.

Note that $f(P'|P)$ can be finite or infinite; the ramification index is always a natural number.

Proposition 3.1.6. *Let F'/K' be an algebraic extension of F/K and let P' be a place of F'/K' lying over $P \in \mathbb{P}_F$. Then*
(a) $f(P'|P) < \infty \iff [F' : F] < \infty$.
(b) If F''/K'' is an algebraic extension of F'/K' and $P'' \in \mathbb{P}_{F''}$ is an extension of P', then

$$e(P''|P) = e(P''|P') \cdot e(P'|P)\,,$$
$$f(P''|P) = f(P''|P') \cdot f(P'|P)\,.$$

Proof. (a) Consider the natural embeddings $K \subseteq F_P \subseteq F'_{P'}$ and $K \subseteq K' \subseteq F'_{P'}$, where $[F_P : K] < \infty$ and $[F'_{P'} : K'] < \infty$. It follows that

$$[F'_{P'} : F_P] < \infty \iff [K' : K] < \infty\,.$$

The latter condition is equivalent to $[F' : F] < \infty$ by Lemma 3.1.2.
(b) The assertion regarding ramification indices follows trivially from the definitions, and $f(P''|P) = f(P''|P') \cdot f(P'|P)$ follows from the inclusions $F_P \subseteq F'_{P'} \subseteq F''_{P''}$. □

Next we investigate the existence of extensions of places in extensions of function fields.

Proposition 3.1.7. *Let F'/K' be an algebraic extension of F/K.*
(a) For each place $P' \in \mathbb{P}_{F'}$ there is exactly one place $P \in \mathbb{P}_F$ such that $P'|P$, namely $P = P' \cap F$.
(b) Conversely, every place $P \in \mathbb{P}_F$ has at least one, but only finitely many extensions $P' \in \mathbb{P}_{F'}$.

Proof. (a) The main step of the proof is the following.

Claim. There is some $z \in F,\ z \neq 0$, with $v_{P'}(z) \neq 0$. (3.3)

Assume this is false. Choose $t \in F'$ with $v_{P'}(t) > 0$. Since F'/F is algebraic, there is an equation

$$c_n t^n + c_{n-1} t^{n-1} + \cdots + c_1 t + c_0 = 0$$

with $c_i \in F$, $c_0 \neq 0$ and $c_n \neq 0$. By assumption we have $v_{P'}(c_0) = 0$ and $v_{P'}(c_i t^i) = v_{P'}(c_i) + i \cdot v_{P'}(t) > 0$ for $i = 1, \ldots, n$, a contradiction to the Strict Triangle Inequality. Thus (3.3) is proved.

We set $\mathcal{O} := \mathcal{O}_{P'} \cap F$ and $P := P' \cap F$. It is obvious by (3.3) that $\mathcal{O}$ is a valuation ring of F/K and that P is the corresponding place. The uniqueness assertion is trivial.

(b) Now a place P of F/K is given. Choose $x \in F \backslash K$ whose only zero is P (this is possible by Proposition 1.6.6). We claim that for $P' \in \mathbb{P}_{F'}$ the following holds:

$$P'|P \iff v_{P'}(x) > 0\,. \tag{3.4}$$

Since x has at least one but only finitely many zeros in F'/K', assertion (b) is an immediate consequence of (3.4).

Now we prove (3.4). If $P'|P$ then $v_{P'}(x) = e(P'|P) \cdot v_P(x) > 0$. Conversely, assume that $v_{P'}(x) > 0$. Let Q denote the place of F/K which lies under P' (here we use (a)). Then $v_Q(x) > 0$, so $Q = P$ since P is the only zero of x in F/K. □

The preceding proposition enables us to define a homomorphism from the divisor group $\mathrm{Div}(F)$ into $\mathrm{Div}(F')$.

Definition 3.1.8. *Let F'/K' be an algebraic extension of F/K. For a place $P \in \mathbb{P}_F$ we define its conorm (with respect to F'/F) as*

$$\mathrm{Con}_{F'/F}(P) := \sum_{P'|P} e(P'|P) \cdot P'\,,$$

where the sum runs over all places $P' \in \mathbb{P}_{F'}$ lying over P. The conorm map is extended to a homomorphism from $\mathrm{Div}(F)$ *to* $\mathrm{Div}(F')$ *by setting*

$$\mathrm{Con}_{F'/F}\Big(\sum n_P \cdot P\Big) := \sum n_P \cdot \mathrm{Con}_{F'/F}(P)\,.$$

The conorm behaves well in towers of function fields $F'' \supseteq F' \supseteq F$; an immediate consequence of Proposition 3.1.6(b) is the formula

$$\mathrm{Con}_{F''/F}(A) = \mathrm{Con}_{F''/F'}(\mathrm{Con}_{F'/F}(A))$$

for every divisor $A \in \mathrm{Div}(F)$.

Another nice property of the conorm is that it sends principal divisors of F to principal divisors of F'. More precisely we have:

Proposition 3.1.9. *Let F'/K' be an algebraic extension of the function field F/K. For $0 \neq x \in F$ let $(x)_0^F$, $(x)_\infty^F$, $(x)^F$ resp. $(x)_0^{F'}$, $(x)_\infty^{F'}$, $(x)^{F'}$ denote the zero, pole, principal divisor of x in $\mathrm{Div}(F)$ resp. in $\mathrm{Div}(F')$. Then*

$$\mathrm{Con}_{F'/F}((x)_0^F) = (x)_0^{F'}, \quad \mathrm{Con}_{F'/F}((x)_\infty^F) = (x)_\infty^{F'}, \text{ and } \mathrm{Con}_{F'/F}((x)^F) = (x)^{F'}.$$

Proof. From the definition of the principal divisor of x follows that

$$\begin{aligned} (x)^{F'} &= \sum_{P' \in \mathbb{P}_{F'}} v_{P'}(x) \cdot P' = \sum_{P \in \mathbb{P}_F} \sum_{P'|P} e(P'|P) \cdot v_P(x) \cdot P' \\ &= \sum_{P \in \mathbb{P}_F} v_P(x) \cdot \mathrm{Con}_{F'/F}(P) = \mathrm{Con}_{F'/F}\Big(\sum_{P \in \mathbb{P}_F} v_P(x) \cdot P \Big) \\ &= \mathrm{Con}_{F'/F}((x)^F). \end{aligned}$$

Considering only the positive (negative) part of the principal divisor, we obtain the corresponding result for the zero (pole) divisor of x. □

By this proposition the conorm induces a homomorphism (again denoted as conorm) of the divisor class groups

$$\mathrm{Con}_{F'/F} : \mathrm{Cl}(F) \to \mathrm{Cl}(F').$$

This map is in general neither injective nor surjective (whereas the map $\mathrm{Con}_{F'/F} : \mathrm{Div}(F) \to \mathrm{Div}(F')$ is trivially injective).

One of our next goals is to find a relation between the degrees of a divisor $A \in \mathrm{Div}(F)$ and of its conorm in $\mathrm{Div}(F')$ in the case of a finite extension F'/F (the general case will be considered in Section 3.6). To this end we first prove a lemma.

Lemma 3.1.10. *Let K'/K be a finite field extension and let x be transcendental over K. Then*

$$[K'(x) : K(x)] = [K' : K].$$

Proof. We can assume that $K' = K(\alpha)$ for some element $\alpha \in K'$. Clearly $[K'(x) : K(x)] \leq [K' : K]$ since $K'(x) = K(x)(\alpha)$. As for the reverse inequality we have to prove that the irreducible polynomial $\varphi(T) \in K[T]$ of α over K remains irreducible over the field $K(x)$. Suppose that this is false, so $\varphi(T) = g(T) \cdot h(T)$ with monic polynomials $g(T), h(T) \in K(x)[T]$ of degree $< \deg \varphi$. Since $\varphi(\alpha) = 0$ we have w.l.o.g. $g(\alpha) = 0$. We write

$$g(T) = T^r + c_{r-1}(x)T^{r-1} + \cdots + c_0(x)$$

with $c_i(x) \in K(x)$ and $r < \deg \varphi$; then

$$\alpha^r + c_{r-1}(x)\alpha^{r-1} + \cdots + c_0(x) = 0 .$$

Multiplying by a common denominator we obtain

$$g_r(x) \cdot \alpha^r + g_{r-1}(x) \cdot \alpha^{r-1} + \cdots + g_0(x) = 0 \tag{3.5}$$

for certain polynomials $g_i(x) \in K[x]$, and we can assume that not all $g_i(x)$ are divisible by x. Setting $x = 0$ in (3.5) yields a non-trivial equation for α over K of degree less than $\deg \varphi$, a contradiction. □

Theorem 3.1.11 (Fundamental Equality). *Let F'/K' be a finite extension of F/K, let P be a place of F/K and let $P_1, \ldots, P_m$ be all the places of F'/K' lying over P. Let $e_i := e(P_i|P)$ denote the ramification index and $f_i := f(P_i|P)$ the relative degree of $P_i|P$. Then*

$$\sum_{i=1}^{m} e_i f_i = [F' : F] .$$

Proof. Choose $x \in F$ such that P is the only zero of x in F/K, and let $v_P(x) =: r > 0$. Then the places $P_1, \ldots, P_m \in \mathbb{P}_{F'}$ are exactly the zeros of x in F'/K' by (3.4). Now we evaluate the degree $[F' : K(x)]$ in two different ways:

$$\begin{aligned}
[F' : K(x)] &= [F' : K'(x)] \cdot [K'(x) : K(x)] \\
&= \left(\sum_{i=1}^{m} v_{P_i}(x) \cdot \deg P_i \right) \cdot [K' : K] \\
&= \sum_{i=1}^{m} (e_i \cdot v_P(x)) \cdot ([F'_{P_i} : K'] \cdot [K' : K]) \\
&= r \cdot \sum_{i=1}^{m} e_i \cdot [F'_{P_i} : F_P] \cdot [F_P : K] \\
&= r \cdot \deg P \cdot \sum_{i=1}^{m} e_i f_i . \qquad (3.6)
\end{aligned}$$

(The second line in the above equations follows from Lemma 3.1.10. The fact that the degree of the zero divisor of x in F'/K' equals $[F' : K'(x)]$ is a consequence of Theorem 1.4.11.) On the other hand,

$$[F' : K(x)] = [F' : F] \cdot [F : K(x)] = [F' : F] \cdot r \cdot \deg P , \tag{3.7}$$

since rP is the zero divisor of x in F/K. Comparing (3.6) and (3.7) one obtains the desired result. □

Corollary 3.1.12. *Let F'/K' be a finite extension of F/K and $P \in \mathbb{P}_F$. Then we have:*

(a) $\big|\{P' \in \mathbb{P}_{F'}\,;\, P' \text{ lies over } P\}\big| \le [F':F]$.

(b) If $P' \in \mathbb{P}_{F'}$ *lies over* P *then* $e(P'|P) \le [F':F]$ *and* $f(P'|P) \le [F':F]$.

According to Corollary 3.1.12 the following definition makes sense:

Definition 3.1.13. *Let* F'/K' *be an extension of* F/K *of degree* $[F':F]=n$ *and let* $P \in \mathbb{P}_F$.

(a) P *splits completely in* F'/F *if there are exactly* n *distinct places* $P' \in \mathbb{P}_{F'}$ *with* $P'|P$.

(b) P *is totally ramified in* F'/F *if there is a place* $P' \in \mathbb{P}_{F'}$ *with* $P'|P$ *and* $e(P'|P) = n$.

By the Fundamental Equality it is clear that a place $P \in \mathbb{P}_F$ splits completely in F'/F if and only if $e(P'|P) = f(P'|P) = 1$ for all places $P'|P$ in F'. If P is totally ramified in F'/F then there is exactly one place $P' \in \mathbb{P}_{F'}$ with $P'|P$.

As a consequence of the Fundamental Equality we obtain:

Corollary 3.1.14. *Let* F'/K' *be a finite extension of* F/K. *Then for each divisor* $A \in \mathrm{Div}(F)$,

$$\deg \mathrm{Con}_{F'/F}(A) = \frac{[F':F]}{[K':K]} \cdot \deg A\,.$$

Proof. It is sufficient to consider a prime divisor $A = P \in \mathbb{P}_F$. We have

$$\begin{aligned}
\deg \mathrm{Con}_{F'/F}(P) &= \deg\Big(\sum_{P'|P} e(P'|P)\cdot P'\Big) \\
&= \sum_{P'|P} e(P'|P)\cdot [F'_{P'} : K'] \\
&= \sum_{P'|P} e(P'|P)\cdot \frac{[F'_{P'} : K]}{[K':K]} \\
&= \frac{1}{[K':K]}\cdot \sum_{P'|P} e(P'|P)\cdot [F'_{P'} : F_P]\cdot [F_P : K] \\
&= \frac{1}{[K':K]}\cdot \Big(\sum_{P'|P} e(P'|P)\cdot f(P'|P)\Big)\cdot \deg P \\
&= \frac{[F':F]}{[K':K]}\cdot \deg P \qquad (\textit{by Theorem } 3.1.11)\,.
\end{aligned}$$

□

The foregoing results can be used to prove a very useful criterion for irreducibility of certain polynomials over a function field. A special case of the following proposition is known as *Eisenstein's Irreducibility Criterion.*

Proposition 3.1.15. *Consider a function field F/K and a polynomial*

$$\varphi(T) = a_n T^n + a_{n-1} T^{n-1} + \ldots + a_1 T + a_0$$

with coefficients $a_i \in F$. Assume that there exists a place $P \in \mathbb{P}_F$ such that one of the following conditions (1) or (2) holds:

(1) $v_P(a_n) = 0$, $v_P(a_i) \geq v_P(a_0) > 0$ for $i = 1, \ldots, n-1$, and $\gcd(n, v_P(a_0)) = 1$.

(2) $v_P(a_n) = 0$, $v_P(a_i) \geq 0$ for $i = 1, \ldots, n-1$, $v_P(a_0) < 0$, and $\gcd(n, v_P(a_0)) = 1$.

Then $\varphi(T)$ is irreducible in $F[T]$. If $F' = F(y)$ where y is a root of $\varphi(T)$, then P has a unique extension $P' \in \mathbb{P}_{F'}$, and we have $e(P'|P) = n$ and $f(P'|P) = 1$ (i.e., P is totally ramified in $F(y)/F$).

Proof. We consider an extension field $F' = F(y)$ with $\varphi(y) = 0$. The degree of F'/F is $[F' : F] \leq \deg \varphi(T) = n$, with equality if and only if $\varphi(T)$ is irreducible in $F[T]$. Choose an extension $P' \in \mathbb{P}_{F'}$ of P. As $\varphi(y) = 0$,

$$-a_n y^n = a_0 + a_1 y + \ldots + a_{n-1} y^{n-1} . \tag{3.8}$$

First we assume (1). From $v_{P'}(a_n) = 0$ and $v_{P'}(a_i) > 0$ for $i = 1, \ldots, n-1$ it follows easily that $v_{P'}(y) > 0$. Setting $e := e(P'|P)$, we have $v_{P'}(a_0) = e \cdot v_P(a_0)$ and $v_{P'}(a_i y^i) = e \cdot v_P(a_i) + i \cdot v_{P'}(y) > e \cdot v_P(a_0)$ for $i = 1, \ldots, n-1$. By the Strict Triangle Inequality, (3.8) implies

$$n \cdot v_{P'}(y) = e \cdot v_P(a_0) .$$

As $\gcd(n, v_P(a_0)) = 1$ by assumption (1), we conclude that $n | e$ and therefore $n \leq e$. On the other hand, $n \geq [F' : F] \geq e$ by Corollary 3.1.12. So we obtain

$$n = e = [F' : F] . \tag{3.9}$$

All assertions of Proposition 3.1.15 follow now immediately from (3.9) and Theorem 3.1.11.

The proof is similar in the case when one assumes (2) instead of (1). □

Before we can proceed further with the theory of extensions of algebraic function fields, we have to study certain subrings of a function field.

3.2 Subrings of Function Fields

As before, F/K denotes a function field with constant field K.

Definition 3.2.1. *A subring of F/K is a ring R such that $K \subseteq R \subseteq F$, and R is not a field.*

In particular, if R is a subring of F/K then $K \subsetneqq R \subsetneqq F$. Here are two typical examples:

(a) $R = \mathcal{O}_P$ for some $P \in \mathbb{P}_F$.

(b) $R = K[x_1, \ldots, x_n]$ where $x_1, \ldots, x_n \in F \backslash K$.

While $\mathcal{O}_P$ is obviously a subring, to see that $K[x_1, \ldots, x_n]$ is also a subring, we have to show that it is not a field. To this end, choose a place $P \in \mathbb{P}_F$ such that $v_P(x_1) \geq 0, \ldots, v_P(x_n) \geq 0$. Let $x = x_1$ and $d := \deg P$. As the residue classes $1, x(P), \ldots, x^d(P) \in \mathcal{O}_P/P$ are linearly dependent over K one can find $\alpha_0, \ldots, \alpha_d \in K$ such that the element $z = \alpha_0 + \alpha_1 x + \ldots + \alpha_d x^d$ is not 0 but $v_P(z) > 0$ (observe that x is transcendental over K since $x \notin K$). Clearly $z \in K[x_1, \ldots, x_n]$ but $z^{-1} \notin K[x_1, \ldots, x_n]$ (since $v_P(y) \geq 0$ for every $y \in K[x_1, \ldots, x_n]$).

A more general class of subrings than (a) is given by the following definition.

Definition 3.2.2. *For $\emptyset \neq S \subsetneqq \mathbb{P}_F$ let*

$$\mathcal{O}_S := \{z \in F \mid v_P(z) \geq 0 \text{ for all } P \in S\}$$

be the intersection of all valuation rings $\mathcal{O}_P$ with $P \in S$. A ring $R \subseteq F$ which is of the form $R = \mathcal{O}_S$ for some $S \subsetneqq \mathbb{P}_F, S \neq \emptyset$ is called a holomorphy ring of F/K.

For instance, the ring $K[x]$ is a holomorphy ring of the rational function field $K(x)/K$. One checks easily that

$$K[x] = \bigcap_{P \neq P_\infty} \mathcal{O}_P \,,$$

where P_∞ denotes the unique pole of x in $K(x)$.

We note some simple consequences of Definition 3.2.2.

Lemma 3.2.3. *(a) Every valuation ring $\mathcal{O}_P$ is a holomorphy ring, namely $\mathcal{O}_P = \mathcal{O}_S$ with $S = \{P\}$.*

(b) Every holomorphy ring $\mathcal{O}_S$ is a subring of F/K.

(c) For $P \in \mathbb{P}_F$ and $\emptyset \neq S \subsetneqq \mathbb{P}_F$ we have

$$\mathcal{O}_S \subseteq \mathcal{O}_P \iff P \in S\,.$$

Consequently, $\mathcal{O}_S = \mathcal{O}_T \iff S = T$.

Proof. (b) Since $\mathcal{O}_S$ is a ring with $K \subseteq \mathcal{O}_S \subseteq F$ we have only to show that it is not a field. Choose a place $P_1 \in S$. As $S \neq \mathbb{P}_F$, the Strong Approximation Theorem yields an element $0 \neq x \in F$ such that

$$v_{P_1}(x) > 0 \quad \textit{and} \quad v_P(x) \geq 0 \quad \textit{for all} \quad P \in S\,.$$

Obviously $x \in \mathcal{O}_S$ but $x^{-1} \notin \mathcal{O}_S$; therefore $\mathcal{O}_S$ is not a field.

(c) Suppose $P \notin S$. By the Strong Approximation Theorem we can find $z \in F$ with

$$v_P(z) < 0 \quad \textit{and} \quad v_Q(z) \geq 0 \quad \textit{for all} \quad Q \in S\,. \tag{3.10}$$

(This is clear if $S \cup \{P\} \neq \mathbb{P}_F$. If however $S \cup \{P\} = \mathbb{P}_F$, choose $z \in \mathcal{O}_S$ which has at least one zero in S; since z must have some pole it follows that $v_P(z) < 0$.) Each element z satisfying (3.10) is in $\mathcal{O}_S$ but not in $\mathcal{O}_P$. Thus we have proved that $P \notin S$ implies $\mathcal{O}_S \not\subseteq \mathcal{O}_P$. The remaining assertions are trivial. □

Definition 3.2.4. *Let R be a subring of F/K.*

(a) An element $z \in F$ is said to be integral over R if $f(z) = 0$ for some monic polynomial $f(X) \in R[X]$; i.e., if there are $a_0, \ldots, a_{n-1} \in R$ such that

$$z^n + a_{n-1}z^{n-1} + \cdots + a_1 z + a_0 = 0\,.$$

Such an equation is called an integral equation for z over R.

(b) The set

$$\mathrm{ic}_F(R) := \{z \in F \mid z \textit{ is integral over } R\}$$

is called the integral closure of R in F.

(c) Let $F_0 \subseteq F$ denote the quotient field of R. The ring R is called integrally closed if $\mathrm{ic}_{F_0}(R) = R$; i.e., every element $z \in F_0$ which is integral over R is already in R.

Proposition 3.2.5. *Let $\mathcal{O}_S$ be a holomorphy ring of F/K. Then*

(a) F is the quotient field of $\mathcal{O}_S$.

(b) $\mathcal{O}_S$ is integrally closed.

Proof. (a) Let $x \in F, x \neq 0$. Choose a place $P_0 \in S$. By the Strong Approximation Theorem there is an element $z \in F$ such that

$$v_{P_0}(z) = \max\{0, v_{P_0}(x^{-1}) \quad \textit{and} \quad v_P(z) \geq \max\{0, v_P(x^{-1})\} \textit{ for all } P \in S\,.$$

Clearly $z \in \mathcal{O}_S$, $z \neq 0$ and $y := zx \in \mathcal{O}_S$, so $x = yz^{-1}$ is in the quotient field of $\mathcal{O}_S$.

(b) Let $u \in F$ be integral over $\mathcal{O}_S$. Choose an integral equation

$$u^n + a_{n-1}u^{n-1} + \cdots + a_0 = 0 \tag{3.11}$$

with $a_i \in \mathcal{O}_S$. We have to show that $v_P(u) \geq 0$ for all $P \in S$. Suppose this is false, so that $v_P(u) < 0$ for some $P \in S$. Since $v_P(a_i) \geq 0$,

$$v_P(u^n) = n \cdot v_P(u) < v_P(a_i u^i)$$

for $i = 0, \ldots, n-1$. Thus the Strict Triangle Inequality yields a contradiction to (3.11). □

Theorem 3.2.6. *Let R be a subring of F/K and*

$$S(R) := \{P \in \mathbb{P}_F \mid R \subseteq \mathcal{O}_P\}.$$

Then the following hold:

(a) $\emptyset \neq S(R) \subsetneqq \mathbb{P}_F$.

(b) The integral closure of R in F is $\mathrm{ic}_F(R) = \mathcal{O}_{S(R)}$. In particular, $\mathrm{ic}_F(R)$ is an integrally closed subring of F/K with quotient field F.

Proof. (a) Since R is not a field we can find a proper ideal $I \subsetneqq R$, and by Theorem 1.1.19 there exists a place $P \in \mathbb{P}_F$ such that $I \subseteq P$ and $R \subseteq \mathcal{O}_P$. Therefore $S(R) \neq \emptyset$. On the other hand, consider an element $x \in R$ which is transcendental over K. Each place $Q \in \mathbb{P}_F$ which is a pole of x is not in $S(R)$, so $S(R) \neq \mathbb{P}_F$.

(b) Since $R \subseteq \mathcal{O}_{S(R)}$ and $\mathcal{O}_{S(R)}$ is integrally closed (by Proposition 3.2.5) it follows immediately that $\mathrm{ic}_F(R) \subseteq \mathcal{O}_{S(R)}$. In order to prove the inclusion in the reverse direction, consider an element $z \in \mathcal{O}_{S(R)}$. We claim:

$$z^{-1} \cdot R[z^{-1}] = R[z^{-1}]. \tag{3.12}$$

Suppose that (3.12) is false; i.e., $z^{-1}R[z^{-1}]$ is a proper ideal in $R[z^{-1}]$. By Theorem 1.1.19 we find a place $Q \in \mathbb{P}_F$ such that

$$R[z^{-1}] \subseteq \mathcal{O}_Q \quad \textit{and} \quad z^{-1} \in Q.$$

It follows that $Q \in S(R)$ and $z \notin \mathcal{O}_Q$ which is a contradiction to $z \in \mathcal{O}_{S(R)}$; thus we have proved (3.12). From (3.12) we obtain a relation

$$1 = z^{-1} \cdot \sum_{i=0}^{s} a_i (z^{-1})^i \tag{3.13}$$

with $a_0, \cdots, a_s \in R$. Multiplying (3.13) by z^{s+1} yields

$$z^{s+1} - \sum_{i=0}^{s} a_i z^{s-i} = 0.$$

This is an integral equation for z over R. □

Remark 3.2.7. Note that in the proof of Theorem 3.2.6 we did not actually use the assumption that K is the full constant field of F. Thus the theorem remains true if we only assume that F/K is a function field – the constant field $\tilde{K}$ of F/K may be larger than K.

An easy consequence of Proposition 3.2.5 and Theorem 3.2.6 is

Corollary 3.2.8. *A subring R of F/K with quotient field F is integrally closed if and only if R is a holomorphy ring.*

Proposition 3.2.9. *Let $\mathcal{O}_S$ be a holomorphy ring of F/K. Then there is a 1–1-correspondence between S and the set of maximal ideals of $\mathcal{O}_S$, given by*

$$P \longmapsto M_P := P \cap \mathcal{O}_S \quad (\textit{for } P \in S)\,.$$

Moreover, the map

$$\varphi : \begin{cases} \mathcal{O}_S/M_P & \longrightarrow \quad F_P = \mathcal{O}_P/P\,, \\ x + M_P & \longmapsto \quad x + P \end{cases}$$

is an isomorphism.

Proof. Consider for $P \in S$ the ring homomorphism

$$\phi : \begin{cases} \mathcal{O}_S & \longrightarrow F_P\,, \\ x & \longmapsto x + P\,. \end{cases}$$

We claim that ϕ is surjective. In fact, let $z + P \in F_P$ with $z \in \mathcal{O}_P$. By the Strong Approximation Theorem there is some $x \in F$ satisfying

$$v_P(x - z) > 0 \quad \textit{and} \quad v_Q(x) \geq 0 \quad \textit{for all} \quad Q \in S \backslash \{P\}\,.$$

Then $x \in \mathcal{O}_S$ and $\phi(x) = z + P$. The kernel of ϕ is $M_P = P \cap \mathcal{O}_S$, hence ϕ induces an isomorphism $\varphi : \mathcal{O}_S/M_P \to F_P$. Since F_P is a field, M_P is a maximal ideal of $\mathcal{O}_S$. If $P \neq Q$, the Strong Approximation Theorem shows that $M_P \neq M_Q$.

It remains to prove that each maximal ideal of $\mathcal{O}_S$ can be written as $P \cap \mathcal{O}_S$ for some $P \in S$. Let $M \subseteq \mathcal{O}_S$ be a maximal ideal. By Theorem 1.1.19 there is a place $P \in \mathbb{P}_F$ with

$$M \subseteq P \quad \textit{and} \quad \mathcal{O}_S \subseteq \mathcal{O}_P\,.$$

Lemma 3.2.3(c) shows that $P \in S$. Since $M \subseteq P \cap \mathcal{O}_S$ and M is a maximal ideal of $\mathcal{O}_S$, we obtain $M = P \cap \mathcal{O}_S$. □

We know by Theorem 1.1.6 that a valuation ring $\mathcal{O}_P$ of F/K is a principal ideal domain (i.e., every ideal of $\mathcal{O}_P$ is principal). In general holomorphy rings are no longer principal ideal domains. However, the following generalization of Theorem 1.1.6 holds:

Proposition 3.2.10. *If $S \subseteq \mathbb{P}_F$ is a non-empty finite set of places of F/K, then $\mathcal{O}_S$ is a principal ideal domain.*

Proof. Let $S = \{P_1, \ldots, P_s\}$ and let $\{0\} \neq I \subseteq \mathcal{O}_S$ be an ideal of $\mathcal{O}_S$. For $i = 1, \ldots, s$ choose $x_i \in I$ such that

$$v_{P_i}(x_i) =: n_i \leq v_{P_i}(u) \quad \textit{for all} \quad u \in I \,.$$

(This is possible since $v_{P_i}(u) \geq 0$ for all $u \in I$ and $I \neq \{0\}$.) By the Approximation Theorem we can find $z_i \in F$ such that

$$v_{P_i}(z_i) = 0 \quad \textit{and} \quad v_{P_j}(z_i) > n_j \quad \textit{for} \quad j \neq i \,.$$

Clearly $z_i \in \mathcal{O}_S$, therefore the element $x := \sum_{i=1}^{s} x_i z_i$ is in I. By the Strict Triangle Inequality we have $v_{P_i}(x) = n_i$ for $i = 1, \ldots, s$. Our proposition will be proved when we can show that $I \subseteq x\mathcal{O}_S$. Consider an element $z \in I$. Set $y := x^{-1}z$, then

$$v_{P_i}(y) = v_{P_i}(z) - n_i \geq 0 \quad \textit{for} \quad i = 1, \ldots, s \,.$$

Consequently $y \in \mathcal{O}_S$ and $z = xy \in x\mathcal{O}_S$. □

3.3 Local Integral Bases

In this section we investigate the integral closure of a subring of F/K in an extension field of F. We consider the following situation:

> *F/K is a function field with constant field K, and $F' \supseteq F$ is a finite field extension (the constant field K' of F' may be larger than K).*

Proposition 3.3.1. *Let R be an integrally closed subring of F/K such that F is the quotient field of R (i.e., R is a holomorphy ring of F/K). For $z \in F'$ let $\varphi(T) \in F[T]$ denote its minimal polynomial over F. Then we have:*

$$z \textit{ is integral over } R \iff \varphi(T) \in R[T] \,.$$

Proof. By definition, $\varphi(T)$ is the unique irreducible monic polynomial with coefficients in F such that $\varphi(z) = 0$. If $\varphi(T) \in R[T]$ then z is clearly integral over R.

The converse is not so evident. In fact, one has to use the assumption that R is integrally closed. So consider an element $z \in F'$ which is integral over R. Choose a monic polynomial $f(T) \in R[T]$ with $f(z) = 0$. Since $\varphi(T)$ is the minimal polynomial of z over F there is some $\psi(T) \in F[T]$ such that $f(T) = \varphi(T) \cdot \psi(T)$. Let $F'' \supseteq F'$ be a finite extension field of F containing all roots of φ, and $R'' = \mathrm{ic}_{F''}(R)$ be the integral closure of R in F''. Since all roots of φ are roots of f as well, they are in R''. The coefficients of $\varphi(T)$ are polynomial expressions of the roots of φ, so $\varphi(T) \in R''[T]$. But $\varphi(T) \in F[T]$ and $F \cap R'' = R$ since R is integrally closed. Therefore $\varphi(T) \in R[T]$. □

Corollary 3.3.2. *Notation as in Proposition* 3.3.1. *Let* $\mathrm{Tr}_{F'/F} : F' \to F$ *denote the trace map from* F' *to* F *and let* $x \in F'$ *be integral over* R. *Then* $\mathrm{Tr}_{F'/F}(x) \in R$.

This corollary follows easily from well-known properties of the trace mapping. Let us briefly recall some of these properties which will be of use in the sequel. We consider a finite field extension M/L of degree n. If M/L is not separable, the trace map $\mathrm{Tr}_{M/L} : M \to L$ is the zero map. Hence we can assume from now on that M/L is separable. In this case, $\mathrm{Tr}_{M/L} : M \to L$ is an L-linear map which is not identically zero. It can be described as follows. Choose an algebraically closed field $\Psi \supseteq L$. An embedding of M/L into Ψ is a field homomorphism $\sigma : M \to \Psi$ such that $\sigma(a) = a$ for all $a \in L$. Since M/L is separable there are exactly n distinct embeddings $\sigma_1, \ldots, \sigma_n$ of M/L into Ψ, and we have for $x \in M$

$$\mathrm{Tr}_{M/L}(x) = \sum_{i=1}^{n} \sigma_i(x) .$$

If $\varphi(T) = T^r + a_{r-1}T^{r-1} + \ldots + a_\circ \in L[T]$ is the minimal polynomial of x over L, then

$$\mathrm{Tr}_{M/L}(x) = -sa_{r-1} , \quad \textit{where} \quad s := [M : L(x)] . \tag{3.14}$$

The trace behaves well in towers of fields, i.e.

$$\mathrm{Tr}_{H/L}(x) = \mathrm{Tr}_{M/L}(\mathrm{Tr}_{H/M}(x)) \tag{3.15}$$

whenever $H \supseteq M \supseteq L$.

Note that Corollary 3.3.2 is an immediate consequence of (3.14) and Proposition 3.3.1.

Proposition 3.3.3. *Let* M/L *be a finite separable field extension, and consider a basis* $\{z_1, \ldots, z_n\}$ *of* M/L. *Then there are uniquely determined elements* $z_1^*, \ldots, z_n^* \in M$, *such that*

$$\mathrm{Tr}_{M/L}(z_i z_j^*) = \delta_{ij} .$$

(δ_{ij} *denotes the Kronecker symbol.) The set* $\{z_1^*, \ldots, z_n^*\}$ *is a basis of* M/L *as well; it is called the dual basis of* $\{z_1, \ldots, z_n\}$ *(with respect to the trace).*

Proof. We consider the dual space $M^\wedge$ of M over L; i.e., $M^\wedge$ is the space of all L-linear maps $\lambda : M \to L$. It is well-known from linear algebra that $M^\wedge$ is an n-dimensional vector space over L. For $z \in M$ and $\lambda \in M^\wedge$ define $z \cdot \lambda \in M^\wedge$ by $(z \cdot \lambda)(w) := \lambda(zw)$. This turns $M^\wedge$ into a vector space over M of dimension one (as $\dim_L(M^\wedge) = [M : L] \cdot \dim_M(M^\wedge)$). Since $\mathrm{Tr}_{M/L}$ is not the zero map, every $\lambda \in M^\wedge$ has a unique representation $\lambda = z \cdot \mathrm{Tr}_{M/L}$. In particular the linear forms $\lambda_j \in M^\wedge$ given by $\lambda_j(z_i) := \delta_{ij}$ $(i = 1, \ldots, n)$ can be written as $\lambda_j = z_j^* \cdot \mathrm{Tr}_{M/L}$ with $z_j^* \in M$. This means that

$$\mathrm{Tr}_{M/L}(z_i z_j^*) = (z_j^* \cdot \mathrm{Tr}_{M/L})(z_i) = \lambda_j(z_i) = \delta_{ij} \,.$$

As $\lambda_1, \ldots, \lambda_n$ are linearly independent over L, the same holds for $z_1^*, \ldots, z_n^*$, and hence they constitute a basis of M/L. □

Our next result holds essentially without the assumption of separability. The proof however is simpler under this additional hypothesis, and later on we shall need the result only for separable extensions.

Theorem 3.3.4. *Let R be an integrally closed subring of F/K with quotient field F, and F'/F be a finite separable extension of degree n. Let $R' = \mathrm{ic}_{F'}(R)$ denote the integral closure of R in F'. Then we have:*

(a) For every basis $\{x_1, \ldots, x_n\}$ of F'/F there are elements $a_i \in R\backslash\{0\}$ such that $a_1x_1, \ldots, a_nx_n \in R'$. Consequently there exist bases of F'/F which are contained in R'.

(b) If $\{z_1, \ldots, z_n\} \subseteq R'$ is a basis of F'/F and $\{z_1^*, \ldots, z_n^*\}$ denotes the dual basis with respect to the trace map then

$$\sum_{i=1}^{n} Rz_i \subseteq R' \subseteq \sum_{i=1}^{n} Rz_i^* \,.$$

(c) If in addition R is a principal ideal domain, then there exists a basis $\{u_1, \ldots, u_n\}$ of F'/F with the property

$$R' = \sum_{i=1}^{n} Ru_i \,.$$

Proof. (a) It must be shown that for every $x \in F'$ there is some element $0 \neq a \in R$ such that ax satisfies an integral equation over R. Since F'/F is algebraic and F is the quotient field of R, there are elements $a_i, b_i \in R$ with $a_i \neq 0$ and

$$x^r + \frac{b_{r-1}}{a_{r-1}}x^{r-1} + \cdots + \frac{b_1}{a_1}x + \frac{b_0}{a_0} = 0 \,.$$

Multiplying this equation by a^r, where $a := a_0 \cdot a_1 \cdot \ldots \cdot a_{r-1}$, we obtain

$$(ax)^r + c_{r-1}(ax)^{r-1} + \ldots + c_1(ax) + c_0 = 0$$

with $c_i \in R$, so $ax \in R'$.

(b) Now $\{z_1, \ldots, z_n\}$ is a basis of F'/F such that all $z_i \in R'$, and $\{z_1^*, \ldots, z_n^*\}$ is the dual basis. In particular each $z \in F'$ can be represented in the form

$$z = e_1 z_1^* + \ldots + e_n z_n^* \quad \textit{with} \quad e_i \in F\,.$$

If $z \in R'$ then $zz_j \in R'$ for $j = 1, \ldots, n$, consequently $\mathrm{Tr}_{F'/F}(zz_j) \in R$ by Corollary 3.3.2. Since

$$\mathrm{Tr}_{F'/F}(zz_j) = \mathrm{Tr}_{F'/F}\left(\sum_{i=1}^{n} e_i z_j z_i^*\right) = \sum_{i=1}^{n} e_i \cdot \mathrm{Tr}_{F'/F}(z_j z_i^*) = e_j\,,$$

we conclude that $e_j \in R$, hence $R' \subseteq \sum_{i=1}^n Rz_i^*$.
(c) Choose a basis $\{w_1, \ldots, w_n\}$ of F'/F with $R' \subseteq \sum_{i=1}^n Rw_i$ (this is possible by (b)). For $1 \le k \le n$ set

$$R_k := R' \cap \sum_{i=1}^{k} Rw_i\,. \tag{3.16}$$

We want to construct recursively $u_1, \ldots, u_n$ such that $R_k = \sum_{i=1}^k Ru_i$. For $k = 1$ (i.e., $R_1 = R' \cap Rw_1$) consider the set

$$I_1 := \{a \in F \,|\, aw_1 \in R'\}\,.$$

It is contained in R, since $R' \subseteq \sum_{i=1}^n Rw_i$. Actually I_1 is an ideal of R, hence $I_1 = a_1 R$ for some $a_1 \in R$ (as R is a principal ideal domain). Setting $u_1 := a_1 w_1$ it is easily verified that $R_1 = Ru_1$.

Suppose now that for $k \ge 2$ we have already found $u_1, \ldots, u_{k-1}$ such that $R_{k-1} = \sum_{i=1}^{k-1} Ru_i$. Let

$$\begin{aligned} I_k := \{a \in F \mid \textit{there are } b_1, \ldots, b_{k-1} \in R \\ \textit{with } b_1 w_1 + \ldots + b_{k-1} w_{k-1} + a w_k \in R'\}\,. \end{aligned}$$

Again I_k is an ideal of R, say $I_k = a_k R$. Choose $u_k \in R'$ with

$$u_k = c_1 w_1 + \ldots + c_{k-1} w_{k-1} + a_k w_k\,.$$

Clearly $R_k \supseteq \sum_{i=1}^k Ru_i$. In order to prove the reverse inclusion, let $w \in R_k$. Write

$$w = d_1 w_1 + \ldots + d_k w_k \quad \textit{with} \quad d_i \in R\,.$$

Then $d_k \in I_k$, hence $d_k = da_k$ with $d \in R$ and

$$w - du_k \in R' \cap \sum_{i=1}^{k-1} Rw_i = R_{k-1} = \sum_{i=1}^{k-1} Ru_i\,.$$

Therefore $w \in \sum_{i=1}^k Ru_i$.

We have proved that $R' = R_n = \sum_{i=1}^n Ru_i$. Since R' contains some basis of F'/F by (a), the elements $u_1, \ldots, u_n$ are linearly independent over F and constitute a basis of F'/F. □

Corollary 3.3.5. *Let F'/F be a finite separable extension of the function field F/K and let $P \in \mathbb{P}_F$ be a place of F/K. Then the integral closure $\mathcal{O}'_P$ of $\mathcal{O}_P$ in F' is*

$$\mathcal{O}'_P = \bigcap_{P'|P} \mathcal{O}_{P'} .$$

There exists a basis $\{u_1, \ldots, u_n\}$ of F'/F such that

$$\mathcal{O}'_P = \sum_{i=1}^{n} \mathcal{O}_P \cdot u_i .$$

Every such basis $\{u_1, \ldots, u_n\}$ is called an integral basis of $\mathcal{O}'_P$ over $\mathcal{O}_P$ (or a local integral basis of F'/F for the place P).

Proof. This is clear by Theorem 3.2.6(b), Remark 3.2.7 and Theorem 3.3.4 (observe that $\mathcal{O}_P$ is a principal ideal domain). □

An important supplement to the existence of local integral bases is given in the next theorem.

Theorem 3.3.6. *Let F/K be a function field, F'/F be a finite separable extension field. Then each basis $\{z_1, \ldots . z_n\}$ of F'/F is an integral basis for almost all (i.e., all but finitely many) places $P \in \mathbb{P}_F$.*

Proof. We consider the dual basis $\{z_1^*, \ldots, z_n^*\}$ of $\{z_1, \ldots, z_n\}$. The minimal polynomials of $z_1, \ldots, z_n, z_1^*, \ldots, z_n^*$ over F involve only finitely many coefficients. Let $S \subseteq \mathbb{P}_F$ be the set of all poles of these coefficients. S is finite, and for $P \notin S$ we have

$$z_1, \ldots, z_n, z_1^*, \ldots, z_n^* \in \mathcal{O}'_P , \tag{3.17}$$

where $\mathcal{O}'_P = \mathrm{ic}_{F'}(\mathcal{O}_P)$. Therefore

$$\sum \mathcal{O}_P \cdot z_i \subseteq \mathcal{O}'_P \subseteq \sum \mathcal{O}_P \cdot z_i^* \subseteq \mathcal{O}'_P \subseteq \sum \mathcal{O}_P \cdot z_i .$$

The first and third of these inclusions are obvious by (3.17), the second and fourth follow immediately from Theorem 3.3.4(b) (note that $\{z_1, \ldots, z_n\}$ is the dual basis of $\{z_1^*, \ldots, z_n^*\}$). Thus $\{z_1, \ldots, z_n\}$ is an integral basis for each $P \notin S$. □

Next we want to describe a method which can often be used to determine all extensions of a place $P \in \mathbb{P}_F$ in F'. For convenience we introduce some notation.

$\bar{F} := F_P$ *is the residue class field of* P.

$\bar{a} := a(P) \in \bar{F}$ *is the residue class of* $a \in \mathcal{O}_P$.

If $\psi(T) = \sum c_i T^i$ *is a polynomial with coefficients* $c_i \in \mathcal{O}_P$, *we set*

$$\bar{\psi}(T) := \sum \bar{c}_i T^i \in \bar{F}[T] .$$

Obviously every polynomial $\gamma(T) \in \bar{F}[T]$ can be represented as $\gamma(T) = \bar{\psi}(T)$ with $\psi(T) \in \mathcal{O}_P[T]$ and $\deg \psi(T) = \deg \gamma(T)$. With these notations we have the following theorem.

Theorem 3.3.7 (Kummer). *Suppose that $F' = F(y)$ where y is integral over $\mathcal{O}_P$, and consider the minimal polynomial $\varphi(T) \in \mathcal{O}_P[T]$ of y over F. Let*

$$\bar{\varphi}(T) = \prod_{i=1}^{r} \gamma_i(T)^{\varepsilon_i}$$

be the decomposition of $\bar{\varphi}(T)$ into irreducible factors over $\bar{F}$ (i.e., the polynomials $\gamma_1(T), \ldots, \gamma_r(T)$ are irreducible, monic, pairwise distinct in $\bar{F}[T]$ and $\varepsilon_i \geq 1$). Choose monic polynomials $\varphi_i(T) \in \mathcal{O}_P[T]$ with

$$\bar{\varphi}_i(T) = \gamma_i(T) \quad \text{and} \quad \deg \varphi_i(T) = \deg \gamma_i(T)\,.$$

Then for $1 \leq i \leq r$, there are places $P_i \in \mathbb{P}_{F'}$ satisfying

$$P_i|P\,, \quad \varphi_i(y) \in P_i \quad \text{and} \quad f(P_i|P) \geq \deg \gamma_i(T)\,.$$

Moreover $P_i \neq P_j$ for $i \neq j$.

Under additional assumptions one can prove more. Suppose that at least one of the following hypotheses $(*)$ resp. $(**)$ is satisfied:

$$\varepsilon_i = 1 \quad \text{for} \quad i = 1, \ldots, r\,, \quad \text{or} \tag{$*$}$$

$$\{1, y, \ldots, y^{n-1}\} \text{ is an integral basis for } P\,. \tag{$**$}$$

Then there exists, for $1 \leq i \leq r$, exactly one place $P_i \in \mathbb{P}_{F'}$ with $P_i|P$ and $\varphi_i(y) \in P_i$. The places $P_1, \ldots, P_r$ are all the places of F' lying over P, and we have

$$\mathrm{Con}_{F'/F}(P) = \sum_{i=1}^{r} \varepsilon_i P_i\,;$$

i.e., $\varepsilon_i = e(P_i|P)$. The residue class field $F'_{P_i} = \mathcal{O}_{P_i}/P_i$ is isomorphic to $\bar{F}[T]/(\gamma_i(T))$, hence $f(P_i|P) = \deg \gamma_i(T)$.

Proof. We set $\bar{F}_i := \bar{F}[T]/(\gamma_i(T))$. Since $\gamma_i(T)$ is irreducible, $\bar{F}_i$ is an extension field of $\bar{F}$ of degree

$$[\bar{F}_i : \bar{F}] = \deg \gamma_i(T)\,. \tag{3.18}$$

Consider the ring $\mathcal{O}_P[y] = \sum_{j=0}^{n-1} \mathcal{O}_P \cdot y^j$ where $n = \deg \varphi(T) = [F' : F]$. There are ring homomorphisms

$$\rho : \begin{cases} \mathcal{O}_P[T] & \longrightarrow \quad \mathcal{O}_P[y]\,, \\ \sum c_j T^j & \longmapsto \quad \sum c_j y^j\,, \end{cases}$$

and

$$\pi_i : \begin{cases} \mathcal{O}_P[T] & \longrightarrow & \bar{F}_i\,, \\ \sum c_j T^j & \longmapsto & \sum \bar{c}_j T^j \bmod \gamma_i(T)\,. \end{cases}$$

The kernel of ρ is the ideal generated by $\varphi(T)$. Since

$$\pi_i(\varphi(T)) = \bar{\varphi}(T) \bmod \gamma_i(T) = 0\,,$$

we have $Ker(\rho) \subseteq Ker(\pi_i)$. We conclude that there is a unique homomorphism $\sigma_i : \mathcal{O}_P[y] \to \bar{F}_i$ with $\pi_i = \sigma_i \circ \rho$. It is explicitly given by

$$\sigma_i : \begin{cases} \mathcal{O}_P[y] & \longrightarrow & \bar{F}_i\,, \\ \sum_{j=0}^{n-1} c_j y^j & \longmapsto & \sum_{j=0}^{n-1} \bar{c}_j T^j \bmod \gamma_i(T)\,. \end{cases}$$

From this it is obvious that σ_i is an epimorphism. We claim that its kernel is

$$\mathrm{Ker}(\sigma_i) = P \cdot \mathcal{O}_P[y] + \varphi_i(y) \cdot \mathcal{O}_P[y]\,. \tag{3.19}$$

The inclusion $\mathrm{Ker}(\sigma_i) \supseteq P \cdot \mathcal{O}_P[y] + \varphi_i(y) \cdot \mathcal{O}_P[y]$ is trivial. In order to show the reverse inclusion, consider an element $\sum_{j=0}^{n-1} c_j y^j \in \mathrm{Ker}(\sigma_i)$. Then $\sum_{j=0}^{n-1} \bar{c}_j T^j = \bar{\varphi}_i(T) \cdot \bar{\psi}(T)$ for some $\psi(T) \in \mathcal{O}_P[T]$, hence

$$\sum_{j=0}^{n-1} c_j T^j - \varphi_i(T) \cdot \psi(T) \in P \cdot \mathcal{O}_P[T]\,.$$

Substituting $T = y$ we obtain that

$$\sum_{j=0}^{n-1} c_j y^j - \varphi_i(y) \cdot \psi(y) \in P \cdot \mathcal{O}_P[y]\,.$$

This proves (3.19).

By Theorem 1.1.19 there exists a place $P_i \in \mathbb{P}_{F'}$ such that $\mathrm{Ker}(\sigma_i) \subseteq P_i$ and $\mathcal{O}_P[y] \subseteq \mathcal{O}_{P_i}$, hence $P_i|P$ and $\varphi_i(y) \in P_i$. The residue class field $\mathcal{O}_{P_i}/P_i$ contains $\mathcal{O}_P[y]/\mathrm{Ker}(\sigma_i)$ which is isomorphic to $\bar{F}_i$ via σ_i. Therefore (by (3.18))

$$f(P_i|P) \geq [\bar{F}_i : \bar{F}] = \deg \gamma_i(T)\,.$$

For $i \neq j$ the polynomials $\gamma_i(T) = \bar{\varphi}_i(T)$ and $\gamma_j(T) = \bar{\varphi}_j(T)$ are relatively prime in $\bar{F}[T]$, so there is a relation

$$1 = \bar{\varphi}_i(T) \cdot \bar{\lambda}_i(T) + \bar{\varphi}_j(T) \cdot \bar{\lambda}_j(T)$$

with $\lambda_i(T), \lambda_j(T) \in \mathcal{O}_P[T]$. This implies

$$\varphi_i(y) \cdot \lambda_i(y) + \varphi_j(y) \cdot \lambda_j(y) - 1 \in P \cdot \mathcal{O}_P[y]\,.$$

We conclude that $1 \in \mathrm{Ker}(\sigma_i) + \mathrm{Ker}(\sigma_j)$, by (3.19). Since $P_i \supseteq \mathrm{Ker}(\sigma_i)$ and $P_j \supseteq \mathrm{Ker}(\sigma_j)$, we have shown that $P_i \neq P_j$ for $i \neq j$.

Now we suppose that hypothesis $(*)$ holds, i.e.

$$\bar{\varphi}(T) = \prod_{i=1}^{r} \gamma_i(T) .$$

Then

$$\begin{aligned}[F' : F] = \deg \varphi(T) &= \sum_{i=1}^{r} \deg \varphi_i(T) \\ &\le \sum_{i=1}^{r} f(P_i|P) \le \sum_{i=1}^{r} e(P_i|P) \cdot f(P_i|P) \\ &\le \sum_{P'|P} e(P'|P) \cdot f(P'|P) = [F' : F] ,\end{aligned}$$

by Theorem 3.1.11. This is possible only if $e(P_i|P) = 1$, $f(P_i|P) = \deg \varphi_i(T)$, and there are no places $P' \in \mathbb{P}_{F'}$ with $P'|P$ other than $P_1, \ldots, P_r$.

Finally we assume hypothesis $(**)$. As before we choose $P_i \in \mathbb{P}_{F'}$ such that $P_i|P$ and $\varphi_i(y) \in P_i$.

Claim: $P_1, \ldots, P_r$ are the only extensions of P in F'.

In fact, let $P' \in \mathbb{P}_{F'}$ with $P'|P$. Since

$$0 = \varphi(y) \equiv \prod_{i=1}^{r} \varphi_i(y)^{\varepsilon_i} \bmod P \cdot \mathcal{O}_P[y] ,$$

we obtain

$$\prod_{i=1}^{r} \varphi_i(y)^{\varepsilon_i} \in P' . \tag{3.20}$$

P' is a prime ideal in $\mathcal{O}_{P'}$, so (3.20) implies $\varphi_i(y) \in P'$ for some $i \in \{1, \ldots, r\}$ and

$$P \cdot \mathcal{O}_P[y] + \varphi_i(y) \cdot \mathcal{O}_P[y] \subseteq P' \cap \mathcal{O}_P[y] . \tag{3.21}$$

The left hand side is a maximal ideal of $\mathcal{O}_P[y]$ by (3.19), therefore equality holds in (3.21). As we also have

$$P \cdot \mathcal{O}_P[y] + \varphi_i(y) \cdot \mathcal{O}_P[y] \subseteq P_i \cap \mathcal{O}_P[y] ,$$

it follows that

$$P' \cap \mathcal{O}_P[y] = P_i \cap \mathcal{O}_P[y] = \varphi_i(y) \cdot \mathcal{O}_P[y] + P \cdot \mathcal{O}_P[y] . \tag{3.22}$$

Since $\mathcal{O}_P[y]$ is the integral closure of $\mathcal{O}_P$ in F' by hypothesis $(**)$, Proposition 3.2.9 shows now that $P' = P_i$, and the claim is proved.

As an immediate consequence of the Claim and Corollary 3.3.5 we see that

$$\mathcal{O}_P[y] = \bigcap_{i=1}^{r} \mathcal{O}_{P_i} \,. \tag{3.23}$$

By the Approximation Theorem one can find elements $t_1, \ldots, t_r \in F'$ satisfying

$$v_{P_i}(t_i) = 1 \ \textit{and} \ v_{P_j}(t_i) = 0 \ \textit{for} \ j \neq i \,.$$

Choose a P-prime element $t \in F$; then

$$t_i \in \mathcal{O}_P[y] \cap P_i = \varphi_i(y) \cdot \mathcal{O}_P[y] + t \cdot \mathcal{O}_P[y]$$

by (3.23) and (3.22). Thus t_i can be written as

$$t_i = \varphi_i(y) \cdot a_i(y) + t \cdot b_i(y) \ \textit{with} \ a_i(y), b_i(y) \in \mathcal{O}_P[y] \,.$$

From this we obtain

$$\prod_{i=1}^{r} t_i^{\varepsilon_i} = a(y) \cdot \prod_{i=1}^{r} \varphi_i(y)^{\varepsilon_i} + t \cdot b(y) \tag{3.24}$$

with $a(y), b(y) \in \mathcal{O}_P[y]$. As

$$\prod_{i=1}^{r} \varphi_i(y)^{\varepsilon_i} \equiv \varphi(y) \bmod t \cdot \mathcal{O}_P[y]$$

and $\varphi(y) = 0$, (3.24) implies that

$$\prod_{i=1}^{r} t_i^{\varepsilon_i} = t \cdot u(y) \quad \textit{for some} \quad u(y) \in \mathcal{O}_P[y] \,. \tag{3.25}$$

Thereby

$$\varepsilon_i = v_{P_i}\left(\prod_{j=1}^{r} t_j^{\varepsilon_j} \right) \geq v_{P_i}(t) = e(P_i|P) \,. \tag{3.26}$$

On the other hand we have

$$f(P_i|P) = \deg \gamma_i(T) \tag{3.27}$$

by (3.18), (3.19), (3.22) and Proposition 3.2.9. It follows, by (3.26), (3.27) and Theorem 3.1.11, that

$$[F' : F] = \sum_{i=1}^{r} e(P_i|P) \cdot f(P_i|P)$$

$$\leq \sum_{i=1}^{r} \varepsilon_i \cdot \deg \gamma_i(T) = \deg \varphi(T) = [F' : F] \,.$$

Hence $\varepsilon_i = e(P_i|P)$ for $i = 1, \ldots, r$. □

We emphasize a special case of Kummer's Theorem which is often particularly useful.

Corollary 3.3.8. *Let $\varphi(T) = T^n + f_{n-1}(x)T^{n-1} + \cdots + f_0(x) \in K(x)[T]$ be an irreducible polynomial over the rational function field $K(x)$. We consider the function field $K(x,y)/K$ where y satsfies the equation $\varphi(y) = 0$, and an element $\alpha \in K$ such that $f_j(\alpha) \neq \infty$ for all j, $0 \leq j \leq n-1$. Denote by $P_\alpha \in \mathbb{P}_{K(x)}$ the zero of $x - \alpha$ in $K(x)$. Suppose that the polynomial*

$$\varphi_\alpha(T) := T^n + f_{n-1}(\alpha)T^{n-1} + \cdots + f_0(\alpha) \in K[T]$$

has the following decomposition in the polynomial ring $K[T]$:

$$\varphi_\alpha(T) = \prod_{i=1}^{r} \psi_i(T)$$

with irreducible, monic, pairwise distinct polynomials $\psi_i(T) \in K[T]$. Then we have:

(a) For every $i = 1, \ldots, r$ there is a uniquely determined place $P_i \in \mathbb{P}_{K(x,y)}$ such that $x - \alpha \in P_i$ and $\psi_i(y) \in P_i$. The element $x - \alpha$ is a prime element of P_i (i.e., $e(P_i|P_\alpha) = 1$), and the residue class field of P_i is K-isomorphic to $K[T]/(\psi_i(T))$. Hence $f(P_i|P_\alpha) = \deg \psi_i(T)$.

(b) If $\deg \psi_i(T) = 1$ for at least one $i \in \{1, \ldots, r\}$, then K is the full constant field of $K(x,y)$.

(c) If $\varphi_\alpha(T)$ has $n = \deg \varphi(T)$ distinct roots β in K, then there is for each β with $\varphi_\alpha(\beta) = 0$ a unique place $P_{\alpha,\beta} \in \mathbb{P}_{K(x,y)}$ such that

$$x - \alpha \in P_{\alpha,\beta} \quad \text{and} \quad y - \beta \in P_{\alpha,\beta}\,.$$

$P_{\alpha,\beta}$ is a place of $K(x,y)$ of degree 1.

Proof. We set $F := K(x)$ and $F' := K(x,y)$. The assumption $f_j(\alpha) \neq \infty$ implies that y is integral over the valuation ring of P_α, and the polynomial $\varphi_\alpha(T)$ is nothing else but $\bar{\varphi}(T)$ (with notation as in Kummer's Theorem). Thus we are in the situation of hypothesis $(*)$ of Kummer's Theorem, and our corollary follows immediately. □

3.4 The Cotrace of Weil Differentials and the Hurwitz Genus Formula

In this section the following situation is considered:

> *F/K is an algebraic function field, F'/F a finite separable extension, K' is the constant field of F'. Clearly, K'/K is a finite separable extension as well.*

Our aim is to associate with each Weil differential of F/K a Weil differential of F'/K'. This will yield a very useful formula for the genus of F', the *Hurwitz Genus Formula*. For this we need to introduce the notion of *different* of an extension F'/F. Note that F'/F is always assumed to be a separable extension, hence the trace map $\mathrm{Tr}_{F'/F}$ is not identically zero.

Definition 3.4.1. *For $P \in \mathbb{P}_F$ let $\mathcal{O}'_P := \mathrm{ic}_{F'}(\mathcal{O}_P)$ denote the integral closure of $\mathcal{O}_P$ in F'. Then the set*

$$\mathcal{C}_P := \{z \in F' \mid \mathrm{Tr}_{F'/F}(z \cdot \mathcal{O}'_P) \subseteq \mathcal{O}_P\}$$

is called the complementary module over $\mathcal{O}_P$.

Proposition 3.4.2. *With notation as in Definition 3.4.1 the following hold:*
(a) $\mathcal{C}_P$ is an $\mathcal{O}'_P$-module and $\mathcal{O}'_P \subseteq \mathcal{C}_P$.
(b) If $\{z_1, \ldots, z_n\}$ is an integral basis of $\mathcal{O}'_P$ over $\mathcal{O}_P$, then

$$\mathcal{C}_P = \sum_{i=1}^{n} \mathcal{O}_P \cdot z_i^* ,$$

where $\{z_1^*, \ldots, z_n^*\}$ is the dual basis of $\{z_1, \ldots, z_n\}$.
(c) There is an element $t \in F'$ (depending on P) such that $\mathcal{C}_P = t \cdot \mathcal{O}'_P$. Moreover,

$$v_{P'}(t) \le 0 \text{ for all } P'|P ,$$

and for every $t' \in F'$ we have:

$$\mathcal{C}_P = t' \cdot \mathcal{O}'_P \iff v_{P'}(t') = v_{P'}(t) \text{ for all } P'|P .$$

(d) $\mathcal{C}_P = \mathcal{O}'_P$ for almost all $P \in \mathbb{P}_F$.

Proof. (a) The assertion that $\mathcal{C}_P$ is an $\mathcal{O}'_P$-module is trivial. Since the trace of an element $y \in \mathcal{O}'_P$ is in $\mathcal{O}_P$, by Corollary 3.3.2, we have $\mathcal{O}'_P \subseteq \mathcal{C}_P$.
(b) First we consider an element $z \in \mathcal{C}_P$. As $\{z_1^*, \ldots, z_n^*\}$ is a basis of F'/F, there are $x_1, \ldots, x_n \in F$ with $z = \sum_{i=1}^n x_i z_i^*$. Since $z \in \mathcal{C}_P$ and $z_1, \ldots z_n \in \mathcal{O}'_P$, it follows that $\mathrm{Tr}_{F'/F}(zz_j) \in \mathcal{O}_P$ for $1 \le j \le n$. Now

$$\begin{aligned} \mathrm{Tr}_{F'/F}(zz_j) &= \mathrm{Tr}_{F'/F}\left(\sum_{i=1}^{n} x_i z_i^* z_j\right) \\ &= \sum_{i=1}^{n} x_i \cdot \mathrm{Tr}_{F'/F}(z_i^* z_j) = x_j , \end{aligned}$$

by the properties of the dual basis. Therefore $x_j \in \mathcal{O}_P$ and $z \in \sum_{i=1}^n \mathcal{O}_P \cdot z_i^*$.

Conversely, let $z \in \sum_{i=1}^n \mathcal{O}_P \cdot z_i^*$ and $u \in \mathcal{O}'_P$, say $z = \sum_{i=1}^n x_i z_i^*$ and $u = \sum_{j=1}^n y_j z_j$ with $x_i, y_j \in \mathcal{O}_P$. Then

$$\begin{aligned}\mathrm{Tr}_{F'/F}(zu) &= \mathrm{Tr}_{F'/F}\left(\sum_{i,j=1}^{n} x_i y_j z_i^* z_j\right)\\ &= \sum_{i,j=1}^{n} x_i y_j \cdot \mathrm{Tr}_{F'/F}(z_i^* z_j) = \sum_{i=1}^{n} x_i y_i \in \mathcal{O}_P .\end{aligned}$$

Hence $z \in \mathcal{C}_P$.

(c) By (b) we know that $\mathcal{C}_P = \sum_{i=1}^{n} \mathcal{O}_P \cdot u_i$ with appropriate elements $u_i \in F'$. Choose $x \in F$ such that

$$v_P(x) \geq -v_{P'}(u_i)$$

for all $P'|P$ and $i = 1, \ldots, n$. Then

$$v_{P'}(xu_i) = e(P'|P) \cdot v_P(x) + v_{P'}(u_i) \geq 0$$

for all $P'|P$ and $i = 1, \ldots, n$, therefore $x \cdot \mathcal{C}_P \subseteq \mathcal{O}'_P$ (observe that we have $\mathcal{O}'_P = \{u \in F' \mid v_{P'}(u) \geq 0 \text{ for all } P'|P\}$, by Corollary 3.3.5). Obviously $x \cdot \mathcal{C}_P$ is an ideal of $\mathcal{O}'_P$. Hence $x \cdot \mathcal{C}_P = y \cdot \mathcal{O}'_P$ for some $y \in \mathcal{O}'_P$, because $\mathcal{O}'_P$ is a principal ideal domain by Proposition 3.2.10. Setting $t := x^{-1}y$ we obtain $\mathcal{C}_P = t \cdot \mathcal{O}'_P$. Since $\mathcal{O}'_P \subseteq \mathcal{C}_P$, it follows immediately that $v_{P'}(t) \leq 0$ for all $P'|P$. Finally we have for $t' \in F'$:

$$\begin{aligned} t \cdot \mathcal{O}'_P = t' \cdot \mathcal{O}'_P &\iff tt'^{-1} \in \mathcal{O}'_P \text{ and } t^{-1}t' \in \mathcal{O}'_P\\ &\iff v_{P'}(tt'^{-1}) \geq 0 \text{ and } v_{P'}(t^{-1}t') \geq 0 \text{ for all } P'|P\\ &\iff v_{P'}(t) = v_{P'}(t') \text{ for all } P'|P .\end{aligned}$$

(d) Choose a basis $\{z_1, \ldots, z_n\}$ of F'/F. By Theorem 3.3.6 $\{z_1, \ldots z_n\}$ and $\{z_1^*, \ldots, z_n^*\}$ are integral bases for almost all $P \in \mathbb{P}_F$. Using (b) we see that $\mathcal{C}_P = \mathcal{O}'_P$ for almost all P. □

Definition 3.4.3. *Consider a place $P \in \mathbb{P}_F$ and the integral closure $\mathcal{O}'_P$ of $\mathcal{O}_P$ in F'. Let $\mathcal{C}_P = t \cdot \mathcal{O}'_P$ be the complementary module over $\mathcal{O}_P$. Then we define for $P'|P$ the different exponent of P' over P by*

$$d(P'|P) := -v_{P'}(t) .$$

By Proposition 3.4.2, *$d(P'|P)$ is well-defined and $d(P'|P) \geq 0$. Moreover $d(P'|P) = 0$ holds for almost all $P \in \mathbb{P}_F$ and $P'|P$, since $\mathcal{C}_P = 1 \cdot \mathcal{O}'_P$ for almost all P. Therefore we can define the divisor*

$$\mathrm{Diff}(F'/F) := \sum_{P \in \mathbb{P}_F} \sum_{P'|P} d(P'|P) \cdot P' .$$

This divisor is called the different of F'/F.

Observe that $\mathrm{Diff}(F'/F)$ is a divisor of F', and $\mathrm{Diff}(F'/F) \geq 0$. Later on we will develop several methods for determining the different in many cases, see for instance Theorem 3.5.1, Theorem 3.5.10 and Theorem 3.8.7.

Remark 3.4.4. We have the following useful characterization of the complementary module $\mathcal{C}_P$, which follows immediately from the definitions. For every element $z \in F'$,

$$z \in \mathcal{C}_P \iff v_{P'}(z) \geq -d(P'|P) \text{ for all } P'|P\,.$$

Let us recall some notions from Chapter 1 which are used in what follows. $\mathcal{A}_F$ is the *adele space* of F/K. For a divisor $A \in \mathrm{Div}(F)$ let

$$\mathcal{A}_F(A) = \{\alpha \in \mathcal{A}_F \mid v_P(\alpha) \geq -v_P(A) \text{ for all } P \in \mathbb{P}_F\}\,.$$

This is a K-subspace of $\mathcal{A}_F$. The field F is embedded into $\mathcal{A}_F$ diagonally. A *Weil differential* ω of F/K is a K-linear mapping $\omega : \mathcal{A}_F \to K$ vanishing on $\mathcal{A}_F(A) + F$ for some divisor $A \in \mathrm{Div}(F)$. If $\omega \neq 0$ is a Weil differential of F/K then its divisor $(\omega) \in \mathrm{Div}(F)$ is defined as

$$(\omega) = \max\{A \in \mathrm{Div}(F) \mid \omega \text{ vanishes on } \mathcal{A}_F(A) + F\}\,.$$

Definition 3.4.5. *Let*

$$\mathcal{A}_{F'/F} := \{\alpha \in \mathcal{A}_{F'} \mid \alpha_{P'} = \alpha_{Q'} \text{ whenever } P' \cap F = Q' \cap F\}\,.$$

This is an F'-subspace of $\mathcal{A}_{F'}$. The trace mapping $\mathrm{Tr}_{F'/F} : F' \to F$ can be extended to an F-linear map (again denoted by $\mathrm{Tr}_{F'/F}$) from $\mathcal{A}_{F'/F}$ to $\mathcal{A}_F$ by setting

$$(\mathrm{Tr}_{F'/F}(\alpha))_P := \mathrm{Tr}_{F'/F}(\alpha_{P'}) \text{ for } \alpha \in \mathcal{A}_{F'/F}\,,$$

where P' is any place of F' lying over P. Observe that $\alpha_{P'} \in \mathcal{O}_{P'}$ for almost all $P' \in \mathbb{P}_{F'}$, therefore $\mathrm{Tr}_{F'/F}(\alpha_{P'}) \in \mathcal{O}_P$ for almost all $P \in \mathbb{P}_F$ by Corollary 3.3.2. Hence $\mathrm{Tr}_{F'/F}(\alpha)$ is an adele of F/K. Clearly the trace of a principal adele $z \in F'$ is the principal adele of $\mathrm{Tr}_{F'/F}(z)$.

For a divisor $A' \in \mathrm{Div}(F')$ we set

$$\mathcal{A}_{F'/F}(A') := \mathcal{A}_{F'}(A') \cap \mathcal{A}_{F'/F}\,.$$

Theorem 3.4.6. *In the above situation, for every Weil differential ω of F/K there exists a unique Weil differential ω' of F'/K' such that*

$$\mathrm{Tr}_{K'/K}(\omega'(\alpha)) = \omega(\mathrm{Tr}_{F'/F}(\alpha)) \tag{3.28}$$

for all $\alpha \in \mathcal{A}_{F'/F}$. This Weil differential is called the cotrace of ω in F'/F, and it is denoted by $\mathrm{Cotr}_{F'/F}(\omega)$. *If $\omega \neq 0$ and $(\omega) \in \mathrm{Div}(F)$ is the divisor of ω, then*

$$(\mathrm{Cotr}_{F'/F}(\omega)) = \mathrm{Con}_{F'/F}((\omega)) + \mathrm{Diff}(F'/F)\,.$$

An important special case of this theorem is:

Corollary 3.4.7. *Let F/K be a function field and $x \in K$ such that the extension $F/K(x)$ is separable. Let η be the Weil differential of the rational function field $K(x)$ whose existence was proved in Proposition* 1.7.4. *Then the divisor of its cotrace in $F/K(x)$ is*

$$(\mathrm{Cotr}_{F/K(x)}(\eta)) = -2(x)_\infty + \mathrm{Diff}(F/K(x)).$$

Remark 3.4.8. Using the notion of local components of a Weil differential (cf. Section 1.7), Equation (3.28) can be replaced by the following *local conditions*: for each $P \in \mathbb{P}_F$ and each $y \in F'$,

$$\omega_P(\mathrm{Tr}_{F'/F}(y)) = \mathrm{Tr}_{K'/K}\left(\sum_{P'|P} \omega'_{P'}(y)\right). \tag{3.29}$$

The equivalence of (3.28) and (3.29) follows after a little thought from Proposition 1.7.2, which says that $\omega(\gamma)$ is the sum of its local components $\omega_P(\gamma_P)$ for each adele $\gamma = (\gamma_P)_{P \in \mathbb{P}_F}$.

In the course of the proof of Theorem 3.4.6 we need two lemmas which will be proved first.

Lemma 3.4.9. *For each $C' \in \mathrm{Div}(F')$ we have $\mathcal{A}_{F'} = \mathcal{A}_{F'/F} + \mathcal{A}_{F'}(C')$.*

Proof. Let $\alpha = (\alpha_{P'})_{P' \in \mathbb{P}_{F'}}$ be an adele of F'. For all $P \in \mathbb{P}_F$ there exists by the Approximation Theorem an element $x_P \in F'$ with

$$v_{P'}(\alpha_{P'} - x_P) \geq -v_{P'}(C') \textit{ for all } P'|P.$$

We set $\beta = (\beta_{P'})_{P' \in \mathbb{P}_{F'}}$ with $\beta_{P'} := x_P$ whenever $P'|P$. Then $\beta \in \mathcal{A}_{F'/F}$ and $\alpha - \beta \in \mathcal{A}_{F'}(C')$. Since $\alpha = \beta + (\alpha - \beta)$, the lemma follows. □

Lemma 3.4.10. *Let M/L be a finite separable field extension, V a vector space over M and $\mu : V \to L$ be an L-linear map. Then there is a unique M-linear map $\mu' : V \to M$ such that $\mathrm{Tr}_{M/L} \circ \mu' = \mu$.*

Proof. As in the proof of Proposition 3.3.3 we consider the space of linear forms $M^\wedge = \{\lambda : M \to L \mid \lambda \text{ is } L\text{-linear}\}$ as a vector space over M by setting $(z \cdot \lambda)(w) = \lambda(z \cdot w)$ for $\lambda \in M^\wedge$ and $z, w \in M$. The dimension of $M^\wedge$ over M is one, hence every $\lambda \in M^\wedge$ has a unique representation $\lambda = z \cdot \mathrm{Tr}_{M/L}$ with $z \in M$.

For a fixed element $v \in V$ define the map $\lambda_v : M \to L$ by $\lambda_v(a) := \mu(av)$; it is clearly L-linear. Therefore $\lambda_v = z_v \cdot \mathrm{Tr}_{M/L}$ with a unique element $z_v \in M$, and we set $\mu'(v) := z_v$. Thus we have

$$\mu(av) = (\mu'(v) \cdot \mathrm{Tr}_{M/L})(a) = \mathrm{Tr}_{M/L}(a \cdot \mu'(v)) \tag{3.30}$$

for all $a \in M$ and $v \in V$, and $\mu'(v)$ is uniquely determined by (3.30). Using this it is easily verified that $\mu' : V \to M$ is M-linear. Setting $a = 1$ in (3.30) we obtain $\mu = \mathrm{Tr}_{M/L} \circ \mu'$, which proves the existence of $\mu' : V \to M$ with the desired properties.

Suppose that there is another $\mu^* : V \to M$ with $\mathrm{Tr}_{M/L} \circ \mu' = \mathrm{Tr}_{M/L} \circ \mu^*$ and $\mu^* \neq \mu'$. Then the image of $\mu' - \mu^*$ is the whole of M, and we have $\mathrm{Tr}_{M/L} \circ (\mu' - \mu^*) = 0$. This is a contradiction, since $\mathrm{Tr}_{M/L}$ is not the zero map. □

Proof of Theorem 3.4.6. First we want to show the existence of a Weil differential ω' such that

$$\mathrm{Tr}_{K'/K}(\omega'(\alpha)) = \omega(\mathrm{Tr}_{F'/F}(\alpha))$$

holds for all $\alpha \in \mathcal{A}_{F'/F}$. For $\omega = 0$ set $\omega' := 0$, therefore we can assume in the sequel that $\omega \neq 0$. For brevity we set

$$W' := \mathrm{Con}_{F'/F}((\omega)) + \mathrm{Diff}(F'/F)\,. \tag{3.31}$$

The construction of ω' is given in three steps.

Step 1. The K-linear mapping $\omega_1 : \mathcal{A}_{F'/F} \to K$ which is defined by setting $\omega_1 := \omega \circ \mathrm{Tr}_{F'/F}$ has the following properties:

(a_1) $\omega_1(\alpha) = 0$ for $\alpha \in \mathcal{A}_{F'/F}(W') + F'$.

(b_1) If $B' \in \mathrm{Div}(F')$ is a divisor with $B' \not\leq W'$, then there is an adele $\beta \in \mathcal{A}_{F'/F}(B')$ with $\omega_1(\beta) \neq 0$.

Proof of Step 1. (a_1) Obviously ω_1 is K-linear, and ω_1 vanishes on F' since ω vanishes on F. Now let $\alpha \in \mathcal{A}_{F'/F}(W')$. In order to prove $\omega_1(\alpha) = 0$ we only have to verify that for all $P \in \mathbb{P}_F$ and $P'|P$ the following holds:

$$v_P(\mathrm{Tr}_{F'/F}(\alpha_{P'})) \geq -v_P(\omega)\,. \tag{3.32}$$

(Observe that ω vanishes on $\mathcal{A}_F((\omega))$ by definition of the divisor (ω).) Choose an element $x \in F$ with $v_P(x) = v_P(\omega)$. Then

$$\begin{aligned} v_{P'}(x\alpha_{P'}) &= v_{P'}(x) + v_{P'}(\alpha_{P'}) \geq e(P'|P) \cdot v_P(\omega) - v_{P'}(W') \\ &= v_{P'}(\mathrm{Con}_{F'/F}((\omega)) - W') = -v_{P'}(\mathrm{Diff}(F'/F)) = -d(P'|P)\,. \end{aligned}$$

(Recall that $d(P'|P)$ denotes the different exponent of $P'|P$.) This implies by Remark 3.4.4 that $x\alpha_{P'} \in \mathcal{C}_P$, the complementary module over $\mathcal{O}_P$, and therefore $v_P(\mathrm{Tr}_{F'/F}(x\alpha_{P'})) \geq 0$. As $\mathrm{Tr}_{F'/F}(x\alpha_{P'}) = x \cdot \mathrm{Tr}_{F'/F}(\alpha_{P'})$ and $v_P(x) = v_P(\omega)$, assertion (3.32) follows.

(b_1) Now there is given a divisor $B' \not\leq W'$; i.e., there is a place $P_0 \in \mathbb{P}_F$ such that

$$v_{P^*}(\mathrm{Con}_{F'/F}((\omega)) - B') < -d(P^*|P_0) \tag{3.33}$$

for some $P^*|P_0$. Let $\mathcal{O}'_{P_0}$ (resp. $\mathcal{C}_{P_0}$) denote the integral closure of $\mathcal{O}_{P_0}$ in F' (resp. the complementary module over $\mathcal{O}_{P_0}$), and consider the set

$$J := \{z \in F' \mid v_{P^*}(z) \geq v_{P^*}(\mathrm{Con}_{F'/F}((\omega)) - B') \textit{ for all } P^*|P_0\}\,.$$

By the Approximation Theorem there is an element $u \in J$ which satisfies $v_{P^*}(u) = v_{P^*}(\mathrm{Con}_{F'/F}((\omega)) - B')$ for all $P^*|P_0$, therefore $J \not\subseteq \mathcal{C}_{P_0}$ by Remark 3.4.4 and (3.33). Since $J \cdot \mathcal{O}'_{P_0} \subseteq J$ it follows that

$$\mathrm{Tr}_{F'/F}(J) \not\subseteq \mathcal{O}_{P_0}\,. \tag{3.34}$$

Choose $t \in F$ with $v_{P_0}(t) = 1$. For some $r \geq 0$ we have $t^r \cdot J \subseteq \mathcal{O}'_{P_0}$ (trivial by definition of J), so $t^r \cdot \mathrm{Tr}_{F'/F}(J) = \mathrm{Tr}_{F'/F}(t^r J) \subseteq \mathcal{O}_{P_0}$. It is easily checked that $t^r \cdot \mathrm{Tr}_{F'/F}(J)$ is an ideal of $\mathcal{O}_{P_0}$, consequently $t^r \cdot \mathrm{Tr}_{F'/F}(J) = t^s \cdot \mathcal{O}_{P_0}$ with $s \geq 0$, and we obtain $\mathrm{Tr}_{F'/F}(J) = t^m \cdot \mathcal{O}_{P_0}$ for some $m \in \mathbb{Z}$. By (3.34), $m \leq -1$ and therefore

$$t^{-1} \cdot \mathcal{O}_{P_0} \subseteq \mathrm{Tr}_{F'/F}(J)\,. \tag{3.35}$$

Recall the notion of the local component of a Weil differential, cf. Section 1.7. By Proposition 1.7.3(a) we can find an element $x \in F$ with

$$v_{P_0}(x) = -v_{P_0}(\omega) - 1 \;\; \textit{and} \;\; \omega_{P_0}(x) \neq 0\,. \tag{3.36}$$

We choose $y \in F$ with $v_{P_0}(y) = v_{P_0}(\omega)$, so $xy \in t^{-1} \cdot \mathcal{O}_{P_0}$. By (3.35) there is some $z \in J$ with $\mathrm{Tr}_{F'/F}(z) = xy$. Consider the adele $\beta \in \mathcal{A}_{F'/F}$ given by

$$\beta_{P'} := \begin{cases} 0 & \textit{if } \; P' \nmid P_0\,, \\ y^{-1}z & \textit{if } \; P' \,|\, P_0\,. \end{cases}$$

It follows from the definition of J that for $P'|P_0$

$$\begin{aligned} v_{P'}(\beta) &= -v_{P'}(y) + v_{P'}(z) \\ &\geq -v_{P'}(\mathrm{Con}_{F'/F}((\omega))) + v_{P'}(\mathrm{Con}_{F'/F}((\omega)) - B') \\ &= -v_{P'}(B')\,. \end{aligned}$$

So $\beta \in \mathcal{A}_{F'/F}(B')$. Finally, we have $\omega_1(\beta) = \omega(\mathrm{Tr}_{F'/F}(\beta)) = \omega_{P_0}(x) \neq 0$ by (3.36). This proves (b_1).

Step 2. We define $\omega_2 : \mathcal{A}_{F'} \to K$ as follows. For $\alpha \in \mathcal{A}_{F'}$ there are adeles $\beta \in \mathcal{A}_{F'/F}$ and $\gamma \in \mathcal{A}_{F'}(W')$ such that $\alpha = \beta + \gamma$, by Lemma 3.4.9. Set

$$\omega_2(\alpha) := \omega_1(\beta)\,.$$

This is well-defined. In fact, if we have two representations $\alpha = \beta + \gamma = \beta_1 + \gamma_1$ with $\beta, \beta_1 \in \mathcal{A}_{F'/F}$ and $\gamma, \gamma_1 \in \mathcal{A}_{F'}(W')$ then

$$\beta_1 - \beta = \gamma - \gamma_1 \in \mathcal{A}_{F'/F} \cap \mathcal{A}_{F'}(W') = \mathcal{A}_{F'/F}(W')\,.$$

Hence $\omega_1(\beta_1) - \omega_1(\beta) = \omega_1(\beta_1 - \beta) = 0$ by (a_1). The mapping ω_2 is obviously K-linear, and by (a_1) and (b_1) it has the following properties:

(a$_2$) $\omega_2(\alpha) = 0$ for $\alpha \in \mathcal{A}_{F'}(W') + F'$.

(b$_2$) If $B' \in \mathrm{Div}(F')$ is a divisor with $B' \not\leq W'$, then there is an adele $\beta \in \mathcal{A}_{F'}(B')$ with $\omega_2(\beta) \neq 0$.

Thus far, we have constructed a K-linear mapping $\omega_2 : \mathcal{A}_{F'} \to K$ vanishing on $\mathcal{A}_{F'}(W') + F'$. However, ω_2 is not a Weil differential of F'/K' if K' is strictly larger than K. Therefore we have to 'lift' ω_2 to a K'-linear map; this is done in the next step.

Step 3. By Lemma 3.4.10 there exists a K'-linear map $\omega' : \mathcal{A}_{F'} \to K'$ such that $\mathrm{Tr}_{K'/K} \circ \omega' = \omega_2$. From the definition of ω_1 and ω_2 we obtain immediately that for $\alpha \in \mathcal{A}_{F'/F}$

$$\mathrm{Tr}_{K'/K}(\omega'(\alpha)) = \omega_2(\alpha) = \omega_1(\alpha) = \omega(\mathrm{Tr}_{F'/F}(\alpha)).$$

This proves (3.28), and it remains to show:

(a$_3$) $\omega'(\alpha) = 0$ for $\alpha \in \mathcal{A}_{F'}(W') + F'$.

(b$_3$) If $B' \in \mathrm{Div}(F')$ is a divisor with $B' \not\leq W'$, then there is some adele $\beta \in \mathcal{A}_{F'}(B')$ with $\omega'(\beta) \neq 0$.

Proof of (a$_3$). Since ω' is K'-linear, the image of $\mathcal{A}_{F'}(W') + F'$ under ω' is either 0 or the whole of K'. In the latter case there is some $\alpha \in \mathcal{A}_{F'}(W') + F'$ such that $\mathrm{Tr}_{K'/K}(\omega'(\alpha)) \neq 0$, since $\mathrm{Tr}_{K'/K} : K' \to K$ is not the zero map. By construction of ω' we have $\omega_2 = \mathrm{Tr}_{K'/K} \circ \omega'$. Hence $\omega_2(\alpha) \neq 0$, which is a contradiction to (a$_2$).

Proof of (b$_3$). By (b$_2$) there exists some adele $\beta \in \mathcal{A}_{F'}(B')$ with the property $\omega_2(\beta) \neq 0$. So $\mathrm{Tr}_{K'/K}(\omega'(\beta)) \neq 0$, and the assertion follows immediately.

We have established the existence of a Weil differential ω' of F'/K' satisfying (3.28), and we have shown that the divisor of ω' is

$$(\omega') = W' = \mathrm{Con}_{F'/F}((\omega)) + \mathrm{Diff}(F'/F).$$

In order to prove uniqueness, suppose that ω^* is another Weil differential of F'/K' with the property (3.28); i.e.,

$$\mathrm{Tr}_{K'/K}(\omega^*(\alpha)) = \mathrm{Tr}_{K'/K}(\omega'(\alpha)) = \omega(\mathrm{Tr}_{F'/F}(\alpha))$$

for all $\alpha \in \mathcal{A}_{F'/F}$. Setting $\eta := \omega^* - \omega'$ we obtain

$$\mathrm{Tr}_{K'/K}(\eta(\alpha)) = 0 \text{ for all } \alpha \in \mathcal{A}_{F'/F}. \tag{3.37}$$

η is a Weil differential of F'/K', hence η vanishes on $\mathcal{A}_{F'}(C')$ for some divisor $C' \in \mathrm{Div}(F')$. By Lemma 3.4.9 and (3.37) it follows that $\mathrm{Tr}_{K'/K}(\eta(\alpha)) = 0$ for all $\alpha \in \mathcal{A}_{F'}$. This implies $\eta = 0$ and $\omega^* = \omega'$. □

We note some formal properties of the cotrace mapping $\omega \mapsto \mathrm{Cotr}_{F'/F}(\omega)$.

Proposition 3.4.11. *(a) If ω, ω_1 and ω_2 are Weil differentials of F/K and $x \in F$, then*

$$\mathrm{Cotr}_{F'/F}(\omega_1 + \omega_2) = \mathrm{Cotr}_{F'/F}(\omega_1) + \mathrm{Cotr}_{F'/F}(\omega_2)$$

and

$$\mathrm{Cotr}_{F'/F}(x\omega) = x \cdot \mathrm{Cotr}_{F'/F}(\omega)\,.$$

(b) Let F''/F' be another finite separable extension. Then

$$\mathrm{Cotr}_{F''/F}(\omega) = \mathrm{Cotr}_{F''/F'}(\mathrm{Cotr}_{F'/F}(\omega))$$

for each Weil differential ω of F/K.

Proof. Keeping in mind the uniqueness assertion in Theorem 3.4.6, it is sufficient to show that

$$\mathrm{Tr}_{K'/K}\Big(\big(\mathrm{Cotr}_{F'/F}(\omega_1) + \mathrm{Cotr}_{F'/F}(\omega_2)\big)(\alpha)\Big) = (\omega_1 + \omega_2)(\mathrm{Tr}_{F'/F}(\alpha)) \quad (3.38)$$

holds for all $\alpha \in \mathcal{A}_{F'/F}$. The proof of (3.38) is straightforward, using the linearity of the trace. In the same manner one proves that $\mathrm{Cotr}_{F'/F}(x\omega) = x \cdot \mathrm{Cotr}_{F'/F}(\omega)$, as well as assertion (b). □

Corollary 3.4.12 (Transitivity of the Different). *If $F'' \supseteq F' \supseteq F$ are finite separable extensions, the following hold:*

(a) $\mathrm{Diff}(F''/F) = \mathrm{Con}_{F''/F'}(\mathrm{Diff}(F'/F)) + \mathrm{Diff}(F''/F')$.

(b) $d(P''|P) = e(P''|P') \cdot d(P'|P) + d(P''|P')$, if P'' (resp. P', P) are places of F'' (resp. F', F) with $P'' \supseteq P' \supseteq P$.

Proof. (b) is merely a reformulation of (a), so we only prove (a). Choose a Weil differential $\omega \neq 0$ of F/K. Then the divisor of $\mathrm{Cotr}_{F''/F}(\omega)$ is

$$(\mathrm{Cotr}_{F''/F}(\omega)) = \mathrm{Con}_{F''/F}((\omega)) + \mathrm{Diff}(F''/F) \quad (3.39)$$

by Theorem 3.4.6. On the other hand, Proposition 3.4.11 yields

$$\begin{aligned}
&(\mathrm{Cotr}_{F''/F}(\omega)) = (\mathrm{Cotr}_{F''/F'}(\mathrm{Cotr}_{F'/F}(\omega))) \\
&= \mathrm{Con}_{F''/F'}((\mathrm{Cotr}_{F'/F}(\omega))) + \mathrm{Diff}(F''/F') \\
&= \mathrm{Con}_{F''/F'}(\mathrm{Con}_{F'/F}((\omega)) + \mathrm{Diff}(F'/F)) + \mathrm{Diff}(F''/F') \\
&= \mathrm{Con}_{F''/F}((\omega)) + \mathrm{Con}_{F''/F'}(\mathrm{Diff}(F'/F)) + \mathrm{Diff}(F''/F')\,. \quad (3.40)
\end{aligned}$$

(We have used the transitivity of the conorm, cf. Definition 3.1.8.) Comparing (3.39) and (3.40) we obtain (a). □

An important consequence of Theorem 3.4.6 is the following result.

Theorem 3.4.13 (Hurwitz Genus Formula). *Let F/K be an algebraic function field of genus g and let F'/F be a finite separable extension. Let K' denote the constant field of F' and g' the genus of F'/K'. Then we have*

$$2g' - 2 = \frac{[F' : F]}{[K' : K]}(2g - 2) + \deg \mathrm{Diff}(F'/F)\,.$$

Proof. Choose a Weil differential $\omega \neq 0$ of F/K. It follows from Theorem 3.4.6 that

$$(\mathrm{Cotr}_{F'/F}(\omega)) = \mathrm{Con}_{F'/F}((\omega)) + \mathrm{Diff}(F'/F)\,. \tag{3.41}$$

Recall that the degree of a canonical divisor is $2g-2$ (resp. $2g'-2$). Then we obtain from (3.41) and Corollary 3.1.14

$$\begin{aligned} 2g' - 2 &= \deg \mathrm{Con}_{F'/F}((\omega)) + \deg \mathrm{Diff}(F'/F) \\ &= \frac{[F' : F]}{[K' : K]}(2g - 2) + \deg \mathrm{Diff}(F'/F)\,. \end{aligned}$$

□

We emphasize a special case of the Hurwitz Genus Formula:

Corollary 3.4.14. *Let F/K be a function field of genus g and let $x \in F \setminus K$ such that the extension $F/K(x)$ is separable. Then*

$$2g - 2 = -2[F : K(x)] + \deg \mathrm{Diff}(F/K(x))\,.$$

Every function field F/K can be regarded as a finite extension of a rational function field $K(x)$ (as we shall prove in Section 3.10, one can choose x in such a way that $F/K(x)$ is separable). Therefore the Hurwitz Genus Formula (resp. Corollary 3.4.14) is a powerful tool that allows determination of the genus of F in terms of the different of $F/K(x)$. Thus far however, we have no methods at hand how to determine this different. The next section addresses this problem.

3.5 The Different

We consider a finite separable extension F'/F where F/K resp. F'/K' are algebraic function fields with constant fields K resp. K'. As always, the field K (hence also K') is assumed to be perfect.

For $P \in \mathbb{P}_F$ and $P' \in \mathbb{P}_{F'}$ with $P'|P$ we have defined the *ramification index* $e(P'|P)$ and the *different exponent* $d(P'|P)$ (Definition 3.1.5 and 3.4.3). There is a close relationship between these two numbers, given by the following theorem:

Theorem 3.5.1 (Dedekind's Different Theorem). *With notation as above we have for all $P'|P$*

(a) $d(P'|P) \geq e(P'|P) - 1$.

(b) $d(P'|P) = e(P'|P) - 1$ *if and only if* $e(P'|P)$ *is not divisible by* $\operatorname{char} K$.

In particular, if $\operatorname{char} K = 0$ *then* $d(P'|P) = e(P'|P) - 1$.

We shall first prove part (a) of Dedekind's Theorem; the proof will require an understanding of the action of automorphisms on the places of a function field. More precisely we need:

Lemma 3.5.2. *Let* F^*/F *be an algebraic extension of function fields,* $P \in \mathbb{P}_F$ *and* $P^* \in \mathbb{P}_{F^*}$ *with* $P^*|P$. *Consider an automorphism* σ *of* F^*/F. *Then* $\sigma(P^*) := \{\sigma(z) \mid z \in P^*\}$ *is a place of* F^*, *and we have*

(a) $v_{\sigma(P^*)}(y) = v_{P^*}(\sigma^{-1}(y))$ *for all* $y \in F^*$.

(b) $\sigma(P^*)|P$.

(c) $e(\sigma(P^*)|P) = e(P^*|P)$ *and* $f(\sigma(P^*)|P) = f(P^*|P)$.

Proof of the Lemma. Clearly $\sigma(\mathcal{O}_{P^*})$ is a valuation ring of F^* and $\sigma(P^*)$ is its maximal ideal; therefore $\sigma(P^*)$ is a place of F^*, and the corresponding valuation ring is $\mathcal{O}_{\sigma(P^*)} = \sigma(\mathcal{O}_{P^*})$. If t^* is a prime element of P^*, i.e. $P^* = t^* \cdot \mathcal{O}_{P^*}$, then $\sigma(P^*) = \sigma(t^*) \cdot \sigma(\mathcal{O}_{P^*})$, so $\sigma(t^*)$ is a prime element for $\sigma(P^*)$.

(a) Let $0 \neq y \in F^*$, say $y = \sigma(z)$. Writing $z = t^{*r}u$ with $r = v_{P^*}(z)$ and $u \in \mathcal{O}_{P^*} \backslash P^*$, we obtain $y = \sigma(t^*)^r \cdot \sigma(u)$ where $\sigma(u) \in \mathcal{O}_{\sigma(P^*)} \backslash \sigma(P^*)$ and $\sigma(t^*)$ is a prime element for $\sigma(P^*)$. Therefore $v_{\sigma(P^*)}(y) = r = v_{P^*}(z) = v_{P^*}(\sigma^{-1}(y))$.

(b) $\sigma(P^*)$ lies over P since $\sigma(P^*) \supseteq \sigma(P) = P$.

(c) Choose a P-prime element $x \in F$. Then

$$e(\sigma(P^*)|P) = v_{\sigma(P^*)}(x) = v_{P^*}(\sigma^{-1}(x)) = v_{P^*}(x) = e(P^*|P)\,.$$

The automorphism σ of F^*/F induces an isomorphism $\bar{\sigma}$ of the residue class field $F^*_{P^*}$ onto $F^*_{\sigma(P^*)}$ given by

$$\bar{\sigma}(z + P^*) := \sigma(z) + \sigma(P^*)\,.$$

$\bar{\sigma}$ is the identity on F_P, hence $f(P^*|P) = f(\sigma(P^*)|P)$. □

Proof of Theorem 3.5.1(a). As before let $\mathcal{O}'_P$ denote the integral closure of $\mathcal{O}_P$ in F', and $\mathcal{C}_P$ the complementary module over $\mathcal{O}_P$. We want to show that

$$\operatorname{Tr}_{F'/F}(t \cdot \mathcal{O}'_P) \subseteq \mathcal{O}_P \tag{3.42}$$

for every element $t \in F'$ which satisfies

$$v_{P'}(t) = 1 - e(P'|P) \;\; \textit{for all} \;\; P'|P\,. \tag{3.43}$$

Note that (3.42) implies $t \in \mathcal{C}_P$, and the characterization of $\mathcal{C}_P$ given in Remark 3.4.4 yields $1 - e(P'|P) \geq -d(P'|P)$, so that $d(P'|P) \geq e(P'|P) - 1$.

In order to prove (3.42), consider a finite Galois extension F^*/F such that $F \subseteq F' \subseteq F^*$, and choose $n := [F' : F]$ automorphisms $\sigma_1, \ldots, \sigma_n$ of F^*/F whose restrictions to F' are pairwise distinct. For $z \in \mathcal{O}'_P$ we have

$$\mathrm{Tr}_{F'/F}(t \cdot z) = \sum_{i=1}^{n} \sigma_i(t \cdot z). \tag{3.44}$$

We fix some place P^* of F^* which lies over P, and set $P_i^* := \sigma_i^{-1}(P^*)$ and $P_i' := P_i^* \cap F'$. Note that $\sigma_i(z)$ is integral over $\mathcal{O}_P$ since $z \in \mathcal{O}'_P$, and therefore $v_{P^*}(\sigma_i(z)) \geq 0$. Then we obtain

$$\begin{aligned} v_{P^*}(\sigma_i(t \cdot z)) &= v_{P^*}(\sigma_i(t)) + v_{P^*}(\sigma_i(z)) \\ &\geq v_{P^*}(\sigma_i(t)) = v_{P_i^*}(t) && (\textit{by Lemma } 3.5.2) \\ &= e(P_i^*|P_i')(1 - e(P_i'|P)) && (\textit{by } (3.43)) \\ &> -e(P_i^*|P_i') \cdot e(P_i'|P) \\ &= -e(P_i^*|P) = -e(P^*|P) && (\textit{by Lemma } 3.5.2). \end{aligned}$$

Using (3.44) we conclude

$$-e(P^*|P) < v_{P^*}(\mathrm{Tr}_{F'/F}(t \cdot z)) = e(P^*|P) \cdot v_P(\mathrm{Tr}_{F'/F}(t \cdot z)).$$

This implies $v_P(\mathrm{Tr}_{F'/F}(t \cdot z)) \geq 0$, hence (3.42). □

The essential step in the proof of part (b) of Theorem 3.5.1 is the following lemma.

Lemma 3.5.3. *Let $P \in \mathbb{P}_F$ and $P_1, \ldots, P_r \in \mathbb{P}_{F'}$ be all the extensions of P in F'/F. Consider the residue class fields $k := \mathcal{O}_P/P$ resp. $k_i := \mathcal{O}_{P_i}/P_i \supseteq k$ and the corresponding residue class maps $\pi : \mathcal{O}_P \to k$ resp. $\pi_i : \mathcal{O}_{P_i} \to k_i$ (for $i = 1, \ldots, r$). Then we have for every $u \in \mathcal{O}'_P$ (the integral closure of $\mathcal{O}_P$ in F')*

$$\pi(\mathrm{Tr}_{F'/F}(u)) = \sum_{i=1}^{r} e(P_i|P) \cdot \mathrm{Tr}_{k_i/k}(\pi_i(u)).$$

Proof of Theorem 3.5.1(b). We maintain the notation of Lemma 3.5.3 and abbreviate $e_i := e(P_i|P)$. Let $P' = P_1$ and $e := e(P'|P)$. It must be shown that

$$d(P'|P) = e - 1 \iff \mathrm{char}\, K \textit{ does not divide } e. \tag{3.45}$$

First assume that e is not divisible by $\mathrm{char}\, K$. Suppose $d(P'|P) \geq e$. Then there exists some $w \in F'$ such that

$$v_{P'}(w) \leq -e \textit{ and } \mathrm{Tr}_{F'/F}(w \cdot \mathcal{O}'_P) \subseteq \mathcal{O}_P. \tag{3.46}$$

Since K is perfect, the extension k_1/k is separable, and we can find $y_0 \in \mathcal{O}_{P'}$ with $\mathrm{Tr}_{k_1/k}(\pi_1(y_0)) \neq 0$. By the Approximation Theorem there is an element $y \in F'$ such that

$$v_{P'}(y - y_0) > 0$$

and

$$v_{P_i}(y) \geq \max\{1, e_i + v_{P_i}(w)\} \quad \textit{for } 2 \leq i \leq r\,. \tag{3.47}$$

Then $y \in \mathcal{O}'_P$ and, by Lemma 3.5.3,

$$\begin{aligned} \pi(\mathrm{Tr}_{F'/F}(y)) &= e \cdot \mathrm{Tr}_{k_1/k}(\pi_1(y)) + \sum_{i=2}^{r} e_i \cdot \mathrm{Tr}_{k_i/k}(\pi_i(y)) \\ &= e \cdot \mathrm{Tr}_{k_1/k}(\pi_1(y_0)) \neq 0\,. \end{aligned}$$

(Here we use the fact that $\mathrm{char}\, K$ does not divide e, hence $e \neq 0$ in k.) We conclude that

$$v_P(\mathrm{Tr}_{F'/F}(y)) = 0\,.$$

Now choose $x \in F$ with $v_P(x) = 1$. Then

$$\mathrm{Tr}_{F'/F}(x^{-1}y) = x^{-1} \cdot \mathrm{Tr}_{F'/F}(y) \notin \mathcal{O}_P\,. \tag{3.48}$$

On the other hand we have $x^{-1}yw^{-1} \in \mathcal{O}'_P$, since

$$v_{P'}(x^{-1}yw^{-1}) = -e + v_{P'}(y) - v_{P'}(w) \geq 0$$

and

$$v_{P_i}(x^{-1}yw^{-1}) = v_{P_i}(y) - (e_i + v_{P_i}(w)) \geq 0$$

for $i = 2, \ldots, r$, by (3.46) and (3.47). It follows that $x^{-1}y \in w \cdot \mathcal{O}'_P$ and $\mathrm{Tr}_{F'/F}(x^{-1}y) \in \mathcal{O}_P$ by (3.46); this contradicts (3.48). So we have proved the implication $\Leftarrow$ of (3.45).

In order to prove the converse, we assume now that $\mathrm{char}\, K$ divides e, and we have to show that $d(P'|P) \geq e$. Choose $u \in F'$ such that

$$v_{P'}(u) = -e \;\; \textit{and} \;\; v_{P_i}(u) \geq -e_i + 1 \; (i = 2, \ldots, r)\,. \tag{3.49}$$

As before, $x \in F$ denotes a P-prime element. For each $z \in \mathcal{O}'_P$ we have

$$v_{P'}(xuz) \geq 0 \;\; \textit{and} \;\; v_{P_i}(xuz) > 0$$

for $i = 2, \ldots, r$. Therefore $xuz \in \mathcal{O}'_P$, and by Lemma 3.5.3,

$$\begin{aligned} \pi(\mathrm{Tr}_{F'/F}(xuz)) &= e \cdot \mathrm{Tr}_{k_1/k}(\pi_1(xuz)) + \sum_{i=2}^{r} e_i \cdot \mathrm{Tr}_{k_i/k}(\pi_i(xuz)) \\ &= e \cdot \mathrm{Tr}_{k_1/k}(\pi_1(xuz)) = 0\,. \end{aligned}$$

We conclude that $x \cdot \mathrm{Tr}_{F'/F}(uz) = \mathrm{Tr}_{F'/F}(xuz) \in P = x\mathcal{O}_P$, and thereby $\mathrm{Tr}_{F'/F}(uz) \in \mathcal{O}_P$ for all $z \in \mathcal{O}'_P$. Thus $u \in \mathcal{C}_P$ and $-e = v_{P'}(u) \geq -d(P'|P)$ by (3.49) and Remark 3.4.4. □

Proof of Lemma 3.5.3. The trace $\mathrm{Tr}_{F'/F}(u)$ can be evaluated as the trace of the F-linear map $\mu : F' \to F'$ which is given by $\mu(z) = u \cdot z$ (see Appendix A). First we show that $\pi(\mathrm{Tr}_{F'/F}(u))$ has an interpretation as the trace of a certain k-linear map $\bar{\mu} : V \to V$ (where V is some k-vector space to be defined below); decomposing V into invariant subspaces will then yield the final result.

Let $t \in F$ be a P-prime element. The quotient $V := \mathcal{O}'_P/t\mathcal{O}'_P$ can be considered as a vector space over k by setting

$$(x+P) \cdot (z + t\mathcal{O}'_P) := xz + t\mathcal{O}'_P \quad (x \in \mathcal{O}_P, z \in \mathcal{O}'_P)\,. \tag{3.50}$$

Note that this scalar multiplication is well-defined. Choose an integral basis $\{z_1, \ldots, z_n\}$ of $\mathcal{O}'_P$ over $\mathcal{O}_P$ (where $n = [F' : F]$, cf. Corollary 3.3.5). Then $\{z_1 + t\mathcal{O}'_P, \ldots, z_n + t\mathcal{O}'_P\}$ constitutes a basis of V over k (the proof is trivial), in particular $\dim_k(V) = n$. We define a k-linear map $\bar{\mu} : V \to V$ by

$$\bar{\mu}(z + t\mathcal{O}'_P) := u \cdot z + t\mathcal{O}'_P\,. \tag{3.51}$$

Let $A = (a_{ij})_{1 \leq i,j \leq n}$ be the matrix which represents μ with respect to the basis $\{z_1, \ldots, z_n\}$. Since this is an integral basis and $u \in \mathcal{O}'_P$, the coefficients a_{ij} are in $\mathcal{O}_P$. Obviously $\bar{A} := (\pi(a_{ij}))_{1 \leq i,j \leq n}$ represents $\bar{\mu}$ with respect to $\{z_1 + t\mathcal{O}'_P, \ldots, z_n + t\mathcal{O}'_P\}$, therefore

$$\pi(\mathrm{Tr}_{F'/F}(u)) = \pi(\mathrm{Tr}\, A) = \mathrm{Tr}\,(\bar{A}) = \mathrm{Tr}\,(\bar{\mu})\,. \tag{3.52}$$

Next we introduce for $1 \leq i \leq r$ the quotients $V_i := \mathcal{O}_{P_i}/P_i^{e_i}$ and the mappings $\mu_i : V_i \to V_i$, defined by

$$\mu_i(z + P_i^{e_i}) := u \cdot z + P_i^{e_i}\,.$$

V_i is considered as a vector space over k in the obvious manner (cf. (3.50)), and μ_i turns out to be k-linear (all this is easily verified). There is a natural isomorphism

$$f : V \longrightarrow \bigoplus_{i=1}^{r} V_i\,,$$

given by

$$f(z + t\mathcal{O}'_P) := (z + P_1^{e_1}, \ldots, z + P_r^{e_r})\,. \tag{3.53}$$

In fact, f is surjective by the Approximation Theorem. In order to prove that f is injective suppose $f(z + t\mathcal{O}'_P) = 0$. Then $v_{P_i}(z) \geq e_i$; i.e., we have $v_{P_i}(z \cdot t^{-1}) \geq 0$ for $i = 1, \ldots, r$. This implies $z \cdot t^{-1} \in \mathcal{O}'_P$, hence $z \in t\mathcal{O}'_P$.

There is a commutative diagram of k-linear mappings

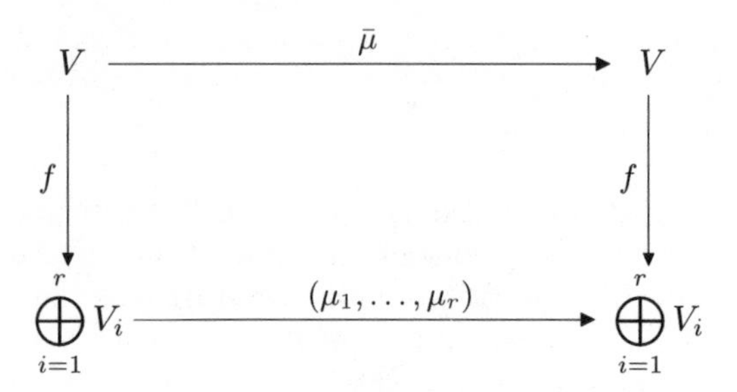

Fig. 3.1.

where $(\mu_1, \ldots, \mu_r)(v_1, \ldots, v_r) := (\mu_1(v_1), \ldots, \mu_r(v_r))$, $v_i \in V_i$. Since f is an isomorphism, it follows that $\mathrm{Tr}(\bar{\mu}) = \mathrm{Tr}((\mu_1, \ldots, \mu_r))$, which is obviously equal to $\sum_{i=1}^{r} \mathrm{Tr}(\mu_i)$. Combining this with (3.52) we obtain

$$\pi(\mathrm{Tr}_{F'/F}(u)) = \sum_{i=1}^{r} \mathrm{Tr}(\mu_i). \tag{3.54}$$

The proof of Lemma 3.5.3 will be completed once we can show that

$$\mathrm{Tr}(\mu_i) = e_i \cdot \mathrm{Tr}_{k_i/k}(\pi_i(u)). \tag{3.55}$$

Consider the chain of k-subspaces

$$V_i = V_i^{(0)} \supseteq V_i^{(1)} \supseteq \ldots \supseteq V_i^{(e_i)} = \{0\},$$

where $V_i^{(j)} := P_i^j / P_i^{e_i} \subseteq V_i$. These spaces are invariant under μ_i, so μ_i induces linear maps

$$\sigma_{ij} : \begin{cases} V_i^{(j)}/V_i^{(j+1)} & \longrightarrow & V_i^{(j)}/V_i^{(j+1)}, \\ [z + P_i^{e_i}] & \longmapsto & [u \cdot z + P_i^{e_i}] \end{cases}$$

for $j = 0, \ldots, e_i - 1$. Here $[z + P_i^{e_i}]$ denotes the residue class of $z + P_i^{e_i}$ in $V_i^{(j)}/V_i^{(j+1)}$. It is easily seen that

$$\mathrm{Tr}(\mu_i) = \sum_{j=0}^{e_i - 1} \mathrm{Tr}(\sigma_{ij}). \tag{3.56}$$

(Represent μ_i by a matrix with respect to a basis of V_i which is composed of bases of $V_i^{(j)}$ modulo $V_i^{(j+1)}$, for $0 \le j \le e_i - 1$.) We know that

$$\mathrm{Tr}_{k_i/k}(\pi_i(u)) = \mathrm{Tr}(\gamma_i), \tag{3.57}$$

where $\gamma_i : k_i \to k_i$ is the k-linear map defined by $\gamma_i(z + P_i) = u \cdot z + P_i$. Now we establish for $0 \le j \le e_i - 1$ an isomorphism $h : k_i \to V_i^{(j)}/V_i^{(j+1)}$ of k-vector spaces such that the following diagram is commutative:

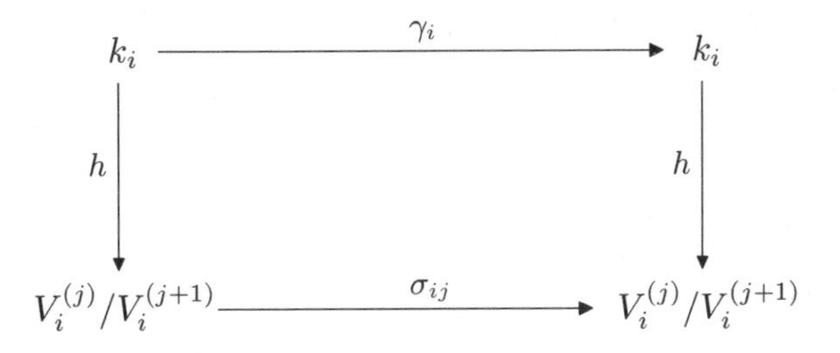

Fig. 3.2.

This diagram yields $\mathrm{Tr}(\gamma_i) = \mathrm{Tr}(\sigma_{ij})$ (because h is an isomorphism), and then (3.55) follows immediately from (3.56) and (3.57).

The map h is defined as follows: Choose a P_i-prime element $t_i \in F'$ and let

$$h(z + P_i) := [t_i^j z + P_i^{e_i}] .$$

Then it is easily verified that h is well-defined, k-linear, bijective and the diagram in Figure 3.2 is commutative. This completes the proof of Lemma 3.5.3 and thereby Theorem 3.5.1. □

We would like to draw some conclusions from Dedekind's Theorem. Recall that $P'|P$ is said to be *ramified* if $e(P'|P) > 1$; otherwise $P'|P$ is *unramified* (cf. Definition 3.1.5).

Definition 3.5.4. *Let F'/F be an algebraic extension of function fields and $P \in \mathbb{P}_F$.*

(a) An extension P' of P in F' is said to be tamely (resp. wildly) ramified if $e(P'|P) > 1$ and the characteristic of K does not divide $e(P'|P)$ (resp. char K *divides $e(P'|P)$).*

(b) We say that P is ramified (resp. unramified) in F'/F if there is at least one $P' \in \mathbb{P}_F$ over P such that $P'|P$ is ramified (resp. if $P'|P$ is unramified for all $P'|P$). The place P is tamely ramified in F'/F if it is ramified in F'/F and no extension of P in F' is wildly ramified. If there is at least one wildly ramified place $P'|P$ we say that P is wildly ramified in F'/F.

(c) P is totally ramified in F'/F if there is only one extension $P' \in \mathbb{P}_{F'}$ of P in F', and the ramification index is $e(P'|P) = [F' : F]$.

(d) F'/F is said to be ramified (resp. unramified) if at least one $P \in \mathbb{P}_F$ is ramified in F'/F (resp. if all $P \in \mathbb{P}_F$ are unramified in F'/F).

(e) F'/F is said to be tame if no place $P \in \mathbb{P}_F$ is wildly ramified in F'/F.

Corollary 3.5.5. *Let F'/F be a finite separable extension of algebraic function fields.*

(a) If $P \in \mathbb{P}_F$ and $P' \in \mathbb{P}_{F'}$ such that $P'|P$, then $P'|P$ is ramified if and only if $P' \leq \operatorname{Diff}(F'/F)$.

If $P'|P$ is ramified, then

$$\begin{aligned} d(P'|P) = e(P'|P) - 1 &\iff P'|P \text{ is tamely ramified }, \\ d(P'|P) \geq e(P'|P) &\iff P'|P \text{ is wildly ramified }. \end{aligned}$$

(b) Almost all places $P \in \mathbb{P}_F$ are unramified in F'/F.

This corollary follows immediately from Dedekind's Theorem. Next we note an important special case of the Hurwitz Genus Formula.

Corollary 3.5.6. *Suppose that F'/F is a finite separable extension of algebraic function fields having the same constant field K. Let g (resp. g') denote the genus of F/K (resp. F'/K). Then*

$$2g' - 2 \geq [F' : F] \cdot (2g - 2) + \sum_{P \in \mathbb{P}_F} \sum_{P'|P} (e(P'|P) - 1) \cdot \deg P' .$$

Equality holds if and only if F'/F is tame (for instance if K is a field of characteristic 0).

Proof. Trivial by Theorems 3.4.13 and 3.5.1. □

Corollary 3.5.7. *Suppose that F'/F is a finite separable extension of function fields having the same constant field. Let g (resp. g') denote the genus of F (resp. F'). Then $g \leq g'$.*

Corollary 3.5.8. *Let $F/K(x)$ be a finite separable extension of the rational function field of degree $[F : K(x)] > 1$ such that K is the constant field of F. Then $F/K(x)$ is ramified.*

Proof. The Hurwitz Genus Formula yields

$$2g - 2 = -2[F : K(x)] + \deg \operatorname{Diff}(F/K(x)) ,$$

where g is the genus of F/K. Therefore

$$\deg \operatorname{Diff}(F/K(x)) \geq 2([F : K(x)] - 1) > 0 .$$

The assertion follows since each place in the support of the different ramifies by Corollary 3.5.5. □

We give another application of the above results.

Proposition 3.5.9 (Lüroth's Theorem). *Every subfield of a rational function field is rational; i.e., if $K \subsetneqq F_0 \subseteq K(x)$ then $F_0 = K(y)$ for some $y \in F_0$.*

Proof. Suppose first that $K(x)/F_0$ is separable. Let g_0 denote the genus of F_0/K. Then

$$-2 = [K(x) : F_0] \cdot (2g_0 - 2) + \deg \mathrm{Diff}(K(x)/F_0)\,,$$

which implies $g_0 = 0$. If P is a place of $K(x)/K$ of degree one then $P_0 = P \cap F_0$ is a place of F_0/K of degree one. Therefore F_0/K is rational by Proposition 1.6.3.

Now assume that $K(x)/F_0$ is not separable. There is an intermediate field $F_0 \subseteq F_1 \subseteq K(x)$ such that F_1/F_0 is separable and $K(x)/F_1$ is purely inseparable. According to what we have proved above, it is sufficient to show that F_1/K is rational. As $K(x)/F_1$ is purely inseparable, $[K(x) : F_1] = q = p^\nu$ where $p = \mathrm{char}\, K > 0$, and $z^q \in F_1$ for each $z \in K(x)$. In particular,

$$K(x^q) \subseteq F_1 \subseteq K(x)\,. \tag{3.58}$$

The degree $[K(x) : K(x^q)]$ is equal to the degree of the pole divisor of x^q in $K(x)/K$ by Theorem 1.4.11, therefore $[K(x) : K(x^q)] = q$. By (3.58), it follows that $F_1 = K(x^q)$, hence F_1/K is rational. □

Next we prove a theorem that is often very useful for evaluating the different of F'/F.

Theorem 3.5.10. *Suppose $F' = F(y)$ is a finite separable extension of a function field F of degree $[F' : F] = n$. Let $P \in \mathbb{P}_F$ be such that the minimal polynomial $\varphi(T)$ of y over F has coefficients in $\mathcal{O}_P$ (i.e., y is integral over $\mathcal{O}_P$), and let $P_1, \ldots, P_r \in \mathbb{P}_{F'}$ be all places of F' lying over P. Then the following hold:*

(a) $d(P_i|P) \le v_{P_i}(\varphi'(y))$ for $1 \le i \le r$.

(b) $\{1, y, \ldots, y^{n-1}\}$ is an integral basis of F'/F at the place P if and only if $d(P_i|P) = v_{P_i}(\varphi'(y))$ for $1 \le i \le r$.

(Here $\varphi'(T)$ denotes the derivative of $\varphi(T)$ in the polynomial ring $F[T]$.)

Proof. The dual basis of $\{1, y, \ldots, y^{n-1}\}$ is closely related to the different exponents $d(P_i|P)$ by Proposition 3.4.2, therefore our first aim is to determine this dual basis. Since $\varphi(y) = 0$, the polynomial $\varphi(T)$ factors in $F'[T]$ as

$$\varphi(T) = (T - y)(c_{n-1}T^{n-1} + \ldots + c_1 T + c_0) \tag{3.59}$$

with $c_0, \ldots, c_{n-1} \in F'$ and $c_{n-1} = 1$. We claim:

$$\left\{ \frac{c_0}{\varphi'(y)}, \ldots, \frac{c_{n-1}}{\varphi'(y)} \right\} \text{ is the dual basis of } \{1, y, \ldots, y^{n-1}\}\,. \tag{3.60}$$

(Note that $\varphi'(y) \neq 0$ since y is separable over F.) By definition of the dual basis, (3.60) is equivalent to

$$\mathrm{Tr}_{F'/F}\left(\frac{c_i}{\varphi'(y)} \cdot y^l\right) = \delta_{il} \quad \textit{for} \quad 0 \leq i, l \leq n-1\,. \tag{3.61}$$

In order to prove (3.61), consider the n distinct embeddings $\sigma_1, \ldots, \sigma_n$ of F'/F into Φ (which denotes, as usual, an algebraically closed extension of F). We set $y_j := \sigma_j(y)$ and obtain

$$\varphi(T) = \prod_{j=1}^{n}(T - y_j)\,.$$

Differentiating this equation and substituting $T = y_\nu$ yields

$$\varphi'(y_\nu) = \prod_{i \neq \nu}(y_\nu - y_i)\,. \tag{3.62}$$

For $0 \leq l \leq n-1$ we consider the polynomial

$$\varphi_l(T) := \left(\sum_{j=1}^{n} \frac{\varphi(T)}{T - y_j} \cdot \frac{y_j^l}{\varphi'(y_j)}\right) - T^l \in \Phi[T]\,.$$

Its degree is at most $n-1$, and for $1 \leq \nu \leq n$ we have

$$\varphi_l(y_\nu) = \left(\prod_{i \neq \nu}(y_\nu - y_i)\right) \cdot \frac{y_\nu^l}{\varphi'(y_\nu)} - y_\nu^l = 0\,,$$

by (3.62). A polynomial of degree $\leq n-1$ having n distinct zeros is the zero polynomial. Hence $\varphi_l(T) = 0$; i.e.,

$$T^l = \sum_{j=1}^{n} \frac{\varphi(T)}{T - y_j} \cdot \frac{y_j^l}{\varphi'(y_j)} \quad \textit{for} \quad 0 \leq l \leq n-1\,. \tag{3.63}$$

The embeddings $\sigma_i : F' \to \Phi$ extend to embeddings $\sigma_i : F'(T) \to \Phi(T)$ by setting $\sigma_i(T) = T$, and we obtain from (3.63)

$$\begin{aligned}
T^l &= \sum_{j=1}^{n} \sigma_j\left(\frac{\varphi(T)}{T - y} \cdot \frac{y^l}{\varphi'(y)}\right) \\
&= \sum_{j=1}^{n} \sigma_j\left(\sum_{i=0}^{n-1} c_i T^i \cdot \frac{y^l}{\varphi'(y)}\right) \qquad \textit{(by (3.59))} \\
&= \sum_{i=0}^{n-1}\left(\sum_{j=1}^{n} \sigma_j\left(\frac{c_i}{\varphi'(y)} \cdot y^l\right)\right) T^i \\
&= \sum_{i=0}^{n-1} \mathrm{Tr}_{F'/F}\left(\frac{c_i}{\varphi'(y)} \cdot y^l\right) T^i\,.
\end{aligned}$$

Comparing coefficients yields (3.61) and thereby (3.60).

Next we want to show:

$$c_j \in \sum_{i=0}^{n-1} \mathcal{O}_P \cdot y^i \quad for \ j = 0, \ldots, n-1 . \tag{3.64}$$

The minimal polynomial $\varphi(T)$ of y over F has the form

$$\varphi(T) = T^n + a_{n-1}T^{n-1} + \ldots + a_1 T + a_0 \tag{3.65}$$

with $a_i \in \mathcal{O}_P$. Thus we have by (3.59) the following recursion formulas:

$$c_{n-1} = 1, \ c_0 y = -a_0, \ and \quad c_i y = c_{i-1} - a_i \quad for \ 1 \le i \le n-1 . \tag{3.66}$$

(3.64) is clearly true for $j = n-1$. Suppose it holds for some $j \in \{1, \ldots, n-1\}$, say

$$c_j = \sum_{i=0}^{n-1} s_i y^i \qquad with \ \ s_i \in \mathcal{O}_P .$$

Then we obtain, using (3.66) and (3.65),

$$\begin{aligned} c_{j-1} &= a_j + c_j y = a_j + \sum_{i=0}^{n-2} s_i y^{i+1} + s_{n-1} y^n \\ &= a_j + \sum_{i=0}^{n-2} s_i y^{i+1} - s_{n-1} \sum_{i=0}^{n-1} a_i y^i \in \sum_{i=0}^{n-1} \mathcal{O}_P \cdot y^i . \end{aligned}$$

Therefore (3.64) is proved. In a similar manner we can show

$$y^j \in \sum_{i=0}^{n-1} \mathcal{O}_P \cdot c_i \quad for \ \ j = 0, \ldots, n-1 . \tag{3.67}$$

In fact, (3.67) holds for $j = 0$; if

$$y^j = \sum_{i=0}^{n-1} r_i c_i \quad with \ \ r_i \in \mathcal{O}_P$$

for some $j \ge 0$, then by (3.66)

$$\begin{aligned} y^{j+1} &= \sum_{i=0}^{n-1} r_i c_i y = \sum_{i=1}^{n-1} r_i (c_{i-1} - a_i) - r_0 a_0 \\ &= \sum_{i=0}^{n-2} r_{i+1} c_i - \left(\sum_{i=0}^{n-1} r_i a_i \right) \cdot c_{n-1} \in \sum_{i=0}^{n-1} \mathcal{O}_P \cdot c_i . \end{aligned}$$

Proof of Theorem 3.5.10(a). As always, $\mathcal{C}_P$ denotes the complementary module and $\mathcal{O}'_P$ the integral closure of $\mathcal{O}_P$ in F'. We have to show that $d(P_i|P) \le v_{P_i}(\varphi'(y))$, which is (by definition of the different exponent) equivalent to the following statement:

$$z \in \mathcal{C}_P \Rightarrow v_{P_i}(z) \ge -v_{P_i}(\varphi'(y)) \quad \textit{for } i = 1, \ldots, r\,. \tag{3.68}$$

The element $z \in \mathcal{C}_P$ can be written as

$$z = \sum_{i=0}^{n-1} r_i \cdot \frac{c_i}{\varphi'(y)} \quad \textit{with } r_i \in F\,.$$

(Note that $\{c_0, \ldots, c_{n-1}\}$ is a basis of F'/F by (3.60).) Since y^l is integral over $\mathcal{O}_P$ and $z \in \mathcal{C}_P$, we have $\mathrm{Tr}_{F'/F}(z \cdot y^l) \in \mathcal{O}_P$. Now

$$\mathrm{Tr}_{F'/F}(z \cdot y^l) = \mathrm{Tr}_{F'/F}\left(\sum_{i=0}^{n-1} r_i \cdot \frac{c_i}{\varphi'(y)} \cdot y^l\right) = r_l$$

by (3.61), so $r_l \in \mathcal{O}_P$. Observing (3.64) we obtain

$$z = \frac{1}{\varphi'(y)} \cdot \sum_{i=0}^{n-1} r_i c_i \in \frac{1}{\varphi'(y)} \cdot \sum_{i=0}^{n-1} \mathcal{O}_P \cdot y^i \subseteq \frac{1}{\varphi'(y)} \cdot \mathcal{O}'_P\,.$$

This implies (3.68) and finishes the proof of part (a) of our theorem.

Proof of Theorem 3.5.10(b). By (3.64) and (3.67) we know that

$$\sum_{i=0}^{n-1} \mathcal{O}_P \cdot y^i = \sum_{i=0}^{n-1} \mathcal{O}_P \cdot c_i\,. \tag{3.69}$$

Suppose now that $\{1, y, \ldots, y^{n-1}\}$ is an integral basis for P. From (3.60) and Proposition 3.4.2 it follows that

$$\begin{aligned} \mathcal{C}_P = \sum_{i=0}^{n-1} \mathcal{O}_P \cdot \frac{c_i}{\varphi'(y)} &= \frac{1}{\varphi'(y)} \cdot \sum_{i=0}^{n-1} \mathcal{O}_P \cdot c_i \\ &= \frac{1}{\varphi'(y)} \cdot \sum_{i=0}^{n-1} \mathcal{O}_P \cdot y^i = \frac{1}{\varphi'(y)} \cdot \mathcal{O}'_P\,. \end{aligned}$$

Consequently $d(P'|P) = v_{P_i}(\varphi'(y))$ by Definition 3.4.3. Conversely, we have to prove that the conditions

$$d(P_i|P) = v_{P_i}(\varphi'(y)) \qquad \textit{for } i = 1, \ldots, r \tag{3.70}$$

imply

$$\mathcal{O}'_P \subseteq \sum_{i=0}^{n-1} \mathcal{O}_P \cdot y^i \,. \tag{3.71}$$

(The inclusion $\mathcal{O}'_P \supseteq \sum_{i=0}^{n-1} \mathcal{O}_P \cdot y^i$ is trivial.) Let $z \in \mathcal{O}'_P$, say

$$z = \sum_{i=0}^{n-1} t_i y^i \quad \textit{with} \;\; t_i \in F \,.$$

Observe that $c_j \in \mathcal{O}'_P$ by (3.64) and $\mathcal{C}_P = \frac{1}{\varphi'(y)} \cdot \mathcal{O}'_P$ by (3.70) and Proposition 3.4.2(c); therefore

$$\mathrm{Tr}_{F'/F}\left(\frac{1}{\varphi'(y)} \cdot c_j \cdot z\right) \in \mathcal{O}_P \,.$$

Since

$$\mathrm{Tr}_{F'/F}\left(\frac{1}{\varphi'(y)} \cdot c_j \cdot z\right) = \mathrm{Tr}_{F'/F}\left(\sum_{i=0}^{n-1} t_i \cdot \frac{c_j}{\varphi'(y)} \cdot y^i\right) = t_j \,,$$

we conclude $t_j \in \mathcal{O}_P$. This proves (3.71). □

Corollary 3.5.11. *Let $F' = F(y)$ be a finite separable extension of function fields of degree $[F' : F] = n$, and let $\varphi(T) \in F[T]$ be the minimal polynomial of y over F. Suppose $P \in \mathbb{P}_F$ satisfies*

$$\varphi(T) \in \mathcal{O}_P[T] \;\; \textit{and} \;\; v_{P'}(\varphi'(y)) = 0$$

for all $P' \in \mathbb{P}_{F'}$ with $P'|P$. Then P is unramified in F'/F, and $\{1, y, \ldots, y^{n-1}\}$ is an integral basis for F'/F at P.

Proof. We have by Theorem 3.5.10

$$0 \le d(P'|P) \le v_{P'}(\varphi'(y)) \le 0$$

for all $P'|P$, hence $v_{P'}(\varphi'(y)) = d(P'|P) = 0$. The corollary follows immediately from Dedekind's Different Theorem and Theorem 3.5.10(b). □

We now give a second simple criterion under which the powers of one element constitute an integral basis at a place P.

Proposition 3.5.12. *Let F'/F be a finite separable extension of function fields, $P \in \mathbb{P}_F$ and $P' \in \mathbb{P}_{F'}$ with $P'|P$. Suppose that $P'|P$ is totally ramified; i.e., $e(P'|P) = [F' : F] = n$. Let $t \in F'$ be a P'-prime element, and consider the minimal polynomial $\varphi(T) \in F[T]$ of t over F. Then $d(P'|P) = v_{P'}(\varphi'(t))$, and $\{1, t, \ldots, t^{n-1}\}$ is an integral basis for F'/F at P.*

Proof. First we claim that $1, t, \ldots, t^{n-1}$ are linearly independent over F. Assume the contrary, so that

$$\sum_{i=0}^{n-1} r_i t^i = 0 \quad \textit{with} \;\; r_i \in F, \textit{ not all } r_i = 0\,.$$

For $r_i \neq 0$ we have

$$v_{P'}(r_i t^i) = v_{P'}(t^i) + e(P'|P) \cdot v_P(r_i) \equiv i \bmod n\,.$$

Therefore $v_{P'}(r_i t^i) \neq v_{P'}(r_j t^j)$ whenever $i \neq j$, $r_i \neq 0$ and $r_j \neq 0$. The Strict Triangle Inequality yields

$$v_{P'}\left(\sum_{i=0}^{n-1} r_i t^i\right) = \min\{v_{P'}(r_i t^i) \mid r_i \neq 0\} < \infty\,,$$

which is a contradiction. Thus $\{1, t, \ldots, t^{n-1}\}$ is a basis of F'/F.

According to the formula $\sum e_i f_i = n$ (Theorem 3.1.11), P' is the only place of F' lying over P, hence $\mathcal{O}_{P'}$ is the integral closure of $\mathcal{O}_P$ in F'. So we have to show that

$$\mathcal{O}_{P'} = \sum_{i=0}^{n-1} \mathcal{O}_P \cdot t^i\,. \tag{3.72}$$

The inclusion $\supseteq$ is trivial, so consider an element $z \in \mathcal{O}_{P'}$. Write

$$z = \sum_{i=0}^{n-1} x_i t^i \quad \textit{with} \;\; x_i \in F\,.$$

Since $0 \leq v_{P'}(z) = \min\{n \cdot v_P(x_i) + i \mid 0 \leq i \leq n-1\}$ by the above argument, we see that $v_P(x_i) \geq 0$ holds for all i. This proves (3.72). The assertion $d(P'|P) = v_{P'}(\varphi'(t))$ now follows from Theorem 3.5.10. □

In later sections we shall give several examples how the above results – combined with the Hurwitz Genus Formula – can be applied to determine the genus of a function field.

3.6 Constant Field Extensions

We consider an algebraic function field F/K with constant field K, where K is assumed to be perfect as always in Chapter 3. *This assumption is essential for the validity of most results in this section.* For example there are counterexamples to most assertions of Proposition 3.6.1 and Theorem 3.6.3, when K is not perfect (see [7]). Recall that $\Phi \supseteq F$ denotes a fixed algebraically closed field.

Let $K' \supseteq K$ be an algebraic extension (with $K' \subseteq \Phi$). The compositum $F' := FK'$ is a function field over K', and its constant field is thereby a finite extension of K' (cf. Corollary 1.1.16). However it is not clear a priori, whether K' is the full constant field of FK'. So we begin this section with the following result:

Proposition 3.6.1. *Let $F' = FK'$ be an algebraic constant field extension of F/K (of finite or infinite degree). Then we have:*
(a) K' is the full constant field of F'.
(b) Each subset of F that is linearly independent over K remains so over K'.
(c) $[F : K(x)] = [F' : K'(x)]$ for every element $x \in F \backslash K$.

For the proof of this proposition we require a simple lemma that generalizes Lemma 3.1.10.

Lemma 3.6.2. *Suppose $\alpha \in \Phi$ is algebraic over K. Then $[K(\alpha) : K] = [F(\alpha) : F]$.*

Proof of the Lemma. The inequality $[F(\alpha) : F] \leq [K(\alpha) : K]$ being trivial, we only have to prove that the minimal polynomial $\varphi(T) \in K[T]$ of α over K remains irreducible in $F[T]$. Assume the contrary, so $\varphi(T) = g(T) \cdot h(T)$ with monic polynomials $g(T), h(T) \in F[T]$ of degree ≥ 1. Each root of $g(T)$ and $h(T)$ in Φ is a root of $\varphi(T)$ as well, hence algebraic over K. Therefore all coefficients of $g(T)$ and $h(T)$ are algebraic over K (observe that they are polynomial expressions of the roots). On the other hand, these coefficients are elements of F. Since K is algebraically closed in F we conclude that $g(T)$, $h(T) \in K[T]$, a contradiction to the irreducibility of $\varphi(T)$ over K. □

Proof of Proposition 3.6.1. (a) Consider an element $\gamma \in F'$ which is algebraic over K'. Then it is algebraic over K, and there are finitely many elements $\alpha_1, \ldots, \alpha_r \in K'$ such that $\gamma \in F(\alpha_1, \ldots, \alpha_r)$. The extension $K(\alpha_1, \ldots, \alpha_r)/K$ is finite and separable, therefore $K(\alpha_1, \ldots, \alpha_r) = K(\alpha)$ for some $\alpha \in K'$ (here we use the assumption that K is perfect). Since γ is algebraic over K, we can find $\beta \in F'$ with $K(\alpha, \gamma) = K(\beta)$. It follows that $F(\beta) = F(\alpha, \gamma) = F(\alpha)$ (as $\gamma \in F(\alpha_1, \ldots, \alpha_r) = F(\alpha)$), and we obtain from Lemma 3.6.2

$$[K(\beta) : K] = [F(\beta) : F] = [F(\alpha) : F] = [K(\alpha) : K].$$

This implies $K(\alpha) = K(\beta)$, hence $\gamma \in K(\alpha) \subseteq K'$.

(b) Let $y_1, \ldots, y_r \in F$ be linearly independent over K, and consider a linear combination

$$\sum_{i=1}^{r} \gamma_i y_i = 0 \qquad \textit{with } \gamma_i \in K'. \tag{3.73}$$

Choose $\alpha \in K'$ such that $\gamma_1, \ldots, \gamma_r \in K(\alpha)$, and write

$$\gamma_i = \sum_{j=0}^{n-1} c_{ij} \alpha^j \quad \textit{with } c_{ij} \in K,\, n = [K(\alpha) : K].$$

From (3.73) we obtain

$$0 = \sum_{i=1}^{r} \Big(\sum_{j=0}^{n-1} c_{ij}\alpha^j \Big) y_i = \sum_{j=0}^{n-1} \Big(\sum_{i=1}^{r} c_{ij} y_i \Big) \alpha^j \tag{3.74}$$

with $\sum c_{ij} y_i \in F$. Since $[F(\alpha) : F] = [K(\alpha) : K]$ by Lemma 3.6.2, the elements $1, \alpha, \ldots, \alpha^{n-1}$ are linearly independent over F, and (3.74) implies

$$\sum_{i=1}^{r} c_{ij} y_i = 0 \ \mathit{for}\ j = 0, \ldots, n-1 \,.$$

As $y_1, \ldots, y_r$ are linearly independent over K, it follows that all $c_{ij} = 0$, and thus (3.73) is the trivial linear combination.

(c) Clearly $[F' : K'(x)] \le [F : K(x)]$. It remains to show that any elements $z_1, \ldots, z_s \in F$, which are linearly independent over $K(x)$, are linearly independent over $K'(x)$ as well. Suppose not, so that

$$\sum_{i=1}^{s} f_i(x) \cdot z_i = 0 \tag{3.75}$$

with $f_i(x) \in K'(x)$, not all $f_i(x) = 0$. Multiplying by a common denominator we can assume that all $f_i(x) \in K'[x]$. Then (3.75) gives a linear dependence of the set $\{x^j z_i \mid 1 \le i \le s,\ j \ge 0\}$ over K'. Part (b) of our proposition implies that this set is then linearly dependent over K as well; so $z_1, \ldots, z_s$ are linearly dependent over $K(x)$, a contradiction. □

Our next theorem contains a summary of the most important properties of constant field extensions.

Theorem 3.6.3. *In an algebraic constant field extension $F' = FK'$ of F/K the following hold:*

(a) F'/F is unramified (i.e., $e(P'|P) = 1$ for all $P \in \mathbb{P}_F$ and all $P' \in \mathbb{P}_{F'}$ with $P'|P$).

(b) F'/K' has the same genus as F/K.

(c) For each divisor $A \in \mathrm{Div}(F)$ we have $\deg \mathrm{Con}_{F'/F}(A) = \deg A$.

(d) For each divisor $A \in \mathrm{Div}(F)$,

$$\ell(\mathrm{Con}_{F'/F}(A)) = \ell(A) \,.$$

More precisely: Every basis of $\mathscr{L}(A)$ is also a basis of $\mathscr{L}(\mathrm{Con}_{F'/F}(A))$. (Note that $\mathscr{L}(A) \subseteq F$ is a K-vector space whereas $\mathscr{L}(\mathrm{Con}_{F'/F}(A))$ is considered as a vector space over K'.)

(e) If W is a canonical divisor of F/K then $\mathrm{Con}_{F'/F}(W)$ is a canonical divisor of F'/K'.

(f) The conorm map $\mathrm{Con}_{F'/F} : \mathrm{Cl}(F) \to \mathrm{Cl}(F')$ from the divisor class group of F/K into that of F'/K' is injective.

(g) The residue class field $F'_{P'}$ of each place $P' \in \mathbb{P}_{F'}$ is the compositum $F_P K'$ of K' and the residue class field F_P, where $P = P' \cap F$.

(h) If K'/K is of finite degree, every basis of K'/K is an integral basis of F'/F for all $P \in \mathbb{P}_F$.

Proof. The proof is organized as follows: first we discuss (a) and (b) in the case of a finite constant field extension, then we prove (h), (a), (c), (b), (d), (e), (f) and (g) in the general case. To begin with, we assume that

$$K' = K(\alpha) \text{ is a finite extension of } K. \tag{3.76}$$

We shall prove (a) and (b) under this additional hypothesis. In this situation, $F' = F(\alpha)$ and the minimal polynomial $\varphi(T)$ of α over K remains irreducible over F by Lemma 3.6.2. Let $P \in \mathbb{P}_F$ and $P' \in \mathbb{P}_{F'}$ with $P'|P$. The different exponent $d(P'|P)$ satisfies

$$0 \leq d(P'|P) \leq v_{P'}(\varphi'(\alpha))$$

by Theorem 3.5.10. Now α is separable over K, and therefore $\varphi'(\alpha) \neq 0$. Since $\varphi'(\alpha) \in K'$, this implies $v_{P'}(\varphi'(\alpha)) = 0$. So we have

$$d(P'|P) = v_{P'}(\varphi'(\alpha)) = 0. \tag{3.77}$$

By Dedekind's Theorem we conclude that $P'|P$ is unramified. The Hurwitz Genus Formula yields

$$2g' - 2 = \frac{[F' : F]}{[K' : K]}(2g - 2) + \deg \operatorname{Diff}(F'/F), \tag{3.78}$$

where g (resp. g') denotes the genus of F/K (resp. F'/K'). In our situation (3.76) we have $[F' : F] = [K' : K]$ by Lemma 3.6.2 and $\operatorname{Diff}(F'/F) = 0$ by (3.77), therefore $2g' - 2 = 2g - 2$ by (3.78). We have shown (a) and (b) in the case of a finite constant field extension.

(h) We can assume that $K' = K(\alpha)$ and set $n := [K' : K]$. From (3.77) and Theorem 3.5.10 we obtain that $\{1, \alpha, \ldots, \alpha^{n-1}\}$ is an integral basis of F'/F for all $P \in \mathbb{P}_F$. Obviously we have for each other basis $\{\gamma_1, \ldots, \gamma_n\}$ of K'/K

$$\sum_{i=0}^{n-1} \mathcal{O}_P \cdot \alpha^i = \sum_{j=1}^{n} \mathcal{O}_P \cdot \gamma_j .$$

So $\{\gamma_1, \ldots, \gamma_n\}$ is an integral basis as well.

From now on, K' is an arbitrary algebraic extension of K (of finite or infinite degree).

(a) Let $P' \in \mathbb{P}_{F'}$ be an extension of P. We choose a P'-prime element $t \in F'$. There exists an intermediate field $K \subseteq K_1 \subseteq K'$ such that the degree $[K_1 : K]$ is finite and $t \in F_1 := FK_1$. Let $P_1 := P' \cap F_1$, then $1 = v_{P'}(t) = e(P'|P_1) \cdot v_{P_1}(t)$ and therefore $e(P'|P_1) = 1$. We have already proved that $e(P_1|P) = 1$, consequently $e(P'|P) = e(P'|P_1) \cdot e(P_1|P) = 1$.

(c) It is sufficient to consider a prime divisor $P \in \mathbb{P}_F$. Choose $x \in F$ such that P is the only zero of x in $\mathbb{P}_F$ (such an element exists by Proposition 1.6.6), so the zero divisor $(x)_0^F$ of x in $\mathrm{Div}(F)$ has the form $(x)_0^F = rP$ with $r > 0$. It follows from Proposition 3.1.9 that

$$(x)_0^{F'} = \mathrm{Con}_{F'/F}((x)_0^F) = r \cdot \mathrm{Con}_{F'/F}(P)\,.$$

Now we use the fact that $[F' : K'(x)] = \deg\,((x)_0^{F'})$ (see Theorem 1.4.11), and we obtain

$$\begin{aligned} r \cdot \deg \mathrm{Con}_{F'/F}(P) &= [F' : K'(x)] \\ &= [F : K(x)] \qquad (\textit{by Proposition } 3.6.1) \\ &= \deg\,((x)_0^F) = r \cdot \deg P\,. \end{aligned}$$

Thus (c) is established.

(b) As a first step, we show that

$$\ell(A) \le \ell(\mathrm{Con}_{F'/F}(A)) \tag{3.79}$$

holds for $A \in \mathrm{Div}(F)$. Indeed, if $\{x_1, \ldots, x_r\}$ is a basis of the space $\mathscr{L}(A)$ then $x_i \in \mathscr{L}(\mathrm{Con}_{F'/F}(A))$ by Proposition 3.1.9, and $x_1, \ldots, x_r$ are linearly independent over K' by Proposition 3.6.1. This proves (3.79).

Let g (resp. g') denote the genus of F/K (resp. F'/K'). Choose a divisor $C \in \mathrm{Div}(F)$ satisfying

$$\deg C \ge \max\{2g - 1,\, 2g' - 1\}\,. \tag{3.80}$$

The Riemann-Roch Theorem states that

$$\ell(C) \;=\; \deg C + 1 - g \tag{3.81}$$

and

$$\ell(\mathrm{Con}_{F'/F}(C)) \;=\; \deg C + 1 - g'\,. \tag{3.82}$$

Here we have used the fact that $\deg \mathrm{Con}_{F'/F}(C) = \deg C$ by (c). Now (3.79), (3.80) and (3.81) imply $g' \ge g$.

In order to prove the reverse inequality $g \le g'$, consider a basis $\{u_1, \ldots, u_s\}$ of $\mathscr{L}(\mathrm{Con}_{F'/F}(C))$. There exists a field $K \subseteq K_0 \subseteq K'$ with $[K_0 : K] < \infty$ and $u_1, \ldots, u_s \in F_0 := FK_0$. Obviously $u_1, \ldots, u_s \in \mathscr{L}(\mathrm{Con}_{F_0/F}(C))$, thus

$$\ell(\mathrm{Con}_{F_0/F}(C)) \ge \ell(\mathrm{Con}_{F'/F}(C))\,. \tag{3.83}$$

We have shown above that F_0/K_0 has genus g (since it is a finite constant field extension of F/K), so the Riemann-Roch Theorem yields

$$\ell(\mathrm{Con}_{F_0/F}(C)) \;=\; \deg C + 1 - g\,. \tag{3.84}$$

Combining (3.82), (3.83) and (3.84) we obtain $g \le g'$, and the proof of (b) is complete.

(d) Suppose first that $\deg A \geq 2g-1$. As $g' = g$ we have by the Riemann-Roch Theorem and (c)

$$\begin{aligned} \ell(\mathrm{Con}_{F'/F}(A)) &= \deg \mathrm{Con}_{F'/F}(A) + 1 - g' \\ &= \deg A + 1 - g = \ell(A) \,. \end{aligned}$$

The argument which we used in the proof of (3.79) shows that every basis of $\mathscr{L}(A)$ is also a basis of $\mathscr{L}(\mathrm{Con}_{F'/F}(A))$.

Now consider an arbitrary divisor $A \in \mathrm{Div}(F)$ and a basis $\{x_1, \ldots, x_r\}$ of $\mathscr{L}(A)$. Since $x_1, \ldots, x_r \in \mathscr{L}(\mathrm{Con}_{F'/F}(A))$, and since they are linearly independent over K', it remains to prove that each $z \in \mathscr{L}(\mathrm{Con}_{F'/F}(A))$ is a K'-linear combination of $x_1, \ldots, x_r$. We choose prime divisors $P_1 \neq P_2$ of F/K and set $A_1 := A + n_1 P_1$ and $A_2 := A + n_2 P_2$ with $n_1, n_2 \geq 0$ such that $\deg A_i \geq 2g - 1$ for $i = 1, 2$. Then

$$A = \min\{A_1, A_2\} \text{ and } \mathscr{L}(A) = \mathscr{L}(A_1) \cap \mathscr{L}(A_2) \,.$$

We extend $\{x_1, \ldots, x_r\}$ to bases $\{x_1, \ldots, x_r, y_1, \ldots, y_m\}$ of $\mathscr{L}(A_1)$ resp. $\{x_1, \ldots, x_r, z_1, \ldots, z_n\}$ of $\mathscr{L}(A_2)$. The elements

$$x_1, \ldots, x_r, y_1, \ldots, y_m, z_1, \ldots, z_n \tag{3.85}$$

are linearly independent over K. Indeed, if

$$\sum_{i=1}^{r} a_i x_i + \sum_{j=1}^{m} b_j y_j + \sum_{k=1}^{n} c_k z_k = 0$$

with a_i, b_j, $c_k \in K$, then

$$\sum_{i=1}^{r} a_i x_i + \sum_{j=1}^{m} b_j y_j = -\sum_{k=1}^{n} c_k z_k \in \mathscr{L}(A_1) \cap \mathscr{L}(A_2) = \mathscr{L}(A) \,.$$

Since $\{x_1, \ldots, x_r\}$ is a basis of $\mathscr{L}(A)$ and $x_1, \ldots, x_r, y_1, \ldots, y_m$ are linearly independent, this implies $b_j = 0\,(j = 1, \ldots, m)$, and then (by the linear independence of $x_1, \ldots, x_r, z_1, \ldots, z_n$) it follows that $a_i = c_k = 0$ for $1 \leq i \leq r$, $1 \leq k \leq n$. Observing that elements of F which are linearly independent over K remain linearly independent over K' by Proposition 3.6.1, we have established that the elements (3.85) are linearly independent over K'.

Now let $z \in \mathscr{L}(\mathrm{Con}_{F'/F}(A))$. Since $\deg A_i \geq 2g-1$, assertion (d) holds for A_i, and we can write

$$z = \sum_{i=1}^{r} d_i x_i + \sum_{j=1}^{m} e_j y_j = \sum_{i=1}^{r} f_i x_i + \sum_{k=1}^{n} g_k z_k \tag{3.86}$$

with d_i, e_j, f_i, $g_k \in K'$. Since $x_1, \ldots, x_r, y_1, \ldots, y_m, z_1, \ldots, z_n$ are linearly independent, the two representations in (3.86) coincide, hence $e_j = g_k = 0$ for $1 \leq j \leq m$, $1 \leq k \leq n$. Thus z is a linear combination of $x_1, \ldots, x_r$ over K'.

(e) If W is a canonical divisor of F/K then $\deg W = 2g-2$ and $\ell(W) = g$. By all that we have proved above, we obtain

$$\deg \mathrm{Con}_{F'/F}(W) = 2g-2 \text{ and } \ell(\mathrm{Con}_{F'/F}(W)) = g\,.$$

These two properties characterize canonical divisors of F'/K' (see Proposition 1.6.2), so $\mathrm{Con}_{F'/F}(W)$ is a canonical divisor of F'/K'.

(f) Since $\mathrm{Con}_{F'/F} : \mathrm{Cl}(F) \to \mathrm{Cl}(F')$ is a homomorphism, we must show that the kernel is trivial. So we consider a divisor $A \in \mathrm{Div}(F)$ whose conorm in F' is principal. This means that

$$\deg \mathrm{Con}_{F'/F}(A) = 0 \text{ and } \ell(\mathrm{Con}_{F'/F}(A)) = 1\,.$$

Therefore $\deg A = 0$ and $\ell(A) = 1$ by (c) and (d), and this implies that A is principal, by Corollary 1.4.12.

(g) Let $z(P') \in F'_{P'}$ where z is an element of $\mathcal{O}_{P'}$. There is an intermediate field $K \subseteq K_1 \subseteq K'$ with $z \in F_1 := FK_1$ and $[K_1 : K] < \infty$. Set $P_1 := P' \cap F_1$ and let $P_2, \ldots, P_r$ be the other places of F_1/K_1 lying over P. Choose $u \in F_1$ such that

$$v_{P_1}(z-u) > 0 \text{ and } v_{P_i}(u) \geq 0 \text{ for } 2 \leq i \leq r\,.$$

Then $z(P') = u(P')$, and u lies in the integral closure of $\mathcal{O}_P$ in F_1 (see Corollary 3.3.5). By (h),

$$u = \sum_{i=1}^{n} \gamma_i x_i \quad \text{with } \gamma_i \in K_1,\, x_i \in \mathcal{O}_P\,.$$

Consequently

$$z(P') = u(P') = \sum_{i=1}^{n} \gamma_i \cdot x_i(P) \in F_P K'\,.$$

□

We can combine Theorem 3.6.3(c) and Corollary 3.1.14 to obtain a formula for the degree of the conorm of a divisor in arbitrary algebraic extensions of function fields (over a perfect constant field).

Corollary 3.6.4. *Let F'/K' be an algebraic extension of F/K (not necessarily a constant field extension). Then we have for each divisor $A \in \mathrm{Div}(F)$,*

$$\deg \mathrm{Con}_{F'/F}(A) = [F' : FK'] \cdot \deg A\,.$$

Proof. By Lemma 3.1.2 we know that $[F' : FK'] < \infty$, and FK'/K' is a constant field extension of F/K. Since

$$\mathrm{Con}_{F'/F}(A) = \mathrm{Con}_{F'/FK'}(\mathrm{Con}_{FK'/F}(A))\,,$$

we obtain that

$$\deg \mathrm{Con}_{F'/F}(A) = [F' : FK'] \cdot \deg \mathrm{Con}_{FK'/F}(A) = [F' : FK'] \cdot \deg A\,,$$

by Corollary 3.1.14 and Theorem 3.6.3. □

The following result is a simple consequence of Theorem 3.6.3(a) and (c).

Corollary 3.6.5. *Let $P \in \mathbb{P}_F$ be a place of F/K of degree r and let $\bar{F} = F\bar{K}$ be the constant field extension of F/K with the algebraic closure $\bar{K}$ of K. Then*

$$\mathrm{Con}_{\bar{F}/F}(P) = \bar{P}_1 + \ldots + \bar{P}_r$$

with pairwise distinct places $\bar{P}_i \in \mathbb{P}_{\bar{F}}$.

We finish this section with a result that often allows us to show that the constant field of a finite extension $F' \supseteq F$ of a function field F/K is no bigger than K.

Proposition 3.6.6. *Let F/K be a function field with constant field K. Suppose that F'/F is a finite extension field, with constant field K'. Let $\bar{K} \subseteq \Phi$ denote the algebraic closure of K. Then*

$$[F' : F] = [F'\bar{K} : F\bar{K}] \cdot [K' : K]. \tag{3.87}$$

In the special case $F' = F(y)$ we obtain: if $\varphi(T) \in F[T]$ is the minimal polynomial of y over F, the following conditions are equivalent:

(1) $K' = K$.

(2) $\varphi(T)$ is irreducible in $F\bar{K}[T]$.

Proof. Since $F \subseteq FK' \subseteq F'$, we have

$$[F' : F] = [F' : FK'] \cdot [FK' : F]. \tag{3.88}$$

The extension K'/K is separable and of finite degree, hence $K' = K(\alpha)$ for some $\alpha \in K'$, and we obtain from Lemma 3.6.2 that

$$[FK' : F] = [K' : K]. \tag{3.89}$$

Proposition 3.6.1(c) shows that for each $x \in F \backslash K$,

$$[FK' : K'(x)] = [F\bar{K} : \bar{K}(x)] \ \text{ and } \ [F' : K'(x)] = [F'\bar{K} : \bar{K}(x)].$$

This implies

$$[F' : FK'] = [F'\bar{K} : F\bar{K}]. \tag{3.90}$$

Substituting (3.89) and (3.90) into (3.88) yields (3.87).

Consider now the case $F' = F(y)$. Observe that $[F' : F] = \deg \varphi(T)$, and $[F'\bar{K} : F\bar{K}]$ equals the degree of the minimal polynomial of y over $F\bar{K}$ (which divides $\varphi(T)$ in $F\bar{K}[T]$). The equivalence of (1) and (2) is therefore an immediate consequence of (3.87). □

Corollary 3.6.7. *Let F/K be a function field and let F'/F be a finite extension such that K is the full constant field of F and of F'. Let L/K be an algebraic extension. Then L is the full constant field of FL and $F'L$, and we have*

$$[F' : F] = [F'L : FL] .$$

Proof. It was already shown in Proposition 3.6.1 that L is the full constant field of FL and of $F'L$. Let $\bar{K} \supseteq L$ be the algebraic closure of K, then we obtain from Proposition 3.6.6

$$[F' : F] = [F'\bar{K} : F\bar{K}]$$

and

$$[F'L : FL] = [F'L\bar{K} : FL\bar{K}] = [F'\bar{K} : F\bar{K}] ,$$

hence $[F' : F] = [F'L : FL]$. □

A polynomial $\varphi(T) \in K(x)[T]$ (over the rational function field $K(x)$) is said to be *absolutely irreducible* if $\varphi(T)$ is irreducible in the polynomial ring $\bar{K}(x)[T]$ (where $\bar{K}$ is the algebraic closure of K). The following corollary is a special case of the Proposition 3.6.6.

Corollary 3.6.8. *Let $F = K(x, y)$ be a function field and $\varphi(T) \in K(x)[T]$ be the minimal polynomial of y over $K(x)$. The following conditions are equivalent:*

(1) K is the full constant field of F.

(2) $\varphi(T)$ is absolutely irreducible.

3.7 Galois Extensions I

In Sections 3.7 and 3.8 we investigate Galois extensions of algebraic function fields. Galois extensions have several useful properties that do not hold in arbitrary finite extensions. Recall that a finite field extension M/L is said to be a *Galois extension* if the automorphism group

$$\begin{aligned}\mathrm{Aut}(M/L) = \{\sigma : M \to M \mid \sigma \ &\textit{is an isomorphism with} \\ &\sigma\,(a) = a \textit{ for all } a \in L\}\end{aligned}$$

has order $[M : L]$. In that case we call $\mathrm{Aut}(M/L)$ the *Galois group* of M/L and write $\mathrm{Gal}(M/L) := \mathrm{Aut}(M/L)$. The main properties of Galois extensions are collected together in Appendix A.

An extension F'/K' of a function field F/K is said to be *Galois* if F'/F is a Galois extension of finite degree.

Let P be a place of F/K. Then $\mathrm{Gal}(F'/F)$ acts on the set of all extensions $\{P' \in \mathbb{P}_{F'} \mid P' \text{ lies over } P\}$ via $\sigma(P') = \{\sigma(x) \mid x \in P'\}$, and we have seen in Lemma 3.5.2 that the corresponding valuation $v_{\sigma(P')}$ is given by

$$v_{\sigma(P')}(y) = v_{P'}(\sigma^{-1}(y)) \quad \textit{for } y \in F'.$$

Theorem 3.7.1. *Let F'/K' be a Galois extension of F/K and $P_1, P_2 \in \mathbb{P}_{F'}$ be extensions of $P \in \mathbb{P}_F$. Then $P_2 = \sigma(P_1)$ for some $\sigma \in \mathrm{Gal}(F'/F)$. In other words, the Galois group acts transitively on the set of extensions of P.*

Proof. Assume that the assertion is false; i.e., $\sigma(P_1) \neq P_2$ for all $\sigma \in G := \mathrm{Gal}(F'/F)$. By the Approximation Theorem there is an element $z \in F'$ such that $v_{P_2}(z) > 0$ and $v_Q(z) = 0$ for all $Q \in \mathbb{P}_{F'}$ with $Q|P$ and $Q \neq P_2$. Let $\mathrm{N}_{F'/F} : F' \to F$ be the *norm map* (see Appendix A). We obtain

$$\begin{aligned} v_{P_1}(\mathrm{N}_{F'/F}(z)) &= v_{P_1}\Big(\prod_{\sigma\in G} \sigma(z)\Big) = \sum_{\sigma\in G} v_{P_1}(\sigma(z)) \\ &= \sum_{\sigma\in G} v_{\sigma^{-1}(P_1)}(z) = \sum_{\sigma\in G} v_{\sigma(P_1)}(z) = 0\,, \end{aligned} \tag{3.91}$$

since P_2 does not occur among the places $\sigma(P_1)$, $\sigma \in G$. On the other hand,

$$v_{P_2}(\mathrm{N}_{F'/F}(z)) = \sum_{\sigma\in G} v_{\sigma(P_2)}(z) > 0\,. \tag{3.92}$$

But $\mathrm{N}_{F'/F}(z) \in F$, therefore

$$v_{P_1}(\mathrm{N}_{F'/F}(z)) = 0 \iff v_P(\mathrm{N}_{F'/F}(z)) = 0 \iff v_{P_2}(\mathrm{N}_{F'/F}(z)) = 0\,.$$

This is a contradiction to (3.91) and (3.92). □

Corollary 3.7.2. *Notation as in Theorem 3.7.1 (in particular F'/F is a Galois extension). Let $P_1, \ldots, P_r$ be all the places of F' lying over P. Then we have:*

(a) $e(P_i|P) = e(P_j|P)$ and $f(P_i|P) = f(P_j|P)$ for all i, j. Therefore we set

$$e(P) := e(P_i|P) \quad \textit{and} \quad f(P) := f(P_i|P)\,,$$

and we call $e(P)$ (resp. $f(P)$) the ramification index (resp. relative degree) of P in F'/F.

(b) $e(P) \cdot f(P) \cdot r = [F' : F]$. In particular $e(P)$, $f(P)$ and r divide the degree $[F' : F]$.

(c) The different exponents $d(P_i|P)$ and $d(P_j|P)$ are the same for all i, j.

Proof. (a) is obvious by Theorem 3.7.1 and Lemma 3.5.2, and (b) is an immediate consequence of (a) and Theorem 3.1.11. As to (c), we have to consider the integral closure

$$\mathcal{O}'_P = \bigcap_{i=1}^{r} \mathcal{O}_{P_i}$$

of $\mathcal{O}_P$ in F', and the complementary module

$$\mathcal{C}_P = \{z \in F' \mid \mathrm{Tr}_{F'/F}(z \cdot \mathcal{O}'_P) \subseteq \mathcal{O}_P\}\,.$$

Let $\sigma \in \mathrm{Gal}(F'/F)$. It is easily seen that $\sigma(\mathcal{O}'_P) = \mathcal{O}'_P$ and $\sigma(\mathcal{C}_P) = \mathcal{C}_P$ (using the fact that $\mathrm{Tr}_{F'/F}(\sigma(u)) = \mathrm{Tr}_{F'/F}(u)$ for $u \in F'$). Writing $\mathcal{C}_P = t \cdot \mathcal{O}'_P$ we obtain $\sigma(t) \cdot \mathcal{O}'_P = \sigma(\mathcal{C}_P) = \mathcal{C}_P = t \cdot \mathcal{O}'_P$, so that

$$-d(P_i|P) = v_{P_i}(t) = v_{P_i}(\sigma(t))$$

for $1 \le i \le r$ (by Proposition 3.4.2(c) and the definition of the different exponent). Consider now two places P_i, P_j lying over P. Choose $\sigma \in \mathrm{Gal}(F'/F)$ such that $\sigma(P_j) = P_i$. Then

$$-d(P_i|P) = v_{P_i}(\sigma(t)) = v_{\sigma^{-1}(P_i)}(t) = v_{P_j}(t) = -d(P_j|P)\,.$$

□

We would like to discuss two special types of Galois extensions of a function field in more detail, namely *Kummer extensions* and *Artin-Schreier extensions.*

Proposition 3.7.3 (Kummer Extensions). *Let F/K be an algebraic function field where K contains a primitive n-th root of unity (with $n > 1$ and n relatively prime to the characteristic of K). Suppose that $u \in F$ is an element satisfying*

$$u \ne w^d \quad \textit{for all } w \in F \textit{ and } d \mid n,\ d > 1\,. \tag{3.93}$$

Let

$$F' = F(y) \quad \textit{with } y^n = u\,. \tag{3.94}$$

Such an extension F'/F is said to be a Kummer extension of F. We have:

(a) The polynomial $\Phi(T) = T^n - u$ is the minimal polynomial of y over F (in particular, it is irreducible over F). The extension F'/F is Galois of degree $[F' : F] = n$; its Galois group is cyclic, and the automorphisms of F'/F are given by $\sigma(y) = \zeta y$, where $\zeta \in K$ is an n-th root of unity.

(b) Let $P \in \mathbb{P}_F$ and $P' \in \mathbb{P}_{F'}$ be an extension of P. Then

$$e(P'|P) = \frac{n}{r_P} \quad \textit{and} \quad d(P'|P) = \frac{n}{r_P} - 1\,,$$

where

$$r_P := \gcd(n, v_P(u)) > 0 \tag{3.95}$$

is the greatest common divisor of n and $v_P(u)$.

(c) If K' denotes the constant field of F' and g (resp. g') the genus of F/K (resp. F'/K'), then

$$g' = 1 + \frac{n}{[K' : K]}\left(g - 1 + \frac{1}{2}\sum_{P \in \mathbb{P}_F}\left(1 - \frac{r_P}{n}\right)\deg P\right),$$

where r_P is defined by (3.95).

We note that every cyclic field extension F'/F of degree n is a Kummer extension, provided that n is relatively prime to the characteristic of F and F contains all n-th roots of unity. This fact is well-known from Galois theory, cf. Appendix A.

The following special case of Proposition 3.7.3 is worth emphasizing.

Corollary 3.7.4. *Let F/K be a function field and $F' = F(y)$ with $y^n = u \in F$, where $n \not\equiv 0 \operatorname{mod}(\operatorname{char} K)$ and K contains a primitive n-th root of unity. Assume there is a place $Q \in \mathbb{P}_F$ such that $\gcd(v_Q(u), n) = 1$. Then K is the full constant field of F', the extension F'/F is cyclic of degree n, and*

$$g' = 1 + n(g-1) + \frac{1}{2}\sum_{P \in \mathbb{P}_F}(n - r_P)\deg P.$$

Proof of Proposition 3.7.3. (a) cf. Appendix A.
(b) *Case* 1. $r_P = 1$. From (3.94) we obtain

$$n \cdot v_{P'}(y) = v_{P'}(y^n) = v_{P'}(u) = e(P'|P) \cdot v_P(u),$$

which implies $e(P'|P) = n$, as n and $v_P(u)$ are relatively prime. Since n is not divisible by $\operatorname{char} K$, Dedekind's Different Theorem yields the different exponent $d(P'|P) = n - 1$.

Case 2. $r_P = n$, say $v_P(u) = l \cdot n$ with $l \in \mathbb{Z}$. We choose $t \in F$ with $v_P(t) = l$ and set

$$y_1 := t^{-1}y, \quad u_1 := t^{-n}u.$$

Then $y_1^n = u_1$, $v_{P'}(y_1) = v_P(u_1) = 0$, and the irreducible polynomial of y_1 over F is

$$\psi(T) = T^n - u_1 \in F[T].$$

Thus y_1 is integral over $\mathcal{O}_P$, and Theorem 3.5.10 yields

$$0 \le d(P'|P) \le v_{P'}(\psi'(y_1)).$$

Now $\psi'(y_1) = n \cdot y_1^{n-1}$, so $v_{P'}(\psi'(y_1)) = (n-1) \cdot v_{P'}(y_1) = 0$ and $d(P'|P) = 0$. By Dedekind's Theorem, $e(P'|P) = 1$, and (b) is established in Case 2.

Case 3. $1 < r_P < n$. Consider the intermediate field

$$F_0 := F(y_0) \ \textit{with} \ y_0 := y^{n/r_P} .$$

Then $[F' : F_0] = n/r_P$ and $[F_0 : F] = r_P$. The element y_0 satisfies the equation

$$y_0^{r_P} = u \tag{3.96}$$

over F. Let $P_0 := P' \cap F_0$. Case 2 applies to F_0/F, and therefore $e(P_0|P) = 1$. By (3.96)

$$v_{P_0}(y_0) = \frac{v_P(u)}{r_P} .$$

This is relatively prime to n/r_P, so Case 1 applies to the extension $F' = F_0(y)$ (note that $y^{n/r_P} = y_0$). Consequently $e(P'|P_0) = n/r_P$ and

$$e(P'|P) = e(P'|P_0) \cdot e(P_0|P) = n/r_P .$$

(c) The degree of the different $\mathrm{Diff}(F'/F)$ is

$$\begin{aligned}
\deg \mathrm{Diff}(F'/F) &= \sum_{P \in \mathbb{P}_F} \sum_{P'|P} d(P'|P) \cdot \deg P' \\
&= \sum_{P \in \mathbb{P}_F} \left(\frac{n}{r_P} - 1 \right) \cdot \sum_{P'|P} \deg P' \qquad \textit{(by (b))}.
\end{aligned} \tag{3.97}$$

Observing that for a fixed place $P \in \mathbb{P}_F$ the ramification index $e(P) = e(P'|P)$ does not depend on the choice of the extension P', we have

$$\begin{aligned}
\sum_{P'|P} \deg P' &= \frac{1}{e(P)} \cdot \deg \left(\sum_{P'|P} e(P'|P) \cdot P' \right) \\
&= \frac{1}{e(P)} \cdot \deg \mathrm{Con}_{F'/F}(P) = \frac{r_P}{n} \cdot \frac{n}{[K' : K]} \cdot \deg P \\
&= \frac{r_P}{[K' : K]} \cdot \deg P ,
\end{aligned}$$

by (b) and Corollary 3.1.14. Substituting this into (3.97) shows that

$$\begin{aligned}
\deg \mathrm{Diff}(F'/F) &= \sum_{P \in \mathbb{P}_F} \frac{n - r_P}{r_P} \cdot \frac{r_P}{[K' : K]} \cdot \deg P \\
&= \frac{n}{[K' : K]} \cdot \sum_{P \in \mathbb{P}_F} \left(1 - \frac{r_P}{n} \right) \deg P .
\end{aligned}$$

Finally, the Hurwitz Genus Formula proves (c). □

Proof of Corollary 3.7.4. From the assumption $\gcd(v_Q(u), n) = 1$ follows easily that u satisfies condition (3.93). It remains to show that the constant field K' of F' is no larger than K; then the corollary follows immediately from Proposition 3.7.3. Choose an extension Q' of Q in F'. Part (b) of the proposition shows that

$$e(Q'|Q) = [F' : F] = n\,. \tag{3.98}$$

Now suppose that $[K' : K] > 1$, and consider the intermediate field $F_1 := FK' \supsetneqq F$ and the place $Q_1 := Q' \cap F_1$. By (3.98) we have that $e(Q_1|Q) = [F_1 : F] > 1$. On the other hand, $e(Q_1|Q) = 1$ since F_1/F is a constant field extension (see Theorem 3.6.3). This contradiction proves that $K' = K$. □

Remark 3.7.5. In the above proofs we never used the assumption that K contains a primitive n-th root of unity. Therefore all assertions of Proposition 3.7.3(b),(c) and Corollary 3.7.4 hold in this more general case, with a single exception: $F(y)/F$ is no longer Galois if K does not contain all n-th roots of unity.

Thus far we have not given explicit examples of function fields of genus $g > 0$. This is easily done now.

Example 3.7.6. Assume $\operatorname{char} K \neq 2$. Let $F = K(x,y)$ with

$$y^2 = f(x) = p_1(x) \cdot \ldots \cdot p_s(x) \in K[x]\,,$$

where $p_1(x), \ldots, p_s(x)$ are distinct irreducible monic polynomials and $s \geq 1$. Let $m = \deg f(x)$. Then K is the constant field of F, and F/K has genus

$$g = \begin{cases} (m-1)/2 & \textit{if} \quad m \equiv 1 \bmod 2\,, \\ (m-2)/2 & \textit{if} \quad m \equiv 0 \bmod 2\,. \end{cases}$$

Proof. We have $F = F_0(y)$ where $F_0 = K(x)$ is the rational function field. Let $P_i \in \mathbb{P}_{K(x)}$ denote the zero of $p_i(x)$ and P_∞ the pole of x in $K(x)$. Then $v_{P_i}(f(x)) = 1$ and $v_{P_\infty}(f(x)) = -m$. From Corollary 3.7.4 we obtain that F/F_0 is cyclic of degree 2 and that K is the constant field of F. The numbers r_P (for $P \in \mathbb{P}_{K(x)}$) are easily seen to be

$$\begin{aligned} r_{P_i} &= 1 && \textit{for } i = 1, \ldots, s\,, \\ r_{P_\infty} &= 1 && \textit{if } m \equiv 1 \bmod 2\,, \\ r_{P_\infty} &= 2 && \textit{if } m \equiv 0 \bmod 2\,. \end{aligned}$$

Now the assertion follows from Corollary 3.7.4. □

We shall return to the previous example in Chapter 6. As a preparation for Artin-Schreier extensions we need a lemma.

Lemma 3.7.7. *Let F/K be an algebraic function field of characteristic $p > 0$. Given an element $u \in F$ and a place $P \in \mathbb{P}_F$, the following holds:*

(a) either there exists an element $z \in F$ such that $v_P(u - (z^p - z)) \geq 0$,

(b) or else, for some $z \in F$,

$$v_P(u - (z^p - z)) = -m < 0 \quad \textit{with } m \not\equiv 0 \bmod p\,.$$

In the latter case the integer m *is uniquely determined by* u *and* P, *namely*

$$-m = \max\{v_P(u - (w^p - w)) \mid w \in F\}\,. \tag{3.99}$$

Proof. We begin by proving the following claim. Assume that $x_1, x_2 \in F\backslash\{0\}$ and $v_P(x_1) = v_P(x_2)$. Then there is some $y \in F$ with

$$v_P(y) = 0 \;\; \textit{and} \;\; v_P(x_1 - y^p x_2) > v_P(x_1)\,. \tag{3.100}$$

Indeed, the residue class $(x_1/x_2)(P) \in \mathcal{O}_P/P$ is not zero, hence $(x_1/x_2)(P) = (y(P))^p$ for some $y \in \mathcal{O}_P\backslash P$ (here the perfectness of $\mathcal{O}_P/P$ is essential). This implies $v_P(y) = 0$ and $v_P((x_1/x_2) - y^p) > 0$, thereby $v_P(x_1 - y^p x_2) > v_P(x_1)$.

Next we show: if $v_P(u - (z_1^p - z_1)) = -lp < 0$, then there is an element $z_2 \in F$ with

$$v_P(u - (z_2^p - z_2)) > -lp\,. \tag{3.101}$$

In order to prove this, we choose $t \in F$ with $v_P(t) = -l$; then

$$v_P(u - (z_1^p - z_1)) = v_P(t^p)\,.$$

By (3.100) we can find $y \in F$ with $v_P(y) = 0$ and

$$v_P(u - (z_1^p - z_1) - (yt)^p) > -lp\,.$$

Since $v_P(yt) = v_P(t) = -l > -lp$,

$$v_P(u - (z_1^p - z_1) - ((yt)^p - yt)) > -lp\,.$$

Setting $z_2 := z_1 + yt$, we have established (3.101).

From (3.101), the existence of an element $z \in F$ such that (a) (resp. (b)) holds, follows immediately. In case (b) we still have to prove the characterization of m given in (3.99). By assumption, we have $v_P(u - (z^p - z)) = -m < 0$ with $m \not\equiv 0 \bmod p$. For every $w \in F$, $p \cdot v_P(w - z) \neq -m$ holds, so we may consider the following cases:

Case 1. $p \cdot v_P(w - z) > -m$. Then $v_P((w - z)^p - (w - z)) > -m$ and $v_P(u - (w^p - w)) = v_P(u - (z^p - z) - ((w - z)^p - (w - z))) = -m$ (we have used the Strict Triangle Inequality).

Case 2. $p \cdot v_P(w - z) < -m$. In this case we obtain $v_P(u - (w^p - w)) = v_P(u - (z^p - z) - ((w - z)^p - (w - z))) < -m$.

In either case $v_P(u - (w^p - w)) \leq -m$, which proves (3.99). □

Proposition 3.7.8 (Artin-Schreier Extensions). *Let F/K be an algebraic function field of characteristic $p > 0$. Suppose that $u \in F$ is an element which satisfies the following condition:*

$$u \neq w^p - w \quad \text{for all } w \in F\,. \tag{3.102}$$

Let

$$F' = F(y) \quad \text{with } y^p - y = u\,. \tag{3.103}$$

Such an extension F'/F is called an Artin-Schreier extension of F. For $P \in \mathbb{P}_F$ we define the integer m_P by

$$m_P := \begin{cases} m & \text{if there is an element } z \in F \text{ satisfying} \\ & v_P(u - (z^p - z)) = -m < 0 \text{ and } m \not\equiv 0 \bmod p\,, \\ -1 & \text{if } v_P(u - (z^p - z)) \geq 0 \text{ for some } z \in F\,. \end{cases}$$

(Observe that m_P is well-defined by Lemma 3.7.7.) We then have:

(a) F'/F is a cyclic Galois extension of degree p. The automorphisms of F'/F are given by $\sigma(y) = y + \nu$, with $\nu = 0, 1, \ldots, p-1$.

(b) P is unramified in F'/F if and only if $m_P = -1$.

(c) P is totally ramified in F'/F if and only if $m_P > 0$. Denote by P' the unique place of F' lying over P. Then the different exponent $d(P'|P)$ is given by

$$d(P'|P) = (p-1)(m_P + 1)\,.$$

(d) If at least one place $Q \in \mathbb{P}_F$ satisfies $m_Q > 0$, then K is algebraically closed in F' and

$$g' = p \cdot g + \frac{p-1}{2}\left(-2 + \sum_{P \in \mathbb{P}_F} (m_P + 1) \cdot \deg P\right),$$

where g' (resp. g) is the genus of F'/K (resp. F/K).

Proof. (a) This is well-known from Galois theory, see Appendix A.
(b) and (c) First we consider the case $m_P = -1$; i.e., $v_P(u - (z^p - z)) \geq 0$ for some $z \in F$. Let $y_1 = y - z$ and $u_1 = u - (z^p - z)$; then $F' = F(y_1)$, and $\varphi_1(T) = T^p - T - u_1$ is the minimal polynomial of y_1 over F. Since $v_P(u_1) \geq 0$, y_1 is integral over the valuation ring $\mathcal{O}_P$, and the different exponent $d(P'|P)$ of an extension P' of P in F satisfies

$$0 \leq d(P'|P) \leq v_{P'}(\varphi_1'(y_1)) = 0\,,$$

since $\varphi_1'(T) = -1$ (see Theorem 3.5.10). Hence $d(P'|P) = 0$, and $P'|P$ is unramified by Dedekind's Different Theorem.

Next we assume $m_P > 0$. Choose $z \in F$ such that $v_P(u - (z^p - z)) = -m_P$. Consider the elements $y_1 = y - z$ and $u_1 = u - (z^p - z)$. As before we have

$F' = F(y_1)$, and $\varphi_1(T) = T^p - T - u_1$ is the minimal polynomial of y_1 over F. Let $P' \in \mathbb{P}_{F'}$ be an extension of P in F'. Since $y_1^p - y_1 = u_1$, we obtain

$$v_{P'}(u_1) = e(P'|P) \cdot v_P(u_1) = -m_P \cdot e(P'|P)$$

and

$$v_{P'}(u_1) = v_{P'}(y_1^p - y_1) = p \cdot v_{P'}(y_1)\,.$$

As p and m_P are relatively prime and $e(P'|P) \le [F' : F] = p$, this implies that

$$e(P'|P) = p \quad \textit{and} \quad v_{P'}(y_1) = -m_P\,.$$

In particular P is totally ramified in F'/F.

Let $x \in F$ be a P-prime element. Choose integers $i, j \ge 0$ such that $1 = ip - jm_P$ (this is possible, as p and m_P are relatively prime). Then the element $t = x^i y_1^j$ is a P'-prime element, since $v_{P'}(t) = i \cdot v_{P'}(x) + j \cdot v_{P'}(y_1) = ip - jm_P = 1$. By Proposition 3.5.12, the different exponent $d(P'|P)$ is

$$d(P'|P) = v_{P'}(\psi'(t))\,,$$

where $\psi(T) \in F[T]$ is the minimal polynomial of t over F. Let $G :=$ Gal(F'/F) be the Galois group of F'/F. Clearly

$$\psi(T) = \prod_{\sigma \in G} (T - \sigma(t)) = (T - t) \cdot h(T)$$

with

$$h(T) = \prod_{\sigma \ne \mathrm{id}} (T - \sigma(t)) \in F'[T]\,.$$

So $\psi'(T) = h(T) + (T - t) \cdot h'(T)$ and $\psi'(t) = h(t)$. We conclude that

$$d(P'|P) = v_{P'}\Bigl(\prod_{\sigma \ne \mathrm{id}} (t - \sigma(t)) \Bigr) = \sum_{\sigma \ne \mathrm{id}} v_{P'}(t - \sigma(t)) \tag{3.104}$$

(the sum runs over all $\sigma \in G$ with $\sigma \ne \mathrm{id}$). Each $\sigma \in G \backslash \{\mathrm{id}\}$ has the form $\sigma(y_1) = y_1 + \mu$ for some $\mu \in \{1, \ldots, p-1\}$, hence

$$t - \sigma(t) = x^i y_1^j - x^i (y_1 + \mu)^j = -x^i \cdot \sum_{l=1}^{j} \binom{j}{l} y_1^{j-l} \mu^l\,.$$

Since $v_{P'}(y_1^{j-1}) < v_{P'}(y_1^{j-l})$ for $l \ge 2$, the Strict Triangle Inequality yields

$$\begin{aligned} v_{P'}(t - \sigma(t)) &= v_{P'}(x^i) + v_{P'}(j \mu y_1^{j-1}) \\ &= ip + (j-1) \cdot (-m_P) = ip - jm_P + m_P = m_P + 1\,. \end{aligned} \tag{3.105}$$

(We have used that $j \ne 0$ in K, which follows from $ip - jm_P = 1$.) Substituting (3.105) into (3.104) gives $d(P'|P) = (p-1)(m_P + 1)$. Thus we have shown (b) and (c).

(d) We assume now that $m_Q > 0$ for at least one place $Q \in \mathbb{P}_F$. From (c) follows that Q is totally ramified in F'/F. That K is the full constant field of F' follows exactly as in the proof of Corollary 3.7.4.

The formula

$$g' = p \cdot g + \frac{p-1}{2}\left(-2 + \sum_{P \in \mathbb{P}_F} (m_P + 1) \cdot \deg P\right)$$

is an immediate consequence of (b), (c) and the Hurwitz Genus Formula. □

Remark 3.7.9. (a) Notation as in Proposition 3.7.8. Suppose that there exists a place $Q \in \mathbb{P}_F$ with

$$v_Q(u) < 0 \quad \textit{and} \quad v_Q(u) \not\equiv 0 \bmod p\,.$$

Then u satisfies condition (3.102) by the Strict Triangle Inequality. Hence Proposition 3.7.8 applies in this case.
(b) All cyclic field extensions F'/F of degree $[F' : F] = p = \operatorname{char} F > 0$ are Artin-Schreier extensions, see Appendix A.

Most of the arguments given in the proof of Proposition 3.7.8 apply in a more general situation. We call a polynomial of the specific form

$$a(T) = a_n T^{p^n} + a_{n-1} T^{p^{n-1}} + \ldots + a_1 T^p + a_0 T \in K[T] \tag{3.106}$$

(where $p = \operatorname{char} K > 0$) an *additive* (or *linearized*) *polynomial* over K. Observe that $a(T)$ is separable if and only if $a(T)$ and its derivative $a'(T)$ have no common factor of degree > 0. Here $a'(T) = a_0$ is constant, so the polynomial (3.106) is separable if and only if $a_0 \neq 0$.

An additive polynomial has the following remarkable property:

$$a(u+v) = a(u) + a(v)$$

for any u, v in some extension field of K. In particular, if $a(T)$ is an additive and separable polynomial over K all of whose roots are in K, then these roots form a subgroup of the additive group of K of order $p^n = \deg a(T)$.

Proposition 3.7.10. *Consider an algebraic function field F/K with constant field K of characteristic $p > 0$, and an additive separable polynomial $a(T) \in K[T]$ of degree p^n which has all its roots in K. Let $u \in F$. Suppose that for each $P \in \mathbb{P}_F$ there is an element $z \in F$ (depending on P) such that*

$$v_P(u - a(z)) \geq 0 \tag{3.107}$$

or

$$v_P(u - a(z)) = -m \quad \textit{with } m > 0 \textit{ and } m \not\equiv 0 \bmod p\,. \tag{3.108}$$

Define $m_P := -1$ *in case* (3.107) *and* $m_P := m$ *in case* (3.108). *Then* m_P *is a well-defined integer. Consider the extension field* $F' = F(y)$ *of* F *where* y *satisfies the equation*

$$a(y) = u\,.$$

If there exists at least one place $Q \in \mathbb{P}_F$ *with* $m_Q > 0$, *the following hold:*

(a) F'/F *is Galois,* $[F' : F] = p^n$, *and the Galois group of* F'/F *is isomorphic to the additive group* $\{\alpha \in K \mid a(\alpha) = 0\}$, *hence isomorphic to* $(\mathbb{Z}/p\mathbb{Z})^n$. *(Such a group is said to be elementary abelian of exponent* p, *hence* F'/F *is called an elementary abelian extension of exponent* p *and degree* p^n.*)*

(b) K *is algebraically closed in* F'.

(c) Each $P \in \mathbb{P}_F$ *with* $m_P = -1$ *is unramified in* F'/F.

(d) Each $P \in \mathbb{P}_F$ *with* $m_P > 0$ *is totally ramified in* F'/F, *and the different exponent* $d(P'|P)$ *of the extension* P' *of* P *in* F' *is*

$$d(P'|P) = (p^n - 1)(m_P + 1)\,.$$

(e) Let g' *(resp.* g*) be the genus of* F' *(resp.* F*). Then*

$$g' = p^n \cdot g + \frac{p^n - 1}{2}\left(-2 + \sum_{P \in \mathbb{P}_F} (m_P + 1) \cdot \deg P\right).$$

The proof of Proposition 3.7.10 can be omitted; it is, with minor modifications, the same as that of Proposition 3.7.8.

3.8 Galois Extensions II

We consider a Galois extension F'/F *of algebraic function fields with Galois group* $G := \mathrm{Gal}(F'/F)$. *Let* P *be a place of* F *and let* P' *be an extension of* P *to* F'.

Definition 3.8.1. *(a)* $G_Z(P'|P) := \{\sigma \in G \mid \sigma(P') = P'\}$ *is called the decomposition group of* P' *over* P.

(b) $G_T(P'|P) := \{\sigma \in G \mid v_{P'}(\sigma z - z) > 0 \text{ for all } z \in \mathcal{O}_{P'}\}$ *is called the inertia group of* $P'|P$.

(c) The fixed field $Z := Z(P'|P)$ *of* $G_Z(P'|P)$ *is called the decomposition field, the fixed field* $T := T(P'|P)$ *of* $G_T(P'|P)$ *is called the inertia field of* P' *over* P.

Clearly $G_T(P'|P) \subseteq G_Z(P'|P)$, *and both are subgroups of* G.

An immediate consequence of the definitions is: for $\tau \in G$ the decomposition group and the inertia group of the place $\tau(P')$ are given by

$$G_Z(\tau(P')|P) = \tau G_Z(P'|P)\tau^{-1}\,,$$
$$G_T(\tau(P')|P) = \tau G_T(P'|P)\tau^{-1}$$

Theorem 3.8.2. *With notation as above the following hold:*

(a) The decomposition group $G_Z(P'|P)$ has order $e(P'|P) \cdot f(P'|P)$.

(b) The inertia group $G_T(P'|P)$ is a normal subgroup of $G_Z(P'|P)$ of order $e(P'|P)$.

(c) The residue class extension $F'_{P'}/F_P$ is a Galois extension. Each automorphism $\sigma \in G_Z(P'|P)$ induces an automorphism $\bar{\sigma}$ of $F'_{P'}$ over F_P by setting $\bar{\sigma}(z(P')) = \sigma(z)(P')$ for $z \in \mathcal{O}_{P'}$. The mapping

$$G_Z(P'|P) \longrightarrow \mathrm{Gal}(F'_{P'}/F_P)\,,$$
$$\sigma \longmapsto \bar{\sigma}\,,$$

is a surjective homomorphism whose kernel is the inertia group $G_T(P'|P)$. In particular, $\mathrm{Gal}(F'_{P'}/F_P)$ is isomorphic to $G_Z(P'|P)/G_T(P'|P)$.

(d) Let P_Z (resp. P_T) denote the restriction of P' to the decomposition field $Z = Z(P'|P)$ (resp. to the inertia field $T = T(P'|P)$). Then the ramification indices and residue degrees of the places $P'|P_T$, $P_T|P_Z$ and $P_Z|P$ are as shown in Figure 3.3 below.

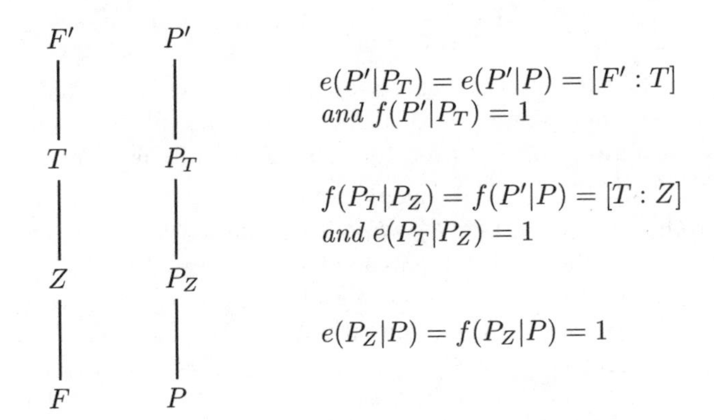

Fig. 3.3.

Proof. (a) By Theorem 3.7.1 G acts transitively on the set of extensions of P in F'. So we can choose $\sigma_1, \ldots, \sigma_r \in G$ such that $\sigma_1(P'), \ldots, \sigma_r(P')$ are all places of F' lying over P and $\sigma_i(P') \neq \sigma_j(P')$ for $i \neq j$. Then $\sigma_1, \ldots, \sigma_r$ are a complete set of coset representatives of G modulo $G_Z(P'|P)$, hence $[F' : F] = \operatorname{ord} G = r \cdot \operatorname{ord} G_Z(P'|P)$. On the other hand, $[F' : F] = e(P'|P) \cdot f(P'|P) \cdot r$ by Corollary 3.7.2(b). This proves (a).

Now we consider the restriction $P_Z = P' \cap Z$ of P' to Z. Obviously the decomposition group of P' over P_Z (with respect to the extension F'/Z) equals $G_Z(P'|P)$, therefore $e(P'|P_Z) \cdot f(P'|P_Z) = \operatorname{ord} G_Z(P'|P) = e(P'|P) \cdot f(P'|P)$

by (a). Since $e(P'|P) = e(P'|P_Z) \cdot e(P_Z|P)$ and $f(P'|P) = f(P'|P_Z) \cdot f(P_Z|P)$, this implies

$$e(P_Z|P) = f(P_Z|P) = 1\,. \tag{3.109}$$

Moreover, P' is the only extension of P_Z in F'.

Next we prove (c). For $z \in \mathcal{O}_{P'}$ let $\bar{z} := z(P') \in F'_{P'}$ denote its residue class mod P', and for $\psi(T) = \sum z_i T^i \in \mathcal{O}_{P'}[T]$ set $\bar{\psi}(T) := \sum \bar{z}_i T^i \in F'_{P'}[T]$. By our general assumption that the constant field of F is perfect, the residue class extension $F'_{P'}/F_P$ is separable, hence $F'_{P'} = F_P(\bar{u})$ for some $u \in \mathcal{O}_{P'}$. We claim that $F'_{P'}$ is the splitting field of some polynomial over F_P (which implies that $F'_{P'}/F_P$ is Galois). The place P' is the only extension of P_Z in F'/Z, so $\mathcal{O}_{P'}$ is the integral closure in F' of the valuation ring $\mathcal{O}_{P_Z}$ of P_Z (see Corollary 3.3.5), and the minimal polynomial $\varphi(T) \in Z[T]$ of u over Z has all its coefficients in $\mathcal{O}_{P_Z}$. As $f(P_Z|P) = 1$ by (3.109), we obtain $\bar{z} \in F_P$ for all $z \in \mathcal{O}_{P_Z}$, thus $\bar{\varphi}(T) \in F_P[T]$. The extension F'/Z being Galois, $\varphi(T)$ splits completely into linear factors $\varphi(T) = \prod (T - u_i)$ with $u_i \in \mathcal{O}_{P'}$, so

$$\bar{\varphi}(T) = \prod (T - \bar{u}_i) \ \ \textit{with}\ \bar{u}_i \in F'_{P'}\,. \tag{3.110}$$

One of the roots of $\bar{\varphi}(T)$ is $\bar{u}$, hence $F'_{P'}$ is the splitting field of $\bar{\varphi}(T)$ over F_P.

Let $\sigma \in G_Z(P'|P)$ and $y, z \in \mathcal{O}_{P'}$ with $\bar{y} = \bar{z}$. Then $y - z \in P'$, hence $\sigma(y) - \sigma(z) = \sigma(y - z) \in \sigma(P') = P'$ and $\sigma(y)(P') = \sigma(z)(P')$. Therefore $\bar{\sigma} : F'_{P'} \to F'_{P'}$ with $\bar{\sigma}(z(P')) := \sigma(z)(P')$ is well-defined, and it is easily verified that the mapping $\sigma \mapsto \bar{\sigma}$ defines a homomorphism from $G_Z(P'|P)$ into the Galois group of $F'_{P'}$ over F_P. The kernel of this homomorphism is just $G_T(P'|P)$, by definition of the inertia group.

An automorphism $\alpha \in \mathrm{Gal}(F'_{P'}/F_P)$ is uniquely determined by $\alpha(\bar{u})$, and $\alpha(\bar{u})$ is a root of the minimal polynomial of $\bar{u}$ over F_P. As this minimal polynomial divides $\bar{\varphi}(T)$, there is some root $u_i \in \mathcal{O}_{P'}$ of $\varphi(T)$ with $\alpha(\bar{u}) = \bar{u}_i$ (by (3.110)). Since $\varphi(T)$ is the minimal polynomial of u over Z and F'/Z is Galois, there is an element $\sigma \in \mathrm{Gal}(F'/Z) = G_Z(P'|P)$ such that $\sigma(u) = u_i$. Clearly $\bar{\sigma} = \alpha$, so our homomorphism from $G_Z(P'|P)$ to $\mathrm{Gal}(F'_{P'}/F_P)$ is surjective. The proof of (c) is now complete.

(b) $G_T(P'|P)$ is a normal subgroup of $G_Z(P'|P)$ since it is the kernel of the homomorphism considered in (c). We have, by (c) and (a),

$$\begin{aligned} f(P'|P) = [F'_{P'} : F_P] &= \mathrm{ord}\,\mathrm{Gal}(F'_{P'}/F_P) \\ &= \mathrm{ord}\, G_Z(P'|P)/\mathrm{ord}\, G_T(P'|P) \\ &= (e(P'|P) \cdot f(P'|P))/\mathrm{ord}\, G_T(P'|P)\,. \end{aligned}$$

Consequently $\mathrm{ord}\, G_T(P'|P) = e(P'|P)$.

(d) It follows from the definition that the inertia group of P' over P_T is equal to $G_T(P'|P)$. Applying (b) first to the extension F'/T and then to F'/F we obtain

$$e(P'|P_T) = \operatorname{ord} G_T(P'|P) = e(P'|P)\,. \tag{3.111}$$

All assertions of (d) are now immediate consequences of (3.109), (3.111) and the multiplicativity of ramification indices and relative degrees in towers of fields (see Proposition 3.1.6(b)). □

There are some useful characterizations of the decomposition field and the inertia field:

Theorem 3.8.3. *Consider a Galois extension F'/F of algebraic function fields, a place $P \in \mathbb{P}_F$ and an extension P' of F in F'. For an intermediate field $F \subseteq M \subseteq F'$ let $P_M := P' \cap M$ denote the restriction of P' to M. Then we have*

(a) $M \subseteq Z(P'|P) \iff e(P_M|P) = f(P_M|P) = 1$.

(b) $M \supseteq Z(P'|P) \iff P'$ is the only place of F' lying over P_M.

(c) $M \subseteq T(P'|P) \iff e(P_M|P) = 1$.

(d) $M \supseteq T(P'|P) \iff P_M$ is totally ramified in F'/M.

Proof. By Theorem 3.8.2(d) all implications $\Rightarrow$ are obvious. Before proving the converse, we remark that the decomposition group of P' over P_M is contained in $G_Z(P'|P)$, and the inertia group of P' over P_M is contained in $G_T(P'|P)$ (this follows immediately from the definition of these groups).

(a) Suppose that $e(P_M|P) = f(P_M|P) = 1$. Then $e(P'|P_M) \cdot f(P'|P_M) = e(P'|P) \cdot f(P'|P)$, so the decomposition group of P' over P_M has the same order as $G_Z(P'|P)$ by Theorem 3.8.2(a). The above remark shows that $G_Z(P'|P)$ is equal to the decomposition group of P' over P_M, in particular $G_Z(P'|P) \subseteq \mathrm{Gal}(F'/M)$. By Galois theory this implies $Z(P'|P) \supseteq M$.

(b), (c), (d) The proofs are similar. □

In what follows, we shall study the phenomenon of wild ramification (see Definition 3.5.4) in a Galois extension more closely.

Definition 3.8.4. *Let F'/F be a Galois extension of algebraic function fields with Galois group $G = \mathrm{Gal}(F'/F)$. Consider a place $P \in \mathbb{P}_F$ and an extension P' of P in F'. For every $i \geq -1$ we define the i-th ramification group of $P'|P$ by*

$$G_i(P'|P) := \{\sigma \in G \mid v_{P'}(\sigma z - z) \geq i+1 \text{ for all } z \in \mathcal{O}_{P'}\}\,.$$

Clearly $G_i(P'|P)$ is a subgroup of G. For abbreviation we write $G_i := G_i(P'|P)$.

Proposition 3.8.5. *With the above notations we have:*

(a) $G_{-1} = G_Z(P'|P)$ *and* $G_0 = G_T(P'|P)$. *In particular,* $\operatorname{ord} G_0 = e(P'|P)$.

(b) $G_{-1} \supseteq G_0 \supseteq \cdots \supseteq G_i \supseteq G_{i+1} \supseteq \cdots$ *and* $G_m = \{\mathrm{id}\}$ *for* m *sufficiently large.*

(c) Let $\sigma \in G_0, i \geq 0$ *and let* t *be a* P'*-prime element; i.e.,* $v_{P'}(t) = 1$. *Then*

$$\sigma \in G_i \iff v_{P'}(\sigma t - t) \geq i + 1 .$$

(d) If $\operatorname{char} F = 0$ *then* $G_i = \{\mathrm{id}\}$ *for all* $i \geq 1$, *and* $G_0 = G_T(P'|P)$ *is cyclic.*

(e) If $\operatorname{char} F = p > 0$ *then* G_1 *is a normal subgroup of* G_0. *The order of* G_1 *is a power of* p, *and the factor group* G_0/G_1 *is cyclic of order relatively prime to* p.

(f) If $\operatorname{char} F = p > 0$ *then* G_{i+1} *is a normal subgroup of* G_i *(for all* $i \geq 1$*), and* G_i/G_{i+1} *is isomorphic to an additive subgroup of the residue class field* $F'_{P'}$. *Hence* G_i/G_{i+1} *is an elementary abelian* p*-group of exponent* p.

Proof. (a) and (b) are obvious.

(c) Consider the inertia field T of P' over P, the restriction $P_T = P' \cap T$ and the corresponding valuation ring $\mathcal{O}_{P_T} = \mathcal{O}_{P'} \cap T$. The elements $1, t, \ldots, t^{e-1}$ (where $e = e(P'|P)$) constitute an integral basis for F'/T at P_T, since $P'|P_T$ is totally ramified (see Proposition 3.5.12). Suppose now that $\sigma \in G_0 = \operatorname{Gal}(F'/T)$ satisfies $v_{P'}(\sigma t - t) \geq i+1$, and let $z \in \mathcal{O}_{P'}$. Writing $z = \sum_{i=0}^{e-1} x_i t^i$ with $x_i \in \mathcal{O}_{P_T}$ we obtain

$$\sigma z - z = \sum_{i=1}^{e-1} x_i((\sigma t)^i - t^i) = (\sigma t - t) \cdot \sum_{i=1}^{e-1} x_i u_i ,$$

where $u_i = ((\sigma t)^i - t^i)/(\sigma t - t) \in \mathcal{O}_{P'}$. This implies $v_{P'}(\sigma z - z) \geq v_{P'}(\sigma t - t) \geq i + 1$, hence $\sigma \in G_i$, and (c) is proved.

We denote by $(F'_{P'})^\times$ (resp. $F'_{P'}$) the multiplicative (resp. additive) group of the residue class field of F' at P', and we shall establish the following facts: There is a homomorphism

$$\chi : G_0 \to (F'_{P'})^\times \ \textit{with} \ \operatorname{Ker}(\chi) = G_1 , \tag{3.112}$$

and for all $i \geq 1$ there is a homomorphism

$$\psi_i : G_i \to F'_{P'} \ \textit{with} \ \operatorname{Ker}(\psi_i) = G_{i+1} . \tag{3.113}$$

The assertions (d), (e) and (f) are easy consequences of (3.112) and (3.113). Indeed, since a finite subgroup of the multiplicative group of a field is cyclic of order prime to the characteristic, G_0/G_1 is a cyclic group by (3.112). If $\operatorname{char} F = 0$, no subgroup of the additive group $F'_{P'}$ is finite, so $G_i = G_{i+1}$ for all $i \geq 1$ by (3.113), and since $G_i = \{\mathrm{id}\}$ for sufficiently large i, (d) follows. If $\operatorname{char} F = p > 0$, each additive subgroup of $F'_{P'}$ is elementary abelian of exponent p, hence (3.113) implies the remaining assertions of (e) and (f).

In order to prove (3.112) and (3.113), we choose a P'-prime element t and set for $\sigma \in G_0$,

$$\chi(\sigma) := \frac{\sigma(t)}{t} + P' \in (F'_{P'})^\times .$$

The definition of $\chi(\sigma)$ does not depend on the specific choice of t: Let $t^* = u \cdot t$ be another prime element for P' (i.e., $u \in F'$ and $v_{P'}(u) = 0$). Then

$$\frac{\sigma(t^*)}{t^*} + P' = \frac{\sigma(t)}{t} + P' , \tag{3.114}$$

since

$$\frac{\sigma(t^*)}{t^*} - \frac{\sigma(t)}{t} = \frac{\sigma(t) \cdot \sigma(u)}{t \cdot u} - \frac{\sigma(t)}{t} = \frac{\sigma(t)}{t} \cdot u^{-1} \cdot (\sigma(u) - u) \in P' .$$

(Observe that $\sigma(u) - u \in P'$, as $\sigma \in G_0$.) For $\sigma, \tau \in G_0$ we have

$$\chi(\sigma\tau) = \frac{(\sigma\tau)(t)}{t} + P' = \frac{\sigma(\tau(t))}{\tau(t)} \cdot \frac{\tau(t)}{t} + P' = \chi(\sigma) \cdot \chi(\tau) .$$

Therefore χ is a homomorphism (here we have used the fact that $\tau(t)$ is a prime element at P', and the definition of $\chi(\sigma)$ is independent of the prime element by (3.114)). An element $\sigma \in G_0$ is in the kernel of χ if and only if $(\sigma(t)/t) - 1 \in P'$; i.e., $v_{P'}(\sigma(t) - t) \geq 2$. Thus $\operatorname{Ker}(\chi) = G_1$ by (c).

It remains to prove (3.113). Let $i \geq 1$ and $\sigma \in G_i$, then $\sigma(t) = t + t^{i+1} \cdot u_\sigma$ for some $u_\sigma \in \mathcal{O}_{P'}$. We define $\psi_i : G_i \to F'_{P'}$ by

$$\psi_i(\sigma) := u_\sigma + P' .$$

(Actually, this definition depends on the choice of t.) For $\tau \in G_i$ we have $\tau(t) = t + t^{i+1} \cdot u_\tau$, hence

$$\begin{aligned}
(\sigma\tau)(t) &= \sigma(t + t^{i+1} u_\tau) = \sigma(t) + \sigma(t)^{i+1} \cdot \sigma(u_\tau) \\
&= t + t^{i+1} \cdot u_\sigma + (t + t^{i+1} \cdot u_\sigma)^{i+1} \cdot (u_\tau + tx) \\
&\textit{(with some } x \in \mathcal{O}_{P'}\textit{; observe that } \sigma(u_\tau) - u_\tau \in P' \textit{ since } \sigma \in G_i) \\
&= t + t^{i+1} \cdot u_\sigma + t^{i+1}(1 + t^i u_\sigma)^{i+1} \cdot (u_\tau + tx) \\
&= t + t^{i+1}(u_\sigma + u_\tau + ty) \;\; \textit{with } y \in \mathcal{O}_{P'} .
\end{aligned}$$

Therefore $u_{\sigma\tau} = u_\sigma + u_\tau + ty$, which implies that

$$\psi_i(\sigma\tau) = \psi_i(\sigma) + \psi_i(\tau) .$$

The kernel of ψ_i is obviously G_{i+1}; this finishes the proof of (3.113). □

As a consequence of Proposition 3.8.5 we have the following supplement to Theorem 3.8.3.

Corollary 3.8.6. *Suppose that F'/F is a Galois extension of algebraic function fields of characteristic $p > 0$. Let $P \in \mathbb{P}_F$ and let $P' \in \mathbb{P}_{F'}$ be an extension of P in F', and consider the fixed field $V_1(P'|P)$ of the first ramification group $G_1(P'|P)$. Denote by P_M the restriction of P' to an intermediate field M, $F \subseteq M \subseteq F'$. Then the following hold:*

(a) $M \subseteq V_1(P'|P) \iff e(P_M|P)$ is prime to p.

(b) $M \supseteq V_1(P'|P) \iff P_M$ is totally ramified in F'/M, and $e(P'|P_M)$ is a power of p.

We omit the proof of this corollary; it is similar to the proof of Theorem 3.8.3.

There is a close relation between the different exponent $d(P'|P)$ and the ramification groups $G_i(P'|P)$.

Theorem 3.8.7 (Hilbert's Different Formula). *Consider a Galois extension F'/F of algebraic function fields, a place $P \in \mathbb{P}_F$ and a place $P' \in \mathbb{P}_{F'}$ lying over P. Then the different exponent $d(P'|P)$ is*

$$d(P'|P) = \sum_{i=0}^{\infty} (\operatorname{ord} G_i(P'|P) - 1)\,.$$

(Note that this is a finite sum, since $G_i(P'|P) = \{\mathrm{id}\}$ for large i.)

Proof. First we assume that $P'|P$ is totally ramified; i.e., $G := \mathrm{Gal}(F'/F) = G_0(P'|P)$. Set $e_i := \operatorname{ord} G_i(P'|P)$ (for $i = 0, 1, \ldots$) and $e := e_0 = [F' : F]$. Choose a P'-prime element t; then $\{1, t, \ldots, t^{e-1}\}$ is an integral basis for F'/F at P by Proposition 3.5.12, and $d(P'|P) = v_{P'}(\varphi'(t))$ where $\varphi(T) \in F[T]$ is the minimal polynomial of t over F. Since F'/F is Galois,

$$\varphi(T) = \prod_{\sigma \in G} (T - \sigma(t))\,.$$

Consequently

$$\varphi'(t) = \pm \prod_{\sigma \neq \mathrm{id}} (\sigma(t) - t)\,.$$

We obtain

$$\begin{aligned} d(P'|P) &= \sum_{\sigma \neq \mathrm{id}} v_{P'}(\sigma t - t) = \sum_{i=0}^{\infty} \sum_{\sigma \in G_i \setminus G_{i+1}} v_{P'}(\sigma t - t) \\ &= \sum_{i=0}^{\infty} (e_i - e_{i+1})(i+1) = \sum_{i=0}^{\infty} (e_i - 1)\,. \end{aligned} \tag{3.115}$$

So our theorem is proved in the totally ramified case. Now we consider the general case. Let T denote the inertia field of $P'|P$ and $P_T := P' \cap T$. Then $P_T|P$ is unramified and $P'|P_T$ is totally ramified. The ramification groups $G_i(P'|P)$ are the same as $G_i(P'|P_T)$, for $i = 0, 1, \ldots$, and the different exponent $d(P'|P)$ is

$$d(P'|P) = e(P'|P_T) \cdot d(P_T|P) + d(P'|P_T) = d(P'|P_T)\,, \tag{3.116}$$

by Corollary 3.4.12(b) (note that $d(P_T|P) = 0$ as $P_T|P$ is unramified). Now the theorem follows from (3.115) and (3.116). □

3.9 Ramification and Splitting in the Compositum of Function Fields

Quite often one considers the following situation: F_1/F and F_2/F are finite extensions of a function field F, and one knows the ramification (resp. splitting) behavior of a place $P \in \mathbb{P}_F$ in both extensions F_1/F and F_2/F. What can then be said about ramification (resp. splitting) of P in the compositum F_1F_2/F? In this section we address this problem.

First we deal with ramification. The main result here is:

Theorem 3.9.1 (Abhyankar's Lemma). *Let F'/F be a finite separable extension of function fields. Suppose that $F' = F_1F_2$ is the compositum of two intermediate fields $F \subseteq F_1, F_2 \subseteq F'$. Let $P' \in \mathbb{P}_{F'}$ be an extension of $P \in \mathbb{P}_F$, and set $P_i := P' \cap F_i$ for $i = 1, 2$. Assume that at least one of the extensions $P_1|P$ or $P_2|P$ is tame. Then*

$$e(P'|P) = \operatorname{lcm}\{e(P_1|P), e(P_2|P)\}\,.$$

For the proof of this theorem we need the following lemma.

Lemma 3.9.2. *Let G be a finite group and $U \subseteq G$ be a normal subgroup such that $\operatorname{ord} U = p^n$ (with either $p = 1$ or else, p a prime number) and G/U is cyclic of order relatively prime to p. Suppose that H_1 is a subgroup of G with $p^n \mid \operatorname{ord} H_1$. Then for every subgroup $H_2 \subseteq G$ we have*

$$\operatorname{ord}(H_1 \cap H_2) = \gcd(\operatorname{ord} H_1, \operatorname{ord} H_2)\,.$$

Proof of the Lemma. Clearly the order of $H_1 \cap H_2$ divides the orders of H_1 and of H_2, thus

$$\operatorname{ord}(H_1 \cap H_2) \mid \gcd(\operatorname{ord} H_1, \operatorname{ord} H_2)\,.$$

We set $\operatorname{ord} H_1 = a_1p^n$ and $\operatorname{ord} H_2 = a_2p^m$ with $(a_1, p) = (a_2, p) = 1$, and $d := \gcd(a_1, a_2)$. Then $\gcd(\operatorname{ord} H_1, \operatorname{ord} H_2) = p^m d$. It is sufficient to prove the following claims:

$$H_1 \cap H_2 \text{ contains a subgroup of order } p^m\,, \text{ and} \tag{3.117}$$

$$H_1 \cap H_2 \text{ contains an element whose order is a multiple of } d\,. \tag{3.118}$$

Let $V \subseteq H_2$ be a p-Sylow subgroup of H_2 (i.e., $\operatorname{ord} V = p^m$). Since U is by assumption a normal subgroup of G, it is the only p-Sylow group of G, therefore $V \subseteq U \subseteq H_1$. This proves (3.117).

Now we consider the canonical homomorphism $\pi : G \to G/U$. The groups $\pi(H_i) \subseteq G/U$ are of order a_i $(i = 1, 2)$, and $\pi(H_1) \cap \pi(H_2)$ is a cyclic group of order $d = \gcd(a_1, a_2)$ (here we use the fact that G/U is cyclic). Choose elements $g_1 \in H_1$ and $g_2 \in H_2$ such that $\pi(g_1) = \pi(g_2)$ is a generator of $\pi(H_1) \cap \pi(H_2)$. Then $g_1^{-1}g_2 =: u \in U \subseteq H_1$, so $g_2 = g_1 u \in H_1 \cap H_2$, and the order of g_2 is a multiple of d. □

Proof of Theorem 3.9.1. Choose a Galois extension F^*/F with $F' \subseteq F^*$ and an extension $P^* \in \mathbb{P}_{F^*}$ of P' in F^*. Then we have the following situation:

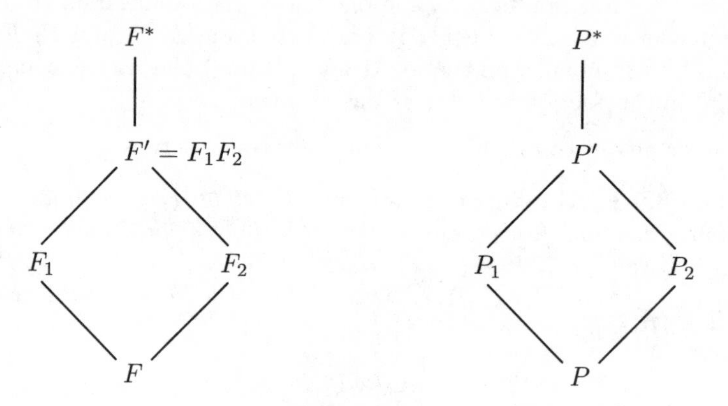

Fig. 3.4.

Consider the group $G := G_T(P^*|P)$ and its subgroups $H_i := G_T(P^*|P_i)$ for $i = 1, 2$. Let $p = \operatorname{char} F$ (in the case of characteristic 0 we set $p = 1$). At least one of the extensions $P_i|P$ is tame, say $\gcd(e(P_1|P), p) = 1$. The groups G, H_1 and H_2 satisfy the hypotheses of Lemma 3.9.2, thus

$$\operatorname{ord}(H_1 \cap H_2) = \gcd(\operatorname{ord} H_1, \operatorname{ord} H_2)\,.$$

The condition $F' = F_1F_2$ shows that $\operatorname{Gal}(F^*/F') = \operatorname{Gal}(F^*/F_1) \cap \operatorname{Gal}(F^*/F_2)$ and $G_T(P^*|P') = G_T(P^*|P_1) \cap G_T(P^*|P_2) = H_1 \cap H_2$. We obtain

$$\begin{aligned} e(P^*|P') &= \operatorname{ord} G_T(P^*|P') = \operatorname{ord}(H_1 \cap H_2) \\ &= \gcd(\operatorname{ord} H_1, \operatorname{ord} H_2) = \gcd(e(P^*|P_1), e(P^*|P_2)) \\ &= \gcd(e(P^*|P') \cdot e(P'|P_1), e(P^*|P') \cdot e(P'|P_2)) \\ &= e(P^*|P') \cdot \gcd(e(P'|P_1), e(P'|P_2))\,. \end{aligned}$$

Therefore

$$\gcd(e(P'|P_1), e(P'|P_2)) = 1 . \tag{3.119}$$

On the other hand we have

$$e(P'|P) = e(P'|P_1) \cdot e(P_1|P) = e(P'|P_2) \cdot e(P_2|P) . \tag{3.120}$$

Equations (3.119) and (3.120) imply that

$$e(P'|P) = \operatorname{lcm}\left(e(P_1|P), e(P_2|P)\right) .$$

(This is a simple fact from elementary number theory: if $ax = by$ with non zero integers a, b, x, y and $\gcd(x, y) = 1$, then the least common multiple of a and b is $\operatorname{lcm}(a, b) = ax = by$.) □

Recall that a place $P \in \mathbb{P}_F$ is said to be *unramified* in a finite extension E/F if $e(Q|P) = 1$ for all places $Q \in \mathbb{P}_E$ with $Q|P$. We have as an immediate consequence of Abhyankar's Lemma:

Corollary 3.9.3. *Let F'/F be a finite separable extension of function fields and let P be a place of F.*
(a) Suppose that $F' = F_1F_2$ is the compositum of two intermediate fields $F \subseteq F_1, F_2 \subseteq F'$. If P is unramified in F_1/F and in F_2/F, then P is unramified in F'/F.
(b) Assume now that F_0 is an intermediate field $F \subseteq F_0 \subseteq F'$ such that F'/F is the Galois closure of F_0/F. If P is unramified in F_0/F then P is unramified in F'/F.

Proof. (a) This is just a special case of Theorem 3.9.1.
(b) The Galois closure F' of F_0/F is the compositum of the fields $\sigma(F_0)$, where σ runs through all embeddings $\sigma : F_0 \to \bar{F}$ over F (where $\bar{F} \supseteq F$ is the algebraic closure of F). As P is unramified in F_0/F, it is also unramified in $\sigma(F_0)/F$. Assertion (b) follows now from (a). □

In the case of wild ramification, Abhyankar's Lemma does not hold in general. With regard to an application in Chapter 7 (see Proposition 7.4.13) we discuss now the simplest case of this situation: we consider a function field F/K of characteristic $p > 0$ and two distinct Galois extensions F_1/F and F_2/F of degree $[F_1 : F] = [F_2 : F] = p$. Let $F' = F_1F_2$ be the compositum of F_1 and F_2; then F'/F is Galois of degree $[F' : F] = p^2$, and also the extensions F'/F_1 and F'/F_2 are Galois of degree p. Let P be a place of F and P' a place of F' lying above P, and denote by $P_i := P' \cap F_i$ the restrictions of P' to F_i, for $i = 1, 2$. If $P_1|P$ or $P_2|P$ is unramified then Abhyankar's Lemma describes how $P'|P_1$ and $P'|P_2$ are ramified. So we assume now that both places $P_1|P$ and $P_2|P$ are ramified. We then have $e(P_1|P) = e(P_2|P) = p$. By Hilbert's Different Formula the different exponents $d(P_i|P)$ satisfy $d(P_i|P) = r_i(p-1)$ with $r_i \geq 2$ for $i = 1, 2$. In the following proposition we consider the special case $r_1 = r_2 = 2$.

Proposition 3.9.4. *In the situation as above we assume that* $d(P_1|P) = d(P_2|P) = 2(p-1)$. *Then one of the following assertions holds:*

(1) $e(P'|P_1) = e(P'|P_2) = 1$, *or*

(2) $e(P'|P_1) = e(P'|P_2) = p$ *and* $d(P'|P_1) = d(P'|P_2) = 2(p-1)$.

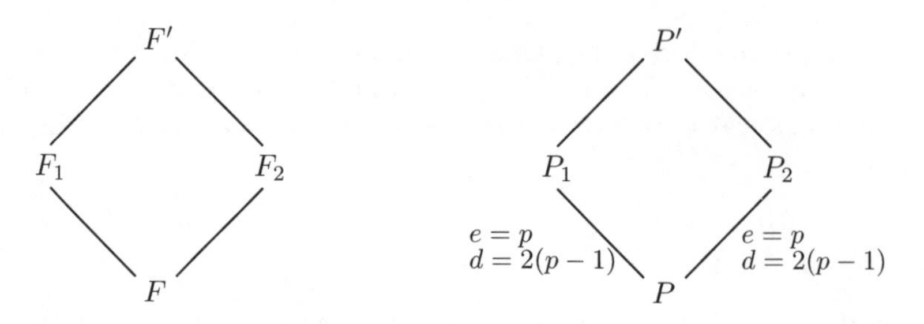

Fig. 3.5.

Proof. It suffices to consider the case $e(P'|P_1) = e(P'|P_2) = p$. Denote by $G_i := G_i(P'|P) \subseteq \mathrm{Gal}(F'/F)$ the i-th ramification group of $P'|P$, then

$$\mathrm{Gal}(F'/F) = G_0 \supseteq G_1 \supseteq \ldots \supseteq G_{s-1} \supsetneqq G_s = \{\mathrm{id}\} \tag{3.121}$$

for some integer $s \geq 2$. For an intermediate field $F \subseteq H \subseteq F'$ we set $Q := P' \cap H$. If $U := \mathrm{Gal}(F'/H)$ is the subgroup of $\mathrm{Gal}(F'/F)$ corresponding to H, the i-th ramification group of $P'|Q$ is (by Definition 3.8.4) $G_i(P'|Q) = U \cap G_i$. By Hilbert's Different Formula the different exponents of $P'|P$ and $P'|Q$ are given by

$$d(P'|P) = \sum_{i=0}^{s-1} (\mathrm{ord}(G_i) - 1)\,, \tag{3.122}$$

$$d(P'|Q) = \sum_{i=0}^{s-1} (\mathrm{ord}(U \cap G_i) - 1)\,. \tag{3.123}$$

We distinguish two cases.

Case 1. $\mathrm{ord}(G_{s-1}) = p^2$. From Equation (3.122) we obtain $d(P'|P) = s(p^2-1)$. We choose $H := F_1$, then (3.123) gives $d(P'|P_1) = s(p-1)$. Since $d(P'|P) = e(P'|P_1) \cdot d(P_1|P) + d(P'|P_1)$ by transitivity of different exponents (Corollary 3.4.12), we conclude that $s(p^2-1) = p \cdot d(P_1|P) + s(p-1)$, hence $d(P_1|P) = s(p-1)$. As $d(P_1|P) = 2(p-1)$ by assumption, we then have $s = 2$ and $d(P'|P_1) = s(p-1) = 2(p-1)$.

Case 2. $\mathrm{ord}(G_{s-1}) = p$. At least one of the fields F_1, F_2 is not the fixed field of the group G_{s-1}, so we can assume w.l.o.g. that $U := \mathrm{Gal}(F'/F_1) \neq G_{s-1}$. We obtain then from (3.123) that $d(P'|P_1) < s(p-1)$, hence

$$\begin{aligned} d(P'|P) &= e(P'|P_1) \cdot d(P_1|P) + d(P'|P_1) \\ &< p \cdot 2(p-1) + s(p-1) \;=\; (2p+s)(p-1)\,. \end{aligned} \tag{3.124}$$

On the other hand, observing that $\mathrm{ord}(G_0) = \mathrm{ord}(G_1) = p^2$ by Proposition 3.8.5 we get from (3.121) and (3.122)

$$d(P'|P) \;\geq\; 2(p^2-1) + (s-2)(p-1) \;=\; (2p+s)(p-1)\,.$$

This inequality contradicts (3.124) and therefore Case 2 cannot occur. □

Now we proceed to splitting places. We recall that a place $P \in \mathbb{P}_F$ *splits completely* in an extension field E/F of degree $[E:F] = n < \infty$, if P has n distinct extensions $Q_1, \ldots, Q_n$ in E. By the Fundamental Equality 3.1.11 this is equivalent to the condition that $e(Q|P) = f(Q|P) = 1$ for all $Q \in \mathbb{P}_E$ lying above P.

Lemma 3.9.5. *Let F_0/F be a finite separable extension of function fields and let $F' \supseteq F_0$ be the Galois closure of F_0/F. Assume that a place $P \in \mathbb{P}_F$ is completely splitting in F_0/F. Then P splits completely in F'/F.*

Proof. Let P' be a place of F' lying above P and consider the decomposition field $Z := Z(P'|P) \subseteq F'$ (see Definition 3.8.1). Set $P_0 := P' \cap F_0$. Since P splits completely in F_0/F we have $e(P_0|P) = f(P_0|P) = 1$, and it follows from Theorem 3.8.3(a) that $F_0 \subseteq Z$. For every embedding $\sigma : F_0 \to F'$ over F, the place P splits completely in $\sigma(F_0)/F$, hence also the field $\sigma(F_0)$ is contained in Z. The Galois closure F' of F/F_0 is the compositum of all these fields $\sigma(F_0)$, so we have that $F' \subseteq Z$ and hence $F' = Z$; we conclude from Theorem 3.8.2 that $e(P'|P) = f(P'|P) = 1$. □

Proposition 3.9.6. *Let F'/F be a finite separable extension of function fields and let F_1, F_2 be intermediate fields of F'/F such that $F' = F_1F_2$ is their compositum.*

(a) Suppose that P is a place of F which splits completely in the extension F_1/F. Then every place Q of F_2 lying above P splits completely in the extension F'/F_2.

(b) If $P \in \mathbb{P}_F$ splits completely in F_1/F and in F_2/F, then P splits completely in F'/F.

Proof. (a) Let E/F be the Galois closure of F_1/F; then P splits completely in the extension E/F by Lemma 3.9.5. We consider the compositum $E' := EF_2$. By Galois theory we know that the extension E'/F_2 is Galois, and the Galois group $\mathrm{Gal}(E'/F_2)$ is isomorphic to a subgroup of $\mathrm{Gal}(E/F)$ under the map $\sigma \mapsto \sigma|_E$ (the restriction of σ to E).

Suppose that there is a place Q of F_2 above P which does not split completely in F'/F_2, so Q does not split completely in E'/F_2. Choose a place Q' of E' lying above Q and set $P' := Q' \cap E$. The situation is shown in Figure 3.6.

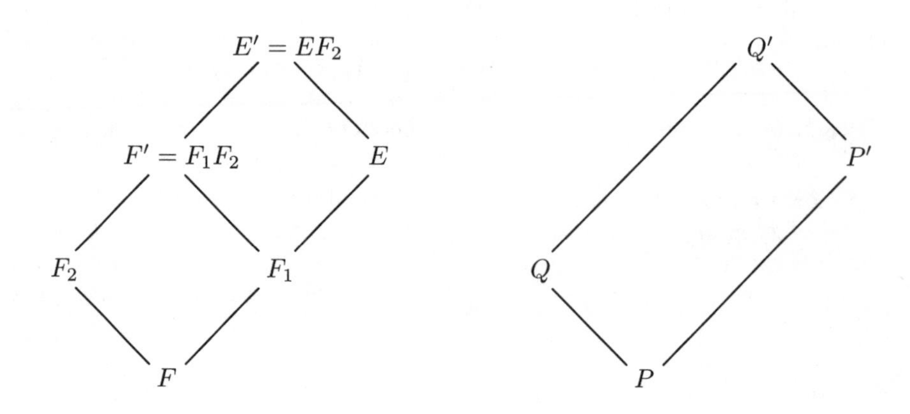

Fig. 3.6.

Since $e(Q'|Q) \cdot f(Q'|Q) > 1$, there is an automorphism $\sigma \in \mathrm{Gal}(E'/F_2)$ with $\sigma(Q') = Q'$ and $\sigma \neq \mathrm{id}$, by Theorem 3.8.2(a). Then the restriction $\sigma' := \sigma|_E \in \mathrm{Gal}(E/F)$ is not the identity on E and $\sigma'(P') = P'$, hence the decomposition group $G_Z(P'|P) \subseteq \mathrm{Gal}(E/F)$ is non-trivial. It follows that $e(P'|P) \cdot f(P'|P) = \mathrm{ord}\, G_Z(P'|P) > 1$, a contradiction to the fact that P splits completely in E/F.
(b) is an immediate consequence of (a). □

Corollary 3.9.7. *Let F/K be a function field whose full constant field is K.*
(a) Suppose that $F' = F_1F_2$ is the compositum of two finite separable extensions F_1/F and F_2/F. Assume that there exists a place $P \in \mathbb{P}_F$ of degree one which splits completely in F_1/F and in F_2/F. Then P splits completely in F'/F, and K is the full constant field of F'.
(b) Suppose that F_0/F is a finite separable extension and $P \in \mathbb{P}_F$ is a place of degree one which splits completely in F_0/F. Let $\tilde{F}/F$ be the Galois closure of F_0/F. Then P splits completely in $\tilde{F}/F$ and K is the full constant field of $\tilde{F}$.

Proof. (a) We only have to show that K is the full constant field of $F' = F_1F_2$; the remaining assertions follow immediately from Proposition 3.9.6. We choose a place P' of F' lying above P, then $f(P'|P) = 1$ and therefore the residue class field $F'_{P'}$ of P' is equal to the residue class field $F_P = K$ of P. Since the full constant field K' of F' satisfies $K \subseteq K' \subseteq F'_{P'}$, we conclude that $K' = K$.

(b) is obvious. □

3.10 Inseparable Extensions

Every algebraic extension F'/F of algebraic function fields can be split into a separable step F_s/F and a purely inseparable step F'/F_s, see Appendix A. Thus far we have mostly studied separable extensions. In the present section, purely inseparable extensions will be investigated. Throughout this section, K is a perfect field of characteristic $p > 0$, and F/K is a function field with constant field K.

Lemma 3.10.1. *Suppose F'/F is a purely inseparable field extension of degree p. Then K is the constant field of F' as well. Every place $P \in \mathbb{P}_F$ has only one extension $P' \in \mathbb{P}_{F'}$, namely*

$$P' = \{z \in F' \mid z^p \in P\}.$$

The corresponding valuation ring is

$$\mathcal{O}_{P'} = \{z \in F' \mid z^p \in \mathcal{O}_P\}.$$

We have $e(P'|P) = p$ and $f(P'|P) = 1$.

Proof. Let $a \in F'$ be algebraic over K. Since F'/F is purely inseparable of degree p, we have $a^p \in F$ and a^p is algebraic over K. As K is the constant field of F this shows that $a^p \in K$. But K is perfect, so $a^p \in K$ implies $a \in K$. Hence K is the constant field of F'.

Next we consider a place $P \in \mathbb{P}_F$. Define

$$R := \{z \in F' \mid z^p \in \mathcal{O}_P\} \ \text{ and } \ M := \{z \in F' \mid z^p \in P\}.$$

Obviously R is a subring of F'/K with $\mathcal{O}_P \subseteq R$, and M is a proper ideal of R containing P. Let $P' \in \mathbb{P}_{F'}$ be an extension of P. For $z \in \mathcal{O}_{P'}$ (resp. $z \in P'$) we have $z^p \in \mathcal{O}_{P'} \cap F = \mathcal{O}_P$ (resp. $z^p \in P' \cap F = P$), hence $\mathcal{O}_{P'} \subseteq R$ and $P' \subseteq M$. Since $\mathcal{O}_{P'}$ is a maximal proper subring of F' (see Theorem 1.1.12(c)) and P' is a maximal ideal of $\mathcal{O}_{P'}$, this implies that $\mathcal{O}_{P'} = R, P' = M$, and P' is the only place of F' lying over P. The residue class field $F'_{P'} = \mathcal{O}_{P'}/P'$ is clearly purely inseparable over $F_P = \mathcal{O}_P/P$, consequently $F'_{P'} = F_P$ (observe that F_P is a finite extension of the perfect field K, thus each algebraic extension of F_P is separable). This proves $f(P'|P) = 1$, and $e(P'|P) = p$ follows now from the formula $\sum_{P_i|P} e(P_i|P) \cdot f(P_i|P) = [F' : F] = p$. □

An element $x \in F$ is called a *separating element* for F/K if $F/K(x)$ is a finite separable extension. F/K is said to be *separably generated* if there exists a separating element for F/K. Next we show, among other things, that every function field F/K is separably generated (this is not true in general if K is not assumed to be perfect).

Proposition 3.10.2. *(a) Assume $z \in F$ satisfies $v_P(z) \not\equiv 0 \bmod p$ for some $P \in \mathbb{P}_F$. Then z is a separating element for F/K. In particular F/K is separably generated.*

(b) There exist $x, y \in F$ such that $F = K(x, y)$.

(c) For each $n \geq 1$ the set $F^{p^n} := \{z^{p^n} \mid z \in F\}$ is a subfield of F. It has the following properties:

- *(1) $K \subseteq F^{p^n} \subseteq F$, and F/F^{p^n} is purely inseparable of degree p^n.*
- *(2) The Frobenius map $\varphi_n : F \to F$, defined by $\varphi_n(z) := z^{p^n}$, is an isomorphism of F onto F^{p^n}. Therefore the function field F^{p^n}/K has the same genus as F/K.*
- *(3) Suppose that $K \subseteq F_0 \subseteq F$ and F/F_0 is purely inseparable of degree $[F : F_0] = p^n$. Then $F_0 = F^{p^n}$.*

(d) An element $z \in F$ is a separating element for F/K if and only if $z \notin F^p$.

Proof. (a) Suppose that z is not separating. The extension $F/K(z)$ is of finite degree since $z \notin K$, hence there is an intermediate field $K(z) \subseteq F_s \subseteq F$ such that F/F_s is purely inseparable of degree p. Let $P_s := P \cap F_s$. By the preceding lemma we have $e(P|P_s) = p$, so $v_P(z) = p \cdot v_{P_s}(z) \equiv 0 \bmod p$.

(b) Choose a separating element $x \in F \backslash K$. Since $F/K(x)$ is a finite separable field extension, there is some $y \in F$ satisfying $F = K(x, y)$ (see Appendix A).

(c) It is easily verified that F^{p^n} is a field, and $K = K^{p^n} \subseteq F^{p^n}$ because K is perfect. The extension F/F^{p^n} is purely inseparable since $z^{p^n} \in F^{p^n}$ for each $z \in F$. We choose $x, y \in F$ such that x is separating and $F = K(x, y)$, and claim that

$$F = K(x, y^{p^n}) \tag{3.125}$$

holds. In fact, $F = K(x, y^{p^n})(y)$ is purely inseparable over $K(x, y^{p^n})$ since y satisfies the equation $T^{p^n} - y^{p^n} = 0$ over $K(x, y^{p^n})$. On the other hand, $K(x) \subseteq K(x, y^{p^n}) \subseteq F$, and therefore the extension $F/K(x, y^{p^n})$ is separable. This proves (3.125).

Now $F^{p^n} = K^{p^n}(x^{p^n}, y^{p^n}) = K(x^{p^n}, y^{p^n})$, and (3.125) implies that $F = F^{p^n}(x)$. Because x is a zero of the polynomial $T^{p^n} - x^{p^n}$ over F^{p^n}, we conclude that

$$[F : F^{p^n}] \leq p^n . \tag{3.126}$$

In order to prove the reverse inequality, choose a place P_0 of F^{p^n}/K and an element $u \in F^{p^n}$ with $v_{P_0}(u) = 1$. Let $P \in \mathbb{P}_F$ be an extension of P_0 in F; then $[F : F^{p^n}] \geq e(P|P_0)$. Writing $u = z^{p^n}$ for some $z \in F$ we obtain

$$p^n \cdot v_P(z) = v_P(z^{p^n}) = v_P(u) = e(P|P_0) \cdot v_{P_0}(u) = e(P|P_0).$$

So

$$p^n \leq e(P|P_0) \leq [F : F^{p^n}]. \tag{3.127}$$

This finishes the proof of (1).

Assertion (2) is trivial, and it remains to prove (3). By assumption, the extension F/F_0 is purely inseparable of degree p^n. Then $z^{p^n} \in F_0$ for each $z \in F$, so $F^{p^n} \subseteq F_0 \subseteq F$. The degree $[F : F^{p^n}]$ is p^n by (1), consequently we have $F^{p^n} = F_0$.

(d) If z is a separating element, $K(z) \not\subseteq F^p$ since F/F^p is purely inseparable of degree > 1. Conversely, if $z \in F \backslash K$ is not separating, there is an intermediate field $K(z) \subseteq F_0 \subseteq F$ such that F/F_0 is purely inseparable of degree p. By (c), $F_0 = F^p$, hence $z \in F^p$. □

In characteristic 0 the situation is of course much simpler: each $x \in F \backslash K$ is a separating element, and therefore we have $F = K(x, y)$ with an appropriate element y.

3.11 Estimates for the Genus of a Function Field

It is often difficult to determine the genus of a function field precisely. Therefore we would like to derive some bounds for the genus in specific cases. As always, F/K is an algebraic function field over the perfect constant field K.

Proposition 3.11.1. *Let F_1/K be a subfield of F/K and $[F : F_1] = n$. Assume that $\{z_1, \ldots, z_n\}$ is a basis of F/F_1 such that all $z_i \in \mathscr{L}(C)$ for some divisor $C \in \mathrm{Div}(F)$. Then*

$$g \leq 1 + n(g_1 - 1) + \deg C,$$

where g (resp. g_1) denotes the genus of F/K (resp. of F_1/K).

Proof. Let A_1 be a divisor of F_1/K of sufficiently large degree such that

$$\ell(A_1) =: t = \deg A_1 + 1 - g_1.$$

Choose a basis $\{x_1, \ldots, x_t\} \subseteq F_1$ of $\mathscr{L}(A_1)$. Set $A := \mathrm{Con}_{F/F_1}(A_1) \in \mathrm{Div}(F)$. The elements

$$x_i z_j \quad (1 \leq i \leq t,\ 1 \leq j \leq n)$$

are in $\mathscr{L}(A + C)$ and they are linearly independent over K. Hence

$$\ell(A + C) \geq n \cdot (\deg A_1 + 1 - g_1). \tag{3.128}$$

We may assume that $\deg\,(A+C)$ is sufficiently large so that the Riemann-Roch Theorem yields

$$\begin{aligned}\ell(A+C) &= \deg(A+C)+1-g \\ &= n\cdot\deg A_1+\deg C+1-g\,. \end{aligned} \tag{3.129}$$

Substituting (3.129) into (3.128) we obtain $g \le 1+n(g_1-1)+\deg C$. □

In the course of the proof of Theorem 3.11.3 we shall need the following lemma:

Lemma 3.11.2. *Assume that K is algebraically closed, and consider a subfield F_1/K of F/K such that F/F_1 is separable of degree $[F:F_1]=n>1$. Let $y\in F$ be an element with $F=F_1(y)$. Then almost all $P\in\mathbb{P}_{F_1}$ have the following properties:*

(a) P splits completely in F/F_1; i.e., it has n distinct extensions $P_1,\dots,P_n$ in F/F_1.

(b) The restrictions $P_1\cap K(y),\dots,P_n\cap K(y)$ are pairwise distinct places of $K(y)$.

Proof. Let $\varphi(T)=T^n+z_{n-1}T^{n-1}+\cdots+z_0\in F_1[T]$ be the minimal polynomial of y over F_1. For almost all $P\in\mathbb{P}_{F_1}$ the following hold:

$$\begin{aligned}&\{1,y,\dots,y^{n-1}\}\ \textit{is an integral basis of}\ F/F_1\ \textit{for}\ P\textit{, and}\\ &P\ \textit{is unramified in}\ F/F_1\,.\end{aligned} \tag{3.130}$$

From now on we assume that P satisfies (3.130). Since K is algebraically closed, P splits completely in F. For $z\in\mathcal{O}_P$ let $\bar z\in\mathcal{O}_P/P=K$ denote its residue class modulo P; then the decomposition of the polynomial $\bar\varphi(T)=T^n+\bar z_{n-1}T^{n-1}+\cdots+\bar z_0\in K[T]$ corresponds to the splitting of P in F (by Kummer's Theorem). Hence we have by (3.130) a decomposition

$$\bar\varphi(T)=\prod_{i=1}^{n}(T-b_i)$$

with pairwise distinct elements $b_i\in K$. For $i=1,\dots,n$ there exists (by Kummer's Theorem) a unique place $P_i\in\mathbb{P}_F$ such that $P_i|P$ and $v_{P_i}(y-b_i)>0$. Since the elements b_i are pairwise distinct, the restrictions $P_i\cap K(y)\in\mathbb{P}_{K(y)}$ are distinct for $i=1,\dots,n$. □

Theorem 3.11.3 (Castelnuovo's Inequality). *Let F/K be a function field with constant field K. Suppose there are given two subfields F_1/K and F_2/K of F/K satisfying*

(1) $F=F_1F_2$ is the compositum of F_1 and F_2, and

(2) $[F:F_i]=n_i$ and F_i/K has genus g_i $(i=1,2)$.

Then the genus g of F/K is bounded by

$$g\le n_1g_1+n_2g_2+(n_1-1)(n_2-1)\,.$$

Proof. We may assume that K is an algebraically closed field. (Otherwise we replace F/K by the constant field extension $F\bar{K}/\bar{K}$ with the algebraic closure $\bar{K} \subseteq \Phi$ of K, and F_i/K by $F_i\bar{K}/\bar{K}$. In a constant field extension, the genera remain unchanged by Theorem 3.6.3, and we also have, by Proposition 3.6.6, that $[F\bar{K} : F_i\bar{K}] = [F : F_i]$.) Moreover we can assume that F/F_1 is a separable extension (if both F/F_1 and F/F_2 were inseparable, then $F_1F_2 \subseteq F^p \subsetneqq F$ by Proposition 3.10.2).

The idea behind the proof of Castelnuovo's Inequality is to find an appropriate divisor $C \in \mathrm{Div}(F)$ of small degree and a basis $\{u_1, \ldots, u_n\} \subseteq \mathscr{L}(C)$ of F/F_1 such that Proposition 3.11.1 will yield the desired inequality.

Since $F = F_1F_2$, there are $y_1, \ldots, y_s \in F_2$ with $F = F_1(y_1, \ldots, y_s)$. The extension F/F_1 is separable, hence we can find $a_1, \ldots, a_s \in K$ such that the element

$$y := \sum_{j=1}^{s} a_j y_j \in F_2$$

is a primitive element of F/F_1; i.e., $F = F_1(y)$ (see Appendix A). By Proposition 1.6.12 there is a divisor $A_0 \in \mathrm{Div}(F_2/K)$ with $A_0 \geq 0$, $\deg A_0 = g_2$ and $\ell(A_0) = 1$. Let $P_0 \in \mathbb{P}_{F_2}$ be a place not in the support of A_0, and set $B_0 := A_0 - P_0$. Since $\mathscr{L}(A_0) = K$, it follows that

$$\deg B_0 = g_2 - 1 \quad \textit{and} \quad \ell(B_0) = 0\,. \tag{3.131}$$

Now we choose a place $P \in \mathbb{P}_{F_1}$ that has n_1 distinct extensions $P_1, \ldots, P_{n_1}$ in F/F_1 such that the restrictions

$$Q_i := P_i \cap F_2 \in \mathbb{P}_{F_2}$$

are pairwise distinct and $Q_i \notin \mathrm{supp}\, B_0$, for $i = 1, \ldots, n_1$. This is possible by Lemma 3.11.2. The Riemann-Roch Theorem yields

$$\ell(B_0 + Q_i) \geq \deg\,(B_0 + Q_i) + 1 - g_2 = 1\,. \tag{3.132}$$

By (3.131) and (3.132) there is an element $u_i \in F_2$ satisfying

$$(u_i) \geq -(B_0 + Q_i) \quad \textit{and} \quad v_{Q_i}(u_i) = -1\,. \tag{3.133}$$

We claim that $\{u_1, \ldots, u_{n_1}\}$ is a basis of F/F_1; so we must show that these elements are linearly independent over F_1. Suppose that

$$\sum_{i=1}^{n_1} x_i u_i = 0 \quad \textit{with} \quad x_i \in F_1$$

is a non-trivial linear combination. Choose $j \in \{1, \ldots, n_1\}$ such that

$$v_P(x_j) \leq v_P(x_i) \quad \textit{for} \quad i = 1, \ldots, n_1\,. \tag{3.134}$$

Then

$$v_{P_j}(x_j u_j) = v_{P_j}(x_j) + v_{P_j}(u_j) \le v_P(x_j) - 1\,.$$

(Observe that $v_{P_j}(x_j) = v_P(x_j)$ since $P_j|P$ is unramified, and $v_{P_j}(u_j) \le -1$ by (3.133).) For $i \ne j$ we have

$$v_{P_j}(x_i u_i) = v_P(x_i) + v_{P_j}(u_i) \ge v_P(x_i) \ge v_P(x_j)\,,$$

by (3.133) and (3.134). Therefore the Strict Triangle Inequality implies

$$v_{P_j}\left(\sum_{i=1}^{n_1} x_i u_i\right) = v_{P_j}(x_j u_j) < \infty\,.$$

This contradiction proves that $\{u_1, \ldots, u_{n_1}\}$ is a basis of F/F_1.

Now we consider the divisor

$$C := \mathrm{Con}_{F/F_2}\left(B_0 + \sum_{i=1}^{n_1} Q_i\right) \in \mathrm{Div}(F)\,.$$

Its degree is

$$\deg C = n_2 \cdot \deg\left(B_0 + \sum_{i=1}^{n_1} Q_i\right) = n_2(g_2 - 1 + n_1)\,.$$

By (3.133) the elements $u_1, \ldots, u_{n_1}$ are in $\mathscr{L}(C)$. Therefore we can apply Proposition 3.11.1 to obtain

$$\begin{aligned} g &\le 1 + n_1(g_1 - 1) + n_2(g_2 - 1 + n_1) \\ &= n_1 g_1 + n_2 g_2 + (n_1 - 1)(n_2 - 1)\,. \end{aligned}$$

□

In the special case $F_1 = K(x)$ and $F_2 = K(y)$, Castelnuovo's Inequality yields:

Corollary 3.11.4 (Riemann's Inequality). *Suppose that $F = K(x, y)$. Then we have the following estimate for the genus g of F/K:*

$$g \le ([F : K(x)] - 1) \cdot ([F : K(y)] - 1)\,.$$

Riemann's Inequality (and therefore also Castelnuovo's Inequality) is often sharp, and it cannot be improved in general. In some situations however, another bound for the genus of $K(x, y)$ is even better.

Proposition 3.11.5. *Consider an algebraic function field $F = K(x,y)$ over K, where the irreducible equation of y over $K(x)$ has the form*

$$y^n + f_1(x)y^{n-1} + \ldots + f_{n-1}(x)y + f_n(x) = 0 \tag{3.135}$$

with $f_j(x) \in K[x]$ and $\deg f_j(x) \le j$ for $j = 1, \ldots, n$. Then the genus g of F/K satisfies the inequality

$$g \le \frac{1}{2}(n-1)(n-2)\,. \tag{3.136}$$

Proof. The proof is similar to that of Proposition 3.11.1. Let $A := (x)_\infty$ denote the pole divisor of x in F. It is a positive divisor of degree n. We claim that

$$v_P(y) \ge -v_P(A) \;\; \text{for all } P \in \mathbb{P}_F\,. \tag{3.137}$$

If P is a place with $v_P(x) \ge 0$ then $v_P(y) \ge 0$ by (3.135), and therefore (3.137) holds for P. Now consider the case $v_P(x) < 0$. Then $v_P(x) = -v_P(A)$, and the hypothesis $\deg f_j(x) \le j$ implies that $v_P(f_j(x)) \ge j \cdot v_P(x)$. Suppose that $v_P(y) < -v_P(A)$. For $j = 1, \ldots, n$ we obtain

$$\begin{aligned} v_P(f_j(x)y^{n-j}) &\ge j \cdot v_P(x) + (n-j) \cdot v_P(y) \\ &> j \cdot v_P(y) + (n-j) \cdot v_P(y) = v_P(y^n)\,. \end{aligned}$$

So Equation (3.135) contradicts the Strict Triangle Inequality, and (3.137) is proved. We conclude that

$$(x) \ge -A \;\; \text{and} \;\; (y) \ge -A\,.$$

It follows that for all $l \ge n$ the elements

$$x^i y^j \;\; \text{with } 0 \le j \le n-1 \text{ and } 0 \le i \le l-j$$

are in $\mathscr{L}(lA)$. They are linearly independent over K as $1, y, \ldots, y^{n-1}$ are linearly independent over $K(x)$. Therefore

$$\begin{aligned} \ell(lA) &\ge \sum_{j=0}^{n-1}(l-j+1) = n(l+1) - \sum_{j=0}^{n-1} j \\ &= n(l+1) - \frac{1}{2}n(n-1)\,. \end{aligned} \tag{3.138}$$

For l sufficiently large, the Riemann-Roch Theorem yields

$$\ell(lA) = l \cdot \deg A + 1 - g = ln + 1 - g\,.$$

We substitute this into (3.138) and obtain $g \le (n-1)(n-2)/2$. □

3.12 Exercises

In all exercises below we assume that K is a perfect field and F/K is a function field with full constant field K.

3.1. Let F'/F be an algebraic extension of F and $A \in \mathrm{Div}(F)$. Show that $\mathscr{L}(\mathrm{Con}_{F'/F}(A)) \cap F = \mathscr{L}(A)$.

3.2. Let $P_1, \ldots, P_r$ be places of F/K ($r \geq 1$). Show that there is an element $x \in F$ with the following properties:

(a) $P_1, \ldots, P_r$ are poles of x, and there are no other poles of x.

(b) The extension $F/K(x)$ is separable.

3.3. Let $R = \mathcal{O}_S$ be a holomorphy ring of F/K, and let R_1 be a subring of F with $R \subseteq R_1 \subsetneqq F$. Show:

(i) For each $x \in R_1$, the ring $R[x]$ is a holomorphy ring.

Hint. Consider the set $T := \{P \in S \mid v_P(x) \geq 0\}$ and show that $R[x] = \mathcal{O}_T$.

(ii) The ring R_1 is a holomorphy ring of F/K.

3.4. Consider an extension field $F' = F(y)$ of degree $[F' : F] = n$. Let $\mathcal{O}_S \subseteq F$ be a holomorphy ring of F and assume that its integral closure in F' is

$$\mathrm{ic}_{F'}(\mathcal{O}_S) = \sum_{i=0}^{n-1} \mathcal{O}_S \cdot y^i \, .$$

Show that $\{1, y, \ldots, y^{n-1}\}$ is an integral basis of F'/F at all places $P \in S$.

3.5. Let $S \subsetneqq \mathbb{P}_F$ such that $\mathbb{P}_F \setminus S$ is finite. Show that there are elements $x_1, \ldots, x_r \in F$ with $\mathcal{O}_S = K[x_1, \ldots, x_r]$.

3.6. We define the ramification locus of a finite separable extension F'/F as $\mathrm{Ram}(F'/F) := \{P \in \mathbb{P}_F \mid \text{there is some place } P' \in \mathbb{P}_{F'} \text{ with } e(P'|P) > 1\}$, and its degree as

$$\deg \mathrm{Ram}(F'/F) := \sum_{P \in \mathrm{Ram}(F'/F)} \deg P \, .$$

Now let $F/K(x)$ be a finite separable extension of the rational function field, having K as its full constant field, and $[F : K(x)] = n > 1$. Show:

(i) $\mathrm{Ram}(F/K(x)) \neq \emptyset$.

(ii) If $\mathrm{char}\, K = 0$ or $\mathrm{char}\, K > n$, then $\deg \mathrm{Ram}(F/K(x)) \geq 2$. If moreover the genus of F is > 0, then $\deg \mathrm{Ram}(F/K(x)) \geq 3$.

3.7. Assume that K is algebraically closed, and $\mathrm{char}\, K = 0$ or $\mathrm{char}\, K > n$. Let $F/K(x)$ be a separable extension of degree n such that $\deg \mathrm{Ram}(F/K(x)) = 2$. Show that there is an element $y \in F$ such that $F = K(y)$ and y satisfies the equation $y^n = (ax+b)/(cx+d)$, with $a, b, c, d \in K$ and $ad \neq bc$.

3.8. Consider the rational function field $F = K(x)$ and a polynomial $f(x) \in K[x]$ of degree $\deg f(x) = n \geq 2$. In case of $\operatorname{char} K = p > 0$ we assume that $f(x) \notin K[x^p]$. We set $z = f(x)$ and consider the field extension $K(x)/K(z)$, which is separable of degree n (cf. Exercise 1.1).

(i) Show that exactly the following places of $K(x)$ are ramified in $K(x)/K(z)$: the pole P_∞ of x and the places which are zeros of the derivative $f'(x)$.

(ii) In the special case $z = f(x) = x^n$ where n is not divisible by the characteristic of K, show that the zero and the pole of z are the only ramified places of $K(z)$ in $K(x)/K(z)$. Calculate the different exponents of the places above them, without using results of Section 3.7.

(iii) Now consider the case $z = x^{p^s} - x$, where $p = \operatorname{char} K > 0$. Without using results from Section 3.7, show that only the pole of z is ramified in $K(x)/K(z)$, with ramification index $e = p^s$ and different exponent $d = 2(p^s - 1)$.

3.9. ($\operatorname{char} K = p > 0$) With notation as in the previous exercise, let $z = f(x) = g(x) + h(x)$ be a polynomial over K of degree n, where

$$g(x) = \sum_{p \nmid i} a_i x^i \;, \quad h(x) = \sum_{p \mid i} a_i x^i \;.$$

Since we assume that $f(x) \notin K[x^p]$, we have that $\deg g(x) \geq 1$. Clearly the pole P of z in $K(z)$ is totally ramified in $K(x)/K(z)$, the only place above it is the pole P_∞ of x in $K(x)$.

Show that the different exponent of P_∞/P is given by

$$d(P_\infty/P) = (n-1) + (n - \deg g(x)) \;.$$

3.10. *(i)* Given an extension $F' = F(y)$ of the function field F/K, where y satisfies the equation

$$y^n = u \in F \;, \quad \textit{with} \;\; (n, \operatorname{char} K) = 1 \;.$$

Let P be a place of F/K and P' a place of F'/K lying above P. Consider F' as the compositum of the fields F and $K(y)$ over the rational function field $K(u)$, and use Exercise 3.8 (ii) and Abhyankar's Lemma to obtain the ramification index of $P'|P$. This gives another proof of Proposition 3.7.3 (ramification in Kummer extensions).

(ii) ($\operatorname{char} K = p > 0$) Let $F' = F(y)$, where y satisfies the equation

$$y^{p^s} - y = u \in F \;.$$

In a similar way as in (i), consider F' as a compositum of two subfields. Use Exercise 3.8 (iii) and Abhyankar's Lemma to show: if there is a pole $P \in \mathbb{P}_F$ of u with $v_P(u) = -m < 0$ and $(m, p) = 1$, then $[F' : F] = p^s$, P is totally ramified in F', and the different exponent of the place $P'|P$ is given by

$$d(P'|P) = (p^s - 1)(m+1) .$$

This gives another proof for ramification and different exponents in Artin-Schreier extensions (Propositions 3.7.8 and 3.7.10).

3.11. (char $K = p > 0$) Determine the genus of the function field $F = K(x, y)$ which is given by the equation

$$y^{rp} + a_{r-1}y^{(r-1)p} + \ldots + a_1 y^p + a_0 y = h(x) \in K[x] ,$$

with $a_i \in K$, $a_0 \neq 0$, $\deg h(x) = m$ and $(m, rp) = 1$.

Hint. Use Exercise 3.9.

3.12. (char $K = p > 0$) Let E/F be a Galois extension of function fields E, F over K. Let P be a place of F and Q a place of E lying above P. Show:

$$e(Q|P) \equiv 0 \bmod p \quad \Rightarrow \quad d(Q|P) \geq (e(Q|P) - 1) + (p-1) .$$

Compare with Exercise 3.9.

3.13. *(i)* Assume that $g \geq 2$. Let $\sigma : F \to F$ be a homomorphism of F to F over K (i.e., $\sigma|_K$ is the identity on K), such that $F/\sigma(F)$ is separable. Show that σ is surjective.

(ii) (char $K \neq 2$) Let $F = K(x, y)$ with $y^2 = x^3 - x$. We know from Example 3.7.6 that K is the full constant field of F, and F has genus $g = 1$. Set

$$u := \frac{(x^2+1)^2}{4y^2} \quad \textit{and} \quad v := \frac{(x^2+1)(y^4 - 4x^4)}{8x^2y^3} .$$

Show that there exists a homomorphism $\sigma : F \to F$ over K with $\sigma(x) = u$ and $\sigma(y) = v$. The extension $F/\sigma(F)$ is separable of degree $[F : \sigma(F)] = 4$.

3.14. In this exercise we assume for simplicity that K is algebraically closed and char$K = 0$. Let $K(x)$ be a rational function field, $F_1 = K(x, y)$, $F_2 = K(x, z)$ and $F = F_1F_2 = K(x, y, z)$, where

$$y^m = f(x) \in K[x] , \quad z^n = g(x) \in K[x] ,$$

$f(x)$ and $g(x)$ are square-free, $\deg f(x) = r$ and $\deg g(x) = s$. Assume that $(m, r) = 1$, $n|s$ and $(f(x), g(x)) = 1$. Determine the genera of F_1, F_2 and F, and show that Castelnuovo's Inequality is sharp in this case.

3.15. *(i)* Consider a Galois extension $F/K(x)$ of degree $[F : K(x)] = \ell$, with a prime number ℓ. Assume that at least $2\ell + 1$ places of $K(x)$ are ramified in $F/K(x)$. Show that $K(x)$ is the only rational subfield of F with $[F : K(x)] = \ell$.

(ii) Assume that $\ell \neq$ char K is a prime number, and $a_1, \ldots, a_\ell, b_1, \ldots, b_\ell$ are distinct elements of K. Set $f(x) = \prod_{1 \leq i \leq \ell}(x - a_i)$ and $g(x) = \prod_{1 \leq i \leq \ell}(x - b_i)$, and consider the function field $F = K(x, y)$ with $y^\ell = f(x)/g(x)$. Show that exactly 2ℓ places of $K(x)$ are ramified in $F/K(x)$, and $K(x)$ is not the only rational subfield of F with $[F : K(x)] = \ell$.

3.16. Let $\sigma \neq \mathrm{id}$ be an automorphism of F/K of finite order, and denote by $F^{\langle\sigma\rangle}$ the fixed field of σ. Let $P \in \mathbb{P}_F$ be a place of degree one. Show:

$$P \text{ is totally ramified in } F/F^{\langle\sigma\rangle} \iff \sigma(P) = P\ .$$

3.17. Let σ be an automorphism of F/K. Assume that there are $2g+3$ distinct places P_i of degree one with $\sigma(P_i) = P_i$. Show that $\sigma = \mathrm{id}$.

3.18. For simplicity we assume that K is an algebraically closed field. Let F/K be a function field of genus $g \geq 2$, and let $G \subseteq \mathrm{Aut}(F/K)$ be a finite group of automorphisms of F/K. Assume that $\gcd(\mathrm{ord}\, G, \mathrm{char}\, K) = 1$. Show that $\mathrm{ord}\, G \leq 84(g-1)$ (this estimate is due to Hurwitz).

Hint. Let $P_1, \ldots, P_r$ be all places of F^G which are ramified in F/F^G. Denote their ramification indices in F/F^G by $e_1, \ldots, e_r$ and assume that $e_1 \leq e_2 \leq \ldots \leq e_r$. Write down the Hurwitz Genus Formula for F/F^G and discuss the possible cases. The case where the field F^G is rational, $r = 3$, $e_1 = 2$, $e_2 = 3$ and $e_3 = 7$, yields the largest possible value for $\mathrm{ord}\, G$, namely $84(g-1)$.

Remark. One can show that the automorphism group $\mathrm{Aut}(F/K)$ is always finite, for all function fields F/K of genus $g \geq 2$ (assuming that K is a perfect field). However, the estimate $\mathrm{ord}\, G \leq 84(g-1)$ does not always hold if $\mathrm{ord}\, G$ is divisible by the characteristic of K.

3.19. Let E/F be a finite extension of F such that $E = F_1F_2$ is the compositum of two intermediate fields $F \subseteq F_i \subseteq E$, $i = 1, 2$. Assume that $[F_1 : F] = [E : F_2]$. Let P_1 be a place of F_1 and P_2 a place of F_2 with $P_1 \cap F = P_2 \cap F$. Show that there exists a place Q of E which satisfies $Q \cap F_1 = P_1$ and $Q \cap F_2 = P_2$.

3.20. Assume that F/K has at least one rational place. Show that there exist $x, y \in F$ such that $F = K(x, y)$ and $K(x) \cap K(y) = K$.

Hint. If $F = K(z)$ is a rational function field, choose $x = z^n(z-1)$ with $n \geq 2$ and $y = z(z-1)h(z)$ with $h(0), h(1) \notin \{0, \infty\}$. Why is $K(x) \cap K(y) = K$? If F is not the rational function field, construct x and y in an analogous way.

3.21. ($\mathrm{char}\, K = p > 0$) Consider a Galois extension F'/F of function fields over K, a place $P \in \mathbb{P}_F$ and a place $P' \in \mathbb{P}_{F'}$ lying over P. Denote by $G_i = G_i(P'|P)$ the i-th ramification group of $P'|P$. An integer $s \geq 1$ is called a jump of $P'|P$ if $G_s \supsetneqq G_{s+1}$.

Assume that the first ramification group G_1 is non-cyclic of order $\mathrm{ord}\, G_1 = p^2$, and that $P'|P$ has two jumps $s < t$, thus

$$G_1 = \ldots = G_s \supsetneqq G_{s+1} = \ldots = G_t \supsetneqq G_{t+1} = \{\mathrm{id}\}\ .$$

Show that $s \equiv t \bmod p$.

Hint. Choose a subgroup $H \subseteq G_1$ with $H \neq G_t$ and $\mathrm{ord}\, H = p$. Let E be the fixed field of H and $Q := P' \cap E$. Calculate the different exponent of $P'|P$ in two ways:

(a) by Hilbert's Different Formula, and

(b) by using transitivity of different exponents for $P' \supset Q \supset P$.

Remark. Exercise 3.20 is the simplest special case of the Hasse-Arf Theorem which states that for every abelian extension F'/F, two consecutive jumps $s < t$ of $P'|P$ satisfy the congruence $s \equiv t \bmod (G_1 : G_t)$.

4

Differentials of Algebraic Function Fields

In the previous chapters we have seen that Weil differentials provide a useful tool for studying algebraic function fields. Now we shall develop the theory of differentials (beginning with a definition of differentials that is closer to analysis) and show how these are related to the notion of Weil differentials.

In this chapter we consider an algebraic function field F/K of one variable. K is the full constant field of F, and K is assumed to be perfect.

4.1 Derivations and Differentials

We begin with some basic notions.

Definition 4.1.1. *Let M be a module (i.e., a vector space) over F. A mapping $\delta : F \to M$ is said to be a derivation of F/K, if δ is K-linear and the product rule*

$$\delta(u \cdot v) = u \cdot \delta(v) + v \cdot \delta(u)$$

holds for all $u, v \in F$.

Some consequences of this definition are listed in the following lemma.

Lemma 4.1.2. *Let $\delta : F \to M$ be a derivation of F/K into M. Then we have:*

(a) $\delta(a) = 0$ for each $a \in K$.

(b) $\delta(z^n) = nz^{n-1} \cdot \delta(z)$ for $z \in F$ and $n \geq 0$.

(c) If $\mathrm{char}\, K = p > 0$, then $\delta(z^p) = 0$ for each $z \in F$.

(d) $\delta(x/y) = (y \cdot \delta(x) - x \cdot \delta(y))/y^2$ for $x, y \in F$ and $y \neq 0$.

The simple proof of this lemma can be omitted.

H. Stichtenoth, *Algebraic Function Fields and Codes*,
Graduate Texts in Mathematics 254,

Before we show that some specific derivations exist, we prove a uniqueness assertion. Recall that an element $x \in F$ is called a *separating element* of F/K if $F/K(x)$ is a separable algebraic extension, cf. Section 3.10.

Lemma 4.1.3. *Suppose that x is a separating element of F/K and that $\delta_1, \delta_2 : F \to M$ are derivations of F/K with $\delta_1(x) = \delta_2(x)$. Then $\delta_1 = \delta_2$.*

Proof. Lemma 4.1.2(b) implies for a polynomial $f(x) = \sum a_i x^i \in K[x]$ that $\delta_j(f(x)) = (\sum i a_i x^{i-1}) \cdot \delta_j(x)$ for $j = 1, 2$, hence $\delta_1(f(x)) = \delta_2(f(x))$. For an arbitrary element $z = f(x)/g(x) \in K(x)$ it follows then, by Lemma 4.1.2(d), that

$$\begin{aligned} \delta_1(z) &= \frac{g(x) \cdot \delta_1(f(x)) - f(x) \cdot \delta_1(g(x))}{g(x)^2} \\ &= \frac{g(x) \cdot \delta_2(f(x)) - f(x) \cdot \delta_2(g(x))}{g(x)^2} = \delta_2(z)\,. \end{aligned}$$

Therefore the restrictions of δ_1 and δ_2 to $K(x)$ are equal. Now we consider an arbitrary element $y \in F$. Let $h(T) = \sum u_i T^i \in K(x)[T]$ be its minimal polynomial over $K(x)$. We apply δ_j $(j = 1, 2)$ to the equation $h(y) = 0$ and obtain

$$\begin{aligned} 0 = \delta_j\left(\sum u_i y^i\right) &= \sum (u_i \cdot \delta_j(y^i) + y^i \cdot \delta_j(u_i)) \\ &= \left(\sum i u_i y^{i-1}\right) \cdot \delta_j(y) + \sum y^i \cdot \delta_j(u_i)\,. \end{aligned}$$

As y is separable over $K(x)$, the derivative $h'(y) = \sum i u_i y^{i-1}$ does not vanish, hence

$$\delta_j(y) = \frac{-1}{h'(y)} \cdot \sum y^i \cdot \delta_j(u_i)$$

for $j = 1, 2$. Since $u_i \in K(x)$, we know already that $\delta_1(u_i) = \delta_2(u_i)$, therefore $\delta_1(y) = \delta_2(y)$. □

Proposition 4.1.4. *(a) Suppose that E/F is a finite separable extension of F and $\delta_0 : F \to N$ is a derivation of F/K into some field $N \supseteq E$. Then δ_0 can be extended to a derivation $\delta : E \to N$. This extension is uniquely determined by δ_0.*

(b) If $x \in F$ is a separating element of F/K and $N \supseteq F$ is some field, then there exists a unique derivation $\delta : F \to N$ of F/K with the property $\delta(x) = 1$.

Proof. (a) Uniqueness follows from the previous lemma. In order to prove the existence of an extension of δ_0, we introduce two mappings s' and s^0 from the polynomial ring $F[T]$ into $N[T]$, namely

$$s(T) = \sum s_i T^i \longmapsto s'(T) := \sum i s_i T^{i-1}$$

and
$$s(T) = \sum s_i T^i \longmapsto s^0(T) := \sum \delta_0(s_i) T^i .$$
Clearly both mappings are K-linear and satisfy the product rule. Now we choose an element $u \in E$ such that $E = F(u)$. Let $f(T) \in F[T]$ be the minimal polynomial of u over F and set $n := [E : F] = \deg f(T)$. Every element $y \in E$ has a unique representation

$$y = h(u) \quad \textit{with} \quad h(T) \in F[T] \quad \textit{and} \quad \deg h(T) < n .$$

We define $\delta : E \to N$ by

$$\delta(y) := h^0(u) - \frac{f^0(u)}{f'(u)} \cdot h'(u) \tag{4.1}$$

and have to verify that δ is a derivation of E which extends δ_0 (observe that $f'(u) \neq 0$ since u is separable over F, hence (4.1) makes sense).

First of all, if $y \in F$ then $h(T) = y, h'(T) = 0$ and $h^0(T) = \delta_0(y)$, hence (4.1) yields $\delta(y) = \delta_0(y)$. The K-linearity of δ is obvious, and it remains to prove the product rule for δ. Consider $y, z \in E$, say $y = h(u), z = g(u)$ with $\deg h(T) < n$ and $\deg g(T) < n$. Write $g(T) \cdot h(T) = c(T) \cdot f(T) + r(T)$ with $c(T), r(T) \in F[T]$ and $\deg r(T) < n$, hence $y \cdot z = c(u) \cdot f(u) + r(u) = r(u)$. Therefore

$$\begin{aligned} \delta(y \cdot z) &= (r^0 - \frac{f^0}{f'} \cdot r')(u) = \frac{1}{f'(u)} \cdot (r^0 f' - f^0 r')(u) \\ &= \frac{1}{f'(u)} \cdot ((gh - cf)^0 \cdot f' - f^0 \cdot (gh - cf)')(u) . \end{aligned} \tag{4.2}$$

We evaluate the terms $(gh - cf)^0$ and $(gh - cf)'$ (using the product rule) and observe that $f(u) = 0$. Then (4.2) is reduced to

$$\delta(y \cdot z) = \frac{1}{f'(u)} \cdot (g^0 h f' + g h^0 f' - f^0 g' h - f^0 g h')(u) . \tag{4.3}$$

On the other hand we obtain from (4.1)

$$\begin{aligned} y \cdot \delta(z) + z \cdot \delta(y) &= h(u) \cdot (g^0 - \frac{f^0}{f'} \cdot g')(u) + g(u) \cdot (h^0 - \frac{f^0}{f'} \cdot h')(u) \\ &= \frac{1}{f'(u)} \cdot (h g^0 f' - h f^0 g' + g h^0 f' - g f^0 h')(u) . \end{aligned}$$

This is in accordance with (4.3).

(b) The uniqueness assertion follows from Lemma 4.1.3. In order to prove the existence of a derivation $\delta : F \to N$ with $\delta(x) = 1$ it is sufficient to show that there is a derivation $\delta_0 : K(x) \to N$ of $K(x)/K$ with $\delta_0(x) = 1$, by (a). We define δ_0 by

$$\delta_0\left(\frac{f(x)}{g(x)}\right) := \frac{g(x)\cdot f'(x) - f(x)\cdot g'(x)}{g(x)^2}\,, \tag{4.4}$$

where $f(x), g(x) \in K[x]$ and $f'(x)$ denotes the formal derivative of $f(x)$ in $K[x]$. Equation (4.4) is well-defined, and it is readily checked that δ_0 is a derivation of $K(x)/K$ with $\delta_0(x) = 1$. □

Definition 4.1.5. *(a) Let x be a separating element of the function field F/K. The unique derivation $\delta_x : F \to F$ of F/K with the property $\delta_x(x) = 1$ is called the derivation with respect to x.*

(b) Let $\mathrm{Der}_F := \{\eta : F \to F \,|\, \eta \text{ is a derivation of } F/K\}$. *For $\eta_1, \eta_2 \in \mathrm{Der}_F$ and $z, u \in F$ we define*

$$(\eta_1 + \eta_2)(z) := \eta_1(z) + \eta_2(z) \quad \text{and} \quad (u\cdot\eta_1)(z) := u\cdot\eta_1(z)\,.$$

It is obvious that $\eta_1 + \eta_2$ and $u\cdot\eta_1$ are derivations of F/K, and Der_F *becomes an F-module in this manner. Hence it is called the module of derivations of F/K.*

Lemma 4.1.6. *Let x be a separating element of F/K. Then the following hold:*

(a) For each derivation $\eta \in \mathrm{Der}_F$ we have $\eta = \eta(x)\cdot\delta_x$. In particular, Der_F *is a one-dimensional F-module.*

(b) (Chain rule) If y is another separating element of F/K, then

$$\delta_y = \delta_y(x)\cdot\delta_x\,. \tag{4.5}$$

(c) For $t \in F$ we have

$$\delta_x(t) \neq 0 \iff t \text{ is a separating element.}$$

Proof. (a) Consider the two derivations η and $\eta(x)\cdot\delta_x$ of F/K into F. Since $(\eta(x)\cdot\delta_x)(x) = \eta(x)\cdot\delta_x(x) = \eta(x)$ and x is separating, Lemma 4.1.3 implies that $\eta(x)\cdot\delta_x = \eta$.

(b) This is a special case of (a).

(c) If t is separating, $1 = \delta_t(t) = \delta_t(x)\cdot\delta_x(t)$ (here we have used the definition of δ_t and the chain rule). Hence $\delta_x(t) \neq 0$. Suppose now that t is not separating. If $\mathrm{char}\,K = 0$ then $t \in K$ and $\delta_x(t) = 0$, since all derivations of F/K vanish on K. If $\mathrm{char}\,K = p > 0$, then $t = u^p$ for some $u \in F$ (see Proposition 3.10.2(d)), and $\delta_x(t) = \delta_x(u^p) = 0$ by Lemma 4.1.2. □

We are now ready to introduce the notion of a *differential* of F/K.

Definition 4.1.7. *(a) On the set* $Z := \{(u,x) \in F \times F \,|\, x \text{ is separating}\}$ *we define a relation* $\sim$ *by*

$$(u,x) \sim (v,y) :\iff v = u \cdot \delta_y(x)\,. \tag{4.6}$$

Using the chain rule (4.5), *it is readily verified that* $\sim$ *is an equivalence relation on* Z.

(b) We denote the equivalence class of $(u,x) \in Z$ *with respect to the above equivalence relation by* $u\,dx$ *and call it a differential of* F/K. *The equivalence class of* $(1,x)$ *is simply denoted by* dx. *Observe that by* (4.6),

$$u\,dx = v\,dy \iff v = u \cdot \delta_y(x)\,. \tag{4.7}$$

(c) Let

$$\Delta_F := \{\, u\,dx \,|\, u \in F, \text{ and } x \in F \text{ is separating}\}$$

be the set of all differentials of F/K. *We define the sum of two differentials* $u\,dx$, $v\,dy \in \Delta_F$ *as follows: Choose a separating element* z*; then*

$$u\,dx = (u \cdot \delta_z(x))\,dz \quad \text{and} \quad v\,dy = (v \cdot \delta_z(y))\,dz\,,$$

by (4.7), *and we set*

$$u\,dx + v\,dy := (u \cdot \delta_z(x) + v \cdot \delta_z(y))\,dz \tag{4.8}$$

This definition (4.8) *is independent of the choice of* z *by the chain rule. Likewise, we define*

$$w \cdot (u\,dx) := (wu)\,dx \in \Delta_F$$

for $w \in F$ *and* $u\,dx \in \Delta_F$. *One checks easily that* Δ_F *becomes an* F*-module in this manner.*

(d) For a non-separating element $t \in F$ *we define* $dt := 0$ *(the zero element of* Δ_F*); thus we obtain a mapping*

$$d : \begin{cases} F \longrightarrow & \Delta_F\,, \\ t \longmapsto & dt\,. \end{cases} \tag{4.9}$$

The pair (Δ_F, d) *is called the differential module of* F/K *(for brevity we shall simply refer to* Δ_F *as the differential module of* F/K*).*

The main properties of the differential module are put together in the following proposition.

Proposition 4.1.8. *(a) Let* $z \in F$ *be separating. Then* $dz \neq 0$, *and every differential* $\omega \in \Delta_F$ *can uniquely be written in the form* $\omega = u\,dz$ *with* $u \in F$. *Hence* Δ_F *is a one-dimensional* F*-module.*

(b) The map $d : F \to \Delta_F$ as defined in (4.9) is a derivation of F/K; i.e.,

$$d(ax) = a\,dx, \quad d(x+y) = dx + dy\,, \quad \textit{and} \quad d(xy) = x\,dy + y\,dx$$

for all $x, y \in F$ and $a \in K$.
(c) For $t \in F$ we have

$$dt \neq 0 \iff t \textit{ is separating.}$$

(d) Suppose that $\delta : F \to M$ is a derivation of F/K into some F-module M. Then there exists a unique F-linear map $\mu : \Delta_F \to M$ such that $\delta = \mu \circ d$.

Proof. (a) The differential $0 = 0\,dz$ is the zero element of Δ_F. By (4.6) we see immediately that $(0, z)$ is not equivalent to $(1, z)$, hence $dz \neq 0$.

Consider now an arbitrary differential $\omega \in \Delta_F$, say $\omega = v\,dy$ with a separating element y. Set $u := v \cdot \delta_z(y)$. Using (4.7) we obtain

$$u\,dz = (v \cdot \delta_z(y))\,dz = v\,dy = \omega\,.$$

The uniqueness of u is evident, since $dz \neq 0$ and Δ_F is a vector space over the field F.
(b) Fix a separating element $z \in F$. For all $t \in F$ we have

$$dt = \delta_z(t)\,dz\,. \tag{4.10}$$

(For separating t this follows from (4.7). If t is not separating, $dt = 0$ by definition, and $\delta_z(t)\,dz = 0$ by Lemma 4.1.6.) Using (4.10), it is easily shown that $d : F \to \Delta_F$ is a derivation of F/K. We prove only the product formula. Since δ_z is a derivation of F/K we get

$$\begin{aligned} d(xy) &= \delta_z(xy)\,dz = (x \cdot \delta_z(y) + y \cdot \delta_z(x))\,dz \\ &= x \cdot (\delta_z(y)\,dz) + y \cdot (\delta_z(x)\,dz) = x\,dy + y\,dx\,. \end{aligned}$$

(c) Clear from the definition of d.
(d) Now there is given a derivation $\delta : F \to M$. By (a), each $\omega \in \Delta_F$ is uniquely written as $\omega = u\,dz$, and we can define $\mu : \Delta_F \to M$ by $\mu(\omega) := u \cdot \delta(z)$. Obviously μ is F-linear. In order to show that $\delta = \mu \circ d$ we have only to prove that

$$\delta(z) = (\mu \circ d)(z) \tag{4.11}$$

(by Lemma 4.1.3). Equation (4.11) holds trivially by definition of μ.

It remains to prove the uniqueness of μ. Suppose that $\nu : \Delta_F \to M$ is F-linear and $\delta = \nu \circ d$. Then

$$\nu(u\,dz) = u \cdot \nu(dz) = u \cdot ((\nu \circ d)(z)) = u \cdot \delta(z) = \mu(u\,dz)\,.$$

Hence $\nu = \mu$. □

Remark 4.1.9. (a) A differential of the specific form $\omega = dx$ (with $x \in F$) is said to be *exact*; the exact differentials form a K-subspace of Δ_F.

(b) Since Δ_F is a one-dimensional F-module, one can define the *quotient* $\omega_1/\omega_2 \in F$ for $\omega_1, \omega_2 \in \Delta_F$ and $\omega_2 \neq 0$ by setting

$$u = \frac{\omega_1}{\omega_2} :\iff \omega_1 = u\omega_2 \,.$$

In particular, if $z \in F$ is separating and $y \in F$, the quotient dy/dz is defined, and we have

$$\delta_z(y) = \frac{dy}{dz} \,,$$

by Equation (4.10). Using this notation, some previous formulas can be written in a more suggestive manner, e.g.

$$u\,dx = v\,dy \iff v = u \cdot \frac{dx}{dy} \iff u = v \cdot \frac{dy}{dx} \tag{4.12}$$

and

$$\frac{dy}{dx} = \frac{dy}{dz} \cdot \frac{dz}{dx} \,, \tag{4.13}$$

if x and z are separating. The first one of these formulas corresponds to (4.7), the second one is the chain rule (4.5).

4.2 The P-adic Completion

The real number field $\mathbb{R}$ is the completion of the rational number field $\mathbb{Q}$ with respect to the ordinary absolute value. That means: (1) the field $\mathbb{Q}$ is dense in $\mathbb{R}$, and (2) every Cauchy sequence in $\mathbb{R}$ is convergent. In the present section we shall consider an analogous situation, namely the completion of a function field F/K with respect to a place $P \in \mathbb{P}_F$. This will provide us with a useful tool for calculating the derivation dz/dt (where t is a P-prime element) and will also enable us to define the *residue* of a differential at the place P. But first we need to generalize slightly some earlier notions.

Definition 4.2.1. *A discrete valuation of a field T is a surjective mapping $v : T \to \mathbb{Z} \cup \{\infty\}$ which satisfies*

(1) $v(x) = \infty \iff x = 0$.

(2) $v(xy) = v(x) + v(y)$ *for all* $x, y \in T$.

(3) $v(x+y) \geq \min\{v(x), v(y)\}$ *for all* $x, y \in T$ *(Triangle Inequality).*

The field T (more precisely: the pair (T, v)) is called a *valued field.* As in Lemma 1.1.11 one can easily prove the *Strict Triangle Inequality*

$$v(x+y) = \min\{v(x), v(y)\} \quad \textit{if } x, y \in T \textit{ and } v(x) \neq v(y) \,.$$

We say that a sequence $(x_n)_{n\geq 0}$ in T is *convergent* if there exists an element $x \in T$ (called the *limit* of the sequence) which satisfies:

$$\text{for every } c \in \mathbb{R} \text{ there is an index } n_0 \in \mathbb{N}$$
$$\text{such that } v(x - x_n) \geq c \text{ whenever } n \geq n_0 \,.$$

A sequence $(x_n)_{n\geq 0}$ is called a *Cauchy sequence* if it has the following property:

$$\text{for every } c \in \mathbb{R} \text{ there is an index } n_0 \in \mathbb{N}$$
$$\text{such that } v(x_n - x_m) \geq c \text{ whenever } n, m \geq n_0 \,.$$

As in analysis one can readily verify the following facts:

(a) If a sequence $(x_n)_{n\geq 0}$ is convergent, then its limit $x \in T$ is unique. Therefore we can write $x = \lim_{n\to\infty} x_n$.

(b) Every convergent sequence is a Cauchy sequence.

In general it is not true that all Cauchy sequences are convergent. Hence we introduce the following notions.

Definition 4.2.2. *(a) A valued field T is said to be complete if every Cauchy sequence in T is convergent.*

(b) Suppose that (T, v) is a valued field. A completion of T is a valued field $(\hat{T}, \hat{v})$ with the following properties:

(1) $T \subseteq \hat{T}$, and v is the restriction of $\hat{v}$ to T.

(2) $\hat{T}$ is complete with respect to the valuation $\hat{v}$.

(3) T is dense in $\hat{T}$; i.e., for each $z \in \hat{T}$ there is a sequence $(x_n)_{n\geq 0}$ in T with $\lim_{n\to\infty} x_n = z$.

Proposition 4.2.3. *For each valued field (T, v) there exists a completion $(\hat{T}, \hat{v})$. It is unique in the following sense: If $(\tilde{T}, \tilde{v})$ is another completion of (T, v) then there is a unique isomorphism $f : \hat{T} \to \tilde{T}$ such that $\hat{v} = \tilde{v} \circ f$. Hence $(\hat{T}, \hat{v})$ is called the completion of (T, v).*

Proof. We give only a sketch of the proof; the tedious details are left to the reader. First of all, we consider the set

$$R := \{(x_n)_{n\geq 0} \,|\, (x_n)_{n\geq 0} \text{ is a Cauchy sequence in } T\}\,.$$

This is a ring if addition and multiplication are defined in the obvious manner via $(x_n) + (y_n) := (x_n + y_n)$ and $(x_n) \cdot (y_n) := (x_n y_n)$. The set

$$I := \{(x_n)_{n\geq 0} \,|\, (x_n)_{n\geq 0} \text{ converges to } 0\}$$

is an ideal in R; actually I is a maximal ideal of R. Therefore the residue class ring

$$\hat{T} := R/I$$

is a field. For $x \in T$ let $\varrho(x) := (x, x, \dots) \in R$ be the constant sequence and $\nu(x) := \varrho(x) + I \in \hat{T}$. It is obvious that $\nu : T \to \hat{T}$ is an embedding, and we can consider T as a subfield of $\hat{T}$ via this embedding.

Now we construct a valuation $\hat{v}$ on $\hat{T}$ as follows. If $(x_n)_{n\geq 0}$ is a Cauchy sequence in T, either

$$\lim_{n\to\infty} v(x_n) = \infty$$

(in this case, $(x_n)_{n\geq 0} \in I$), or there is an integer $n_0 \geq 0$ such that

$$v(x_n) = v(x_m) \quad \textit{for all} \quad m, n \geq n_0 .$$

This follows easily from the Strict Triangle Inequality. In any case the limit $\lim_{n\to\infty} v(x_n)$ exists in $\mathbb{Z} \cup \{\infty\}$. Moreover, if $(x_n) - (y_n) \in I$ then we have $\lim_{n\to\infty} v(x_n) = \lim_{n\to\infty} v(y_n)$. Hence we can define the function $\hat{v} : \hat{T} \to \mathbb{Z} \cup \{\infty\}$ by

$$\hat{v}((x_n)_{n\geq 0} + I) := \lim_{n\to\infty} v(x_n) .$$

Using the corresponding properties of v, it is easily verified that $\hat{v}$ is a valuation of $\hat{T}$ and $\hat{v}(x) = v(x)$ for $x \in T$.

Next we consider a Cauchy sequence $(z_m)_{m\geq 0}$ in $\hat{T}$, say

$$z_m = (x_{mn})_{n\geq 0} + I \quad \textit{with} \quad (x_{mn})_{n\geq 0} \in R .$$

Then the diagonal sequence $(x_{nn})_{n\geq 0}$ is a Cauchy sequence in T and

$$\lim_{n\to\infty} z_n = (x_{nn})_{n\geq 0} + I \in \hat{T} .$$

Thus $\hat{T}$ is complete with respect to $\hat{v}$.

Now let $z = (x_n)_{n\geq 0} + I$ be an element of $\hat{T}$. Upon checking, one finds that $z = \lim_{n\to\infty} x_n$, hence T is dense in $\hat{T}$.

Thus far we have shown that a completion $(\hat{T}, \hat{v})$ of (T, v) exists. Suppose that $(\tilde{T}, \tilde{v})$ is another completion of (T, v). For the moment, we denote by $\hat{v}$-lim (resp. $\tilde{v}$-lim) the limit of a sequence in $\hat{T}$ (resp. $\tilde{T}$). Then we can construct a mapping $f : \hat{T} \to \tilde{T}$ as follows: if $z \in \hat{T}$ is represented as

$$z = \hat{v}\text{-}\lim_{n\to\infty} x_n \quad \textit{with} \quad x_n \in T ,$$

we define

$$f(z) := \tilde{v}\text{-}\lim_{n\to\infty} x_n .$$

It turns out that f is a well-defined isomorphism of $\hat{T}$ onto $\tilde{T}$ with the additional property $\hat{v} = \tilde{v} \circ f$. □

It is often more convenient to consider convergent series instead of sequences. Let $(z_n)_{n\geq 0}$ be a sequence in a valued field (T, v) and $s_m := \sum_{i=0}^{m} z_i$. We say that the infinite series $\sum_{i=0}^{\infty} z_i$ is convergent if the sequence of its partial sums $(s_m)_{m\geq 0}$ is convergent; in this case we write, as usual,

$$\sum_{i=0}^{\infty} z_i := \lim_{m\to\infty} s_m .$$

In a complete field there is a very simple criterion for convergence of an infinite series.

Lemma 4.2.4. *Let $(z_n)_{n\geq 0}$ be a sequence in a complete valued field (T, v). Then we have: The infinite series $\sum_{i=0}^{\infty} z_i$ is convergent if and only if the sequence $(z_n)_{n\geq 0}$ converges to* 0.

Proof. Suppose that $(z_n)_{n\geq 0}$ converges to 0. Consider the m-th partial sum $s_m := \sum_{i=0}^{m} z_i$. For $n > m$ we have

$$v(s_n - s_m) = v\left(\sum_{i=m+1}^{n} z_i\right) \geq \min\{v(z_i)\,|\, m < i \leq n\} \geq \min\{v(z_i)\,|\, i > m\} .$$

Since $v(z_i) \to \infty$ for $i \to \infty$, this shows that the sequence $(s_n)_{n\geq 0}$ is a Cauchy sequence in T, hence convergent.

The converse statement is easy; its proof is the same as in analysis. □

Now we specialize the above results to the case of an algebraic function field F/K.

Definition 4.2.5. *Let P be a place of F/K. The completion of F with respect to the valuation v_P is called the P-adic completion of F. We denote this completion by $\hat{F}_P$ and the valuation of $\hat{F}_P$ by v_P.*

Theorem 4.2.6. *Let $P \in \mathbb{P}_F$ be a place of degree one and let $t \in F$ be a P-prime element. Then every element $z \in \hat{F}_P$ has a unique representation of the form*

$$z = \sum_{i=n}^{\infty} a_i t^i \quad \textit{with} \quad n \in \mathbb{Z} \quad \textit{and} \quad a_i \in K . \tag{4.14}$$

This representation is called the P-adic power series expansion of z with respect to t.

Conversely, if $(c_i)_{i\geq n}$ is a sequence in K, then the series $\sum_{i=n}^{\infty} c_i t^i$ converges in $\hat{F}_P$, and we have

$$v_P\left(\sum_{i=n}^{\infty} c_i t^i\right) = \min\{i\,|\, c_i \neq 0\} .$$

Proof. First we prove the existence of a representation of the form in (4.14). Given $z \in \hat{F}_P$ we choose $n \in \mathbb{Z}$ with $n \leq v_P(z)$. There is an element $y \in F$ with $v_P(z-y) > n$ (since F is dense in $\hat{F}_P$). By the Triangle Inequality it follows that $v_P(y) \geq n$, hence $v_P(yt^{-n}) \geq 0$. As P is a place of degree one, there is an element $a_n \in K$ with $v_P(yt^{-n} - a_n) > 0$, and

$$v_P(z - a_n t^n) = v_P\big((z-y) + (y - a_n t^n)\big) > n\,.$$

In the same manner, we find $a_{n+1} \in K$ such that

$$v_P(z - a_n t^n - a_{n+1} t^{n+1}) > n+1\,.$$

Iterating this construction, we obtain an infinite sequence $a_n, a_{n+1}, a_{n+2}, \ldots$ in K such that

$$v_P\left(z - \sum_{i=n}^{m} a_i t^i\right) > m$$

for all $m \geq n$. This shows that

$$z = \sum_{i=n}^{\infty} a_i t^i\,.$$

In order to prove uniqueness we consider another sequence $(b_i)_{i \geq m}$ in K which satisfies

$$z = \sum_{i=n}^{\infty} a_i t^i = \sum_{i=m}^{\infty} b_i t^i\,.$$

We can assume that $n = m$ (otherwise, if $n < m$, define $b_i := 0$ for $n \leq i < m$). Suppose there is some j with $a_j \neq b_j$. We choose j minimal with this property and obtain for all $k > j$

$$v_P\left(\sum_{i=n}^{k} a_i t^i - \sum_{i=n}^{k} b_i t^i\right) = v_P\left((a_j - b_j)t^j + \sum_{i=j+1}^{k} (a_i - b_i)t^i\right) = j \tag{4.15}$$

(since $v_P\big((a_j - b_j)t^j\big) = j$, the Strict Triangle Inequality applies). On the other hand,

$$v_P\left(\sum_{i=n}^{k} a_i t^i - \sum_{i=n}^{k} b_i t^i\right) = v_P\left(\sum_{i=n}^{k} a_i t^i - z + z - \sum_{i=n}^{k} b_i t^i\right)$$

$$\geq \min\left\{ v_P\left(z - \sum_{i=n}^{k} a_i t^i\right), v_P\left(z - \sum_{i=n}^{k} b_i t^i\right)\right\}. \tag{4.16}$$

For $k \to \infty$, (4.16) tends to infinity. This is a contradiction to (4.15) and proves that the representation (4.14) is unique.

Finally we consider an arbitrary sequence $(c_i)_{i\geq n}$ in K. As $v_P(c_it^i) \geq i$ for all i, the sequence $(c_it^i)_{i\geq n}$ converges to 0. Hence by Lemma 4.2.4 the series $\sum_{i=n}^{\infty} c_it^i$ is convergent in $\hat{F}_P$, say

$$\sum_{i=n}^{\infty} c_it^i =: y \in \hat{F}_P .$$

Set $j_0 := \min\{i \,|\, c_i \neq 0\}$. If $j_0 = \infty$ then all $c_i = 0$, hence $y = 0$ and $v_P(y) = \infty$ as well. In the case $j_0 < \infty$ we have for all $k \geq j_0$

$$v_P\left(\sum_{i=n}^{k} c_it^i\right) = j_0$$

by the Strict Triangle Inequality. Since

$$v_P\left(y - \sum_{i=n}^{k} c_it^i\right) > j_0$$

for all sufficiently large k, this implies

$$\begin{aligned} v_P(y) &= v_P\left(y - \sum_{i=n}^{k} c_it^i + \sum_{i=n}^{k} c_it^i\right) \\ &= \min\left\{v_P\left(y - \sum_{i=n}^{k} c_it^i\right), v_P\left(\sum_{i=n}^{k} c_it^i\right)\right\} = j_0 . \end{aligned}$$

□

We continue to consider a place P of F/K of degree one and a P-prime element t. By Proposition 3.10.2, t is a separating element of F/K, and thus one can speak of the derivation $\delta_t : F \to F$ with respect to t (cf. Definition 4.1.5). Using the P-adic power series expansion, we can easily calculate $dz/dt = \delta_t(z)$ for $z \in F$ (the notation dz/dt is explained in Remark 4.1.9).

Proposition 4.2.7. *Let P be a place of F/K of degree one and let $t \in F$ be a P-prime element. If $z \in F$ has the P-adic expansion $z = \sum_{i=n}^{\infty} a_it^i$ with coefficients $a_i \in K$, then*

$$\frac{dz}{dt} = \sum_{i=n}^{\infty} ia_it^{i-1} .$$

Proof. We define a mapping $\delta : \hat{F}_P \to \hat{F}_P$ by

$$\delta\left(\sum_{i=m}^{\infty} c_it^i\right) := \sum_{i=m}^{\infty} ic_it^{i-1} .$$

This mapping is obviously K-linear and satisfies the product rule $\delta(u \cdot v) = u \cdot \delta(v) + v \cdot \delta(u)$ for all $u, v \in \hat{F}_P$ (the verification of the product rule is a bit technical but straightforward). Moreover $\delta(t) = 1$ holds. Therefore $\delta(z) = \delta_t(z) = dz/dt$ for each $z \in F$, by Proposition 4.1.4(b) and Definition 4.1.5. □

Our next goal is to introduce the *residue* of a differential $\omega \in \Delta_F$ at a place P. For this we need some background.

Definition 4.2.8. *Suppose that P is a place of F/K of degree one and $t \in F$ is a P-prime element. If $z \in F$ has the P-adic expansion $z = \sum_{i=n}^{\infty} a_i t^i$ with $n \in \mathbb{Z}$ and $a_i \in K$ we define its residue with respect to P and t by*

$$\mathrm{res}_{P,t}(z) := a_{-1} \,.$$

Clearly, $\mathrm{res}_{P,t} : F \longrightarrow K$ *is a K-linear map and*

$$\mathrm{res}_{P,t}(z) = 0 \quad \text{if} \quad v_P(z) \geq 0 \,. \tag{4.17}$$

The residue satisfies the following transformation formula:

Proposition 4.2.9. *Let $s, t \in F$ be P-prime elements (where P is a place of degree one). Then*

$$\mathrm{res}_{P,s}(z) = \mathrm{res}_{P,t}\Big(z \cdot \frac{ds}{dt}\Big)$$

for all $z \in F$.

Proof. The power series expansion of s with respect to t has the following form (see Theorem 4.2.6):

$$s = \sum_{i=1}^{\infty} c_i t^i \quad \text{with} \quad c_1 \neq 0 \,.$$

Proposition 4.2.7 yields

$$\frac{ds}{dt} = c_1 + \sum_{i=2}^{\infty} i c_i t^{i-1} \,. \tag{4.18}$$

Now we distinguish several cases.

Case 1. $v_P(z) \geq 0$. Then $v_P(z \cdot ds/dt) \geq 0$ as well (by (4.18)), and from (4.17) it follows that

$$\mathrm{res}_{P,s}(z) = \mathrm{res}_{P,t}\Big(z \cdot \frac{ds}{dt}\Big) = 0 \,.$$

Case 2. $z = s^{-1}$. Clearly we have $\mathrm{res}_{P,s}(s^{-1}) = 1$. We determine the power series expansion of s^{-1} with respect to t:

$$
\begin{aligned}
s^{-1} &= \frac{1}{c_1t + c_2t^2 + \cdots} = \frac{1}{c_1t} \cdot \left(1 + \frac{c_2}{c_1}t + \frac{c_3}{c_1}t^2 + \cdots\right)^{-1} \\
&= \frac{1}{c_1t} \cdot \left(1 + \sum_{r=1}^{\infty}(-1)^r \left(\frac{c_2}{c_1}t + \frac{c_3}{c_1}t^2 + \cdots\right)^r\right) \\
&= \frac{1}{c_1t}\left(1 + \frac{f_2(c_2)}{c_1}t + \frac{f_3(c_2,c_3)}{c_1^2}t^2 + \cdots\right)
\end{aligned}
\tag{4.19}
$$

with certain polynomials $f_j(X_2,\ldots,X_j) \in \mathbb{Z}[X_2,\ldots,X_j]$. Therefore

$$
s^{-1} \cdot \frac{ds}{dt} = \frac{1}{t} + y \quad \textit{with} \quad v_P(y) \geq 0
$$

from (4.18) and (4.19), and we obtain

$$
\mathrm{res}_{P,t}\left(s^{-1} \cdot \frac{ds}{dt}\right) = 1 + \mathrm{res}_{P,t}(y) = 1
$$

by Case 1.

Case 3. $z = s^{-n}$ with $n \geq 2$. Here $\mathrm{res}_{P,s}(s^{-n}) = 0$. To begin with, we calculate $\mathrm{res}_{P,t}(s^{-n} \cdot ds/dt)$ in the case when $\mathrm{char}\, K = 0$. Then

$$
s^{-n} \cdot \frac{ds}{dt} = \frac{1}{-n+1} \cdot \frac{d(s^{-n+1})}{dt}\,.
$$

We write

$$
s^{-n+1} = \sum_{i=k}^{\infty} d_i t^i
$$

with $k = -n+1$ and $d_i \in K$ and obtain

$$
\frac{d(s^{-n+1})}{dt} = \sum_{i=k}^{\infty} i d_i t^{i-1}\,.
$$

Hence

$$
\mathrm{res}_{P,t}\left(s^{-n} \cdot \frac{ds}{dt}\right) = \frac{1}{-n+1} \cdot \mathrm{res}_{P,t}\left(\sum_{i=k}^{\infty} i d_i t^{i-1}\right) = 0\,. \tag{4.20}
$$

Next we consider Case 3 in arbitrary characteristic. By (4.18) and (4.19) we find

$$
\begin{aligned}
s^{-n} \cdot \frac{ds}{dt} &= \frac{1}{c_1^n t^n}(c_1 + 2c_2t + \cdots) \cdot \left(1 + \frac{f_2(c_2)}{c_1}t + \frac{f_2(c_2,c_3)}{c_1^2}t^2 + \cdots\right)^n \\
&= \frac{1}{c_1^n t^n}\left(c_1 + \frac{g_2(c_1,c_2)}{c_1}t + \frac{g_3(c_1,c_2,c_3)}{c_1^2}t^2 + \cdots\right)
\end{aligned}
$$

with polynomials $g_j(X_1,\ldots,X_j) \in \mathbb{Z}[X_1,\ldots,X_j]$. These polynomials are independent of the characteristic of K, and we have

$$\mathrm{res}_{P,t}\left(s^{-n} \cdot \frac{ds}{dt}\right) = \frac{1}{c_1^{2n-1}} \cdot g_n(c_1, \cdots, c_n)\,.$$

From (4.20) it follows that $g_n(c_1, \ldots, c_n) = 0$ for any elements $c_1 \neq 0, c_2, \ldots, c_n$ in a field of characteristic zero, so $g_n(X_1, \ldots, X_n)$ must be the zero polynomial in $\mathbb{Z}[X_1, \ldots, X_n]$. Thus the equality

$$\mathrm{res}_{P,t}\left(s^{-n} \cdot \frac{ds}{dt}\right) = 0 = \mathrm{res}_{P,s}(s^{-n})$$

holds for a field of arbitrary characteristic (for $n \geq 2$).

Case 4. Finally let z be an arbitrary element of F with $v_P(z) < 0$, say

$$z = \sum_{i=-n}^{\infty} a_i s^i \quad \textit{with} \quad n \geq 1 \quad \textit{and} \quad a_i \in K\,.$$

Then $\mathrm{res}_{P,s}(z) = a_{-1}$ and $z = a_{-n}s^{-n} + \cdots + a_{-1}s^{-1} + y$ with $v_P(y) \geq 0$. Using the results of Cases 1, 2 and 3 we get

$$\mathrm{res}_{P,t}\left(z \cdot \frac{ds}{dt}\right) = \sum_{i=-n}^{-1} a_i \cdot \mathrm{res}_{P,t}\left(s^i \cdot \frac{ds}{dt}\right) + \mathrm{res}_{P,t}\left(y \cdot \frac{ds}{dt}\right)$$

$$= a_{-1} \cdot \mathrm{res}_{P,t}\left(s^{-1} \cdot \frac{ds}{dt}\right) = a_{-1} = \mathrm{res}_{P,s}(z)\,.$$

□

Definition 4.2.10. *Let $\omega \in \Delta_F$ be a differential and let $P \in \mathbb{P}_F$ be a place of degree one. Choose a P-prime element $t \in F$ and write $\omega = u\,dt$ with $u \in F$. Then we define the residue of ω at P by*

$$\mathrm{res}_P(\omega) := \mathrm{res}_{P,t}(u)\,.$$

This definition is independent of the specific choice of the prime element t. In fact, if s is another P-prime element and $\omega = u\,dt = z\,ds$, then $u = z \cdot ds/dt$, and Proposition 4.2.9 yields

$$\mathrm{res}_{P,s}(z) = \mathrm{res}_{P,t}\left(z \cdot \frac{ds}{dt}\right) = \mathrm{res}_{P,t}(u)\,.$$

We will show in the following section that the residue of a differential at a place P of degree one has an interpretation as the local component of a specific Weil differential at this place.

4.3 Differentials and Weil Differentials

The goal in this section is to establish a relationship between the notions of differentials and Weil differentials of an algebraic function field. As always, K is assumed to be perfect.

To begin with, we recall some notations and results from previous chapters (in particular Sections 1.5, 1.7 and 3.4). $\mathcal{A}_F$ denotes the *adele space* of F/K; its elements are *adeles* $\alpha = (\alpha_P)_{P\in\mathbb{P}_F}$ where the *P-component* α_P of α is an element of F and $v_P(\alpha) := v_P(\alpha_P) \geq 0$ for almost all $P \in \mathbb{P}_F$. The field F is considered as a subspace of $\mathcal{A}_F$ via the diagonal embedding $F \hookrightarrow \mathcal{A}_F$. If A is a divisor of F/K one considers the space $\mathcal{A}_F(A) = \{\alpha \in \mathcal{A}_F \mid v_P(\alpha) \geq -v_P(A)$ for all $P \in \mathbb{P}_F\}$. A *Weil differential* of F is a K-linear map $\omega : \mathcal{A}_F \to K$ that vanishes on $\mathcal{A}_F(A) + F$ for some divisor A. The Weil differentials constitute a one-dimensional F-module Ω_F. For $0 \neq \omega \in \Omega_F$ there exists a uniquely determined divisor $W = (\omega) \in \mathrm{Div}(F)$ such that ω vanishes on $\mathcal{A}_F(W)$ but not on $\mathcal{A}_F(B)$ for each divisor $B > W$. Such a divisor (ω) is called a *canonical* divisor of F/K. For each $P \in \mathbb{P}_F$ we have another embedding $\iota_P : F \to \mathcal{A}_F$ where $\iota_P(z)$ is the adele whose P-component is z, and all other components of $\iota_P(z)$ are 0. The *local component* of the Weil differential ω at the place P is the mapping $\omega_P : F \to K$ given by $\omega_P(z) := \omega(\iota_P(z))$.

If F'/F is a finite separable extension of function fields we have defined the *cotrace* $\omega' := \mathrm{Cotr}_{F'/F}(\omega)$ of a Weil differential $\omega \in \Omega_F$; this is a Weil differential of F', and if F' and F have the same constant field, ω' is characterized by the condition

$$\omega_P(\mathrm{Tr}_{F'/F}(y)) = \sum_{P'|P} \omega'_{P'}(y) \tag{4.21}$$

for all $P \in \mathbb{P}_F$ and $y \in F'$, cf. Theorem 3.4.6 and Remark 3.4.8. In Proposition 1.7.4 we have shown the existence of a specific Weil differential η of the rational function field $K(x)/K$ which is uniquely determined by the following properties:

$$\textit{the divisor of } \eta \textit{ is } (\eta) = -2P_\infty\,, \textit{ and } \eta_{P_\infty}(x^{-1}) = -1\,. \tag{4.22}$$

(P_∞ is the pole of x in $K(x)$, and η_{P_∞} is the local component of η at P_∞.)

Definition 4.3.1. *Let F/K be an algebraic function field. We define a mapping*

$$\delta : \begin{cases} F \to & \Omega_F\,, \\ x \mapsto & \delta(x) \end{cases}$$

as follows: if $x \in F\backslash K$ is a separating element of F/K we set

$$\delta(x) := \mathrm{Cotr}_{F/K(x)}(\eta)\,,$$

where $\eta \in \Omega_{K(x)}$ is the Weil differential of $K(x)/K$ characterized by (4.22). *For a non-separating element $x \in F$ we define $\delta(x) := 0$. We call $\delta(x)$ the Weil differential of F/K associated with x.*

Note that $\delta(x) \neq 0$ if x is separating, hence each Weil differential $\omega \in \Omega_F$ can be written as $\omega = z \cdot \delta(x)$ with $z \in F$. Now we state the main results of this section.

Theorem 4.3.2. *Suppose that F/K is an algebraic function field over a perfect field K, and let $x \in F$ be a separating element.*

(a) The map $\delta : F \to \Omega_F$ given by Definition 4.3.1 is a derivation of F/K.

(b) For every $y \in F$ we have

$$\delta(y) = \frac{dy}{dx} \cdot \delta(x)\,.$$

(c) The map

$$\mu : \begin{cases} \Delta_F & \to \quad \Omega_F\,, \\ z\,dx & \mapsto z \cdot \delta(x) \end{cases}$$

is an isomorphism of the differential module Δ_F onto Ω_F. This isomorphism is compatible with the derivations $d : F \to \Delta_F$ and $\delta : F \to \Omega_F$, that means $\mu \circ d = \delta$.

(d) If $P \in \mathbb{P}_F$ is a place of F/K of degree one and $\omega = z \cdot \delta(x) \in \Omega_F$, the local component of ω at P is given by

$$(z \cdot \delta(x))_P(u) = \mathrm{res}_P(uz\; dx)\,.$$

In particular,

$$(z \cdot \delta(x))_P(1) = \mathrm{res}_P(z\; dx)\,.$$

(e) If $\omega = z \cdot \delta(t) \in \Omega_F$ and t is a prime element at the place P, then we have $v_P(\omega) = v_P(z)$.

An immediate consequence of this theorem is

Corollary 4.3.3 (Residue Theorem). *Let F/K be an algebraic function field over an algebraically closed field, and let $\omega \in \Delta_F$ be a differential of F/K. Then $\mathrm{res}_P(\omega) = 0$ for almost all places $P \in \mathbb{P}_F$, and*

$$\sum_{P \in \mathbb{P}_F} \mathrm{res}_P(\omega) = 0\,.$$

Proof of the Corollary. Write $\omega = z\; dx$ with $z \in F$ and a separating element $x \in F$. By Theorem 4.3.2(d) we have $\mathrm{res}_P(\omega) = (z \cdot \delta(x))_P(1)$. Now Proposition 1.7.2 yields the desired result. □

The proof of Theorem 4.3.2 is rather tedious. First we shall assume that the constant field is algebraically closed. The case of an arbitrary perfect constant field K will then be reduced to this special case by considering the constant field extension $\bar{F} = F\bar{K}$ of F with the algebraic closure $\bar{K}$ of K. We begin with some preliminaries.

Lemma 4.3.4. *Suppose that F/F_0 is a finite separable extension of algebraic function fields over an algebraically closed field K. Let $\psi \in \Omega_{F_0}$ be a Weil differential of F_0/K and $\omega := \mathrm{Cotr}_{F/F_0}(\psi)$. Consider a place $P_0 \in \mathbb{P}_{F_0}$ which is unramified in F/F_0, and a place $P \in \mathbb{P}_F$ which lies over P_0. Then we have*

$$\omega_P(z) = \psi_{P_0}(z) \quad \textit{for all } z \in F_0 .$$

Proof. We can assume that $\psi \neq 0$. Let $P_1, \ldots, P_n \in \mathbb{P}_F$ be the places of F lying over P_0, say $P_1 = P$. Since P_0 is unramified in F/F_0 and K is algebraically closed, $n = [F : F_0]$ by Theorem 3.1.11. By the Approximation Theorem we can find an element $z' \in F$ satisfying

$$\begin{aligned} v_P(z' - z) &\geq -v_{P_0}(\psi) , \\ v_{P_i}(z') &\geq -v_{P_0}(\psi) \qquad \textit{for } i = 2, \ldots, n . \end{aligned} \tag{4.23}$$

(Recall that the integer $v_Q(\psi)$ is defined by $v_Q(\psi) := v_Q(W)$ where $W = (\psi)$ denotes the divisor of ψ.) Since P_0 is unramified in F/F_0 and $\omega = \mathrm{Cotr}_{F/F_0}(\psi)$,

$$v_{P_i}(\omega) = v_{P_0}(\psi) \quad \textit{for } i = 1, \ldots, n \tag{4.24}$$

by Theorem 3.4.6 and Dedekind's Different Theorem. Consider the adele $\alpha = (\alpha_Q)_{Q \in \mathbb{P}_F}$ with

$$\begin{aligned} \alpha_P &:= z' - z , \\ \alpha_{P_i} &:= z' \quad \textit{for } i = 2, \ldots, n , \textit{ and} \\ \alpha_Q &:= 0 \quad \textit{for } Q \neq P_1, \ldots, P_n . \end{aligned}$$

Then α is $\mathcal{A}_F((\omega))$ by (4.23) and (4.24), hence

$$\begin{aligned} \omega_P(z) &= \omega_P(z) + \omega(\alpha) = \omega_P(z) + \omega_P(z' - z) + \sum_{i=2}^{n} \omega_{P_i}(z') \\ &= \sum_{i=1}^{n} \omega_{P_i}(z') = \psi_{P_0}(\mathrm{Tr}_{F/F_0}(z')) . \end{aligned} \tag{4.25}$$

(The last equality in (4.25) follows from (4.21).) By (4.25), the proof of our lemma will be finished when we show that

$$\psi_{P_0}(\mathrm{Tr}_{F/F_0}(z')) = \psi_{P_0}(z) . \tag{4.26}$$

The trace Tr_{F/F_0} can be evaluated by using the embeddings of F/F_0 into an extension field of F. We proceed as follows: Choose a Galois closure $E \supseteq F$ of F/F_0 (i.e., $F_0 \subseteq F \subseteq E$, the extension E/F_0 is Galois, and E is minimal with this property). Then P_0 is unramified in E/F_0 by Corollary 3.9.3. Choose places $Q_1 = Q, Q_2, \ldots, Q_n \in \mathbb{P}_E$ with $Q_i \mid P_i$. Since E/F_0 is Galois, there exist automorphisms $\sigma_1, \ldots, \sigma_n \in \mathrm{Gal}(E/F_0)$ such that

$$\sigma_i^{-1}(Q) = Q_i \quad \textit{for } i = 1, \ldots, n\,, \tag{4.27}$$

cf. Theorem 3.7.1. We assert that the restrictions $\sigma_i|_F$ $(i = 1, \ldots, n)$ of σ_i to F are pairwise distinct. In fact, if $\sigma_i|_F = \sigma_j|_F$, we have for each $u \in F$

$$v_{P_i}(u) = v_{Q_i}(u) = v_{\sigma_i^{-1}(Q)}(u) = v_Q(\sigma_i(u)) = v_Q(\sigma_j(u)) = v_{P_j}(u)$$

(we have used (4.27) and Lemma 3.5.2(a)). Therefore $i = j$, and the embeddings $\sigma_i|_F \colon F \to E$ are pairwise distinct. Hence

$$\mathrm{Tr}_{F/F_0}(u) = \sum_{i=1}^{n} \sigma_i(u)$$

for $u \in F$. Now we can verify (4.26). By (4.23) and (4.27),

$$\begin{aligned} &v_Q(z' - z) \geq -v_{P_0}(\psi) \quad \textit{and} \\ &v_Q(\sigma_i(z')) = v_{Q_i}(z') = v_{P_i}(z') \geq -v_{P_0}(\psi) \end{aligned}$$

for $i = 2, \ldots, n$. Thus

$$\begin{aligned} v_{P_0}(\mathrm{Tr}_{F/F_0}(z') - z) &= v_Q(\mathrm{Tr}_{F/F_0}(z') - z) \\ &= v_Q\left((z' - z) + \sum_{i=2}^{n} \sigma_i(z')\right) \geq -v_{P_0}(\psi)\,. \end{aligned}$$

Using Proposition 1.7.3 we obtain

$$\psi_{P_0}(\mathrm{Tr}_{F/F_0}(z') - z) = 0\,,$$

hence (4.26) follows. □

Lemma 4.3.5. *Let F be an algebraic function field over an algebraically closed field K. Suppose that x is a separating element of F/K and $P_0 \in \mathbb{P}_{K(x)}$ satisfies the following conditions:*

(1) P_0 is unramified in $F/K(x)$.

(2) P_0 is not the pole of x in $K(x)$.

If $\delta(x) \in \Omega_F$ denotes the Weil differential associated with x (as defined in Definition 4.3.1) and $u \in F$, then

$$\delta(x)_P(u) = \mathrm{res}_P(u\,dx) \tag{4.28}$$

holds for all $P \in \mathbb{P}_F$ with $P|P_0$.

Proof. By (1) and (2) there exists an element $a \in K$ such that $t := x - a$ is a P-prime element. As always, P_∞ denotes the pole of x in $K(x)$. Consider the Weil differential $\eta \in \Omega_{K(x)}$ which is given by (4.22); then $\delta(x) = \mathrm{Cotr}_{F/K(x)}(\eta)$.

First we evaluate the left-hand side of (4.28) for $u = t^k$, $k \in \mathbb{Z}$. By the previous lemma and Proposition 1.7.4(c) we obtain

$$\delta(x)_P(t^k) = \eta_{P_0}(t^k) = \begin{cases} 0 & \text{for } k \neq -1\,, \\ 1 & \text{for } k = -1\,. \end{cases} \tag{4.29}$$

An arbitrary element $u \in F$ can be written as

$$u = \sum_{\nu=m}^{l-1} a_\nu t^\nu + u'$$

with $a_\nu \in K$, $l \geq \max\{0, -v_P(\delta(x))\}$ and $v_P(u') \geq l$; this follows easily from the P-adic power series expansion of u with respect to t (Theorem 4.2.6). We have $\delta(x)_P(u') = 0$ by Proposition 1.7.3, so (4.29) implies

$$\delta(x)_P(u) = \sum_{\nu=m}^{l-1} a_\nu \cdot \delta(x)_P(t^\nu) = a_{-1}\,.$$

On the other hand, $dt = d(x-a) = dx$, hence

$$\mathrm{res}_P(u\,dx) = \mathrm{res}_P(u\,dt) = \mathrm{res}_{P,t}(u) = a_{-1}\,.$$

This proves (4.28). □

Proof of Theorem 4.3.2 *under the additional hypothesis that K is algebraically closed.* We begin with (b). If $y \in F$ is not separating, $\delta(y) = 0$ and $dy/dx = 0$ (see Definition 4.3.1 and Proposition 4.1.8). So $\delta(y) = (dy/dx) \cdot \delta(x)$ in this case. Henceforth we can assume that y is separating. As $\delta(x) \neq 0$ and Ω_F is a one-dimensional F-module (Proposition 1.5.9), $\delta(y) = z \cdot \delta(x)$ for some $z \in F$. Only finitely many places of F are ramified over $K(x)$ or $K(y)$ (cf. Corollary 3.5.5), and we can find a place $P \in \mathbb{P}_F$ such that the restrictions of P to $K(x)$ (resp. $K(y)$) are unramified in $F/K(x)$ (resp. $F/K(y)$), and P is neither a pole of x nor a pole of y. For every $u \in F$,

$$\delta(y)_P(u) = (z \cdot \delta(x))_P(u) = \delta(x)_P(zu)$$

holds. On the other hand, Lemma 4.3.5 (applied to the extensions $F/K(y)$ and $F/K(x)$) yields

$$\delta(y)_P(u) = \mathrm{res}_P(u\,dy) = \mathrm{res}_P\left(u\frac{dy}{dx}\,dx\right) = \delta(x)_P\left(u\frac{dy}{dx}\right)\,.$$

Hence

$$\delta(x)_P\left(u(z - \frac{dy}{dx})\right) = 0 \quad \text{for all} \quad u \in F\,.$$

This implies $z = dy/dx$ (by Proposition 1.7.3), and proves (b).

(a) We use (b) and Proposition 4.1.8(b) and obtain for $y_1, y_2 \in F$ and $a \in K$

$$\delta(y_1 + y_2) = \frac{d(y_1 + y_2)}{dx} \cdot \delta(x) = \left(\frac{dy_1}{dx} + \frac{dy_2}{dx}\right) \cdot \delta(x) = \delta(y_1) + \delta(y_2),$$

$$\delta(ay_1) = \frac{d(ay_1)}{dx} \cdot \delta(x) = a \cdot \frac{dy_1}{dx} \cdot \delta(x) = a \cdot \delta(y_1),$$

and the product rule

$$\delta(y_1 y_2) = \frac{d(y_1 y_2)}{dx} \cdot \delta(x) = \left(y_1 \frac{dy_2}{dx} + y_2 \frac{dy_1}{dx}\right) \cdot \delta(x) = y_1 \cdot \delta(y_2) + y_2 \cdot \delta(y_1).$$

(c) By Proposition 4.1.8(d) there exists a uniquely determined F-linear map $\mu : \Delta_F \to \Omega_F$ such that $\delta(y) = (\mu \circ d)(y) = \mu(dy)$ for all $y \in F$. Since μ is F-linear, $\mu(z\,dx) = z \cdot \mu(dx) = z \cdot \delta(x)$. As dx generates Δ_F and $\delta(x)$ generates Ω_F (as F-modules), μ is bijective.

(d) This is a generalization of Lemma 4.3.5, and we want to reduce the proof of (d) to this lemma. We choose sufficiently many pairwise distinct places $P_1 := P, P_2, \ldots, P_r$ such that $\mathscr{L}(P_1 + P_2 + \ldots + P_r)$ is strictly larger than $\mathscr{L}(P_2 + \ldots + P_r)$. Each element

$$t_1 \in \mathscr{L}(P_1 + P_2 + \ldots + P_r) \backslash \mathscr{L}(P_2 + \ldots + P_r)$$

has only simple poles in F, and $P = P_1$ is one of them. Setting $t := t_1^{-1}$ and $P_0 := P \cap K(t) \in \mathbb{P}_{K(t)}$ we have seen that P_0 is unramified in the separable extension $F/K(t)$ and that P_0 is not the pole of t in $K(t)$. Now (b) and Lemma 4.3.5 yield

$$\begin{aligned} (z \cdot \delta(x))_P(u) &= \left(z \cdot \frac{dx}{dt} \cdot \delta(t)\right)_P (u) = \delta(t)_P \left(uz \frac{dx}{dt}\right) \\ &= \operatorname{res}_P \left(uz \frac{dx}{dt}\, dt\right) = \operatorname{res}_P(uz\, dx). \end{aligned}$$

(e) Since t is a P-prime element, the extension $F/K(t)$ is separable (Proposition 3.10.2), and P is unramified in $F/K(t)$. The divisor of the Weil differential $\delta(t)$ is given by

$$(\delta(t)) = -2(t)_\infty + \operatorname{Diff}(F/K(t)),$$

where $(t)_\infty$ is the pole divisor of t in F, by (4.22) and Theorem 3.4.6. As P is not a pole of t and P does not occur in the different of $F/K(t)$ (by Dedekind's Different Theorem), we have $v_P(\delta(t)) = 0$. Hence

$$v_P(z \cdot \delta(t)) = v_P(z) + v_P(\delta(t)) = v_P(z).$$

□

Now we consider a function field F/K where K is an arbitrary perfect field. Our aim is to prove Theorem 4.3.2 also in this case. Let $\bar{K} \supseteq K$ be an algebraic closure of K and let $\bar{F} := F\bar{K}$ be the constant field extension of F/K with $\bar{K}$. Each place of F/K has finitely many extensions in $\bar{F}$; conversely, each place of $\bar{F}/\bar{K}$ is an extension of some place of F/K. So we have a natural embedding $\mathcal{A}_F \hookrightarrow \mathcal{A}_{\bar{F}}$ of the adele spaces and principal adeles of F/K are mapped to principal adeles of $\bar{F}/\bar{K}$ under this embedding. Hence we shall regard $\mathcal{A}_F$ as a subspace of $\mathcal{A}_{\bar{F}}$.

We fix a separating element $x \in F \backslash K$; then x is separating for $\bar{F}/\bar{K}$ as well, and $[F : K(x)] = [\bar{F} : \bar{K}(x)]$, cf. Proposition 3.6.1. For $y \in F$ let $\delta(y) \in \Omega_F$ (resp. $\bar{\delta}(y) \in \Omega_{\bar{F}}$) be the Weil differential of F/K (resp. $\bar{F}/\bar{K}$) associated with y. The restriction of a Weil differential $\bar{\omega} \in \Omega_{\bar{F}}$ to the space $\mathcal{A}_F \subseteq \mathcal{A}_{\bar{F}}$ is denoted by $\bar{\omega}|_{\mathcal{A}_F}$.

Proposition 4.3.6. *With the above notation we have*

$$\bar{\delta}(x)|_{\mathcal{A}_F} = \delta(x)\,. \tag{4.30}$$

Proof. Recall that $\delta(x) = \mathrm{Cotr}_{F/K(x)}(\eta)$ and $\bar{\delta}(x) = \mathrm{Cotr}_{\bar{F}/\bar{K}(x)}(\bar{\eta})$ with specific Weil differentials $\eta \in \Omega_{K(x)}$ (resp. $\bar{\eta} \in \Omega_{\bar{K}(x)}$), cf. Definition 4.3.1. The first step in the proof of Proposition 4.3.6 is the following assertion:

$$\bar{\eta}|_{\mathcal{A}_{K(x)}} = \eta\,. \tag{4.31}$$

In fact, the map $\bar{\eta}\;|_{\mathcal{A}_{K(x)}} : \mathcal{A}_{K(x)} \to \bar{K}$ is K-linear, and $\bar{\eta}$ vanishes on $\mathcal{A}_{K(x)}(-2P_\infty) + K(x)$ (as usual, P_∞ is the pole of x in $K(x)$). Consider the adele $\gamma := \iota_{P_\infty}(x^{-1}) \in \mathcal{A}_{K(x)}$ which is given by $\gamma_{P_\infty} := x^{-1}$ and $\gamma_Q := 0$ for $Q \neq P_\infty$. Then $\bar{\eta}(\gamma) = \eta(\gamma) = -1$ by (4.22). Observe that

$$\mathcal{A}_{K(x)} = \mathcal{A}_{K(x)}(-2P_\infty) + K(x) + K \cdot \gamma\,,$$

since $\gamma \notin \mathcal{A}_{K(x)}(-2P_\infty) + K(x)$ and $\dim_K(\mathcal{A}_{K(x)}/\mathcal{A}_{K(x)}(-2P_\infty) + K(x)) = 1$, cf. Theorem 1.5.4. So each adele $\beta \in \mathcal{A}_{K(x)}$ can be written as $\beta = \beta_0 + c\gamma$ with $c \in K$ and $\eta(\beta_0) = \bar{\eta}(\beta_0) = 0$, and we obtain $\bar{\eta}(\beta) = \bar{\eta}(\beta_0) + c \cdot \bar{\eta}(\gamma) = -c = \eta(\beta)$. This proves (4.31).

In the second step we evaluate the local component $\delta(x)_P(u)$ for a place $P \in \mathbb{P}_F$ and $u \in F$. Let $Q := P \cap K(x)$ and let $P_1, \ldots, P_r \in \mathbb{P}_F$ be the places of F lying over Q, say $P = P_1$. Let $\bar{P}_{ij} \in \mathbb{P}_{\bar{F}}$ be the extensions of P_i in $\bar{F}$ $(1 \leq i \leq r;\ 1 \leq j \leq s_i)$ and $\bar{Q}_1, \ldots, \bar{Q}_s \in \mathbb{P}_{\bar{K}(x)}$ the extensions of Q in $\bar{K}(x)$. There exists an element $z \in F$ such that

$$\begin{aligned}
\delta(x)_P(z-u) &= 0\,, \\
\bar{\delta}(x)_{\bar{P}_{1j}}(z-u) &= 0 \quad \textit{for } 1 \leq j \leq s_1\,, \\
\delta(x)_{P_i}(z) &= 0 \quad \textit{for } 2 \leq i \leq r\,, \\
\bar{\delta}(x)_{\bar{P}_{ij}}(z) &= 0 \quad \textit{for } 2 \leq i \leq r,\ 1 \leq j \leq s_i\,.
\end{aligned} \tag{4.32}$$

This follows from the Approximation Theorem and the fact that the local component $\omega_R(y)$ of a Weil differential ω at a place R vanishes if $v_R(y)$ is sufficiently large. We obtain

$$\begin{aligned}\delta(x)_P(u) &= \delta(x)_P(z) = \sum_{i=1}^{r} \delta(x)_{P_i}(z) && \textit{by } (4.32)\\ &= \eta_Q(\mathrm{Tr}_{F/K(x)}(z)) && \textit{by } (4.21)\\ &= \sum_{l=1}^{s} \bar{\eta}_{\bar{Q}_l}(\mathrm{Tr}_{F/K(x)}(z)) && \textit{by } (4.31).\end{aligned}$$

Noting that $\mathrm{Tr}_{F/K(x)}(z) = \mathrm{Tr}_{\bar{F}/\bar{K}(x)}(z)$ (since $[F : K(x)] = [\bar{F} : \bar{K}(x)]$) and applying (4.21) once again, we get

$$\sum_{l=1}^{s} \bar{\eta}_{\bar{Q}_l}(\mathrm{Tr}_{F/K(x)}(z)) = \sum_{i=1}^{r} \sum_{j=1}^{s_i} \bar{\delta}(x)_{\bar{P}_{ij}}(z) = \sum_{j=1}^{s_1} \bar{\delta}(x)_{\bar{P}_{1j}}(u)\,.$$

Hence

$$\delta(x)_P(u) = \sum_{\bar{P}|P} \bar{\delta}(x)_{\bar{P}}(u) \tag{4.33}$$

for all $u \in F$ (where $\bar{P}$ runs over all places of $\bar{F}/\bar{K}$ lying over P).

Finally we consider an arbitrary adele $\alpha = (\alpha_P)_{P \in \mathbb{P}_F}$ of F/K. Then

$$\begin{aligned}\delta(x)(\alpha) &= \sum_{P \in \mathbb{P}_F} \delta(x)_P(\alpha_P) && \textit{by Proposition } 1.7.2\\ &= \sum_{P \in \mathbb{P}_F} \sum_{\bar{P}|P} \bar{\delta}(x)_{\bar{P}}(\alpha_P) && \textit{by } (4.33)\\ &= \bar{\delta}(x)(\alpha) && \textit{by Proposition } 1.7.2.\end{aligned}$$

□

As before we consider the constant field extension $\bar{F} = F\bar{K}$ of F/K with the algebraic closure $\bar{K}$ of K. Let $(\Delta_{\bar{F}}, \bar{d})$ be the differential module of $\bar{F}/\bar{K}$. There is an F-linear map $\mu : \Delta_F \to \Delta_{\bar{F}}$ given by $\mu(z\,dx) = z\,\bar{d}x$ (cf. Proposition 4.1.8(d)) , and we can regard Δ_F as a submodule of $\Delta_{\bar{F}}$ via this embedding μ. Since $\bar{d}y = dy$ for $y \in F$, we denote the derivation $\bar{d} : \bar{F} \to \Delta_{\bar{F}}$ by d as well (i.e., $\bar{F}/\bar{K}$ has the differential module $(\Delta_{\bar{F}}, d)$).

Proof of Theorem 4.3.2 *for an arbitrary constant field* K. The idea is to reduce the theorem to the case of an algebraically closed constant field by considering the constant field extension $\bar{F} = F\bar{K}$. For $y \in F$ and a separating element $x \in F$ we have

$$\delta(y) = \bar{\delta}(y)\,|_{\mathcal{A}_F} = (\frac{dy}{dx} \cdot \bar{\delta}(x))\,|_{\mathcal{A}_F} = \frac{dy}{dx} \cdot \bar{\delta}(x)\,|_{\mathcal{A}_F} = \frac{dy}{dx} \cdot \delta(x)$$

(here we have used Proposition 4.3.6 and Theorem 4.3.2(b) for $\bar{F}/\bar{K}$). Hence (b) is proved. Parts (a), (c) and (e) follow now precisely as in the case of an algebraically closed constant field.

(d) A place $P \in \mathbb{P}_F$ of degree one has exactly one extension $\bar{P} \in \mathbb{P}_{\bar{F}}$ by Corollary 3.6.5. Thus

$$(z \cdot \delta(x))_P(u) = \delta(x)_P(zu) = \bar{\delta}(x)_{\bar{P}}(zu) = \text{res}_{\bar{P}}(zu\,dx) = \text{res}_P(zu\,dx)\,.$$

(The last equation follows from the observation that the residue of a differential is defined by means of the power series expansion with respect to a prime element; since $\bar{P} \mid P$ is unramified, every prime $t \in F$ is a $\bar{P}$-prime element.) □

Remark 4.3.7. (a) As a consequence of Theorem 4.3.2 we identify the differential module Δ_F with the module Ω_F of Weil differentials of F/K. This means that a differential $\omega = z\,dx \in \Delta_F$ is the same as the Weil differential $\omega = z \cdot \delta(x) \in \Omega_F$ (where $x \in F$ is separating and $z \in F$). In other words,

$$\Delta_F = \Omega_F \quad \textit{and} \quad z\,dx = z \cdot \delta(x)\,. \tag{4.34}$$

(b) If $0 \neq \omega \in \Delta_F$ and t is a prime element at the place $P \in \mathbb{P}_F$, we can write $\omega = z\,dt$ with $z \in F$, and we define

$$v_P(\omega) := v_P(z) \quad \textit{and} \quad (\omega) := \sum_{P \in \mathbb{P}_F} v_P(\omega) P\,. \tag{4.35}$$

Theorem 4.3.2(e) shows that the definition of $v_P(\omega)$ is independent of the choice of the prime element, and it is compatible with the identification (4.34) of Δ_F and Ω_F. Hence (ω) is just the divisor of the corresponding Weil differential ω as defined in Section 1.5.

(c) As an important special case of Theorem 3.4.6 we obtain the following formula for the divisor of a differential $\omega = z\,dx \neq 0$:

$$(z\,dx) = (z) + (dx) = (z) - 2(x)_\infty + \text{Diff}(F/K(x))\,. \tag{4.36}$$

A particular case of this formula is

$$(dx) = -2(x)_\infty + \text{Diff}(F/K(x))\,. \tag{4.37}$$

(d) Once again we consider the constant field extension $\bar{F} = F\bar{K}$ of F/K with the algebraic closure of K. We have identified the differential module Δ_F with a submodule of $\Delta_{\bar{F}}$, hence we obtain a corresponding embedding of Ω_F into $\Omega_{\bar{F}}$ which is given by

$$\omega = z \cdot \delta(x) \longmapsto \bar{\omega} := z \cdot \bar{\delta}(x)$$

(we use the notation of Proposition 4.3.6). By Proposition 4.3.6, ω is the restriction of $\bar{\omega}$ to $\mathcal{A}_F$. This observation yields a formula for the local component

ω_P of ω at a place P of degree $f \geq 1$: by Corollary 3.6.5 there are exactly $f = \deg P$ places $\bar{P}_1, \ldots, \bar{P}_f \in \mathbb{P}_{\bar{F}}$ lying over P, hence Theorem 4.3.2(d) implies

$$\omega_P(u) = \sum_{i=1}^{f} \operatorname{res}_{\bar{P}_i}(u\omega) \tag{4.38}$$

for $\omega \in \Omega_F$ and $u \in F$. In particular we have

$$\omega_P(u) = \operatorname{res}_P(u\omega)\,, \tag{4.39}$$

if $\deg P = 1$.

4.4 Exercises

In all exercises below we assume that K is a perfect field and F/K is a function field with full constant field K.

4.1. Let E/F be a finite extension field of F. Let $P \in \mathbb{P}_F$ and $Q \in \mathbb{P}_E$ such that Q is an extension of P.

(i) Show that the completion $\hat{F}_P$ of F at P can be considered in a natural way as a subfield of $\hat{E}_Q$.

(ii) Show that $\hat{F}_P = \hat{E}_Q$ if and only if $e(Q|P) = f(Q|P) = 1$.

4.2. Let P be a place of F/K of degree one, and let $t \in F$ be a P-prime element. Determine the P-adic power series expansions with respect to t for the following elements of F:

$$\frac{1}{1-t^r} \quad \textit{and} \quad \frac{d}{dt}\Big(\frac{1}{1-t^r}\Big), \quad \textit{with } 0 \neq r \in \mathbb{Z},$$

$$\frac{1+t}{1-t} \quad \textit{and} \quad \frac{1-t}{1+t}.$$

4.3. ($\operatorname{char} K \neq 2$) Consider the function field $F = K(x,y)$ with defining equation $y^2 = x^3 + x$ (see Example 3.7.6). The element x has a unique pole in F, call it P, and then $t := x/y$ is a P-prime element. Determine the P-adic power series of the elements x and y with respect to t.

4.4. Let $x \in F$ be a separating element of F/K. Show that there exists a unique derivation $\delta : F \to F$ of F/K such that $\delta(x) = x$.

4.5. Let $x, y \in F$ and $c \in K$. Show that the following conditions are equivalent:

(a) $dy = c\,dx$.

(b) $y = cx + z$, where $z \in F$ is a non-separating element of F/K (i.e., $z \in K$ if the characteristic of K is 0, and $z = u^p$ with some element $u \in F$ if $\operatorname{char} K = p > 0$).

4.6. (char $K \neq 2$) Consider the function field $F = K(x,y)$, given by the equation

$$y^2 = \prod_{i=1}^{2m+1} (x - a_i),$$

where $m \geq 0$ and $a_1, \ldots, a_{2m+1} \in K$ are distinct elements of K.

(i) Determine the divisor of the differential dx/y and obtain a new proof that the genus of F/K is $g = m$ (cf. Example 3.7.6).

(ii) Show that the differentials $x^i\, dx/y$ with $0 \leq i \leq m-1$ are a basis of the space $\Omega_F(0)$ (the space of regular differentials of F/K).

4.7. Let x be a separating element of F/K. Show that

$$\mathrm{res}_P\left(\frac{dx}{x}\right) = v_P(x),$$

for every place P of degree one.

As pointed out in Remark 4.3.7, we identify the differential module Δ_F and the module of Weil differentials Ω_F. Therefore it makes sense to define for a divisor $A \in \mathrm{Div}(F)$ the K-vector space

$$\Delta_F(A) := \{\omega \in \Delta_F \,|\, \omega = 0 \text{ or } (\omega) \geq A\},$$

which corresponds to $\Omega_F(A)$ under the identification of Δ_F and Ω_F. We define further:

(1) A differential $\omega \in \Delta_F$ is called regular (or holomorphic, or a differential of the first kind), if $\omega = 0$ or $(\omega) \geq 0$.

(2) A differential $\omega \in \Delta_F$ is called exact, if $\omega = dx$ for some element $x \in F$.

(3) A differential $\omega \in \Delta_F$ is called residue-free, if $\mathrm{res}_P(\omega) = 0$ for all rational places $P \in \mathbb{P}_F$.

(4) A differential $\omega \in \Delta_F$ is called a differential of the second kind, if for all places $P \in \mathbb{P}_F$ there exists an element $u \in F$ (depending on P) such that $v_P(\omega - du) \geq 0$.

Then we consider the following subsets of Δ_F:

$$\begin{aligned}
\Delta_F^{(1)} &:= \Delta_F(0) = \{\omega \in \Delta_F \,|\, \omega \text{ is regular}\} \\
\Delta_F^{(\mathrm{ex})} &:= \{\omega \in \Delta_F \,|\, \omega \text{ is exact}\} \\
\Delta_F^{(\mathrm{rf})} &:= \{\omega \in \Delta_F \,|\, \omega \text{ is residue-free}\} \\
\Delta_F^{(2)} &:= \{\omega \in \Delta_F \,|\, \omega \text{ is of the second kind}\}
\end{aligned}$$

4.8. Prove:

(i) $\Delta_F^{(1)}$, $\Delta_F^{(\mathrm{ex})}$, $\Delta_F^{(\mathrm{rf})}$ and $\Delta_F^{(2)}$ are K-subspaces of Δ_F.

(ii) $\Delta_F^{(\mathrm{ex})} \subseteq \Delta_F^{(2)} \subseteq \Delta_F^{(\mathrm{rf})}$.

(iii) If $\operatorname{char} K = 0$ and K is algebraically closed, then $\Delta_F^{(2)} = \Delta_F^{(\mathrm{rf})}$.

4.9. (The result of this exercise will be used in Exercise 4.10.) Assume that F/K has genus $g > 0$. Then there exists a non-special divisor B of degree g such that $B = P_1 + \ldots + P_g$, with pairwise distinct places $P_1, \ldots, P_g$.

4.10. (K algebraically closed and $\operatorname{char} K = 0$) The aim of this exercise is to show that

$$\dim_K(\Delta_F^{(2)} \,/\, \Delta_F^{(\mathrm{ex})}) = 2g\,,$$

if F/K is a function field of genus g over an algebraically closed field K with $\operatorname{char} K = 0$.

(i) Prove the claim in the case $g = 0$.

Hence we will assume now that $g > 0$, and we fix a non-special divisor $B = P_1 + \ldots + P_g$ with distinct places $P_1, \ldots, P_g$.

(ii) Show that for every $\omega \in \Delta_F^{(2)}$ there exists a unique differential $\omega^* \in \Omega_F(-2B)$ such that $\omega - \omega^*$ is exact.

(iii) Show that the map $f : \omega \mapsto \omega^*$, where ω^* is defined as in (ii), is a K-linear surjective map from $\Delta_F^{(2)}$ onto $\Delta_F^{(2)} \cap \Delta_F(-2B)$, with kernel $\Delta_F^{(\mathrm{ex})}$. Conclude that

$$\dim_K(\Delta_F^{(2)} \,/\, \Delta_F^{(\mathrm{ex})}) = \dim_K(\Delta_F^{(2)} \cap \Delta_F(-2B))\,.$$

(iv) For $i = 1, \ldots, g$ we fix a P_i-prime element $t_i \in F$. Then every $\omega \in \Delta_F^{(2)} \cap \Delta_F(-2B)$ has a P_i-adic expansion

$$\omega = \Big(a_{-2}^{(i)} t_i^{-2} + a_0^{(i)} + \sum_{j \geq 1} a_j^{(i)} t_i^j\Big)\, dt_i$$

with $a_j^{(i)} \in K$, for $1 \leq i \leq g$. Show that the map

$$\omega \mapsto (a_{-2}^{(i)}, a_0^{(i)})_{i=1,\ldots,g}$$

defines an isomorphism of $\Delta_F^{(2)} \cap \Delta_F(-2B)$ onto K^{2g}.

(v) Conclude that $\dim_K(\Delta_F^{(2)} \,/\, \Delta_F^{(\mathrm{ex})}) = 2g$.

4.11. ($\operatorname{char} K = p > 0$) Recall that there is a unique subfield $M \subseteq F$ such that F/M is purely inseparable of degree $[F : M] = p$, namely $M = F^p$. In what follows, we fix a separating element x of F/K.

(i) Show that every element $z \in F$ has a unique representation of the form

$$z = \sum_{i=0}^{p-1} u_i^p x^i\,, \quad \textit{with} \;\; u_0, \ldots, u_{p-1} \in F\,.$$

(ii) If $z \in F$ is given as above, then

$$\frac{dz}{dx} = \sum_{i=1}^{p-1} i\, u_i^p x^{i-1}\,.$$

(iii) For a derivation η of F/K and $n \geq 1$ we define $\eta^n := \eta \circ \eta \circ \ldots \circ \eta$ (n times). Show that $\delta_x^p = 0$.

(iv) Is $\delta^p = 0$ for every derivation $\delta \in \mathrm{Der}_F$?

4.12. (char $K = p > 0$) Fix a separating element $x \in F$; then every differential $\omega \in \Delta_F$ has a unique representation of the form

$$\omega = \left(u_0^p + u_1^p x + \ldots + u_{p-1}^p x^{p-1}\right) dx$$

with $u_0, \ldots, u_{p-1} \in F$, cf. Exercise 4.11. We define the map $C : \Delta_F \to \Delta_F$ by

$$C(\omega) := u_{p-1}\, dx\,.$$

This map C is called the Cartier operator. Its definition seems to depend on the element x. We will show in Exercise 4.13 that it is actually independent of the choice of x.

Prove the following properties of C, for all $\omega, \omega_1, \omega_2 \in \Delta_F$ and all $z \in F$:

(i) $C(\omega_1 + \omega_2) = C(\omega_1) + C(\omega_2)$.

(ii) $C(z^p\omega) = z\, C(\omega)$.

(iii) $C : \Delta_F \to \Delta_F$ is surjective.

(iv) $C(\omega) = 0$ if and only if ω is exact.

(v) If $0 \neq z \in F$, then $C(dz/z) = dz/z$.

Hint. The proofs of (i) - (iv) are easy. In order to prove (v), one may proceed as follows. Observe that $C(dz/z) = dz/z$ if and only if $C(z^{p-1}dz) = dz$, by (ii). Therefore consider the set

$$M := \{z \in F \,|\, C(z^{p-1}dz) = dz\}$$

and show:

(a) $F^p \subseteq M$ and $x \in M$.

(b) If $0 \neq z \in M$, then $z^{-1} \in M$.

(c) If $z \in M$, then $z + 1 \in M$.

(d) If $z_1, z_2 \in M$, then $z_1 z_2 \in M$ and $z_1 + z_2 \in M$.

(e) Conclude that $M = F$, which finishes the proof of (v).

4.13. (char $K = p > 0$) Suppose that $C^* : \Delta_F \to \Delta_F$ is a map with the following properties: for all $\omega, \omega_1, \omega_2 \in \Delta_F$ and all $z \in F$,

(i) $C^*(\omega_1 + \omega_2) = C^*(\omega_1) + C^*(\omega_2)$.

(ii) $C^*(z^p \omega) = z\, C^*(\omega)$.

(iv) $C^*(dz) = 0$.

(v) If $0 \neq z \in F$, then $C^*(dz/z) = dz/z$.

Show that $C^* = C$, where C is defined as in the previous exercise. This implies in particular, that the definition of the Cartier operator does not depend on the choice of the separating element x.

4.14. (char $K = p > 0$) Prove the following properties of the Cartier divisor. For all $\omega \in \Delta_F$ and all places $P \in \mathbb{P}_F$,

(i) $v_P(\omega) \geq 0 \Longrightarrow v_P(C(\omega)) \geq 0$.

(ii) $v_p(\omega) = -1 \Longrightarrow v_P(C(\omega)) = -1$.

(iii) $v_P(\omega) < -1 \Longrightarrow v_P(C(\omega)) > v_P(\omega)$.

(iv) $C(\Delta_F^{(1)}) \subseteq \Delta_F^{(1)}$.

(v) $\mathrm{res}_P(\omega) = (\mathrm{res}_P(C(\omega))^p$, if P is a place of F/K of degree one.

4.15. (K algebraically closed and char $K = p > 0$)

(i) Show that $\dim_K(\Delta_F^{(2)} / \Delta_F^{(\mathrm{ex})}) = g$.

(ii) Show that $\dim_K(\Delta_F^{(\mathrm{rf})} / \Delta_F^{(\mathrm{ex})}) = \infty$.

Compare these results with Exercise 4.10.

Hint. Use the fact that $\Delta_F^{(\mathrm{ex})}$ is the kernel of the Cartier operator.

4.16. (char $K = p > 0$) A differential $\omega \in \Omega_F$ is called logarithmic, if $\omega = dx/x$ for some element $0 \neq x \in F$. We define

$$\Lambda := \{\omega \in \Omega_F \mid (\omega) \geq 0 \;\; \textit{and } \omega \textit{ is logarithmic}\}\,.$$

Show:

(i) Λ is an additive subgroup of $\Delta_F(0)$, i.e., we can consider Λ as a vector space over the prime field $\mathbb{F}_p$.

(ii) If $\omega_1, \ldots, \omega_m \in \Lambda$ are linearly independent over $\mathbb{F}_p$, then they are linearly independent over K.

(iii) Λ is a finite group of order p^s, with $0 \leq s \leq g$ (as usual, g denotes the genus of F/K).

4.17. (char $K = p > 0$) Show that a differential $\omega \in \Omega_F$ is logarithmic if and only if $C(\omega) = \omega$.

4.18. ($\operatorname{char} K = p > 0$) Consider the divisor class group $\mathrm{Cl}(F)$ of F/K and define the subgroup

$$\mathrm{Cl}(F)(p) := \{[A] \in \mathrm{Cl}(F) \,|\, p\,[A] = 0\}\,.$$

We want to define a map $f : \mathrm{Cl}(F)(p) \to \Delta_F$. For $[A] \in \mathrm{Cl}(F)(p)$ there exists an element $0 \neq x \in F$ such that $pA = (x)$, by definition of $\mathrm{Cl}(F)(p)$. Then we set $f([A]) := dx/x$.

Prove:

(i) f is a well-defined group homomorphism from $\mathrm{Cl}(F)(p)$ to Λ (with Λ as in the previous exercise).

(ii) f is an isomorphism from $\mathrm{Cl}(F)(p)$ onto Λ. Conclude that $\mathrm{Cl}(F)(p)$ is a finite group of order p^s with $0 \leq s \leq g$.

Remark. If K is algebraically closed, the number s above is called the p-rank (or the Hasse-Witt rank) of F/K. The function field is called regular (resp. singular) if $s = g$ (resp. $s < g$); it is called supersingular if $s = 0$.

5

Algebraic Function Fields over Finite Constant Fields

In the previous chapters we developed the theory of algebraic function fields over an arbitrary perfect constant field K. We would now like to consider in greater detail the case of a finite constant field. Observe that a finite field is perfect, so that all results from Chapters 3 and 4 apply. We will mainly be interested in the places of degree one of a function field over a finite field. Their number is finite and can be estimated by the Hasse-Weil Bound (see Theorem 5.2.3). This bound has many number-theoretical implications, and it plays a crucial role in the applications of algebraic function fields to coding theory, cf. Chapter 8 and 9.

Throughout this chapter, F denotes an algebraic function field of genus g whose constant field is the finite field $\mathbb{F}_q$.

5.1 The Zeta Function of a Function Field

As in Chapter 1, $\mathrm{Div}(F)$ denotes the divisor group of the function field $F/\mathbb{F}_q$. A divisor $A = \sum_{P \in \mathbb{P}_F} a_P P$ is *positive* if all $a_P \geq 0$; we write $A \geq 0$.

Lemma 5.1.1. *For every $n \geq 0$ there exist only finitely many positive divisors of degree n.*

Proof. A positive divisor is a sum of prime divisors. Hence it is sufficient to prove that the set $S := \{P \in \mathbb{P}_F \mid \deg P \leq n\}$ is finite. We choose an element $x \in F \backslash \mathbb{F}_q$ and consider the set $S_0 := \{P_0 \in \mathbb{P}_{\mathbb{F}_q(x)} \mid \deg P_0 \leq n\}$. Obviously $P \cap \mathbb{F}_q(x) \in S_0$ for all $P \in S$, and each $P_0 \in S_0$ has only finitely many extensions in F. Therefore we have only to show that S_0 is finite. Since the places of $\mathbb{F}_q(x)$ (except the pole of x) correspond to irreducible monic polynomials $p(x) \in \mathbb{F}_q[x]$ of the same degree (cf. Section 1.2), the finiteness of S_0 follows readily. $\square$

H. Stichtenoth, *Algebraic Function Fields and Codes*,
Graduate Texts in Mathematics 254,

Let us recall some notation from previous chapters: $\mathrm{Princ}(F)$ denotes the subgroup of $\mathrm{Div}(F)$ consisting of all principal divisors $(x) = \sum_{P \in \mathbb{P}_F} v_P(x) \cdot P$ (with $0 \neq x \in F$). The factor group $\mathrm{Cl}(F) = \mathrm{Div}(F)/\mathrm{Princ}(F)$ is called the *divisor class group* of $F/\mathbb{F}_q$. Two divisors $A, B \in \mathrm{Div}(F)$ are equivalent (written $A \sim B$) if $B = A + (x)$ for some principal divisor $(x) \in \mathrm{Princ}(F)$. The class of A in the divisor class group $\mathrm{Cl}(F)$ is denoted by $[A]$, hence $A \sim B \iff A \in [B] \iff [A] = [B]$. Equivalent divisors have the same degree and the same dimension, so the integers

$$\deg [A] := \deg A \quad \textit{and} \quad \ell([A]) := \ell(A)$$

are well-defined for a divisor class $[A] \in \mathrm{Cl}(F)$.

Definition 5.1.2. *The set*

$$\mathrm{Div}^0(F) := \{A \in \mathrm{Div}(F) \mid \deg A = 0\}\,,$$

which is obviously a subgroup of $\mathrm{Div}(F)$*, is called the group of divisors of degree zero, and*

$$\mathrm{Cl}^0(F) := \{[A] \in \mathrm{Cl}(F) \mid \deg [A] = 0\}$$

is called the group of divisor classes of degree zero.

Proposition 5.1.3. $\mathrm{Cl}^0(F)$ *is a finite group. Its order* $h = h_F$ *is called the class number of* $F/\mathbb{F}_q$*; i.e.,*

$$h := h_F := \mathrm{ord}\, \mathrm{Cl}^0(F)\,.$$

Proof. Choose a divisor $B \in \mathrm{Div}(F)$ of degree $\geq g$, say $n := \deg B$, and consider the set of divisor classes

$$\mathrm{Cl}^n(F) := \{[C] \in \mathrm{Cl}(F) \mid \deg [C] = n\}\,.$$

The map

$$\begin{cases} \mathrm{Cl}^0(F) & \longrightarrow & \mathrm{Cl}^n(F)\,, \\ [A] & \longmapsto & [A+B] \end{cases}$$

is bijective (this is trivial), so we only have to verify that $\mathrm{Cl}^n(F)$ is finite. We claim:

$$\begin{array}{l} \textit{for each divisor class } [C] \in \mathrm{Cl}^n(F) \textit{ there} \\ \textit{exists a divisor } A \in [C] \textit{ with } A \geq 0\,. \end{array} \tag{5.1}$$

In fact, since $\deg C = n \geq g$, we have

$$\ell([C]) \geq n + 1 - g \geq 1 \tag{5.2}$$

by the Riemann-Roch Theorem, and (5.1) follows from (5.2) and Remark 1.4.5(b). There are only finitely many divisors $A \geq 0$ of degree n (by Lemma 5.1.1), so (5.1) implies the finiteness of $\mathrm{Cl}^n(F)$. □

We define the integer $\partial > 0$ by

$$\partial := \min\{\deg A \mid A \in \mathrm{Div}(F) \textit{ and } \deg A > 0\}. \tag{5.3}$$

The image of the degree mapping $\deg : \mathrm{Div}(F) \to \mathbb{Z}$ is the subgroup of $\mathbb{Z}$ generated by ∂, and the degree of each divisor of $F/\mathbb{F}_q$ is a multiple of ∂.

In what follows, we would like to study the numbers

$$A_n := \big| \{A \in \mathrm{Div}(F) \mid A \geq 0 \textit{ and } \deg A = n\} \big|. \tag{5.4}$$

For instance, $A_0 = 1$, and A_1 is the number of places $P \in \mathbb{P}_F$ of degree one.

Lemma 5.1.4. *(a) $A_n = 0$ if $\partial \nmid n$.*
(b) For a fixed divisor class $[C] \in \mathrm{Cl}(F)$ we have

$$\big| \{A \in [C] \mid A \geq 0\} \big| = \frac{1}{q-1}\big(q^{\ell([C])} - 1\big).$$

(c) For each integer $n > 2g - 2$ with $\partial | n$ we have

$$A_n = \frac{h}{q-1}\big(q^{n+1-g} - 1\big).$$

Proof. (a) is trivial.
(b) The conditions $A \in [C]$ and $A \geq 0$ are equivalent to

$$A = (x) + C \quad \textit{for some} \quad x \in F \quad \textit{with} \quad (x) \geq -C;$$

i.e., $x \in \mathscr{L}(C)\backslash\{0\}$. There exist exactly $q^{\ell([C])} - 1$ elements $x \in \mathscr{L}(C)\backslash\{0\}$, and two of them yield the same divisor if and only if they differ by a constant factor $0 \neq \alpha \in \mathbb{F}_q$. This proves (b).
(c) There are $h = h_F$ divisor classes of degree n, say $[C_1], \ldots, [C_h]$. By (b) and the Riemann-Roch Theorem,

$$\big| \{A \in [C_j] | A \geq 0\} \big| = \frac{1}{q-1}\big(q^{\ell(C_j)} - 1\big) = \frac{1}{q-1}\big(q^{n+1-g} - 1\big).$$

Each divisor of degree n lies in exactly one of the divisor classes $[C_1], \ldots, [C_h]$, hence

$$A_n = \sum_{j=1}^{h} \big| \{A \in [C_j]\,;\, A \geq 0\} \big| = \frac{h}{q-1}\big(q^{n+1-g} - 1\big).$$

□

Definition 5.1.5. *The power series*

$$Z(t) := Z_F(t) := \sum_{n=0}^{\infty} A_n t^n \in \mathbb{C}[[t]]$$

(where A_n is defined by (5.4)) is called the Zeta function of $F/\mathbb{F}_q$.

Observe that we regard t here as a complex variable, and $Z(t)$ is a power series over the field of complex numbers (rather than the P-adic power series that we considered in Chapter 4). We shall show now that this power series converges in a neighbourhood of 0.

Proposition 5.1.6. *The power series $Z(t) = \sum_{n=0}^{\infty} A_n t^n$ is convergent for $|t| < q^{-1}$. More precisely, we have for $|t| < q^{-1}$:*
(a) If $F/\mathbb{F}_q$ has genus $g = 0$ then

$$Z(t) = \frac{1}{q-1}\left(\frac{q}{1-(qt)^{\partial}} - \frac{1}{1-t^{\partial}}\right).$$

(b) If $g \geq 1$, then $Z(t) = F(t) + G(t)$ with

$$F(t) = \frac{1}{q-1} \sum_{0 \leq \deg[C] \leq 2g-2} q^{\ell([C])} \cdot t^{\deg[C]},$$

(where $[C]$ runs over all divisor classes $[C] \in \mathrm{Cl}(F)$ with $0 \leq \deg[C] \leq 2g-2$) and

$$G(t) = \frac{h}{q-1}\left(q^{1-g}(qt)^{2g-2+\partial}\frac{1}{1-(qt)^{\partial}} - \frac{1}{1-t^{\partial}}\right).$$

Proof. (a) $g = 0$. To begin with, we show that a function field of genus zero has class number $h = 1$; i.e., every divisor A of degree 0 is principal. This fact follows easily from the Riemann-Roch Theorem: as $0 > 2g-2$, we have $\ell(A) = \deg A + 1 - g = 1$, and we can therefore find an element $x \neq 0$ with $(x) \geq -A$. Both divisors are of degree 0, hence $A = -(x) = (x^{-1})$ is principal. Now we apply Lemma 5.1.4 and obtain

$$\begin{aligned}
\sum_{n=0}^{\infty} A_n t^n &= \sum_{n=0}^{\infty} A_{\partial n} t^{\partial n} \\
&= \sum_{n=0}^{\infty} \frac{1}{q-1}(q^{\partial n+1} - 1)t^{\partial n} \\
&= \frac{1}{q-1}\left(q\sum_{n=0}^{\infty}(qt)^{\partial n} - \sum_{n=0}^{\infty} t^{\partial n}\right) \\
&= \frac{1}{q-1}\left(\frac{q}{1-(qt)^{\partial}} - \frac{1}{1-t^{\partial}}\right)
\end{aligned}$$

for $|qt| < 1$.
(b) For $g \geq 1$ the calculation is quite similar. We obtain

$$\sum_{n=0}^{\infty} A_n t^n = \sum_{\deg[C]\geq 0} \left|\{A \in [C]\,;\, A \geq 0\}\right| \cdot t^{\deg[C]} = \sum_{\deg[C]\geq 0} \frac{q^{\ell([C])}-1}{q-1} \cdot t^{\deg[C]}$$

$$= \frac{1}{q-1} \sum_{0\leq \deg[C]\leq 2g-2} q^{\ell([C])} \cdot t^{\deg[C]} + \frac{1}{q-1} \sum_{\deg[C]>2g-2} q^{\deg[C]+1-g} \cdot t^{\deg[C]}$$

$$-\frac{1}{q-1} \sum_{\deg[C]\geq 0} t^{\deg[C]} = F(t) + G(t)\,,$$

with

$$F(t) = \frac{1}{q-1} \sum_{0\leq \deg[C]\leq 2g-2} q^{\ell([C])} \cdot t^{\deg[C]}$$

and

$$\begin{aligned}(q-1)G(t) &= \sum_{n=((2g-2)/\partial)+1}^{\infty} hq^{n\partial+1-g} \cdot t^{n\partial} - \sum_{n=0}^{\infty} ht^{n\partial} \\ &= hq^{1-g}(qt)^{2g-2+\partial}\frac{1}{1-(qt)^{\partial}} - h\frac{1}{1-t^{\partial}}\,.\end{aligned}$$

□

Corollary 5.1.7. *$Z(t)$ can be extended to a rational function on $\mathbb{C}$; it has a simple pole at $t = 1$.*

Proof. Obvious, since $1/(1-t^{\partial})$ has a simple pole at $t = 1$. □

In order to study the behavior of the Zeta function of $F/\mathbb{F}_q$ under finite constant field extensions, it is convenient to have a second representation of $Z(t)$ as an infinite product. Recall that an infinite product $\prod_{i=1}^{\infty}(1+a_i)$ (with complex numbers $a_i \neq -1$) is said to be *convergent* with limit $a \in \mathbb{C}$ if $\lim_{n\to\infty}\prod_{i=1}^{n}(1+a_i) = a \neq 0$. The product is called *absolutely convergent* if $\sum_{i=1}^{\infty}|a_i| < \infty$. From analysis it is well-known that absolute convergence implies convergence of the product, and the limit of an absolutely convergent product is independent of the order of the factors. Moreover, if the product $\prod_{i=1}^{\infty}(1+a_i) = a$ is absolutely convergent, then $\prod_{i=1}^{\infty}(1+a_i)^{-1}$ converges absolutely, too, and $\prod_{i=1}^{\infty}(1+a_i)^{-1} = a^{-1}$.

Proposition 5.1.8 (Euler Product). *For $|t| < q^{-1}$ the Zeta function can be represented as an absolutely convergent product*

$$Z(t) = \prod_{P\in\mathbb{P}_F} (1-t^{\deg P})^{-1}\,. \tag{5.5}$$

In particular $Z(t) \neq 0$ for $|t| < q^{-1}$.

Proof. The right hand side of (5.5) converges absolutely for $|t| < q^{-1}$, since $\sum_{P \in \mathbb{P}_F} |t|^{\deg P} \leq \sum_{n=0}^{\infty} A_n |t|^n < \infty$ by Proposition 5.1.6. Each factor of (5.5) can be written as a geometric series, and we obtain

$$\begin{aligned} \prod_{P \in \mathbb{P}_F} (1 - t^{\deg P})^{-1} &= \prod_{P \in \mathbb{P}_F} \sum_{n=0}^{\infty} t^{\deg (nP)} \\ &= \sum_{A \in \mathrm{Div}(F); A \geq 0} t^{\deg A} = \sum_{n=0}^{\infty} A_n t^n = Z(t) . \end{aligned}$$

□

In the following we fix an algebraic closure $\bar{\mathbb{F}}_q$ of $\mathbb{F}_q$ and consider the constant field extension $\bar{F} = F\bar{\mathbb{F}}_q$ of $F/\mathbb{F}_q$. For each $r \geq 1$ there exists exactly one extension $\mathbb{F}_{q^r}/\mathbb{F}_q$ of degree r with $\mathbb{F}_{q^r} \subseteq \bar{\mathbb{F}}_q$, and we set

$$F_r := F\mathbb{F}_{q^r} \subseteq \bar{F} .$$

Lemma 5.1.9. *(a) F_r/F is a cyclic extension of degree r (i.e., F_r/F is Galois with a cyclic Galois group of order r). The Galois group* $\mathrm{Gal}(F_r/F)$ *is generated by the Frobenius automorphism σ which acts on $\mathbb{F}_{q^r}$ by $\sigma(\alpha) = \alpha^q$.*

(b) $\mathbb{F}_{q^r}$ is the full constant field of F_r.

(c) $F_r/\mathbb{F}_{q^r}$ has the same genus as $F/\mathbb{F}_q$.

(d) Let $P \in \mathbb{P}_F$ be a place of degree m. Then $\mathrm{Con}_{F_r/F}(P) = P_1 + \ldots + P_d$ *with $d :=$* $\gcd(m, r)$ *pairwise distinct places $P_i \in \mathbb{P}_{F_r}$ and* $\deg P_i = m/d$.

Proof. (a) It is well-known that $\mathbb{F}_{q^r}/\mathbb{F}_q$ is cyclic of degree r, and its Galois group is generated by the Frobenius map $\alpha \mapsto \alpha^q$. Since $[F_r : F] = [\mathbb{F}_{q^r} : \mathbb{F}_q]$ by Lemma 3.6.2, assertion (a) follows immediately.

(b) and (c) See Proposition 3.6.1 and Theorem 3.6.3.

(d) P is unramified in F_r/F, cf. Theorem 3.6.3. Consider some place $P' \in \mathbb{P}_{F_r}$ lying over P. The residue class field of P' is the compositum of $\mathbb{F}_{q^r}$ with the residue class field F_P of P, by Theorem 3.6.3(g). Set $l := \mathrm{lcm}(m, r)$. As $F_P = \mathbb{F}_{q^m}$, this compositum is

$$\mathbb{F}_{q^m} \cdot \mathbb{F}_{q^r} = \mathbb{F}_{q^l} .$$

Therefore

$$\deg P' = [\mathbb{F}_{q^l} : \mathbb{F}_{q^r}] = m/d .$$

Since $\deg(\mathrm{Con}_{F_r/F}(P)) = \deg P = m$ (cf. Theorem 3.6.3(c)), we conclude that $\mathrm{Con}_{F_r/F}(P) = P_1 + \ldots + P_d$ with places P_i of degree m/d. □

In the proof of our next proposition we need a simple polynomial identity: if $m \geq 1$ and $r \geq 1$ are integers and $d = \gcd(m, r)$ then

$$(X^{r/d} - 1)^d = \prod_{\zeta^r=1} (X - \zeta^m), \tag{5.6}$$

where ζ runs over all r-th roots of unity in $\mathbb{C}$. In fact, both sides of (5.6) are monic polynomials of the same degree, and each (r/d)-th root of unity is a d-fold root of them. Hence the polynomials are equal. We substitute $X = t^{-m}$ in (5.6), multiply by t^{mr} and obtain

$$(1 - t^{mr/d})^d = \prod_{\zeta^r=1} (1 - (\zeta t)^m). \tag{5.7}$$

Proposition 5.1.10. *Let $Z(t)$ (resp. $Z_r(t)$) denote the Zeta function of F (resp. of $F_r = F\mathbb{F}_{q^r}$). Then*

$$Z_r(t^r) = \prod_{\zeta^r=1} Z(\zeta t) \tag{5.8}$$

for all $t \in \mathbb{C}$ (ζ runs over the r-th roots of unity).

Proof. It is sufficient to prove (5.8) for $|t| < q^{-1}$. In this region the Euler product representation yields

$$Z_r(t^r) = \prod_{P \in \mathbb{P}_F} \prod_{P'|P} (1 - t^{r \cdot \deg P'})^{-1}. \tag{5.9}$$

For a fixed place $P \in \mathbb{P}_F$ we set $m := \deg P$ and $d := \gcd(r, m)$; then

$$\begin{aligned} \prod_{P'|P} (1 - t^{r \cdot \deg P'}) &= (1 - t^{rm/d})^d \\ &= \prod_{\zeta^r=1} (1 - (\zeta t)^m) = \prod_{\zeta^r=1} (1 - (\zeta t)^{\deg P}), \end{aligned}$$

by (5.7) and Lemma 5.1.9. Now we obtain from (5.9)

$$Z_r(t^r) = \prod_{\zeta^r=1} \prod_{P \in \mathbb{P}_F} (1 - (\zeta t)^{\deg P})^{-1} = \prod_{\zeta^r=1} Z(\zeta t).$$

□

Corollary 5.1.11 (F.K. Schmidt). $\partial = 1$.

Proof. For $\zeta^\partial = 1$ we have

$$Z(\zeta t) = \prod_{P \in \mathbb{P}_F} (1 - (\zeta t)^{\deg P})^{-1} = \prod_{P \in \mathbb{P}_F} (1 - t^{\deg P})^{-1} = Z(t),$$

since ∂ divides the degree of P for every $P \in \mathbb{P}_F$. Therefore $Z_\partial(t^\partial) = Z(t)^\partial$ by Proposition 5.1.10. The rational function $Z_\partial(t^\partial)$ has a simple pole at $t = 1$, by Corollary 5.1.7, and $Z(t)^\partial$ has a pole of order ∂ at $t = 1$. Hence $\partial = 1$. □

Corollary 5.1.12. *(a) Every function field $F/\mathbb{F}_q$ of genus 0 is rational, and its Zeta function is*

$$Z(t) = \frac{1}{(1-t)(1-qt)}\,.$$

(b) If $F/\mathbb{F}_q$ has genus $g \geq 1$, its Zeta function can be written in the form $Z(t) = F(t) + G(t)$ with

$$F(t) = \frac{1}{q-1} \sum_{0 \leq \deg[C] \leq 2g-2} q^{\ell([C])} \cdot t^{\deg[C]}$$

and

$$G(t) = \frac{h}{q-1}\left(q^g t^{2g-1} \frac{1}{1-qt} - \frac{1}{1-t}\right).$$

Proof. A function field of genus 0 having a divisor of degree 1 is rational, cf. Proposition 1.6.3. The remaining assertions follow from Proposition 5.1.6 and $\partial = 1$. □

Proposition 5.1.13 (Functional Equation of the Zeta Function). *The Zeta function of $F/\mathbb{F}_q$ satisfies the functional equation*

$$Z(t) = q^{g-1} t^{2g-2} Z\left(\frac{1}{qt}\right).$$

Proof. For $g = 0$ this is obvious from Corollary 5.1.12(a). For $g \geq 1$ we write $Z(t) = F(t) + G(t)$ as in Corollary 5.1.12(b). Let W be a canonical divisor of F; then

$$\begin{aligned}
(q-1)F(t) &= \sum_{0 \leq \deg[C] \leq 2g-2} q^{\ell([C])} \cdot t^{\deg[C]} \\
&= \sum_{0 \leq \deg[C] \leq 2g-2} q^{\deg[C]+1-g+\ell([W-C])} \cdot t^{\deg[C]} \\
&= q^{g-1} t^{2g-2} \sum_{0 \leq \deg[C] \leq 2g-2} q^{\deg[C]-(2g-2)+\ell([W-C])} \cdot t^{\deg[C]-(2g-2)} \\
&= q^{g-1} t^{2g-2} \sum_{0 \leq \deg[C] \leq 2g-2} q^{\ell([W-C])} \cdot \left(\frac{1}{qt}\right)^{\deg[W-C]} \\
&= q^{g-1} t^{2g-2} (q-1) F\left(\frac{1}{qt}\right).
\end{aligned} \tag{5.10}$$

We have used that $\deg[W] = 2g-2$ and, if $[C]$ runs over all divisor classes with $0 \leq \deg[C] \leq 2g-2$, so does $[W-C]$. For the function $G(t)$ we obtain

$$q^{g-1}t^{2g-2}G(\frac{1}{qt}) = \frac{h}{q-1}q^{g-1}t^{2g-2}\left(q^g(\frac{1}{qt})^{2g-1}\frac{1}{1-q\frac{1}{qt}} - \frac{1}{1-\frac{1}{qt}}\right)$$

$$= \frac{h}{q-1}\left(\frac{1}{t}\frac{1}{1-\frac{1}{t}} - \frac{q^g t^{2g-1}}{qt(1-\frac{1}{qt})}\right) = G(t)\,. \qquad (5.11)$$

Adding (5.10) and (5.11) yields the functional equation for $Z(t)$. □

Definition 5.1.14. *The polynomial $L(t) := L_F(t) := (1-t)(1-qt)Z(t)$ is called the L-polynomial of $F/\mathbb{F}_q$.*

By Corollary 5.1.12 it is obvious that $L(t)$ is a polynomial of degree $\le 2g$. Observe that $L(t)$ contains all information about the numbers $A_n (n \ge 0)$ since

$$L(t) = (1-t)(1-qt)\sum_{n=0}^{\infty} A_n t^n\,. \qquad (5.12)$$

Theorem 5.1.15. *(a) $L(t) \in \mathbb{Z}[t]$ and $\deg L(t) = 2g$.*

(b) (Functional Equation) $L(t) = q^g t^{2g} L(1/qt)$.

(c) $L(1) = h$, the class number of $F/\mathbb{F}_q$.

(d) We write $L(t) = a_0 + a_1 t + \cdots + a_{2g}t^{2g}$. Then the following hold:

(1) $a_0 = 1$ and $a_{2g} = q^g$.

(2) $a_{2g-i} = q^{g-i}a_i$ for $0 \le i \le g$.

(3) $a_1 = N - (q+1)$ where N is the number of places $P \in \mathbb{P}_F$ of degree one.

(e) $L(t)$ factors in $\mathbb{C}[t]$ in the form

$$L(t) = \prod_{i=1}^{2g}(1-\alpha_i t)\,. \qquad (5.13)$$

The complex numbers $\alpha_1, \ldots, \alpha_{2g}$ are algebraic integers, and they can be arranged in such a way that $\alpha_i \alpha_{g+i} = q$ holds for $i = 1, \ldots, g$. (We note that a complex number α is called an algebraic integer if it satisfies an equation $\alpha^m + c_{m-1}\alpha^{m-1} + \cdots + c_1\alpha + c_0 = 0$ with coefficients $c_i \in \mathbb{Z}$.)

(f) If $L_r(t) := (1-t)(1-q^r t)Z_r(t)$ denotes the L-polynomial of the constant field extension $F_r = F\mathbb{F}_{q^r}$, then

$$L_r(t) = \prod_{i=1}^{2g}(1-\alpha_i^r t)\,,$$

where the α_i are given by (5.13).

Proof. All assertions are trivial for $g = 0$, hence we can assume from now on that $g \geq 1$.

(a) We have already remarked that $L(t)$ is a polynomial of degree $\leq 2g$. In (d) we shall prove that its leading coefficient is q^g, so $\deg L(t) = 2g$. The assertion $L(t) \in \mathbb{Z}[t]$ follows from (5.12) by comparing coefficients.

(b) is nothing but the functional equation for the Zeta function, see Proposition 5.1.13.

(c) With the notation of Corollary 5.1.12(b) we have

$$L(t) = (1-t)(1-qt)F(t) + \frac{h}{q-1}\left(q^g t^{2g-1}(1-t) - (1-qt)\right).$$

Hence $L(1) = h$.

(d) Write $L(t) = a_0 + a_1 t + \cdots + a_{2g}t^{2g}$. The functional equation (b) yields

$$L(t) = q^g t^{2g} L(\frac{1}{qt}) = \frac{a_{2g}}{q^g} + \frac{a_{2g-1}}{q^{g-1}}t + \cdots + q^g a_0 t^{2g}.$$

Therefore $a_{2g-i} = q^{g-i}a_i$ for $i = 0, \ldots, g$, and (2) is proved. Comparing the coefficients of t^0 and t^1 in (5.12) shows that $a_0 = A_0$ and $a_1 = A_1 - (q+1)A_0$. Since $A_0 = 1$ and $A_1 = N$ (this is trivial by the definition of A_n, see (5.4)), we obtain $a_0 = 1$ and $a_1 = N - (q+1)$. Finally $a_{2g} = q^g a_0 = q^g$ by (2).

(e) We consider the reciprocal polynomial

$$L^{\perp}(t) := t^{2g} L(\frac{1}{t}) = a_0 t^{2g} + a_1 t^{2g-1} + \cdots + a_{2g} = t^{2g} + a_1 t^{2g-1} + \cdots + q^g. \quad (5.14)$$

$L^{\perp}(t)$ is a monic polynomial with coefficients in $\mathbb{Z}$, so its roots are algebraic integers. We write

$$L^{\perp}(t) = \prod_{i=1}^{2g}(t - \alpha_i) \quad \textit{with} \quad \alpha_i \in \mathbb{C},$$

therefore

$$L(t) = t^{2g} L^{\perp}(\frac{1}{t}) = \prod_{i=1}^{2g}(1 - \alpha_i t).$$

Observe that $L(\alpha_i^{-1}) = 0$ for $i = 1, ..., 2g$, and

$$\prod_{i=1}^{2g} \alpha_i = q^g.$$

Substituting $t = qu$ and using the functional equation (b) we obtain

$$\begin{aligned}
\prod_{i=1}^{2g}(t-\alpha_i) &= L^{\perp}(t) = t^{2g}L(\frac{1}{t}) \\
&= q^{2g}u^{2g}L(\frac{1}{qu}) = q^g L(u) = q^g \cdot \prod_{j=1}^{2g}(1-\alpha_j u) \\
&= q^g \cdot \prod_{j=1}^{2g}(1-\frac{\alpha_j}{q}t) = q^g \cdot \prod_{j=1}^{2g}\frac{\alpha_j}{q} \cdot \prod_{j=1}^{2g}(t-\frac{q}{\alpha_j}) \\
&= \prod_{j=1}^{2g}(t-\frac{q}{\alpha_j}) .
\end{aligned}$$

So we can arrange the roots of $L^{\perp}(t)$ as

$$\alpha_1, \frac{q}{\alpha_1}, \ldots, \alpha_k, \frac{q}{\alpha_k}, q^{1/2}, \ldots, q^{1/2}, -q^{1/2}, \ldots, -q^{1/2},$$

where $q^{1/2}$ occurs m times and $-q^{1/2}$ occurs n times. By (5.14),

$$\alpha_1 \cdot \frac{q}{\alpha_1} \cdot \ldots \cdot \alpha_k \cdot \frac{q}{\alpha_k} \cdot (q^{1/2})^m \cdot (-q^{1/2})^n = q^g .$$

Therefore n is even. Since $n+m+2k = 2g$, m is also even, and we can rearrange $\alpha_1, \ldots, \alpha_{2g}$ such that $\alpha_i\alpha_{g+i} = q$ holds for $i = 1, \ldots, g$.
(f) We use Proposition 5.1.10 and obtain

$$\begin{aligned}
L_r(t^r) &= (1-t^r)(1-q^r t^r)Z_r(t^r) = (1-t^r)(1-q^r t^r) \prod_{\zeta^r=1} Z(\zeta t) \\
&= (1-t^r)(1-q^r t^r) \prod_{\zeta^r=1} \frac{L(\zeta t)}{(1-\zeta t)(1-q\zeta t)} = \prod_{\zeta^r=1} L(\zeta t) \\
&= \prod_{i=1}^{2g} \prod_{\zeta^r=1} (1-\alpha_i \zeta t) = \prod_{i=1}^{2g}(1-\alpha_i^r t^r) .
\end{aligned}$$

Hence $L_r(t) = \prod_{i=1}^{2g}(1-\alpha_i^r t)$. □

The above theorem shows that the number

$$N(F) := N = \left| \{P \in \mathbb{P}_F \, ; \, \deg P = 1\} \right| \tag{5.15}$$

can easily be calculated if the L-polynomial $L(t)$ of $F/\mathbb{F}_q$ is known. More generally we consider for $r \geq 1$ the number

$$N_r := N(F_r) = \left| \{P \in \mathbb{P}_{F_r} \, ; \, \deg P = 1\} \right| , \tag{5.16}$$

where $F_r = F\mathbb{F}_{q^r}$ is the constant field extension of $F/\mathbb{F}_q$ of degree r. In Section 5.2 the following result will play an essential role.

Corollary 5.1.16. *For all* $r \geq 1$,

$$N_r = q^r + 1 - \sum_{i=1}^{2g} \alpha_i^r ,$$

where $\alpha_1, \ldots, \alpha_{2g} \in \mathbb{C}$ *are the reciprocals of the roots of* $L(t)$. *In particular, since* $N_1 = N(F)$, *we have*

$$N(F) = q + 1 - \sum_{i=1}^{2g} \alpha_i .$$

Proof. By Theorem 5.1.15(d), $N_r - (q^r + 1)$ is the coefficient of t in the L-polynomial $L_r(t)$. On the other hand, since $L_r(t) = \prod_{i=1}^{2g}(1 - \alpha_i^r t)$, this coefficient is $-\sum_{i=1}^{2g} \alpha_i^r$. □

Conversely, if the numbers N_r are known for sufficiently many r, one can calculate the coefficients of $L(t)$ as follows.

Corollary 5.1.17. *Let* $L(t) = \sum_{i=0}^{2g} a_i t^i$ *be the L-polynomial of* $F/\mathbb{F}_q$, *and* $S_r := N_r - (q^r + 1)$. *Then we have:*

(a) $L'(t)/L(t) = \sum_{r=1}^{\infty} S_r t^{r-1}$.

(b) $a_0 = 1$ *and*

$$i a_i = S_i a_0 + S_{i-1} a_1 + \ldots + S_1 a_{i-1} \tag{5.17}$$

for $i = 1, \ldots, g$.

Hence, given $N_1, \ldots, N_g$ *we can determine* $L(t)$ *from* (5.17) *and the equations* $a_{2g-i} = q^{g-i} a_i$ *(for* $i = 0, \ldots, g$*).*

Proof. (a) Write $L(t) = \prod_{i=1}^{2g}(1 - \alpha_i t)$ as in (5.13). Then

$$\begin{aligned} L'(t)/L(t) &= \sum_{i=1}^{2g} \frac{-\alpha_i}{(1 - \alpha_i t)} = \sum_{i=1}^{2g} (-\alpha_i) \cdot \sum_{r=0}^{\infty} (\alpha_i t)^r \\ &= \sum_{r=1}^{\infty} \left(\sum_{i=1}^{2g} -\alpha_i^r \right) t^{r-1} = \sum_{r=1}^{\infty} S_r t^{r-1} , \end{aligned}$$

by Corollary 5.1.16 and the definition of S_r.

(b) We know that $a_0 = 1$, by Theorem 5.1.15. From (a) it follows that

$$a_1 + 2a_2 t + \ldots + 2g a_{2g} t^{2g-1} = (a_0 + a_1 t + \ldots + a_{2g} t^{2g}) \cdot \sum_{r=1}^{\infty} S_r t^{r-1} .$$

Comparing the coefficients of $t^0, t^1, \ldots, t^{g-1}$ yields (5.17). □

5.2 The Hasse-Weil Theorem

We retain all notation of Section 5.1. Thus $F/\mathbb{F}_q$ is a function field of genus $g(F) = g$ over the finite field $\mathbb{F}_q$,

$$
\begin{aligned}
&Z_F(t) = L_F(t)/(1-t)(1-qt) \quad \textit{is its Zeta function,} \\
&\alpha_1, \ldots, \alpha_{2g} \quad \textit{are the reciprocals of the roots of} \quad L_F(t), \\
&N(F) = \left| \{P \in \mathbb{P}_F \, ; \, \deg P = 1\} \right|, \\
&F_r = F\mathbb{F}_{q^r} \quad \textit{is the constant field extension of degree } r, \textit{ and} \\
&N_r = N(F_r).
\end{aligned}
$$

The main result of this section is

Theorem 5.2.1 (Hasse-Weil). *The reciprocals of the roots of $L_F(t)$ satisfy*

$$|\alpha_i| = q^{1/2} \quad \textit{for} \quad i = 1, \ldots, 2g\,.$$

Remark 5.2.2. The Hasse-Weil Theorem is often referred to as the *Riemann Hypothesis for Function Fields.* Let us briefly explain this notation. One can regard the Zeta function $Z_F(t)$ of a function field $F/\mathbb{F}_q$ as an analogue of the classical Riemann ζ-function

$$\zeta(s) := \sum_{n=1}^{\infty} n^{-s} \tag{5.18}$$

(where $s \in \mathbb{C}$ and $\mathrm{Re}(s) > 1$) in the following manner. Define the *absolute norm* of a divisor $A \in \mathrm{Div}(F)$ by

$$\mathcal{N}(A) := q^{\deg A}\,.$$

For instance, the absolute norm $\mathcal{N}(P)$ of a prime divisor $P \in \mathbb{P}_F$ is the cardinality of its residue class field F_P. Then the function

$$\zeta_F(s) := Z_F(q^{-s})$$

can be written as

$$\zeta_F(s) = \sum_{n=0}^{\infty} A_n q^{-sn} = \sum_{A \in \mathrm{Div}(F), A \geq 0} \mathcal{N}(A)^{-s}\,,$$

which is the appropriate analogue to (5.18). It is well-known from number theory that the Riemann ζ-function (5.18) has an analytic continuation as a meromorphic function on $\mathbb{C}$. The classical Riemann Hypothesis states that – besides the so-called trivial zeros $s = -2, -4, -6, \ldots$ – all zeros of $\zeta(s)$ lie on the line $\mathrm{Re}(s) = 1/2$.

In the function field case, the Hasse-Weil Theorem states that

$$\zeta_F(s) = 0 \Rightarrow Z_F(q^{-s}) = 0 \Rightarrow |q^{-s}| = q^{-1/2}.$$

Since $|q^{-s}| = q^{-\mathrm{Re}(s)}$, this means that

$$\zeta_F(s) = 0 \Rightarrow \mathrm{Re}(s) = 1/2.$$

Therefore Theorem 5.2.1 can be viewed as an analogue of the classical Riemann Hypothesis.

Before proving it, we draw an important conclusion from the Hasse-Weil Theorem.

Theorem 5.2.3 (Hasse-Weil Bound). *The number $N = N(F)$ of places of $F/\mathbb{F}_q$ of degree one satisfies the inequality*

$$|N - (q+1)| \le 2gq^{1/2}.$$

Proof. Corollary 5.1.16 yields

$$N - (q+1) = -\sum_{i=1}^{2g} \alpha_i.$$

Hence the Hasse-Weil Bound is an immediate consequence of the Hasse-Weil Theorem. □

Note that Theorem 5.2.3, applied to the function field $F_r/\mathbb{F}_{q^r}$, gives

$$|N_r - (q^r + 1)| \le 2gq^{r/2} \tag{5.19}$$

for all $r \ge 1$.

Our proof of the Hasse-Weil Theorem is due to E. Bombieri. The proof is divided into several steps. The first step is almost trivial.

Lemma 5.2.4. *Let $m \ge 1$. Then the Hasse-Weil Theorem holds for $F/\mathbb{F}_q$ if and only if it holds for the constant field extension $F_m/\mathbb{F}_{q^m}$.*

Proof. The reciprocals of the roots of $L_F(t)$ are $\alpha_1, \ldots, \alpha_{2g}$. By Theorem 5.1.15(f), the reciprocals of the roots of $L_m(t)$ are $\alpha_1^m, \ldots, \alpha_{2g}^m$ (as in Theorem 5.1.15, we denote by $L_m(t)$ the L-polynomial of F_m). Our lemma follows immediately since $|\alpha_i| = q^{1/2} \iff |\alpha_i^m| = (q^m)^{1/2}$. □

The next step reduces the proof of the Hasse-Weil Theorem to an assertion that is closely related to (5.19).

Lemma 5.2.5. *Assume there is a constant $c \in \mathbb{R}$ such that for all $r \geq 1$,*

$$|N_r - (q^r + 1)| \leq cq^{r/2}. \tag{5.20}$$

Then the Hasse-Weil Theorem holds for $F/\mathbb{F}_q$.

Proof. Corollary 5.1.16 states that $N_r - (q^r + 1) = -\sum_{i=1}^{2g} \alpha_i^r$, hence (5.20) yields

$$\left|\sum_{i=1}^{2g} \alpha_i^r\right| \leq cq^{r/2} \tag{5.21}$$

for all $r \geq 1$. Consider the meromorphic function

$$H(t) := \sum_{i=1}^{2g} \frac{\alpha_i t}{1 - \alpha_i t}. \tag{5.22}$$

Let $\varrho := \min\{|\alpha_i^{-1}|;\, 1 \leq i \leq 2g\}$. The convergence radius of the power series expansion of $H(t)$ around $t = 0$ is precisely ϱ (since $\alpha_1^{-1}, \ldots, \alpha_{2g}^{-1}$ are the only singularities of $H(t)$). On the other hand we obtain for $|t| < \varrho$

$$H(t) = \sum_{i=1}^{2g} \sum_{r=1}^{\infty} (\alpha_i t)^r = \sum_{r=1}^{\infty} \Big(\sum_{i=1}^{2g} \alpha_i^r\Big) t^r.$$

By (5.21) this series converges for $|t| < q^{-1/2}$, hence $q^{-1/2} \leq \varrho$. This implies $q^{1/2} \geq |\alpha_i|$ for $i = 1, \ldots, 2g$. Since $\prod_{i=1}^{2g} \alpha_i = q^g$ (by Theorem 5.1.15(e)), we conclude that $|\alpha_i| = q^{1/2}$. □

Observe that the inequalities (5.20) are equivalent to an upper bound and a lower bound for N_r: there exist constants $c_1 > 0$ and $c_2 > 0$ such that

$$N_r \leq q^r + 1 + c_1 q^{r/2} \tag{5.23}$$

and

$$N_r \geq q^r + 1 - c_2 q^{r/2} \tag{5.24}$$

for all $r \geq 1$. By Lemma 5.2.4 the Hasse-Weil Theorem holds for $F/\mathbb{F}_q$ if it holds for some constant field extension of F. Therefore it is sufficient to prove (5.23) and (5.24) under additional assumptions which can be realized in an appropriate finite constant field extension.

Proposition 5.2.6. *Suppose that $F/\mathbb{F}_q$ satisfies the following assumptions:*

(1) q is a square, and (2) $q > (g + 1)^4$.

Then the number $N = N(F)$ of places of $F/\mathbb{F}_q$ of degree one can be estimated by

$$N < (q + 1) + (2g + 1)q^{1/2}.$$

Proof. We can assume that there exists a place $Q \in \mathbb{P}_F$ of degree one (otherwise $N = 0$, and the proposition is trivial). Set

$$q_0 := q^{1/2}\,,\ m := q_0 - 1 \quad \textit{and} \quad n := 2g + q_0\,.$$

One checks easily that

$$r := q - 1 + (2g+1)q^{1/2} = m + nq_0\,. \tag{5.25}$$

Let $T := \{i \mid 0 \le i \le m, \text{ and } i \text{ is a pole number of } Q\}$. (Recall that i is called a pole number of Q if there exists an element $x \in F$ with pole divisor $(x)_\infty = iQ$, cf. Definition 1.6.7.) For every $i \in T$ choose an element $u_i \in F$ with pole divisor iQ. Then the set $\{u_i \mid i \in T\}$ is a basis of $\mathscr{L}(mQ)$. We consider the space

$$\mathscr{L} := \mathscr{L}(mQ) \cdot \mathscr{L}(nQ)^{q_0} \subseteq \mathscr{L}(rQ)\,.$$

(By definition, $\mathscr{L}$ consists of all finite sums $\sum x_\nu y_\nu^{q_0}$ with $x_\nu \in \mathscr{L}(mQ)$ and $y_\nu \in \mathscr{L}(nQ)$; obviously $\mathscr{L}$ is a vector space over $\mathbb{F}_q$, and $\mathscr{L} \subseteq \mathscr{L}(rQ)$ by (5.25).) Our aim is to construct an element $0 \ne x \in \mathscr{L}$ such that

$$x(P) = 0 \quad \textit{for all } P \in \mathbb{P}_F \quad \textit{with } \deg P = 1 \textit{ and } P \ne Q\,. \tag{5.26}$$

Suppose for a moment that we have found such an element x. Then all places of degree one (except Q) are zeros of x, and the zero divisor $(x)_0$ has degree

$$\deg\,(x)_0 \ge N - 1\,.$$

As $x \in \mathscr{L} \subseteq \mathscr{L}(rQ)$,

$$\deg\,(x)_0 = \deg\,(x)_\infty \le r = q - 1 + (2g+1)q^{1/2}\,.$$

Combining these inequalities, we get $N \le q + (2g+1)q^{1/2}$, which proves the proposition.

Claim 1. Every element $y \in \mathscr{L}$ can be written uniquely in the form

$$y = \sum_{i \in T} u_i z_i^{q_0} \quad \textit{with} \quad z_i \in \mathscr{L}(nQ)\,, \tag{5.27}$$

where $\{u_i \mid i \in T\}$ is the above-mentioned basis of $\mathscr{L}(mQ)$.

The existence of a representation (5.27) follows almost immediately from the definition of $\mathscr{L}$. In order to prove uniqueness, we assume that there is an equation

$$\sum_{i \in T} u_i x_i^{q_0} = 0 \tag{5.28}$$

with $x_i \in \mathscr{L}(nQ)$, not all $x_i = 0$. For each index $i \in T$ with $x_i \ne 0$ we have

$$v_Q(u_i x_i^{q_0}) \equiv v_Q(u_i) \equiv -i \bmod q_0\,.$$

Since $m = q_0 - 1$, the numbers $i \in T$ are pairwise distinct modulo q_0. Therefore the Strict Triangle Inequality yields

$$v_Q\Big(\sum_{i\in T} u_i x_i^{q_0}\Big) = \min\{v_Q(u_i x_i^{q_0}) \mid i \in T\} \neq \infty .$$

This contradiction to (5.28) proves Claim 1.

Next we consider the mapping $\lambda : \mathscr{L} \to \mathscr{L}((q_0 m + n)Q)$ given by

$$\lambda\Big(\sum_{i\in T} u_i z_i^{q_0}\Big) := \sum_{i\in T} u_i^{q_0} z_i$$

(with $z_i \in \mathscr{L}(nQ)$). By Claim 1 this map is well-defined. Observe that λ is not $\mathbb{F}_q$-linear, but it is a homomorphism of the additive group of $\mathscr{L}$ into $\mathscr{L}((q_0 m + n)Q)$.

Claim 2. The kernel of λ is not $\{0\}$.

As λ is a homomorphism from $\mathscr{L}$ into $\mathscr{L}((q_0 m + n)Q)$, it is sufficient to show that

$$\dim \mathscr{L} > \dim \mathscr{L}((q_0 m + n)Q) \tag{5.29}$$

(where dim denotes the dimension as a vector space over $\mathbb{F}_q$). We have, by Claim 1 and the Riemann-Roch Theorem,

$$\dim \mathscr{L} = \ell(mQ) \cdot \ell(nQ) \geq (m + 1 - g)(n + 1 - g) .$$

On the other hand, since

$$q_0 m + n = q_0(q_0 - 1) + (2g + q_0) = 2g + q ,$$

we obtain

$$\dim \mathscr{L}((q_0 m + n)Q) = (2g + q) + 1 - g = g + q + 1 .$$

Hence (5.29) follows if we can prove that

$$(m + 1 - g)(n + 1 - g) > g + q + 1 . \tag{5.30}$$

Consider the following equivalences:

$$\begin{aligned} &(m + 1 - g)(n + 1 - g) > g + q + 1 \\ \iff\ & (q_0 - g)(2g + q_0 + 1 - g) > g + q + 1 \\ \iff\ & q - g^2 + q_0 - g > g + q + 1 \\ \iff\ & q_0 > g^2 + 2g + 1 = (g + 1)^2 \\ \iff\ & q > (g + 1)^4 . \end{aligned}$$

As we assumed that $q > (g + 1)^4$ (see assumption (2) of Proposition 5.2.6), (5.30) is established.

Claim 3. Let $0 \neq x \in \mathscr{L}$ be an element in the kernel of λ and let $P \neq Q$ be a place of degree one. Then $x(P) = 0$.

Note that $y(P) \neq \infty$ for all $y \in \mathscr{L}$ because Q is the only pole of y. Moreover, since $\mathbb{F}_q$ is the residue class field of P, we have $y(P)^q = y(P)$. Now we consider an element $x \in \mathscr{L}$ with $\lambda(x) = 0$. We write $x = \sum_{i \in T} u_i z_i^{q_0}$ and obtain

$$\begin{aligned} x(P)^{q_0} &= \Big(\sum_{i\in T} u_i(P) \cdot z_i(P)^{q_0}\Big)^{q_0} \\ &= \sum_{i\in T} u_i^{q_0}(P) \cdot z_i(P)^q \\ &= \Big(\sum_{i\in T} u_i^{q_0} z_i\Big)(P) = \lambda(x)(P) = 0\,. \end{aligned}$$

So we have proved Claim 3. As we mentioned above, this implies the assertion of Proposition 5.2.6. □

The previous proposition provides an upper bound (5.23) for the numbers N_r (after an appropriate constant field extension). Next we would like to obtain a lower bound. We begin with a lemma from group theory.

Lemma 5.2.7. *Let G' be a group which is the direct product*

$$G' = \langle\sigma\rangle \times G \tag{5.31}$$

of a cyclic subgroup $\langle\sigma\rangle$ and a subgroup $G \subseteq G'$ such that $\operatorname{ord} G = m$, $\operatorname{ord}(\sigma) = n$ *and m divides n. Suppose that $H \subseteq G'$ is a subgroup of G' with*

$$\operatorname{ord} H = ne \quad \textit{and} \quad \operatorname{ord}(H \cap G) = e\,. \tag{5.32}$$

Then there exist exactly e subgroups $U \subseteq H$ with the following properties:

$$U \textit{ is cyclic of order } n, \textit{ and } U \cap G = \{1\}\,. \tag{5.33}$$

Proof. For $\tau \in G$ we consider the cyclic subgroup $\langle\sigma\tau\rangle \subseteq G'$. Since $\sigma\tau = \tau\sigma$ (by (5.31)), $\operatorname{ord}(\sigma) = n$ and $\operatorname{ord}(\tau) | m$ we conclude that $\operatorname{ord}(\sigma\tau) = n$. Moreover, $\langle\sigma\tau\rangle \cap G = \{1\}$ and $\langle\sigma\tau\rangle \neq \langle\sigma\tau'\rangle$ for $\tau \neq \tau'$ (all this follows immediately from (5.31) because the elements $\lambda \in G'$ have a unique representation $\lambda = \sigma^i \varrho$ with $0 \leq i < n$ and $\varrho \in G$). Thus we have found $m = \operatorname{ord} G$ distinct subgroups $U \subseteq G'$ with the properties (5.33).

The subgroup $G \subseteq G'$ is a normal subgroup, hence $H/H \cap G \simeq HG/G$. By (5.32) this implies that $HG = G'$, and $H/H \cap G \simeq G'/G$ is cyclic of order n. Choose an element $\lambda_0 \in H$ whose order modulo $H \cap G$ is n, and write $\lambda_0 = \sigma^a \tau'$ with $\tau' \in G$ and $a \in \mathbb{Z}$. The exponent a is relatively prime to n (otherwise there would be an integer $1 \leq d < n$ with $\sigma^{ad} = 1$, hence

$\lambda_0^d = \tau'^d \in H \cap G$; so the order of λ_0 modulo $H \cap G$ would be less than n). Therefore a suitable power $\lambda = \lambda_0^t$ has a representation $\lambda = \sigma\tau_0$ with $\tau_0 \in G$. Let $H \cap G = \{\psi_1, \ldots, \psi_e\}$. We define

$$U^{(j)} := \langle \sigma\tau_0\psi_j \rangle \quad \textit{for} \quad j = 1, \ldots, e\,.$$

The subgroups $U^{(j)} \subseteq H$ are cyclic of order n, they are pairwise distinct, and $U^{(j)} \cap G = \{1\}$.

It remains to show that H contains no other cyclic subgroup U of order n with $U \cap G = \{1\}$. In fact, let $U \subseteq H$ be a subgroup satisfying (5.33). As above we find a generator of U of the specific form $\sigma\tau_1$ with $\tau_1 \in G$. Since $\sigma\tau_1 \in H$ and $\sigma\tau_0 \in H$,

$$\tau_0^{-1}\tau_1 = (\sigma\tau_0)^{-1}(\sigma\tau_1) \in H \cap G = \{\psi_1, \ldots, \psi_e\}\,.$$

Hence $\tau_1 = \tau_0\psi_j$ for some j, and $U = \langle \sigma\tau_1 \rangle = \langle \sigma\tau_0\psi_j \rangle = U^{(j)}$. □

The next proposition is the essential step in the proof of a lower bound for N_r. We consider the following situation: E/L is a Galois extension of function fields of degree $[E : L] = m$, and it is assumed that $\mathbb{F}_q$ is the full constant field of E and L. We choose an integer $n > 0$ with $m|n$ and let $E' := E\mathbb{F}_{q^n}$ (resp. $L' := L\mathbb{F}_{q^n} \subseteq E'$) be the corresponding constant field extension of degree n. Then E'/L is Galois with Galois group $G' = \langle \sigma \rangle \times G$, where $G := \mathrm{Gal}(E'/L') \simeq \mathrm{Gal}(E/L)$ and σ is the Frobenius automorphism of E'/E (i.e., $\sigma(z) = z$ for $z \in E$ and $\sigma(\alpha) = \alpha^q$ for $\alpha \in \mathbb{F}_{q^n}$). By the previous lemma, G' contains exactly m cyclic subgroups $U \subseteq G'$ with ord $U = n$ and $U \cap G = \{1\}$, say $U_1, \ldots, U_m$. We can assume that $U_1 = \langle \sigma \rangle$.

Let E_i be the fixed field of U_i $(i = 1, \ldots, m)$. Then $E_1 = E$, and we have the situation as shown in Figure 5.1:

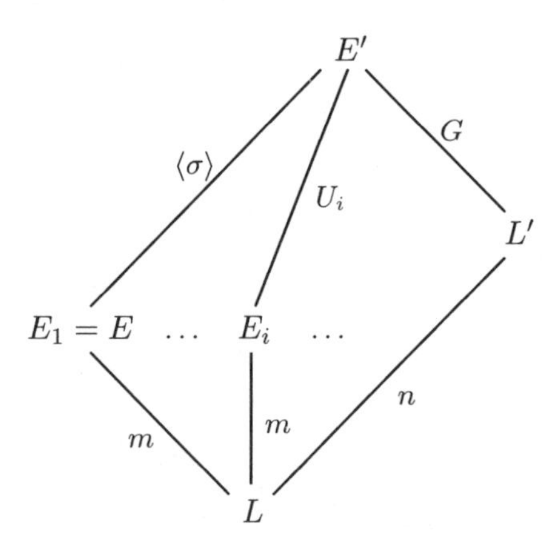

Fig. 5.1.

Denote by $g(E_i)$ the genus of E_i, and by $N(E_i)$ (resp. $N(L)$) the number of places of degree one of E_i (resp. L).

Proposition 5.2.8. *Under the above assumptions the following hold:*
(a) $\mathbb{F}_q$ *is the full constant field of* E_i*, for* $1 \le i \le m$.
(b) $E' = E_i\mathbb{F}_{q^n}$ *and* $g(E_i) = g(E)$*, for* $i = 1, \ldots, m$.
(c) $m \cdot N(L) = \sum_{i=1}^{m} N(E_i)$.

Proof. (a), (b) Note that $U_i \cap G = \{1\}$. By Galois theory, E' is then the compositum of E_i and L', hence $E' = E_iL' = E_iL\mathbb{F}_{q^n} = E_i\mathbb{F}_{q^n}$ is the constant field extension of E_i with $\mathbb{F}_{q^n}$. Since $[E' : E_i] = \operatorname{ord} U_i = n$, this implies that $\mathbb{F}_q$ is the full constant field of E_i. The genus is invariant under constant field extensions, so $g(E_i) = g(E') = g(E)$ for $i = 1, \ldots, m$.
(c) We consider the sets $X := \{P \in \mathbb{P}_L \,|\, \deg P = 1\}$ and, for $i = 1, \ldots, m$, $X_i := \{Q \in \mathbb{P}_{E_i} \,|\, \deg Q = 1\}$, and we have to prove the following assertion:

$$\left| \bigcup_{i=1}^{m} X_i \right| = m \cdot |X| \,. \tag{5.34}$$

Let $P \in X$. Choose a place $P' \in \mathbb{P}_{E'}$ lying over P, and set $P_1 := P' \cap E$. The relative degree $f(P_1|P)$ divides m, as E/L is Galois. Hence $f(P_1|P)$ divides n, and the residue class field of P' is $\mathbb{F}_{q^n}$ by Theorem 3.6.3(g). This means that the relative degree of $P'|P$ is $f(P'|P) = n$. Denote by $e := e(P'|P)$ the ramification index of P in E'/L and by r the number of places in $\mathbb{P}_{E'}$ lying over P (as E'/L is Galois, e depends only on P). We obtain

$$m \cdot n = [E' : L] = e(P'|P) \cdot f(P'|P) \cdot r = e \cdot n \cdot r \,.$$

Hence $m = e \cdot r$, and (5.34) is reduced to the following claims:

Claim 1. For every $Q \in X_i$ with $Q|P$ there is exactly one place $Q' \in \mathbb{P}_{E'}$ which lies above Q.

Claim 2. For every place $Q' \in \mathbb{P}_{E'}$ with $Q'|P$ there are exactly e distinct places $Q \in \bigcup_{i=1}^{m} X_i$ such that $Q'|Q$.

Proof of Claim 1. If $Q' \in \mathbb{P}_{E'}$ lies above the place $Q \in X_i$ and $Q|P$, then

$$f(Q'|Q) = f(Q'|Q) \cdot f(Q|P) = f(Q'|P) = n \,.$$

(Observe that $f(Q|P) = 1$ as $Q \in X_i$.) Thus $f(Q'|Q) = [E' : E_i]$, which implies that Q' is the only extension of Q in $E'|E_i$.

Proof of Claim 2. Here a place $Q' \in \mathbb{P}_{E'}$ with $Q'|P$ is given. Let $H \subseteq \operatorname{Gal}(E'/L)$ be the decomposition group of Q' over P, $Z \subseteq E'$ the fixed field of H and $P_Z := Q' \cap Z$. Then

$$\operatorname{ord} H = e(Q'|P) \cdot f(Q'|P) = e \cdot n$$

and $f(P_Z|P) = 1$, by Theorem 3.8.2. It follows in particular that

$$\mathbb{F}_q \textit{ is the full constant field of } Z\,. \tag{5.35}$$

By Galois theory, the fixed field of $H \cap G$ is the compositum of Z and L'. We have $ZL' = ZL\mathbb{F}_{q^n} = Z\mathbb{F}_{q^n}$ and $[Z\mathbb{F}_{q^n} : Z] = n$ (by (5.35)), hence

$$\operatorname{ord}(H \cap G) = [E' : Z]/[ZL' : Z] = ne/n = e\,.$$

As P_Z is unramified in $ZL' = Z\mathbb{F}_{q^n}$, it also follows that $T := ZL'$ is the inertia field and $H \cap G$ is the inertia group of $Q'|P$ (cf. Theorem 3.8.3).

Now we apply Lemma 5.2.7 once again: Exactly e of the cyclic groups $U_1, \ldots, U_m \subseteq \mathrm{Gal}(E'/L)$ of order n with $U_i \cap G = \{1\}$ are contained in H, say $U_{i_1}, \ldots, U_{i_e}$. Let $Q_{i_j} := Q' \cap E_{i_j}$. Because E_{i_j} contains the decomposition field of Q' over P, Q' is the only place of E' lying over Q_{i_j}. On the other hand, $e(Q'|Q_{i_j}) = 1$ since E' is a constant field extension of E_{i_j} (by (b)). This implies $f(Q'|Q_{i_j}) = [E' : E_{i_j}] = n = f(Q'|P)$, hence $\deg Q_{i_j} = 1$. In this manner we have constructed e distinct places $Q_{i_j} \in \bigcup_{i=1}^m X_i$ such that $Q'|Q_{i_j}$.

Conversely, suppose that $Q \in X_i$ for some $i \in \{1, \ldots, m\}$ and $Q'|Q$. Then $f(Q'|Q) = n$. So $U_i = \mathrm{Gal}(E'|E_i)$ is contained in the decomposition group H of Q' over P, i.e. U_i is one of the above groups U_{i_j}, and Q is the corresponding place Q_{i_j} ($j \in \{1, \ldots, e\}$). This proves Claim 2 and finishes the proof of Proposition 5.2.8. □

End of the Proof of the Hasse-Weil Theorem 5.2.1. As we mentioned above, it remains to establish a lower bound (5.24) for $N_r = N(F_r)$. We proceed as follows. Choose a rational subfield $F_0 = \mathbb{F}_q(t) \subseteq F$ such that F/F_0 is separable, and a finite extension $E \supseteq F$ such that E/F_0 is Galois (observe that there exists a separating element t by Proposition 3.10.2). It is possible that the constant field of E is a proper extension $\mathbb{F}_{q^d}$ of $\mathbb{F}_q$. In this case we consider the fields $F\mathbb{F}_{q^d}$ and $F_0\mathbb{F}_{q^d} = \mathbb{F}_{q^d}(t)$ instead of F and F_0. The extension $E/F_0\mathbb{F}_{q^d}$ is Galois, and it is sufficient to prove the Hasse-Weil Theorem for $F\mathbb{F}_{q^d}/\mathbb{F}_{q^d}$ (by Lemma 5.2.4). So we may change notation and assume from the beginning that $\mathbb{F}_q$ is also the full constant field of E. Moreover we can assume that

$$q \textit{ is a square and } q > (g(E)+1)^4\,. \tag{5.36}$$

Let $m := [E : F]$ and $n := [E : F_0]$, and consider the constant field extensions $E' = E\mathbb{F}_{q^n}$, $F' := F\mathbb{F}_{q^n}$ and $F_0' := F_0\mathbb{F}_{q^n}$. By Lemma 5.2.7 there exist exactly m distinct cyclic subgroups $V_1, \ldots, V_m \subseteq \mathrm{Gal}(E'/F)$ of order n such that $V_i \cap \mathrm{Gal}(E'/F') = \{1\}$. On the other hand there are n cyclic subgroups $U_1, \ldots, U_n \subseteq \mathrm{Gal}(E'/F_0)$ with the property $\operatorname{ord}(U_j) = n$ and $U_j \cap \mathrm{Gal}(E'/F_0') = \{1\}$. It is easily seen that $V_i \cap \mathrm{Gal}(E'/F_0') = \{1\}$ (by showing that E' is the compositum of F_0' with the fixed field of V_i), hence we can assume that $V_i = U_i$ for $i = 1, \ldots, m$. Denote by E_i the fixed field of U_i, for $i = 1, \ldots, n$. Proposition 5.2.8 yields

$$m \cdot N(F) = \sum_{i=1}^{m} N(E_i) \tag{5.37}$$

and

$$n \cdot N(F_0) = \sum_{i=1}^{n} N(E_i)\,. \tag{5.38}$$

As we assumed (5.36), the upper bound

$$N(E_i) \leq q + 1 + (2g(E) + 1)q^{1/2}$$

holds for $1 \leq i \leq n$, by Proposition 5.2.6. The places of $F_0 = \mathbb{F}_q(t)$ of degree one are the pole of t and, for each $\alpha \in \mathbb{F}_q$, the zero of $t - \alpha$; thus $N(F_0) = q+1$. We combine this with (5.37) and (5.38) and obtain

$$\begin{aligned} m \cdot N(F) &= n \cdot N(F_0) + \sum_{i=1}^{m} N(E_i) - \sum_{i=1}^{n} N(E_i) \\ &= n(q+1) - \sum_{i=m+1}^{n} N(E_i) \\ &\geq n(q+1) - (n-m)(q + 1 + (2g(E)+1)q^{1/2}) \\ &= m(q+1) - (n-m)(2g(E)+1)q^{1/2}\,. \end{aligned}$$

Therefore

$$N(F) \geq q + 1 - \frac{n-m}{m}(2g(E)+1)q^{1/2}\,.$$

Observe that the numbers m, n and $g(E)$ are invariant under constant field extensions; so we have established a lower bound

$$N_r \geq q^r + 1 - c_2 q^{r/2} \tag{5.24}$$

with a constant $c_2 > 0$. This finishes the proof of the Hasse-Weil Theorem. □

Using the Hasse-Weil Bound, one can also give an estimate for the number of places of a fixed degree r. Given a function field $F/\mathbb{F}_q$ of genus g, we define

$$B_r := B_r(F) := \big|\{P \in \mathbb{P}_F\,;\, \deg P = r\}\big|\,. \tag{5.39}$$

Observe that $B_1 = N(F)$. There is a close relationship between the numbers B_r and N_s (the number of places of degree one in the constant field extension $F_s = F\mathbb{F}_{q^s}$), namely

$$N_r = \sum_{d|r} d \cdot B_d \tag{5.40}$$

(the sum runs over all integers $d \geq 1$ that divide r). This formula follows easily from Lemma 5.1.9(d): every place $P \in \mathbb{P}_F$ of degree $d|r$ decomposes into d places of degree one in $\mathbb{P}_{F_r}$, and the extensions P' of P in F_r/F have degree $\deg P' > 1$ if $\deg P \nmid r$. The *Möbius inversion formula* (cf. [24]) converts (5.40) into

$$r \cdot B_r = \sum_{d|r} \mu\Big(\frac{r}{d}\Big) \cdot N_d \,. \tag{5.41}$$

Here $\mu : \mathbb{N} \to \{0, -1, 1\}$ denotes the *Möbius function* which is defined by

$$\mu(n) = \begin{cases} 1 & \textit{if } n = 1\,, \\ 0 & \textit{if there is an integer } k > 1 \textit{ with } k^2|n\,, \\ (-1)^l & \textit{if } n \textit{ is the product of } l \textit{ distinct primes}\,. \end{cases}$$

We set

$$S_r := -\sum_{i=1}^{2g} \alpha_i^r \,, \tag{5.42}$$

where $\alpha_1, \ldots, \alpha_{2g} \in \mathbb{C}$ are the reciprocals of the roots of $L_F(t)$ (for $g = 0$ we set $S_r := 0$). Then

$$N_r = q^r + 1 + S_r \,,$$

by Corollary 5.1.16. Substitute this into (5.41) and observe that

$$\sum_{d|r} \mu(\frac{r}{d}) = 0 \quad \textit{for} \quad r > 1$$

(cf. [24]). The result is:

Proposition 5.2.9. *For all $r \geq 2$,*

$$B_r = \frac{1}{r} \cdot \sum_{d|r} \mu(\frac{r}{d})(q^d + S_d)\,.$$

Corollary 5.2.10. *(a) The estimate*

$$|B_r - \frac{q^r}{r}| \leq \Big(\frac{q}{q-1} + 2g\frac{q^{1/2}}{q^{1/2}-1}\Big) \cdot \frac{q^{r/2}-1}{r} < (2+7g) \cdot \frac{q^{r/2}}{r}$$

holds for all $r \geq 1$.
(b) If $g = 0$ then $B_r > 0$ for all $r \geq 1$.
(c) For each r such that $2g + 1 \leq q^{(r-1)/2}(q^{1/2} - 1)$ there exists at least one place of degree r. In particular, if $r \geq 4g + 3$ then $B_r \geq 1$.

Proof. (a) For $r = 1$ we have $B_1 = N$, and the assertion follows easily from the Hasse-Weil Bound. For $r \geq 2$ Proposition 5.2.9 yields

$$B_r - \frac{q^r}{r} = \frac{1}{r} \sum_{d|r,\, d<r} \mu\left(\frac{r}{d}\right) q^d + \frac{1}{r} \sum_{d|r} \mu\left(\frac{r}{d}\right) S_d \,.$$

Setting $l := [r/2]$ (the integer part of $r/2$) and observing that $|S_d| = |\sum_{i=1}^{2g} \alpha_i^d| \leq 2gq^{d/2}$, we obtain

$$\begin{aligned}
|B_r - \frac{q^r}{r}| &\leq \frac{1}{r} \sum_{d=1}^{l} q^d + \frac{2g}{r} \sum_{d=1}^{r} q^{d/2} \\
&= \frac{q}{r} \cdot \frac{q^l - 1}{q - 1} + \frac{2gq^{1/2}}{r} \cdot \frac{q^{r/2} - 1}{q^{1/2} - 1} \\
&\leq \left(\frac{q}{q-1} + 2g \frac{q^{1/2}}{q^{1/2} - 1} \right) \cdot \frac{q^{r/2} - 1}{r} \\
&< (2 + 7g) \cdot \frac{q^{r/2}}{r} \,.
\end{aligned}$$

(b),(c) From (a) it follows that $B_r > 0$ whenever

$$\frac{q^r}{r} > \left(\frac{q}{q-1} + 2g \frac{q^{1/2}}{q^{1/2} - 1} \right) \cdot \frac{q^{r/2} - 1}{r} \,. \tag{5.43}$$

In the case $g = 0$, (5.43) holds for all $r \geq 1$. This implies (b), and we can assume $g \geq 1$ from now on. A simple calculation shows that (5.43) is equivalent to

$$2g + \frac{1}{1 + q^{-1/2}} < \frac{q^r (q^{1/2} - 1)}{q^{1/2} (q^{r/2} - 1)} \,. \tag{5.44}$$

The inequalities

$$2g + \frac{1}{1 + q^{-1/2}} < 2g + 1 \quad \textit{and} \quad q^{(r-1)/2} (q^{1/2} - 1) < \frac{q^r (q^{1/2} - 1)}{q^{1/2} (q^{r/2} - 1)}$$

are trivial, hence our assumption $2g + 1 \leq q^{(r-1)/2} (q^{1/2} - 1)$ implies (5.44) and therefore $B_r > 0$. If $r \geq 4g + 3$ then

$$2g + 1 < 2^{2g+1} (2^{1/2} - 1) \leq 2^{(r-1)/2} (2^{1/2} - 1) \leq q^{(r-1)/2} (q^{1/2} - 1) \,.$$

This completes the proof of (c). □

5.3 Improvements of the Hasse-Weil Bound

In general the Hasse-Weil Bound $|N - (q + 1)| \leq 2gq^{1/2}$ is sharp. There are examples of function fields $F/\mathbb{F}_q$ such that $N = q + 1 + 2gq^{1/2}$ (resp.

$N = q + 1 - 2gq^{1/2}$). We shall present several examples in Chapter 6. Under certain assumptions however, the bound can be improved. For instance, if q is not a square, we have the trivial improvement

$$|N - (q + 1)| \leq [2gq^{1/2}], \tag{5.45}$$

where $[a]$ denotes the integer part of the real number a. The bound (5.45) can be ameliorated substantially as follows:

Theorem 5.3.1 (Serre Bound). *For a function field $F/\mathbb{F}_q$ of genus g the number of places of degree one is bounded by*

$$|N - (q + 1)| \leq g\,[2q^{1/2}]. \tag{5.46}$$

Proof. Let $\mathbb{A} \subseteq \mathbb{C}$ be the set of algebraic integers; i.e., a complex number α is in $\mathbb{A}$ if and only if α satisfies an equation $\alpha^m + b_{m-1}\alpha^{m-1} + \cdots + b_1\alpha + b_0 = 0$ with coefficients $b_i \in \mathbb{Z}$. It is an elementary fact from algebraic number theory that

$$\mathbb{A} \quad \textit{is a subring of} \quad \mathbb{C}, \quad \textit{and} \quad \mathbb{A} \cap \mathbb{Q} = \mathbb{Z}. \tag{5.47}$$

For the proof of the Serre Bound we can assume that $g > 0$. We consider the L-polynomial $L(t) = \prod_{i=1}^{2g}(1 - \alpha_i t)$ of $F/\mathbb{F}_q$. The complex numbers $\alpha_1, \ldots, \alpha_{2g}$ are algebraic integers with $|\alpha_i| = q^{1/2}$ (Theorem 5.1.15 and Theorem 5.2.1). They can be ordered such that $\alpha_i \alpha_{g+i} = q$, hence

$$\bar{\alpha}_i = \alpha_{g+i} = q/\alpha_i \qquad \textit{for} \ \ 1 \leq i \leq g.$$

(We denote by $\bar{\alpha}$ the complex conjugate of α.) Set

$$\begin{aligned} \gamma_i &:= \alpha_i + \bar{\alpha}_i + [2q^{1/2}] + 1, \\ \delta_i &:= -(\alpha_i + \bar{\alpha}_i) + [2q^{1/2}] + 1. \end{aligned}$$

By (5.47), γ_i and δ_i are real algebraic integers and, as $|\alpha_i| = q^{1/2}$, they satisfy

$$\gamma_i > 0 \quad \textit{and} \quad \delta_i > 0. \tag{5.48}$$

Each embedding $\sigma : \mathbb{Q}(\alpha_1, \ldots, \alpha_{2g}) \longrightarrow \mathbb{C}$ permutes $\alpha_1, \ldots, \alpha_{2g}$ because $\prod_{i=1}^{2g}(t - \alpha_i) = L^{\perp}(t) \in \mathbb{Z}[t]$, cf. (5.14). Moreover, if $\sigma(\alpha_i) = \alpha_j$ then

$$\sigma(\bar{\alpha}_i) = \sigma(q/\alpha_i) = q/\sigma(\alpha_i) = \overline{\sigma(\alpha_i)} = \bar{\alpha}_j.$$

Therefore σ acts as a permutation on the sets $\{\gamma_1, \ldots, \gamma_g\}$ and $\{\delta_1, \ldots, \delta_g\}$. Define

$$\gamma := \prod_{i=1}^{g} \gamma_i \quad \textit{and} \quad \delta := \prod_{i=1}^{g} \delta_i.$$

γ and δ are algebraic integers which are invariant under all embeddings of $\mathbb{Q}(\alpha_1, \ldots, \alpha_{2g})$ into $\mathbb{C}$. Hence $\gamma, \delta \in \mathbb{Q} \cap \mathbb{A} = \mathbb{Z}$. By (5.48), $\gamma > 0$ and $\delta > 0$, so we have

$$\prod_{i=1}^{g} \gamma_i \geq 1 \quad \textit{and} \quad \prod_{i=1}^{g} \delta_i \geq 1 \,.$$

The well-known inequality between the arithmetic and geometric mean yields now

$$\frac{1}{g}\sum_{i=1}^{g} \gamma_i \geq \left(\prod_{i=1}^{g} \gamma_i\right)^{1/g} \geq 1 \,.$$

Hence

$$\begin{aligned} g &\leq \left(\sum_{i=1}^{g} (\alpha_i + \bar{\alpha}_i)\right) + g[2q^{1/2}] + g \\ &= \sum_{i=1}^{2g} \alpha_i + g[2q^{1/2}] + g \,. \end{aligned}$$

Observing that $\sum_{i=1}^{2g} \alpha_i = (q+1) - N$ by Corollary 5.1.16, we obtain

$$N \leq q + 1 + g[2q^{1/2}] \,.$$

In the same manner, the inequality

$$\frac{1}{g}\sum_{i=1}^{g} \delta_i \geq \left(\prod_{i=1}^{g} \delta_i\right)^{1/g} \geq 1$$

implies that

$$N \geq q + 1 - g[2q^{1/2}] \,.$$

□

One is often interested in function fields which have many places of degree one. So we introduce the following notion:

Definition 5.3.2. *A function field $F/\mathbb{F}_q$ of genus g is said to be maximal if $N = q + 1 + 2gq^{1/2}$.*

Obviously maximal function fields over $\mathbb{F}_q$ can exist only if q is a square. Our next result is due to Y. Ihara. It shows that $F/\mathbb{F}_q$ cannot be maximal if the genus is large with respect to q.

Proposition 5.3.3 (Ihara). *Suppose that $F/\mathbb{F}_q$ is a maximal function field. Then $g \leq (q - q^{1/2})/2$.*

Proof. Let $\alpha_1, \ldots, \alpha_{2g}$ be the reciprocals of the roots of $L(t)$. Since

$$N = q + 1 - \sum_{i=1}^{2g} \alpha_i \quad \textit{and} \quad |\alpha_i| = q^{1/2}$$

(by Corollary 5.1.16 and Theorem 5.2.1), the assumption $N = q + 1 + 2gq^{1/2}$ implies

$$\alpha_i = -q^{1/2} \quad \textit{for} \quad i = 1, \ldots, 2g\,. \tag{5.49}$$

Next we consider the number N_2 of places of degree one in the constant field extension $F\mathbb{F}_{q^2}/\mathbb{F}_{q^2}$. We have $N_2 \geq N$ and

$$N_2 = q^2 + 1 - \sum_{i=1}^{2g} \alpha_i^2 = q^2 + 1 - 2gq\,,$$

by Corollary 5.1.16 and (5.49). So

$$q + 1 + 2gq^{1/2} \leq q^2 + 1 - 2gq\,.$$

The inequality $g \leq (q - q^{1/2})/2$ follows immediately. □

Ihara's estimate for the genus of a maximal function field cannot be improved in general: in Chapter 6 we shall show that there exist maximal function fields of genus $g = (q - q^{1/2})/2$ (if q is a square).

One can refine the proof of Proposition 5.3.3 in order to obtain other bounds for the number of places of degree one. This method was developed by J.-P. Serre. We proceed as follows: Let

$$N_r = N(F_r) = \left| \{ P \in \mathbb{P}_{F_r}\,;\, \deg P = 1 \} \right|,$$

where $F_r = F\mathbb{F}_{q^r}$ is the constant field extension of F of degree r. We write for $i = 1, \ldots, 2g$,

$$\omega_i := \alpha_i q^{-1/2}\,, \tag{5.50}$$

where $\alpha_1, \ldots, \alpha_{2g}$ are the reciprocals of the roots of $L_F(t)$. Then $|\omega_i| = 1$ by the Hasse-Weil Theorem, and we can assume that

$$\omega_{g+i} = \bar{\omega}_i = \omega_i^{-1} \quad \textit{for} \quad i = 1, \ldots, g\,. \tag{5.51}$$

(This follows from Theorem 5.1.15(e).) By Corollary 5.1.16,

$$N_r q^{-r/2} = q^{r/2} + q^{-r/2} - \sum_{i=1}^{g} (\omega_i^r + \omega_i^{-r})\,. \tag{5.52}$$

Given real numbers $c_1, c_2, \ldots$ we multiply (5.52) by c_r and obtain

$$N_1 c_r q^{-r/2} = c_r q^{r/2} + c_r q^{-r/2} - \sum_{i=1}^{g} c_r(\omega_i^r + \omega_i^{-r}) - (N_r - N_1) c_r q^{-r/2}\,. \tag{5.53}$$

Summing up the equations (5.53) for $r = 1, \ldots, m$ gives

$$N_1 \cdot \lambda_m(q^{-1/2}) = \lambda_m(q^{1/2}) + \lambda_m(q^{-1/2}) + g - \sum_{i=1}^{g} f_m(\omega_i)$$
$$- \sum_{r=1}^{m} (N_r - N_1) c_r q^{-r/2}, \qquad (5.54)$$

where

$$\lambda_m(t) := \sum_{r=1}^{m} c_r t^r \qquad (5.55)$$

and

$$f_m(t) := 1 + \lambda_m(t) + \lambda_m(t^{-1}) \qquad (5.56)$$

for $t \in \mathbb{C}$, $t \neq 0$. Note that $f_m(t) \in \mathbb{R}$ for $|t| = 1$. For specific choices of the constants c_r, Equation (5.54) yields good estimates for N. For instance we show:

Proposition 5.3.4 (Serre's Explicit Formulas). *Suppose that $c_1, \ldots, c_m \in \mathbb{R}$ satisfy the following conditions:*

(1) $c_r \geq 0$ for $r = 1, \ldots, m$, and not all $c_r = 0$.

(2) $f_m(t) \geq 0$ for all $t \in \mathbb{C}$ with $|t| = 1$ (where $f_m(t)$ is defined by (5.56)).

Then the number of rational places of $F/\mathbb{F}_q$ is bounded by

$$N \leq \frac{g}{\lambda_m(q^{-1/2})} + \frac{\lambda_m(q^{1/2})}{\lambda_m(q^{-1/2})} + 1, \qquad (5.57)$$

with $\lambda_m(t)$ as in (5.55).

Proof. We have $N = N_1 \leq N_r$ for all $r \geq 1$. So (5.54) and the assumptions (1) and (2) imply

$$N \cdot \lambda_m(q^{-1/2}) \leq \lambda_m(q^{1/2}) + \lambda_m(q^{-1/2}) + g.$$

Dividing this inequality by $\lambda_m(q^{-1/2})$ yields (5.57) (note that $\lambda_m(q^{-1/2}) > 0$ by assumption (1)). □

In Chapter 7 we will use Proposition 5.3.4 to prove an asymptotic bound for $N(F)$ when the genus of F tends to infinity.

5.4 Exercises

In all exercises below we assume that $F/\mathbb{F}_q$ is a function field with full constant field $\mathbb{F}_q$, and $N = N_1$ denotes the number of rational places of $F/\mathbb{F}_q$. The constant field extension of F of degree r is denoted by $F_r = F\mathbb{F}_{q^r}$, and N_r is the number of rational places of $F_r/\mathbb{F}_{q^r}$. Moreover, $L(t)$ denotes the L-polynomial of $F/\mathbb{F}_q$.

5.1. Let F be a function field of genus 1 over $\mathbb{F}_5$ having $N = 10$ rational places.

(i) Determine the L-polynomial $L(t)$ and the class number $h = h_F$.

(ii) Determine N_2 and N_3.

(iii) How many divisors $A \geq 0$ are there with $\deg A = 1, 2, 3$ and 4?

(iv) Determine the number of places of $F/\mathbb{F}_5$ of degree 2, 3 and 4.

5.2. Show that over every finite field $\mathbb{F}_q$ there exists a function field without rational places.

5.3. Using results of this chapter, find a formula for the number of irreducible monic poynomials $p(x) \in \mathbb{F}_q[x]$ of degree $\deg p(x) = n$.

5.4. If $F/\mathbb{F}_q$ is a function field over a finite field, show that its automorphism group $\mathrm{Aut}\,(F/\mathbb{F}_q)$ is finite.

5.5. Assume that $q = \ell^2$ is a square and the genus of F is $g \geq 1$.

(i) Show that $F/\mathbb{F}_q$ is maximal if and only if $L(t) = (1 + \ell t)^{2g}$.

(ii) If $F/\mathbb{F}_q$ is maximal, show that $F_r/\mathbb{F}_{q^r}$ is maximal if and only if r is odd.

5.6. Assume that the number of rational places of $F/\mathbb{F}_q$ attains the upper Serre Bound $N = q + 1 + g[2q^{1/2}]$. Determine the L-polynomial $L(t)$.

Hint. When is the arithmetic mean equal to the geometric mean?

5.7. *(i)* Let P be a rational place of $F/\mathbb{F}_q$. Show that all integers k with $1 \leq k \leq (N-2)/q$ are gap numbers of P.

(ii) Conclude from (i) the estimate $N \leq q + 1 + qg$.

Note. This simple estimate is in general much weaker than the Hasse-Weil Bound. But note that it is better than Hasse-Weil for $q = 2$ and $q = 3$, where it coincides with the Serre Bound.

5.8. Assume that $F/\mathbb{F}_q$ is a non-rational function field. In this exercise we will provide a lower bound for the class number $h = h_F$.

(i) Show that the number of positive divisors of $F/\mathbb{F}_q$ of degree $2g$ is equal to $h \cdot (q^{g+1} - 1)/(q - 1)$.

(ii) Consider the constant field extension F_{2g} of degree $2g$. Let Q be a rational place of F_{2g} and let $P = Q \cap F_q$ be the restriction of Q to F. Then $a := 2g/\deg P$ is an integer, hence we obtain a positive divisor $a\,P$ of $F/\mathbb{F}_q$. Show that in this way one constructs at least $N_{2g}/2g$ distinct positive divisors of $F/\mathbb{F}_q$ of degree $2g$.

(iii) Using (i), (ii) and the Hasse-Weil Bound for N_{2g}, show that

$$h \geq \frac{q-1}{2} \cdot \frac{q^{2g} + 1 - 2gq^g}{g(q^{g+1} - 1)} \,.$$

(iv) Prove that $h > (q-1)/4$.

(v) Given a real number $M > 0$, show that there are - up to isomorphism - only finitely many non-rational function fields over finite fields having class number $h \le M$. In particular, there exist only finitely many non-rational function fields over finite fields with class number one (an analogous result does not hold for algebraic number fields).

5.9. Let $F/\mathbb{F}_q$ be a non-rational function field with class number $h = 1$. Prove:

(i) $q \le 4$.

(ii) If $q = 4$, then $g = 1$.

(iii) If $q = 3$, then $g \le 2$.

(iv) If $q = 2$, then $g \le 4$.

Remark. One can show that there is no function field of class number 1 with $q = 3$, $g = 2$ or $q = 2$, $g = 4$.

5.10. For this exercise we first recall some notations and introduce a few others. Let $S, T \subseteq \mathbb{P}_F$ be non-empty subsets of $\mathbb{P}_F$ such that $S \cup T = \mathbb{P}_F$ and $S \cap T = \emptyset$. Let $\mathcal{O}_S = \bigcap_{P\in S} \mathcal{O}_P$ be the corresponding holomorphy ring, cf. Section 3.2. For $0 \ne x \in F$, its S-divisor $(x)_S$ is defined as

$$(x)_S := \sum_{P\in S} v_P(x) P .$$

We consider the following groups:

$\mathrm{Div}(F)$, *the divisor group of* F

$\mathrm{Div}^0(F)$, *the group of divisors of degree* 0

$\mathrm{Princ}(F)$, *the group of principal divisors of* F

$\mathrm{Cl}^0(F) = \mathrm{Div}^0(F)/\mathrm{Princ}(F)$, *the group of divisor classes of degree* 0

Div_S, *the subgroup of* $\mathrm{Div}(F)$ *which is generated by all* $P \in S$

$\mathrm{Princ}_S := \{(x)_S \,|\, 0 \ne x \in F\}$

$\mathrm{Cl}_S = \mathrm{Div}_S/\mathrm{Princ}_S$, *the* S*-class group of* F

Finally we define

$h := h_F$, *the class number of* F

$h_S := \mathrm{ord}\,(\mathrm{Cl}_S)$, *the* S*-class number of* F

$r_S := \mathrm{ord}\,(\mathrm{Div}_T \cap \mathrm{Div}^0(F)) \,/\, (\mathrm{Div}_T \cap \mathrm{Princ}(F))$, *the regulator of* $\mathcal{O}_S$

$u_S := \gcd\{\deg P \,|\, P \in T\}$

(i) Show that Cl_S is isomorphic to $\mathrm{Div}(F)/(\mathrm{Princ}(F) + \mathrm{Div}_T)$.

(ii) Show that $u_S = \mathrm{ord}(\mathrm{Div}(F)/(\mathrm{Div}^0(F) + \mathrm{Div}_T))$.

(iii) Show that there is an exact sequence

$$0 \longrightarrow \frac{\mathrm{Div}_T \cap \mathrm{Div}^0(F)}{\mathrm{Div}_T \cap \mathrm{Princ}(F)} \longrightarrow \mathrm{Cl}^0(F) \longrightarrow \frac{\mathrm{Div}^0(F) + \mathrm{Div}_T}{\mathrm{Princ}(F) + \mathrm{Div}_T} \longrightarrow 0 .$$

(iv) Conclude that h_S and r_S are finite, $r_S | h$ and

$$h_S = u_S \cdot \frac{h}{r_S} .$$

Remark. We specialize the situation of this exercise as follows: F is an extension of the rational function field $\mathbb{F}_q(x)$ of degree $[F : \mathbb{F}_q(x)] = 2$. Choose $S := \{P \in \mathbb{P}_F \mid v_P(x) \geq 0\}$, then $\mathcal{O}_S$ is the integral closure of $\mathbb{F}_q[x]$ in F (see Section 3.2). Now we distinguish 3 cases.

Case 1. $(x)_\infty = P_1 + P_2$ with $P_1 \neq P_2$. Then $u_S = 1$, and r_S is the order of the divisor class $[P_2 - P_1]$ in the class group $\mathrm{Cl}(F)$, hence $h_S = h/\mathrm{ord}([P_2 - P_1])$.

Case 2. $(x)_\infty = 2P$. Now $u_S = r_S = 1$ and $h_S = h$.

Case 3. $(x)_\infty = P$ with $\deg P = 2$. Now $u_S = 2, r_S = 1$ and $h_S = 2h$.

In analogy with algebraic number theory, in case 1 the function field F is called real-quadratic, and in the cases 2 and 3 it is called imaginary quadratic (with respect to x).

There are more exercises about function fields over finite fields at the end of Chapter 6.

6

Examples of Algebraic Function Fields

Thus far we have encountered very few explicit examples of algebraic function fields, namely the rational function field $K(x)/K$ (cf. Section 1.2) and some quadratic extensions of the rational function field (Example 3.7.6). Now we would like to discuss some other examples in detail. These examples serve as an illustration of the general theory of algebraic function fields developed in Chapters 1, 3, 4 and 5. Some of the examples will be used in Chapter 8 for the construction of algebraic geometry codes.

Throughout this chapter K denotes a perfect field.

This assumption is not essential; actually most results of Chapter 6 hold for an arbitrary constant field K, with minor modifications.

6.1 Elliptic Function Fields

The rational function field $K(x)$ has genus 0. Conversely, if F/K is a function field of genus 0 which has a divisor $A \in \mathrm{Div}(F)$ of degree one then F/K is rational, see Proposition 1.6.3. Therefore the simplest non-rational function fields are fields of genus one.

Definition 6.1.1. *An algebraic function field F/K (where K is the full constant field of F) is said to be an elliptic function field if the following conditions hold:*

(1) the genus of F/K is $g = 1$, and

(2) there exists a divisor $A \in \mathrm{Div}(F)$ with $\deg A = 1$.

There are numerous connections between elliptic function fields and other branches of mathematics (such as number theory and complex analysis), and there exists an extensive literature on the theory of elliptic function fields, cf. [38]. Here we present only some basic facts on the subject.

H. Stichtenoth, *Algebraic Function Fields and Codes*,
Graduate Texts in Mathematics 254,

Proposition 6.1.2. *Let F/K be an elliptic function field.*
(a) If $\operatorname{char} K \neq 2$, *there exist $x, y \in F$ such that $F = K(x, y)$ and*

$$y^2 = f(x) \in K[x] \tag{6.1}$$

with a square-free polynomial $f(x) \in K[x]$ of degree 3.
(b) If $\operatorname{char} K = 2$, *there exist $x, y \in F$ such that $F = K(x, y)$ and*

$$y^2 + y = f(x) \in K[x] \quad \textit{with} \quad \deg f = 3\,, \tag{6.2}$$

or

$$y^2 + y = x + \frac{1}{ax + b} \quad \textit{with} \quad a, b \in K \quad \textit{and} \quad a \neq 0\,. \tag{6.3}$$

Proof. Choose a divisor A of degree one. By the Riemann-Roch Theorem, $\ell(A) = \deg A + 1 - g = 1$ (note that $\deg A > 2g - 2$). Hence A is equivalent to a positive divisor A_1, cf. Remark 1.4.5. As $\deg A_1 = 1$ we conclude that A_1 is a prime divisor $A_1 = P \in \mathbb{P}_F$. So we have shown that an elliptic function field F/K has at least one place $P \in \mathbb{P}_F$ with $\deg P = 1$.

We consider the spaces $K = \mathscr{L}(0) \subseteq \mathscr{L}(P) \subseteq \ldots \subseteq \mathscr{L}(nP) \subseteq \ldots$. Because $2g - 2 = 0$, the Riemann-Roch Theorem gives $\dim \mathscr{L}(iP) = i$ for all $i > 0$, hence $\mathscr{L}(P) = K$ and $\mathscr{L}((i+1)P) \supsetneqq \mathscr{L}(iP)$ for $i > 0$. We choose elements $x_1 \in \mathscr{L}(2P) \backslash K$ and $y_1 \in \mathscr{L}(3P) \backslash \mathscr{L}(2P)$. Their pole divisors are

$$(x_1)_\infty = 2P \quad \textit{and} \quad (y_1)_\infty = 3P\,.$$

As $[F : K(x_1)] = 2$ and $[F : K(y_1)] = 3$ (by Theorem 1.4.11), it follows that $F = K(x_1, y_1)$.

The seven elements $1, x_1, y_1, x_1^2, x_1 y_1, x_1^3, y_1^2$ are in the space $\mathscr{L}(6P)$. Since $\ell(6P) = 6$, there is a non-trivial relation

$$\alpha_1 y_1^2 + \beta_1 x_1 y_1 + \gamma_1 y_1 = \delta_1 x_1^3 + \varepsilon_1 x_1^2 + \lambda_1 x_1 + \mu_1 \tag{6.4}$$

with $\alpha_1, \beta_1, \ldots \in K$. The coefficient α_1 does not vanish; otherwise (6.4) would give an equation for y_1 over $K(x_1)$ of degree one (which is impossible as $F = K(x_1, y_1)$ and $[F : K(x_1)] = 2$). In the same manner we see that $\delta_1 \neq 0$. Multiply (6.4) by $\alpha_1^3 \delta_1^2$; then

$$\alpha_1^4 \delta_1^2 y_1^2 + \cdots = \alpha_1^3 \delta_1^3 x_1^3 + \cdots\,.$$

Setting $y_2 := \alpha_1^2 \delta_1 y_1$ and $x_2 := \alpha_1 \delta_1 x_1$ we obtain $F = K(x_2, y_2)$ and

$$y_2^2 + (\beta_2 x_2 + \gamma_2) y_2 = x_2^3 + \varepsilon_2 x_2^2 + \lambda_2 x_2 + \mu_2 \tag{6.5}$$

with $\beta_2, \gamma_2, \ldots \in K$. Now we must distinguish the cases $\operatorname{char} K \neq 2$ and $\operatorname{char} K = 2$.

(a) char $K \neq 2$. We set $y := y_2 + (\beta_2 x_2 + \gamma_2)/2$ and $x := x_2$; then $F = K(x, y)$ and

$$y^2 = x^3 + \varepsilon x^2 + \lambda x + \mu = f(x) \in K[x] \tag{6.1}$$

with $\varepsilon, \lambda, \mu \in K$. It remains to show that $f(x)$ is square-free. Suppose the contrary; i.e., $f(x) = (x - \zeta)^2(x - \eta)$ with $\zeta, \eta \in K$. Consider the element $z := y/(x - \zeta)$. Then $z^2 = x - \eta$, and $F = K(x, y) = K(x, z) = K(z)$. So F/K is rational, a contradiction.

(b) char $K = 2$. We have already shown that $F = K(x_2, y_2)$ with

$$y_2^2 + (\beta_2 x_2 + \gamma_2) y_2 = x_2^3 + \varepsilon_2 x_2^2 + \lambda_2 x_2 + \mu_2 \,. \tag{6.5}$$

We claim that $\beta_2 x_2 + \gamma_2 \neq 0$. In fact, if $\beta_2 x_2 + \gamma_2 = 0$ then $y_2^2 \in K(x_2)$; i.e., the extension $F/K(x_2)$ is purely inseparable of degree $p = 2$. By Proposition 3.10.2 the only intermediate field $K \subseteq F_0 \subseteq F$ such that F/F_0 is purely inseparable of degree p is the field $F_0 = F^p$, so $K(x_2) = F^p$. However, the genus of $K(x_2)/K$ is zero, and the genus of F^p/K is one (cf. Proposition 3.10.2(c)). This contradiction proves the claim.

We set $y_3 := y_2(\beta_2 x_2 + \gamma_2)^{-1}$; then $F = K(x_2, y_3)$ and

$$y_3^2 + y_3 = (\beta_2 x_2 + \gamma_2)^{-2}(x_2^3 + \varepsilon_2 x_2^2 + \lambda_2 x_2 + \mu_2) \,. \tag{6.6}$$

If $\beta_2 = 0$, the right hand side of (6.6) is a polynomial $f(x_2) \in K[x_2]$ of degree 3, and we are in the situation of (6.2).

If $\beta_2 \neq 0$, the right hand side of (6.6) can be written in the form

$$\nu x_2 + \varrho + \frac{\sigma}{(\beta_2 x_2 + \gamma_2)^2} + \frac{\tau}{\beta_2 x_2 + \gamma_2}$$

with $\nu, \varrho, \sigma, \tau \in K$ and $\nu \neq 0$. As K is perfect, $\sigma = \sigma_1^2$ for some $\sigma_1 \in K$, and the element $y_4 := y_3 + \sigma_1(\beta_2 x_2 + \gamma_2)^{-1}$ satisfies an equation

$$y_4^2 + y_4 = \nu_2 x_2 + \varrho_2 + \frac{\tau_2}{\beta_2 x_2 + \gamma_2} \tag{6.7}$$

with $\nu_2, \varrho_2, \tau_2 \in K$ and $\nu_2 = \nu \neq 0$. Also, the coefficient τ_2 does not vanish (otherwise $F = K(x_2, y_4)$ would be rational by (6.7)). We set $y := y_4$ and $x := \nu_2 x_2 + \varrho_2$ and obtain $F = K(x, y)$ with

$$y^2 + y = x + \frac{1}{ax + b} \tag{6.8}$$

($a, b \in K$ and $a \neq 0$). □

Next we show that each of the above equations (6.1), (6.2) and (6.3) defines an elliptic function field.

Proposition 6.1.3. *(a)* char $K \neq 2$. *Suppose that* $F = K(x,y)$ *with*

$$y^2 = f(x) \in K[x]\,, \tag{6.1}$$

where $f(x)$ *is a square-free polynomial of degree* 3. *Consider the decomposition* $f(x) = c\prod_{i=1}^{r} p_i(x)$ *of* $f(x)$ *into irreducible monic polynomials* $p_i(x) \in K[x]$ *with* $0 \neq c \in K$. *Denote by* $P_i \in \mathbb{P}_{K(x)}$ *the place of* $K(x)$ *corresponding to* $p_i(x)$, *and by* $P_\infty \in \mathbb{P}_{K(x)}$ *the pole of* x. *Then the following hold:*

(1) K *is the full constant field of* F, *and* F/K *is an elliptic function field.*

(2) The extension $F/K(x)$ *is cyclic of degree* 2. *The places* $P_1,\ldots,P_r$ *and* P_∞ *are ramified in* $F/K(x)$; *each of them has exactly one extension in* F, *say* $Q_1,\ldots,Q_r$ *and* Q_∞, *and we have* $e(Q_j|P_j) = e(Q_\infty|P_\infty) = 2$, $\deg Q_j = \deg P_j$ *and* $\deg Q_\infty = 1$.

(3) $P_1,\ldots,P_r$ *and* P_∞ *are the only places of* $K(x)$ *which are ramified in* $F/K(x)$, *and the different of* $F/K(x)$ *is*

$$\mathrm{Diff}(F/K(x)) = Q_1 + \cdots + Q_r + Q_\infty\,.$$

(b) char $K = 2$. *Suppose that* $F = K(x,y)$ *with*

$$y^2 + y = f(x) \in K[x] \quad \text{and} \quad \deg f(x) = 3 \tag{6.2}$$

or

$$y^2 + y = x + \frac{1}{ax+b} \quad \text{with} \quad a,b \in K \quad \text{and} \quad a \neq 0\,. \tag{6.3}$$

Denote by $P_\infty \in \mathbb{P}_{K(x)}$ *the pole of* x *in* $K(x)$ *and by* $P' \in \mathbb{P}_{K(x)}$ *the zero of* $ax+b$ *in* $K(x)$ *(in Case* (6.3)*). Then the following hold:*

(1) K *is the full constant field of* F, *and* F/K *is an elliptic function field.*

(2) The extension $F/K(x)$ *is cyclic of degree* 2. *The only places of* $K(x)$ *which ramify in* $F/K(x)$ *are*

$$\begin{array}{l} P_\infty\,, \ \textit{in Case}\ \ (6.2)\,, \\ P_\infty \ \textit{and}\ P', \ \textit{in Case}\ \ (6.3)\,. \end{array}$$

Let Q_∞ *(resp.* Q' *in Case* (6.3)*) be the place of* F/K *lying over* P_∞ *(resp.* P'*). Then* $\deg Q_\infty = \deg Q' = 1$ *and*

$$\mathrm{Diff}(F/K(x)) = \begin{cases} 4Q_\infty & \textit{in Case}\ (6.2)\,, \\ 2Q_\infty + 2Q' & \textit{in Case}\ (6.3)\,. \end{cases}$$

Proof. In case char $K \neq 2$, all assertions follow easily from Proposition 3.7.3 (see also Corollary 3.7.4 and Example 3.7.6). For the case of char $K = 2$ apply Proposition 3.7.8. □

For an elliptic function field F/K the zero divisor is a canonical divisor since

$$\ell(0) = 1 = g \quad and \quad \deg(0) = 0 = 2g - 2$$

(see Proposition 1.6.2). In each of the cases (6.1), (6.2) and (6.3) one can easily write down a differential $\omega \in \Omega_F$ with $(\omega) = 0$, namely

$$\omega = \begin{cases} y^{-1}dx & \text{in Case (6.1)}, \\ dx & \text{in Case (6.2)}, \\ (ax+b)^{-1}dx & \text{in Case (6.3)}. \end{cases}$$

The proof of this assertion is left to the reader. (Hint: Calculate the divisor of the differential dx by using Remark 4.3.7(c).)

Example 6.1.4. Let us briefly describe the classical example of an elliptic function field (without giving proofs). Consider a *lattice* $\Gamma \subseteq \mathbb{C}$; i.e.,

$$\Gamma = \mathbb{Z}\gamma_1 \oplus \mathbb{Z}\gamma_2$$

with $\gamma_1, \gamma_2 \in \mathbb{C}\backslash\{0\}$ and $\gamma_1/\gamma_2 \notin \mathbb{R}$. An *elliptic function* (with respect to Γ) is a meromorphic function $f(z)$ on $\mathbb{C}$ satisfying

$$f(z+\gamma) = f(z) \quad \text{for all} \quad \gamma \in \Gamma .$$

The elliptic functions form a subfield $\mathcal{M}(\Gamma)$ of the field of all meromorphic functions on $\mathbb{C}$, and $\mathbb{C} \subseteq \mathcal{M}(\Gamma)$ (we consider a complex number as a constant function). Two specific non-constant elliptic functions are the *Weierstrass $\wp$-function* which is defined by

$$\wp(z) := \frac{1}{z^2} + \sum_{0 \neq \gamma \in \Gamma} \left(\frac{1}{(z-\gamma)^2} - \frac{1}{\gamma^2} \right) ,$$

and its derivative $\wp'(z)$. It is not difficult to prove the following facts:

(1) $\mathcal{M}(\Gamma) = \mathbb{C}(\wp(z), \wp'(z))$, and

(2) $\wp'(z)^2 = 4\wp(z)^3 - g_2 \cdot \wp(z) - g_3$

with constants $g_2, g_3 \in \mathbb{C}$, where the polynomial $f(T) = 4T^3 - g_2T - g_3 \in \mathbb{C}[T]$ is square-free. Hence $\mathcal{M}(\Gamma)/\mathbb{C}$ is an elliptic function field by Proposition 6.1.3. For $\alpha \in \mathbb{C}$, every function $0 \neq f \in \mathcal{M}(\Gamma)$ has a Laurent series expansion

$$f(z) = \sum_{i=i_0}^{\infty} a_i (z-\alpha)^i$$

with $a_i \in \mathbb{C}$, $i_0 \in \mathbb{Z}$ and $a_{i_0} \neq 0$. Setting $v_\alpha(f) := i_0$ we define a discrete valuation v_α, hence a place P_α of $\mathcal{M}(\Gamma)/\mathbb{C}$. It is obvious that $P_\alpha = P_\beta$ if and only if $\alpha \equiv \beta \bmod \Gamma$. In this manner one obtains all places of the elliptic function field $\mathcal{M}(\Gamma)$.

Example 6.1.5. We would like to investigate some elliptic function fields F over the field $\mathbb{F}_2$. Let N denote the number of places of $F/\mathbb{F}_2$ of degree one. The Serre Bound states that

$$N \le 2 + 1 + g \cdot [2\sqrt{2}] = 5\,.$$

Let us show that (up to isomorphism) there exists exactly one elliptic function field $F/\mathbb{F}_2$ with $N = 5$. By Proposition 6.1.2(b) we can write $F = \mathbb{F}_2(x,y)$ with

$$y^2 + y = x + \frac{1}{x+b}\,, \qquad b \in \mathbb{F}_2 \tag{6.8}$$

or

$$y^2 + y = f(x) \in \mathbb{F}_2[x]\,, \quad \deg f(x) = 3\,, \tag{6.9}$$

Observe that there are exactly 3 places of $\mathbb{F}_2(x)$ of degree one, and every place of F of degree one must lie over one of them. In Case (6.8) two places of $\mathbb{F}_2(x)$ of degree one ramify in $F/\mathbb{F}_2(x)$, so $N \le 4$. In Case (6.9) we can assume that $f(x) = x^3 + bx + c$ with $b, c \in \mathbb{F}_2$ (if $f(x) = x^3 + x^2 + bx + c$, we replace y by $z := y + x$; then $z^2 + z = x^3 + b_1 x + c_1$). It remains to consider the four cases $f(x) = x^3, x^3 + x, x^3 + 1, x^3 + x + 1$. Using Kummer's Theorem (or its Corollary 3.3.8) one can easily calculate N in each of these cases; the result is

$$N = \begin{cases} 1 & \textit{for } y^2 + y = x^3 + x + 1\,, \\ 3 & \textit{for } y^2 + y = x^3 \textit{ or } x^3 + 1\,, \\ 5 & \textit{for } y^2 + y = x^3 + x\,. \end{cases}$$

So the only elliptic function field $F/\mathbb{F}_2$ with $N = 5$ is

$$F = \mathbb{F}_2(x,y) \quad \textit{with} \quad y^2 + y = x^3 + x\,. \tag{6.10}$$

Now we determine the L-polynomial $L_F(t)$ of (6.10). We know by Theorem 5.1.15 that $L_F(t) = a_0 + a_1 t + a_2 t^2$ with $a_0 = 1$, $a_2 = 2$ and $a_1 = N - (2+1) = 2$. Hence

$$L_F(t) = 1 + 2t + 2t^2 = (1 - \alpha t)(1 - \bar{\alpha} t) \tag{6.11}$$

with $\alpha = -1 + i = \omega\sqrt{2}$, $\omega = \exp(3\pi i/4)$ (here, $i = \sqrt{-1} \in \mathbb{C}$, and $\bar{\alpha}$ is the complex conjugate of α). Consider the constant field extension $F_r := F\mathbb{F}_{2^r}$ of degree r. The number N_r of places of degree one of the function field $F_r/\mathbb{F}_{2^r}$ is given by

$$N_r = 2^r + 1 - (\alpha^r + \bar{\alpha}^r) \tag{6.12}$$

(cf. Corollary 5.1.16). We obtain $N_r = 2^r + 1 - 2 \cdot 2^{r/2} \cdot \mathrm{Re}\,(\omega^r)$, hence

$$N_r = \begin{cases} 2^r + 1 & \textit{for } r \equiv 2, 6 \bmod 8\,, \\ 2^r + 1 + 2 \cdot 2^{r/2} & \textit{for } r \equiv 4 \bmod 8\,, \\ 2^r + 1 - 2 \cdot 2^{r/2} & \textit{for } r \equiv 0 \bmod 8\,, \\ 2^r + 1 + 2^{(r+1)/2} & \textit{for } r \equiv 1, 7 \bmod 8\,, \\ 2^r + 1 - 2^{(r+1)/2} & \textit{for } r \equiv 3, 5 \bmod 8\,. \end{cases}$$

We see that for $r \equiv 4 \bmod 8$ the upper Hasse-Weil Bound $q+1+2gq^{1/2}$ is attained; for $r \equiv 0 \bmod 8$ the lower Hasse-Weil Bound $q+1-2gq^{1/2}$ is attained; for $r=1$ the upper Serre Bound $q+1+g\cdot[2q^{1/2}]$ is attained.

We finish this section with a result that is fundamental to the theory of elliptic function fields F/K (over an arbitrary field K). Let us recall some notation: $\mathrm{Cl}(F)$ is the divisor class group of F/K, and $\mathrm{Cl}^0(F) \subseteq \mathrm{Cl}(F)$ is the subgroup consisting of the divisor classes of degree zero. For a divisor $B \in \mathrm{Div}(F)$, $[B] \in \mathrm{Cl}(F)$ denotes the corresponding divisor class. $A \sim B$ means that the divisors A, B are equivalent.

Proposition 6.1.6. *Let F/K be an elliptic function field. Define*

$$\mathbb{P}_F^{(1)} := \{P \in \mathbb{P}_F \mid \deg P = 1\}.$$

Then the following hold:
(a) For each divisor $A \in \mathrm{Div}(F)$ with $\deg A = 1$ there exists a unique place $P \in \mathbb{P}_F^{(1)}$ with $A \sim P$. In particular $\mathbb{P}_F^{(1)} \neq \emptyset$.
(b) Fix a place $P_0 \in \mathbb{P}_F^{(1)}$. Then the mapping

$$\Phi : \begin{cases} \mathbb{P}_F^{(1)} & \longrightarrow & \mathrm{Cl}^0(F)\,, \\ P & \longmapsto & [P-P_0] \end{cases} \tag{6.13}$$

is bijective.

Proof. (a) Let $A \in \mathrm{Div}(F)$ and $\deg A = 1$. We show the existence of a place $P \in \mathbb{P}_F^{(1)}$ with $A \sim P$ as in the proof of Proposition 6.1.2; since $\ell(A) = \deg A + 1 - g > 0$, there is a divisor $A_1 \sim A$ with $A_1 > 0$, and from $\deg A = 1$ follows immediately that $A_1 = P \in \mathbb{P}_F^{(1)}$.

Next we prove uniqueness. Suppose that $A \sim P$ and $A \sim Q$ for $P, Q \in \mathbb{P}_F^{(1)}$ and $P \neq Q$. Then $P \sim Q$; i.e., $P - Q = (x)$ for some $x \in F$. By Theorem 1.4.11 we have $[F : K(x)] = \deg\,(x)_\infty = \deg Q = 1$, hence $F = K(x)$. This is a contradiction as F/K is elliptic.
(b) First we show that Φ is surjective: Let $[B] \in \mathrm{Cl}^0(F)$. The divisor $B + P_0$ has degree one. By (a) we find a place $P \in \mathbb{P}_F^{(1)}$ with $B + P_0 \sim P$. Then $[B] = [P - P_0] = \Phi(P)$, and Φ is surjective.

Suppose now that $\Phi(P) = \Phi(Q)$ for $P, Q \in \mathbb{P}_F^{(1)}$. Then $P - P_0 \sim Q - P_0$, hence $P \sim Q$. By the uniqueness assertion of (a) it follows that $P = Q$. □

The bijection Φ of the foregoing proposition can be used to carry over the group structure of $\mathrm{Cl}^0(F)$ to the set $\mathbb{P}_F^{(1)}$. That means, we define for $P, Q \in \mathbb{P}_F^{(1)}$

$$P \oplus Q := \Phi^{-1}(\Phi(P) + \Phi(Q))\,. \tag{6.14}$$

Some consequences of this definition are put together in the following proposition.

Proposition 6.1.7. *Let F/K be an elliptic function field. Then:*

(a) $\mathbb{P}_F^{(1)}$ is an abelian group with respect to the operation $\oplus$ as defined in (6.14).

(b) The place P_0 is the zero element of the group $\mathbb{P}_F^{(1)}$.

(c) For $P, Q, R \in \mathbb{P}_F^{(1)}$ the following holds:

$$P \oplus Q = R \iff P + Q \sim R + P_0 \,.$$

(d) The map $\Phi : \mathbb{P}_F^{(1)} \to \mathrm{Cl}^0(F)$ given by (6.13) *is a group isomorphism.*

Proof. (a), (b) and (d) are obvious.
(c) By (6.14) we have the following equivalences:

$$\begin{aligned} P \oplus Q = R &\iff \Phi(R) = \Phi(P) + \Phi(Q) \\ &\iff R - P_0 \sim (P - P_0) + (Q - P_0) \\ &\iff P + Q \sim R + P_0 \,. \end{aligned}$$

□

We note that the group law on $\mathbb{P}_F^{(1)}$ depends on the choice of the place P_0. However, the group-theoretical structure of $\mathbb{P}_F^{(1)}$ is independent of this choice, since $\mathbb{P}_F^{(1)}$ is isomorphic to $\mathrm{Cl}^0(F)$ in any case. If F is represented in the form $F = K(x, y)$ as in Proposition 6.1.3, one usually chooses $P_0 := Q_\infty$, the pole of x.

6.2 Hyperelliptic Function Fields

In this section we discuss another important class of non-rational function fields over K.

Definition 6.2.1. *A hyperelliptic function field over K is an algebraic function field F/K of genus $g \geq 2$ which contains a rational subfield $K(x) \subseteq F$ with $[F : K(x)] = 2$.*

Lemma 6.2.2. *(a) A function field F/K of genus $g \geq 2$ is hyperelliptic if and only if there exists a divisor $A \in \mathrm{Div}(F)$ with $\deg A = 2$ and $\ell(A) \geq 2$.*
(b) Every function field F/K of genus 2 is hyperelliptic.

Proof. (a) Suppose that F/K is hyperelliptic. Choose an element $x \in F$ such that $[F : K(x)] = 2$, and consider the divisor $A := (x)_\infty$. Then $\deg A = 2$ and the elements $1, x \in \mathscr{L}(A)$ are linearly independent over K, hence $\ell(A) \geq 2$.

Conversely, assume that F/K has genus $g \geq 2$ and that A is a divisor of degree 2 with $\ell(A) \geq 2$. There is a divisor $A_1 \geq 0$ with $A_1 \sim A$, so $\deg A_1 = 2$ and $\ell(A_1) \geq 2$, and we can find an element $x \in \mathscr{L}(A_1) \backslash K$. Then $(x)_\infty \leq A_1$ and therefore $[F : K(x)] = \deg\,(x)_\infty \leq 2$. Since F/K is not rational, we conclude that $[F : K(x)] = 2$.

(b) Now a function field F/K of genus $g = 2$ is given. For each canonical divisor $W \in \mathrm{Div}(F)$ we have $\deg W = 2g - 2 = 2$ and $\ell(W) = g = 2$ by Corollary 1.5.16. This implies that F/K is hyperelliptic, by (a). □

If F/K is hyperelliptic and $K(x)$ is a subfield of F with $[F : K(x)] = 2$, the extension $F/K(x)$ is separable (if $F/K(x)$ were purely inseparable, F itself would be rational, by Proposition 3.10.2). Hence $F/K(x)$ is a cyclic extension of degree 2, and we can use Proposition 3.7.3 (resp. 3.7.8) to provide an explicit description of F/K (analogous to the description of elliptic function fields given in Section 6.1). For simplicity we restrict ourselves to the case $\mathrm{char}\, K \neq 2$.

Proposition 6.2.3. *Assume that* $\mathrm{char}\, K \neq 2$.
(a) Let F/K be a hyperelliptic function field of genus g. Then there exist $x, y \in F$ such that $F = K(x, y)$ and

$$y^2 = f(x) \in K[x] \tag{6.15}$$

with a square-free polynomial $f(x)$ of degree $2g + 1$ or $2g + 2$.
(b) Conversely, if $F = K(x, y)$ and $y^2 = f(x) \in K[x]$ with a square-free polynomial $f(x)$ of degree $m > 4$, then F/K is hyperelliptic of genus

$$g = \begin{cases} (m-1)/2 & \text{if } m \equiv 1 \bmod 2\,, \\ (m-2)/2 & \text{if } m \equiv 0 \bmod 2\,. \end{cases}$$

(c) Let $F = K(x, y)$ with $y^2 = f(x)$ as in (6.15). Then the places $P \in \mathbb{P}_{K(x)}$ which ramify in $F/K(x)$ are the following :

all zeros of $f(x)$, if $\deg f(x) \equiv 0 \bmod 2$,
all zeros of $f(x)$ and the pole of x, if $\deg f(x) \equiv 1 \bmod 2$.

Hence, if $f(x)$ decomposes into linear factors, exactly $2g + 2$ places of $K(x)$ are ramified in $F/K(x)$.

Proof. (b) and (c) are special cases of Proposition 3.7.3 (cf. also Example 3.7.6).
(a) As $F/K(x)$ is cyclic of degree 2 and $\mathrm{char}\, K \neq 2$, there exists an element $z \in F$ such that $F = K(x, z)$ and $z^2 = u(x) \in K(x)$. Write

$$u(x) = c \cdot \prod p_i(x)^{r_i} \quad , 0 \neq c \in K\,,$$

with pairwise distinct irreducible monic polynomials $p_i(x) \in K[x]$ and $r_i \in \mathbb{Z}$. Let $r_i = 2s_i + \varepsilon_i$, $s_i \in \mathbb{Z}$ and $\varepsilon_i \in \{0, 1\}$. Set

$$y := z \cdot \prod p_i^{-s_i}\,.$$

Then $F = K(x, y)$ and $y^2 = f(x)$ with a square-free polynomial $f(x) \in K[x]$. Now Example 3.7.6 implies that $\deg f = 2g + 1$ or $2g + 2$. □

In the case of $\operatorname{char} K = 2$, all places of F which are ramified in the quadratic extension $F/K(x)$ are wildly ramified, so their different exponent in $\operatorname{Diff}(F/K(x))$ is at least 2 (by Dedekind's Different Theorem or Proposition 3.7.8(c)). It follows that the number s of ramified places lies in the interval $1 \le s \le g+1$. One can easily construct examples for each s in this range. For example, the hyperelliptic field $F = K(x,y)$ with

$$y^2 + y = f(x) \in K[x], \quad \deg f(x) = 2g+1 \tag{6.16}$$

has genus g and exactly one ramified place in $F/K(x)$ (the pole of x). On the other hand, if

$$y^2 + y = \sum_{i=1}^{g+1} (x + a_i)^{-1} \tag{6.17}$$

with pairwise distinct $a_i \in K$, the genus of F/K is g, and we have exactly $g+1$ ramified places in $F/K(x)$. All this follows immediately from Proposition 3.7.8.

Thus far the condition $g \ge 2$ was not essential. Our previous results on hyperelliptic function fields hold for elliptic function fields as well. However, the next proposition is false in the case of an elliptic function field.

We recall that the space Ω_F of differentials of F/K is a one-dimensional F-module. Hence for $\omega_1, \omega_2 \in \Omega_F$ and $\omega_2 \ne 0$, the quotient $\omega_1/\omega_2 \in F$ is defined. $\Omega_F(0) = \{\omega \in \Omega_F \mid (\omega) \ge 0\}$ is the space of regular differentials of F/K.

Proposition 6.2.4. *Consider a hyperelliptic function field F/K of genus g and a rational subfield $K(x) \subseteq F$ with $[F : K(x)] = 2$. Then the following hold:*

(a) All rational subfields $K(z) \subseteq F$ with $[F : K(z)] \le g$ are contained in $K(x)$. In particular, $K(x)$ is the only rational subfield of F with $[F : K(x)] = 2$.

(b) $K(x)$ is the subfield of F which is generated by the quotients of regular differentials of F/K.

Proof. (a) Suppose that $[F : K(z)] \le g$ but $z \notin K(x)$. Then $F = K(x,z)$, and Riemann's Inequality (Theorem 3.11.4) yields the contradiction

$$g \le ([F : K(x)] - 1) \cdot ([F : K(z)] - 1) \le g - 1 .$$

(b) First we claim that the divisor $W := (g-1) \cdot (x)_\infty \in \operatorname{Div}(F)$ is a canonical divisor of F/K. This follows from Proposition 1.6.2 since $\deg W = 2g - 2$ (obvious) and $\ell(W) \ge g$ (the elements $1, x, \ldots, x^{g-1}$ are in $\mathscr{L}(W)$). Choose a differential $\omega \in \Omega_F$ with $(\omega) = W$; then the differentials $x^i\omega$, $0 \le i \le g-1$ are in $\Omega_F(0)$. As $\Omega_F(0)$ is a g-dimensional vector space over K (Remark 1.5.12), this implies

$$\Omega_F(0) = \{f(x) \cdot \omega \mid f(x) \in K[x] \text{ and } \deg f(x) \le g-1\}.$$

Thus $K(x)$ is the subfield of F generated by the quotients of regular differentials. □

We mention without proof that if F/K is a non-hyperelliptic function field of genus $g \ge 2$, the quotients of regular differentials generate F (under a weak additional assumption, e.g. the existence of a divisor of degree one), cf. [6].

6.3 Tame Cyclic Extensions of the Rational Function Field

We study function fields $F = K(x,y)$ which are defined by an equation

$$y^n = a \cdot \prod_{i=1}^{s} p_i(x)^{n_i} \tag{6.18}$$

with $s > 0$ pairwise distinct irreducible monic polynomials $p_i(x) \in K[x]$, $0 \ne a \in K$ and $0 \ne n_i \in \mathbb{Z}$. Throughout this section we will assume that the following conditions hold:

$$\operatorname{char} K \nmid n, \quad \textit{and} \quad \gcd(n, n_i) = 1 \quad \textit{for} \quad 1 \le i \le s. \tag{6.19}$$

Note that hyperelliptic function fields of characteristic $\ne 2$ are special cases of (6.18).

Proposition 6.3.1. *Suppose that $F = K(x,y)$ is defined by (6.18) and (6.19). Then we have:*

(a) K is the full constant field of F, and $[F : K(x)] = n$. If K contains a primitive n-th root of unity, $F/K(x)$ is a cyclic field extension.

(b) Let P_i (resp. P_∞) denote the zero of $p_i(x)$ (resp. the pole of x) in $K(x)$. The places $P_1, \ldots, P_s$ are totally ramified in $F/K(x)$. All places $Q_\infty \in \mathbb{P}_F$ with $Q_\infty \mid P_\infty$ have ramification index $e(Q_\infty \mid P_\infty) = n/d$ where

$$d := \gcd\left(n, \sum_{i=1}^{s} n_i \cdot \deg p_i(x)\right). \tag{6.20}$$

No places $P \in \mathbb{P}_{K(x)}$ other than $P_1, \ldots, P_s, P_\infty$ ramify in $F/K(x)$.

(c) The genus of F/K is

$$g = \frac{n-1}{2}\left(-1 + \sum_{i=1}^{s} \deg p_i(x)\right) - \frac{d-1}{2},$$

with d as in (6.20).

Proof. All assertions follow immediately from Proposition 3.7.3, Corollary 3.7.4 and Remark 3.7.5. □

Now we consider some special cases of Proposition 6.3.1.

Example 6.3.2. Let $F = K(x, y)$ with

$$y^n = (x^m - b)/(x^m - c)\,,$$

where $b, c \in K\backslash\{0\}$, $b \neq c$ and $\operatorname{char} K \nmid mn$. Then (6.18) and (6.19) hold, and we obtain from Proposition 6.3.1

$$g = (n-1)(m-1) = \big([F : K(x)] - 1\big)\big([F : K(y)] - 1\big)\,.$$

Thus Riemann's Inequality (Corollary 3.11.4) is sharp in this case.

Example 6.3.3. The function field $F = K(x, y)$ with defining equation

$$ax^m + by^n = c, \qquad a, b, c \in K\backslash\{0\}\,, \ \operatorname{char} K \nmid mn$$

has genus

$$g = \frac{1}{2}\big((n-1)(m-1) + 1 - \gcd(m, n)\big)\,.$$

Example 6.3.4. A function field $F = K(x, y)$ with

$$ax^n + by^n = c, \qquad a, b, c \in K\backslash\{0\}, \ \operatorname{char} K \nmid n$$

is said to be of *Fermat type*. Its genus is $g = (n-1)(n-2)/2$ by the previous example. This shows that the estimate for the genus given in Proposition 3.11.5 cannot be improved in general.

Example 6.3.5. Let $K = \mathbb{F}_{q^2}$ be the finite field of cardinality q^2, where q is a power of a prime number. Consider the function field $F = K(x, y)$ with

$$ax^{q+1} + by^n = c\,, \quad a, b, c \in \mathbb{F}_q\backslash\{0\}\,, \quad n \,|\, (q+1)\,. \tag{6.21}$$

We want to determine the number of rational places

$$N = N(F/\mathbb{F}_{q^2}) = \big|\{P \in \mathbb{P}_F\,;\, \deg P = 1\}\big|\,.$$

First we substitute $x_1 := \gamma x$, $y_1 := \delta y$ with $\gamma^{q+1} = a/c$ and $\delta^n = -b/c$, and we obtain $F = K(x_1, y_1)$ with $y_1^n = x_1^{q+1} - 1$ (observe that $\gamma, \delta \in \mathbb{F}_{q^2}$ since all elements of $\mathbb{F}_q$ are $(q+1)$-th powers of elements of $\mathbb{F}_{q^2}$). So we can assume from the beginning that $F = K(x, y)$ with

$$y^n = x^{q+1} - 1 \quad \textit{and} \quad n \,|\, (q+1)\,. \tag{6.22}$$

Let $P_\alpha \in \mathbb{P}_{K(x)}$ (resp. P_∞) denote the zero of $x - \alpha$ (resp. the pole of x) in $K(x)$. Each place $P \in \mathbb{P}_F$ of degree one lies over P_∞ or some P_α (with $\alpha \in K$), hence we have to study the decomposition of P_α and P_∞ in $F/K(x)$.

Case 1. $\alpha \in K$ and $\alpha^{q+1} = 1$. In this case α is a simple root of the polynomial $T^{q+1} - 1 \in K[T]$, and P_α is fully ramified in $F/K(x)$ by Proposition 6.3.1(b). So P_α has a unique extension $P \in \mathbb{P}_F$, and $\deg P = 1$.

Case 2. $\alpha \in K$ and $\alpha^{q+1} \neq 1$. We use Kummer's Theorem (resp. Corollary 3.3.8) to determine the decomposition of P_α in $F/K(x)$. The minimal polynomial of y over $K(x)$ is $\varphi(T) = T^n - (x^{q+1} - 1) \in K(x)[T]$, and

$$\varphi_\alpha(T) := T^n - (\alpha^{q+1} - 1) \in K[T]$$

has n distinct roots $\beta \in K = \mathbb{F}_{q^2}$ (here we use that $\alpha^{q+1} - 1 \in \mathbb{F}_q \setminus \{0\}$ and $n \mid (q+1)$). For any such β there is a unique place $P_{\alpha,\beta} \in \mathbb{P}_F$ with $P_{\alpha,\beta} \mid P_\alpha$ and $y - \beta \in P_{\alpha,\beta}$, and $P_{\alpha,\beta}$ is of degree one. So P_α has n distinct extensions $P \in \mathbb{P}_F$ with $\deg P = 1$.

Case 3. $\alpha = \infty$. In this case Kummer's Theorem does not apply directly as not all coefficients of the minimal polynomial of y over $K(x)$ are in the valuation ring $\mathcal{O}_\infty$ of P_∞. So we consider the element $z := y/x^{(q+1)/n}$ which satisfies the equation

$$z^n = 1 - (1/x)^{q+1}.$$

As $T^n - 1$ has n distinct roots in K, we see that P_∞ has n distinct extensions $P \in \mathbb{P}_F$, all of degree one, by Kummer's Theorem.

There are $q+1$ elements $\alpha \in \mathbb{F}_{q^2}$ belonging to Case 1, and $q^2 - (q+1)$ elements α that fall under Case 2. Summing up we find that $F/\mathbb{F}_{q^2}$ has

$$N = (q+1) + n(q^2 - (q+1)) + n = q + 1 + n(q^2 - q)$$

places of degree one. By Example 6.3.3 the genus of F is $g = (n-1)(q-1)/2$, hence

$$q^2 + 1 + 2gq = q^2 + 1 + q(n-1)(q-1) = q + 1 + n(q^2 - q).$$

We see that the function fields $F/\mathbb{F}_{q^2}$ which are defined by (6.21) are maximal; i.e., they attain the upper Hasse-Weil Bound

$$N = q^2 + 1 + 2gq \tag{6.23}$$

(over the constant field $\mathbb{F}_{q^2}$). Now one can easily determine the L-polynomial $L_F(t)$ of $F/\mathbb{F}_{q^2}$: if $\alpha_1, \ldots, \alpha_{2g} \in \mathbb{C}$ are the reciprocals of the roots of $L_F(t)$ then

$$N = q^2 + 1 - \sum_{i=1}^{2g} \alpha_i, \tag{6.24}$$

by Corollary 5.1.16. On the other hand, $|\alpha_i| = q$ by the Hasse-Weil Theorem. By (6.23) and (6.24) this implies $\alpha_i = -q$ for $i = 1, \ldots, 2g$, and thus

$$L_F(t) = (1 + qt)^{2g}. \tag{6.25}$$

The above proof shows that Equation (6.25) holds for all maximal function fields over $\mathbb{F}_{q^2}$.

Example 6.3.6. The special case $H := \mathbb{F}_{q^2}(x,y)$ with

$$x^{q+1} + y^{q+1} = 1 \tag{6.26}$$

is called the *Hermitian function field* over $\mathbb{F}_{q^2}$. It is a maximal function field by (6.23), so it provides an example of a maximal function field of genus $g = q(q-1)/2$ and shows that Proposition 5.3.3 cannot be improved. The number of places of degree one is $N = 1 + q^3$.

The Hermitian function field has some other remarkable properties. For instance, its automorphism group $\mathrm{Aut}(H/\mathbb{F}_{q^2})$ is very large, cf. [41] and Exercise 6.10. In Section 6.4 we shall give a second description of the Hermitian function field.

Remark 6.3.7. Once again we consider a function field $F = \mathbb{F}_{q^2}(u,v)$ as in Example 6.3.5; i.e., $au^{q+1} + bv^n = c$ with $a,b,c \in \mathbb{F}_q \setminus \{0\}$ and $n \mid (q+1)$. By (6.22) we can replace u,v by elements w,t such that $F = \mathbb{F}_{q^2}(t,w)$ and

$$w^n = t^{q+1} - 1\,. \tag{6.27}$$

On the other hand, let $H = \mathbb{F}_{q^2}(x,y)$ be the Hermitian function field given by $x^{q+1} + y^{q+1} = 1$. Write $q+1 = sn$, choose $\zeta \in \mathbb{F}_{q^2}$ with $\zeta^n = -1$ and set $z := \zeta y^s \in H$. Then

$$z^n = \zeta^n y^{q+1} = x^{q+1} - 1\,.$$

Thus $F = \mathbb{F}_{q^2}(u,v)$ is isomorphic to the subfield $\mathbb{F}_{q^2}(x,z) \subseteq H$. In other words, all function fields which were considered in Example 6.3.5 can be regarded as subfields of the Hermitian function field H. More generally, it is easily seen that the function fields $F = \mathbb{F}_{q^2}(u,v)$ with

$$au^n + bv^m = c\,, \qquad a,b,c \in \mathbb{F}_q \setminus \{0\}\,,\ m \mid (q+1) \text{ and } n \mid (q+1) \tag{6.28}$$

can be regarded as subfields of H. One can show by direct computation that all function fields (6.28) are maximal.

Example 6.3.8. We consider the function field $F = K(y,z)$ defined by

$$z^3 + y^3 z + y = 0\,. \tag{6.29}$$

F is called the function field of the *Klein Quartic.* The polynomial $T^3 + y^3 T + y \in K(y)[T]$ is absolutely irreducible (by Proposition 3.1.15), so K is the full constant field of F (see Corollary 3.6.8), and $[F : K(y)] = 3$.

It is convenient to choose other generators of F/K. We multiply (6.29) by y^6, set $x := -y^2 z$ and obtain $F = K(x,y)$ with

$$y^7 = x^3/(1-x)\,. \tag{6.30}$$

If $\mathrm{char}\, K = 7$, $F/K(x)$ is purely inseparable; therefore F/K is rational in this case (Proposition 3.10.2(c)). In case $\mathrm{char}\, K \neq 7$ we can apply Proposition

6.3.1. Exactly three places of $K(x)$ ramify in $F/K(x)$, namely the pole P_∞ of x, the zero P_0 of x and the zero P_1 of $x-1$. All these places have ramification index $e=7$ in the extension $F/K(x)$, and the genus of F/K is $g=3$.

Now we specialize to $K=\mathbb{F}_2$. As in Chapter 5, N_r denotes the number of places of degree one in the constant field extension $F_r = F\mathbb{F}_{2^r}$. We claim that

$$N_1 = 3\,,\ N_2 = 5 \ \textit{and}\ N_3 = 24\,. \tag{6.31}$$

$N_1 = 3$ is obvious since the three places of $\mathbb{F}_2(x)$ of degree one are fully ramified in F, so each of them has a unique extension of degree one in $\mathbb{P}_F$. For $r=2$ the constant field is $\mathbb{F}_4 = \{0,1,\alpha,\alpha+1\}$ where $\alpha^2+\alpha+1=0$. Let $P_\gamma \in \mathbb{P}_{K(x)}$ be the zero of $x-\gamma$. We determine the decomposition of P_γ in $\mathbb{F}_4(x,y)/\mathbb{F}_4(x)$ for $\gamma \in \{\alpha,\alpha+1\}$. In order to apply Kummer's Theorem we have to study the polynomials

$$\varphi_\alpha(T) = T^7 + \frac{\alpha^3}{1+\alpha} = T^7+\alpha \ \textit{and}$$
$$\varphi_{\alpha+1}(T) = T^7 + \frac{(\alpha+1)^3}{1+(1+\alpha)} = T^7+\alpha+1\,.$$

Both polynomials have only simple irreducible factors in $\mathbb{F}_4[T]$ as they are relatively prime to their derivative, and α (resp. $\alpha+1$) is the only root of $\varphi_\alpha(T)$ (resp. $\varphi_{\alpha+1}(T)$) in $\mathbb{F}_4$. Hence there is exactly one place of degree one lying over P_α (resp. $P_{\alpha+1}$), the other extensions of P_α and $P_{\alpha+1}$ are of degree >1. Summing up, we have found exactly 5 places of $\mathbb{F}_4(x,y)/\mathbb{F}_4$ of degree one, so $N_2=5$.

Next we consider the constant field $\mathbb{F}_8 = \mathbb{F}_2(\beta)$ where $\beta^3+\beta+1=0$. One has to study the decomposition of the polynomials

$$\varphi_\gamma(T) = T^7 + \frac{\gamma^3}{1+\gamma} \in \mathbb{F}_8[T]$$

for $\gamma \in \mathbb{F}_8 \setminus \{0,1\}$. For $\gamma \in \{\beta,\beta^2,\beta^4\}$ we have $\varphi_\gamma(T) = T^7+1$ which decomposes into seven distinct linear factors in $\mathbb{F}_8[T]$. For $\gamma \in \{\beta^3,\beta^5,\beta^6\}$, $\varphi_\gamma(T)$ has no root in $\mathbb{F}_8$. Hence $N_3 = 3+3\cdot 7 = 24$, and the proof of (6.31) is finished.

Now it is easy to determine the L-polynomial of the Klein Quartic over $\mathbb{F}_2$. With the notations as in Corollary 5.1.17 we find $S_1 = S_2 = 0$ and $S_3 = 24-(8+1) = 15$, hence $a_0=1$, $a_1=a_2=0$, $a_3=5$, $a_4=a_5=0$ and $a_6=8$. Thus

$$L_F(t) = 1+5t^3+8t^6\,.$$

The Klein Quartic over $\mathbb{F}_8$ provides an example where the upper Serre Bound $N = q+1+g\cdot[2q^{1/2}]$ is attained (Theorem 5.3.1), since we have $N = 24 = 8+1+3\cdot[2\sqrt{8}]$.

6.4 Some Elementary Abelian p-Extensions of $K(x)$, char $K = p > 0$

In this section K is a field of characteristic $p > 0$.

The function fields to be discussed in this section have interesting applications in coding theory, cf. Chapter 8 and 9.

Proposition 6.4.1. *Consider a function field $F = K(x,y)$ with*

$$y^q + \mu y = f(x) \in K[x]\,, \tag{6.32}$$

where $q = p^s > 1$ is a power of p and $0 \neq \mu \in K$. Assume that $\deg f =: m > 0$ is prime to p, and that all roots of $T^q + \mu T = 0$ are in K. Then the following hold:

(a) $[F : K(x)] = q$, and K is the full constant field of F.

(b) $F/K(x)$ is Galois. The set $A := \{\gamma \in K \mid \gamma^q + \mu\gamma = 0\}$ is a subgroup of order q of the additive group of K. For all $\sigma \in \mathrm{Gal}(F/K(x))$ there is a unique $\gamma \in A$ such that $\sigma(y) = y + \gamma$, and the map

$$\begin{cases} \mathrm{Gal}(F/K(x)) & \longrightarrow \; A\,, \\ \qquad \sigma & \longmapsto \; \gamma \end{cases}$$

is an isomorphism of $\mathrm{Gal}(F/K(x))$ onto A.

(c) The pole $P_\infty \in \mathbb{P}_{K(x)}$ of x in $K(x)$ has a unique extension $Q_\infty \in \mathbb{P}_F$, and $Q_\infty | P_\infty$ is totally ramified (i.e., $e(Q_\infty/P_\infty) = q$). Hence Q_∞ is a place of F/K of degree one.

(d) P_∞ is the only place of $K(x)$ which ramifies in $F/K(x)$.

(e) The genus of F/K is $g = (q-1)(m-1)/2$.

(f) The divisor of the differential dx is

$$(dx) = (2g-2)Q_\infty = \big((q-1)(m-1)-2\big)Q_\infty\,.$$

(g) The pole divisor of x (resp. y) is $(x)_\infty = qQ_\infty$ (resp. $(y)_\infty = mQ_\infty$).

(h) Let $r \geq 0$. Then the elements $x^i y^j$ with

$$0 \leq i\,, \quad 0 \leq j \leq q-1\,, \quad qi + mj \leq r$$

form a basis of the space $\mathscr{L}(rQ_\infty)$ over K.

(i) For all $\alpha \in K$ one of the following cases holds:

Case 1. The equation $T^q + \mu T = f(\alpha)$ has q distinct roots in K.

Case 2. The equation $T^q + \mu T = f(\alpha)$ has no root in K.

In Case 1, for each β with $\beta^q + \mu\beta = f(\alpha)$ there exists a unique place $P_{\alpha,\beta} \in \mathbb{P}_F$ such that $P_{\alpha,\beta}|P_\alpha$ and $y(P_{\alpha,\beta}) = \beta$. Hence P_α has q distinct extensions in $F/K(x)$, each of degree one.

In Case 2, all extensions of P_α in F have degree > 1.

Proof. Equation (6.32) is a special case of the situation which was considered in Proposition 3.7.10, so (a) - (e) hold.

(g) $(x)_\infty = qQ_\infty$ follows from (c). The elements x and y have the same poles, hence Q_∞ is the only pole of y as well. Since $q \cdot v_{Q_\infty}(y) = v_{Q_\infty}(y^q + y) = v_{Q_\infty}(f(x)) = -mq$ we obtain $(y)_\infty = mQ_\infty$.

(f) The different of $F/K(x)$ is $\mathrm{Diff}(F/K(x)) = (q-1)(m+1)Q_\infty$, by Proposition 3.7.10(d). So Remark 4.3.7(c) yields

$$(dx) = -2(x)_\infty + \mathrm{Diff}(F/K(x)) = \big((q-1)(m-1) - 2\big)Q_\infty = (2g-2)Q_\infty\,.$$

(h) The elements $1, y, \ldots, y^{q-1}$ form an integral basis of $F/K(x)$ at all places $P \in \mathbb{P}_{K(x)}$ different from P_∞. This follows from Theorem 3.5.10(b), as the minimal polynomial $\varphi(T) = T^q + \mu T - f(x)$ of y over $K(x)$ is in $\mathcal{O}_P[T]$ and for all $Q|P$,

$$v_Q(\varphi'(y)) = v_Q(\mu) = 0 = d(Q|P)\,.$$

Let $z \in \mathscr{L}(rQ_\infty)$. As Q_∞ is the only pole of z, z is integral over $\mathcal{O}_P$ for all $P \in \mathbb{P}_{K(x)}$, $P \neq P_\infty$, thus $z = \sum_{j=0}^{q-1} z_j y^j$ with $z_j \in K(x)$, and z_j has no poles other than P_∞. Hence z_j is a polynomial in $K[x]$; i.e.,

$$z = \sum_{j=0}^{q-1} \sum_{i \geq 0} a_{ij} x^i y^j \quad \textit{with} \quad a_{ij} \in K\,. \tag{6.33}$$

The elements $x^i y^j$ with $0 \leq j \leq q-1$ have pairwise distinct pole orders because $v_{Q_\infty}(x) = -q$, $v_{Q_\infty}(y) = -m$ and m and q are relatively prime. Therefore the Strict Triangle Inequality implies

$$v_{Q_\infty}(z) = \min\{-iq - jm \mid a_{ij} \neq 0\}\,.$$

This proves (h).

(i) Suppose there is some $\beta \in K$ such that $\beta^q + \mu\beta = f(\alpha)$. It follows that $(\beta + \gamma)^q + \mu(\beta + \gamma) = f(\alpha)$ for all γ with $\gamma^q + \mu\gamma = 0$, so

$$T^q + \mu T - f(\alpha) = \prod_{j=1}^{q} (T - \beta_j)$$

with pairwise distinct elements $\beta_j \in K$. By Corollary 3.3.8(c) there exists for $j = 1, \ldots, q$, a unique place $P_j \in \mathbb{P}_F$ such that $P_j|P_\alpha$ and $y - \beta_j \in P_j$, and the degree of P_j is one.

In Case 2 the polynomial $T^q + \mu T - f(\alpha) \in K[T]$ splits into pairwise distinct irreducible factors of degree > 1. By Corollary 3.3.8(a), all places $P \in \mathbb{P}_F$ with $P|P_\alpha$ have degree > 1. □

Example 6.4.2. We consider a special case of the previous proposition, namely

$$F = \mathbb{F}_{q^2}(x,y) \quad \textit{with} \quad y^q + y = x^m \quad \textit{and} \quad m|(q+1)\,. \tag{6.34}$$

The genus of F is $g = (q-1)(m-1)/2$. We claim that $F/\mathbb{F}_{q^2}$ has

$$N = 1 + q\big(1 + (q-1)m\big) \tag{6.35}$$

places of degree one. The pole Q_∞ of x is one of them. The other places of degree one are extensions of some place $P_\alpha \in \mathbb{P}_{K(x)}$. Hence, by Proposition 6.4.1(i), we have to count the elements $\alpha \in \mathbb{F}_{q^2}$ such that the equation

$$T^q + T = \alpha^m \tag{6.36}$$

has a root $\beta \in \mathbb{F}_{q^2}$. The map $\beta \mapsto \beta^q + \beta$ is the trace mapping from $\mathbb{F}_{q^2}$ to $\mathbb{F}_q$, and therefore it is surjective (cf. Appendix A). Thus (6.36) has a root in $\mathbb{F}_{q^2}$ if and only if $\alpha^m \in \mathbb{F}_q$. Let $U \subseteq \mathbb{F}_{q^2}^*$ be the subgroup of order $(q-1)m$ (here we use the assumption $m|(q+1)$). Then for $\alpha \in \mathbb{F}_{q^2}$,

$$\alpha^m \in \mathbb{F}_q \iff \alpha \in U \cup \{0\}\,.$$

Hence $N = 1 + q((q-1)m+1)$ by Proposition 6.4.1(i). This proves (6.35).

Because $1 + q((q-1)m+1) = 1 + q^2 + 2gq$, the fields which are defined by (6.34) provide other examples of maximal function fields over $\mathbb{F}_{q^2}$.

Example 6.4.3. The Hermitian function field H which was studied in Example 6.3.6 is given by

$$H = \mathbb{F}_{q^2}(u,v) \quad \textit{with} \quad u^{q+1} + v^{q+1} = 1\,. \tag{6.37}$$

We choose $a, b, c \in \mathbb{F}_{q^2}$ such that

$$a^{q+1} = -1\,, \quad b^q + b = 1 \quad \textit{and} \quad c = -ab^q\,;$$

then it follows that

$$\begin{aligned} ab^q + c &= 0\,, \\ a^q b + c^q &= (ab^q + c)^q = 0\,, \\ ac^q + a^q c &= a(-a^q b) + a^q(-ab^q) = -a^{q+1}(b + b^q) = 1\,. \end{aligned} \tag{6.38}$$

We set

$$x = \frac{1}{u + av} \quad \textit{and} \quad y = \frac{bu + cv}{u + av}\,.$$

Then $H = \mathbb{F}_{q^2}(x,y)$, and we obtain

$$(u + av)^{q+1} \cdot x^{q+1} = 1 \tag{6.39}$$

and

$$\begin{aligned}&(u+av)^{q+1}\cdot(y^q+y)\\&=(u+av)(bu+cv)^q+(u+av)^q(bu+cv)\\&=(b^q+b)u^{q+1}+(b^qa+c)u^qv+(c^q+ba^q)uv^q+(ac^q+a^qc)v^{q+1}\\&=u^{q+1}+v^{q+1}=1 \qquad (6.40)\end{aligned}$$

(we have used (6.38)). Comparing Equations (6.39) and (6.40) we see that

$$H=\mathbb{F}_{q^2}(x,y) \quad \textit{and} \quad y^q+y=x^{q+1}\,.$$

So the Hermitian function field H can be regarded as a special case of the function fields which were considered in Example 6.4.2. This representation of H is particularly useful, because we have a simple explicit description of the canonical divisor $W=(dx)$, the spaces $\mathscr{L}(rQ_\infty)$ and all places of degree one (by Proposition 6.4.1 and Example 6.4.2). For applications to coding theory in Section 8.3 we put together these results in a lemma.

Lemma 6.4.4. *The Hermitian function field over $\mathbb{F}_{q^2}$ can be defined by*

$$H=\mathbb{F}_{q^2}(x,y) \quad \textit{with} \quad y^q+y=x^{q+1}\,. \qquad (6.41)$$

It has the following properties:

(a) The genus of H is $g=q(q-1)/2$.

(b) H has q^3+1 places of degree one over $\mathbb{F}_{q^2}$, namely

(1) the common pole Q_∞ of x and y, and

(2) for each $\alpha\in\mathbb{F}_{q^2}$ there are q elements $\beta\in\mathbb{F}_{q^2}$ such that $\beta^q+\beta=\alpha^{q+1}$, and for all such pairs (α,β) there is a unique place $P_{\alpha,\beta}\in\mathbb{P}_H$ of degree one with $x(P_{\alpha,\beta})=\alpha$ and $y(P_{\alpha,\beta})=\beta$.

(c) $H/\mathbb{F}_{q^2}$ is a maximal function field.

(d) The divisor of the differential dx is $(dx)=(q(q-1)-2)Q_\infty$.

(e) For $r\geq 0$, the elements x^iy^j with $0\leq i$, $0\leq j\leq q-1$ and $iq+j(q+1)\leq r$ form a basis of $\mathscr{L}(rQ_\infty)$.

Remark 6.4.5. One can show that the Hermitian function field is - up to isomorphism - the only maximal function field over $\mathbb{F}_{q^2}$ of genus $g=q(q-1)/2$, see [34].

6.5 Exercises

6.1. Consider the function field $F=\mathbb{F}_2(x,y)$ over $\mathbb{F}_2$ which is defined by the equation $y^2=f(x)$. For each of the following choices of $f(x)\in\mathbb{F}_2(x)$, determine the L-polynomial $L(t)$.

(i) $f(x) = x^3 + 1$,
(ii) $f(x) = x^3 + x$,
(iii) $f(x) = x^3 + x + 1$,
(iv) $f(x) = (x^2 + x)/(x^3 + x + 1)$.

6.2. Let $F = \mathbb{F}_3(x, y)$ be the elliptic function field over $\mathbb{F}_3$ which is defined by the equation $y^2 = x^3 - x$. Determine N_r for all $r \geq 1$.

6.3. Show that there exists an elliptic function field over $F/\mathbb{F}_5$ having 10 rational places, cf. Exercise 5.1. Is F unique (up to isomorphism)?

6.4. Construct a function field of genus $g = 2$ over $\mathbb{F}_3$ with $N = 8$ rational places.

6.5. Consider a function field $E = \mathbb{F}_2(x, y, z)$ and its subfield $F = \mathbb{F}_2(x, y)$, where x, y, z satisfy the equations

$$y^2 x + y + x^2 + 1 = 0 \quad \textit{and} \quad z^2 y + z + y^2 + 1 = 0\,.$$

(i) Show that $[E : F] = [F : \mathbb{F}_2(x)] = 2$, and that the pole of x is totally ramified in $E/\mathbb{F}_2(x)$.

(ii) Determine the genus and the number of rational places of $F/\mathbb{F}_2$. Compare with Example 6.1.5.

(iii) Determine the genus and the number of rational places of $E/\mathbb{F}_2$.

6.6. (p prime, $p \equiv 1 \bmod 4$) Consider the function field $F = \mathbb{F}_p(x, y)$ with the defining equation

$$y^p - y = x^{p+1}\,.$$

It is clear from Proposition 6.4.1 that $F/\mathbb{F}_p(x)$ is a Galois extension of degree $[F : \mathbb{F}_p(x)] = p$, and that x has a unique pole Q_∞ in F.

(i) Show that the Riemann-Roch space $\mathscr{L}(pQ_\infty)$ has dimension 2 and is generated by the elements 1 and x.

(ii) Determine all rational places of $F/\mathbb{F}_p$.

(iii) Show that the automorphism group $\mathrm{Aut}\,(F/\mathbb{F}_p)$ acts transitively on the places of degree one.

Hint. Choose $\alpha \in \mathbb{F}_p$ with $\alpha^2 = -1$ (only here the assumption $p \equiv 1 \bmod 4$ is used). Show that there is an automorphism $\sigma \in \mathrm{Aut}\,(F/\mathbb{F}_p)$ with $\sigma(y) = 1/y$ and $\sigma(x) = \alpha x/y$. This automorphism permutes the zero and the pole of y.

6.7. (p prime, $p \equiv 1 \bmod 4$) Consider the function fields $E = \mathbb{F}_p(s, t)$ with the defining equation

$$t^p + t = s^{p+1}\,,$$

and $F = \mathbb{F}_p(x, y)$ with

$$y^p - y = x^{p+1}$$

(as in the previous exercise).

(i) Show that the extension $E/\mathbb{F}_p(s)$ is not Galois, and determine all rational places of $E/\mathbb{F}_p$.

(ii) It follows that E and F have the same number of rational places, namely $N = p+1$. Show that E and F are not isomorphic.

(iii) Consider the constant field extensions $E_2 := E\mathbb{F}_{p^2}$ and $F_2 := F\mathbb{F}_{p^2}$ of degree 2. Show that E_2 is isomorphic to F_2.

The subsequent exercises are related to the automorphism group $\mathrm{Aut}\,(F/K)$ *of a function field* F/K*. Recall that we have already shown the following facts:*

(1) The automorphism group of a rational function field $K(x)/K$ *is isomorphic to* $\mathrm{PGL}_2(K)$ *(see Exercise* 1.2.*)*

(2) If K *is a finite field, then the automorphism group of* F/K *is finite (see Exercise* 5.4*)*

(3) Let K *be an algebraically closed field and* G *a finite subgroup of* $\mathrm{Aut}\,(F/K)$*. Assume that the order of* G *is not divisible by the characteristic of* K *and that the genus of* F/K *is* $g \geq 2$*. Then the order of* G *satisfies the bound* $\mathrm{ord}\, G \leq 84(g-1)$ *(see Exercise* 3.18*).*

We mention without proof that $\mathrm{Aut}\,(F/K)$ *is always a finite group if* K *is algebraically closed and the genus of* F *is* ≥ 2*. Most proofs of this fact use the theory of Weierstrass points.*

6.8. Let F/K be a function field with exact constant field K, and consider the constant field extension FL/L with some algebraic extension field $L \supseteq K$.

(i) Let σ be an automorphism of F/K. Show that there is a unique automorphism $\tilde{\sigma}$ of FL/L whose restriction to F is σ.

(ii) Let $\mathrm{Aut}\,(F/K)$ be the group of automorphisms of F over K. With notation as in (i), show that the map $\sigma \mapsto \tilde{\sigma}$ is a monomorphism of $\mathrm{Aut}\,(F/K)$ into $\mathrm{Aut}\,(FL/L)$.

As a consequence of (ii), one can consider the automorphism group $\mathrm{Aut}\,(F/K)$ as a subgroup of $\mathrm{Aut}\,(\bar{F}/\bar{K})$, where $\bar{F}$ is the constant field extension of F with the algebraic closure $\bar{K}$ of K.

6.9. Consider the rational function field $F := \mathbb{F}_q(x)$ and its automorphism group $G := \mathrm{Aut}\,(F/\mathbb{F}_q)$. We know that $G \simeq \mathrm{PGL}_2(\mathbb{F}_q)$, see Exercise 1.2.

(i) Determine the order of G.

(ii) Let $U := \{\sigma \in G \,|\, \sigma(x) = x + c \text{ with } c \in \mathbb{F}_q\}$. Show that U is a p-Sylow subgroup of G (with $p := \mathrm{char}\, \mathbb{F}_q$). Find an element $z \in F$ such that the fixed field of U is $F^U = \mathbb{F}_q(z)$. Describe all ramified places, their ramification indices and different exponents in F/F^U.

(iii) Let $V := \{\sigma \in G \,|\, \sigma(x) = ax + c \text{ with } a, c \in \mathbb{F}_q \text{ and } a \neq 0\}$. Find an element $v \in F$ such that the fixed field of V is $F^V = \mathbb{F}_q(v)$. Describe all ramified places, their ramification indices and different exponents in F/F^V.

(iv) Show that exactly 2 places of F^G are ramified in F/F^G, and both of them are rational places of F^G. Determine their relative degrees, ramification indices and different exponents in F/F^G. Show that all places of F of degree 2 are conjugate under the group G.

(v) Find an element $t \in F$ such that the fixed field of G is $F^G = \mathbb{F}_q(t)$.

6.10. Let $K = \mathbb{F}_{q^2}$ and consider the Hermitian function field $H = K(x,y)$ with defining equation

$$y^q + y = x^{q+1} ,$$

cf. Lemma 6.4.4. The element x has a unique pole in H that will be denoted by Q_∞.

(i) Show that for each pair $(d,e) \in K \times K$ with $e^q + e = d^{q+1}$ there is an automorphism $\sigma \in \mathrm{Aut}\,(H/K)$ with $\sigma(x) = x + d$ and $\sigma(y) = y + d^q x + e$. These automorphisms form a subgroup $V \subseteq \mathrm{Aut}\,(H/K)$ of order q^3.

(ii) Show that for each element $c \in K^\times$ there is an automorphism $\tau \in \mathrm{Aut}\,(H/K)$ with $\tau(x) = cx$ and $\tau(y) = c^{q+1}y$. These automorphisms form a cyclic subgroup $W \subseteq \mathrm{Aut}\,(H/K)$ of order $q^2 - 1$.

(iii) Let $U \subseteq \mathrm{Aut}\,(H/K)$ be the group which is generated by V and W. Prove:

- *(a)* $\mathrm{ord}\, U = q^3(q^2-1)$, and V is a normal subgroup of U.
- *(b)* For every $\rho \in U$ holds $\rho(Q_\infty) = Q_\infty$.
- *(c)* The group U acts transitively on the set $S := \{Q \mid Q$ is a rational place of H/K and $Q \neq Q_\infty\}$.

(iv) Show that every automorphism $\lambda \in \mathrm{Aut}\,(H/K)$ with $\lambda(Q_\infty) = Q_\infty$ lies in U.

Hint. Observe that the elements $1, x, y$ form a K-basis of $\mathscr{L}((q+1)Q_\infty)$, by Lemma 6.4.4.

(v) Show that there is an automorphism $\mu \in \mathrm{Aut}\,(H/K)$ with $\mu(x) = x/y$ and $\mu(y) = 1/y$. This automorphism maps the place Q_∞ to the common zero of x and y.

(vi) Let $G \subseteq \mathrm{Aut}\,(H/K)$ be the group which is generated by U and μ. Prove:

- *(a)* G acts transitively on the set of all rational places of H/K.
- *(b)* $G = \mathrm{Aut}\,(H/K)$ and $\mathrm{ord}\, G = q^3(q^3+1)(q^2-1)$.
- *(c)* If $g = g(H)$ denotes the genus of H/K, then $\mathrm{ord}\, G > 16g^4 > 84(g-1)$.

6.11. We assume for simplicity that K is an algebraically closed field. Let F/K be a hyperelliptic function field of genus $g \geq 2$, and let $K(x) \subseteq F$ be its unique rational subfield with $[F : K(x)] = 2$ (cf. Proposition 6.2.4). We set $S := \{P \in \mathbb{P}_F \mid P$ is ramified in $F/K(x)\}$.

(i) Recall that $|S| = 2g+2$ if $\mathrm{char}\, K \neq 2$, and $1 \leq |S| \leq g+1$ if $\mathrm{char}\, K = 2$.

(ii) Show that $\mathrm{Aut}\,(F/K)$ acts on S (i.e., if $\sigma \in \mathrm{Aut}\,(F/K)$ and $P \in S$, then $\sigma(P) \in S$).

(iii) Fix a place $P_0 \in S$ and consider $U := \{\sigma \in \mathrm{Aut}\,(F/K) \,|\, \sigma(P_0) = P_0\}$. Show that U is a subgroup of $\mathrm{Aut}\,(F/K)$ of finite index.

(iv) Show that the subgroup U as above is a finite group.

(v) Conclude that the automorphism group of a hyperelliptic function field (over an arbitrary perfect constant field) is finite.

6.12. Let F/K be a function field over an algebraically closed field. For a place $P \in \mathbb{P}_F$ we consider its *Weierstrass semigroup*

$$W(P) := \{r \in \mathbb{N}_0 \,|\, r \textit{ is a pole number of } P\},$$

see Definition 1.6.7. Assume now that F/K is hyperelliptic of genus g, and $K(x) \subseteq F$ is the unique rational subfield of F with $[F : K(x)] = 2$. Let $S \subseteq \mathbb{P}_F$ be the set of ramified places in $F/K(x)$ (as in the previous exercise). Prove:

(i) If $P \in S$, then $W(P) = \{0, 2, 4, \ldots, 2g-2, 2g, 2g+1, 2g+2, \ldots\} = \mathbb{N}_0 \setminus \{1, 3, \ldots, 2g-1\}$.

(ii) If $P \notin S$, then $W(P) = \{0, g+1, g+2, \ldots\} = \mathbb{N}_0 \setminus \{1, 2, \ldots, g\}$.

These results show that the Weierstrass points (see Remark 1.6.9) of a hyperelliptic function field F/K are exactly the places which are ramified in the extension $F/K(x)$, where $K(x) \subseteq F$ is the unique rational subfield of degree $[F : K(x)] = 2$.

6.13. Let K be a perfect field of characteristic $\neq 2$ and $F = K(x,y)$ a hyperelliptic function field of genus $g \geq 2$ with the defining equation

$$y^2 = \prod_{i=1}^{2g+1} (x - a_i)\,,$$

with pairwise distinct elements $a_1, \ldots, a_{2g+1} \in K$. Let $P_i \in \mathbb{P}_F$ be the unique zero of $x - a_i$ ($i = 1, \ldots, 2g+1$), and $P_\infty \in \mathbb{P}_F$ the unique pole of x in F. Let $A_i := P_i - P_\infty$ and $[A_i] \in \mathrm{Cl}(F)$ be the divisor class of A_i (cf. Definition 1.4.3). We study the subgroup of the divisor class group of F/K which is generated by the classes $[A_1], [A_2], \ldots, [A_{2g+1}]$. Show:

(i) $[A_i] \neq [0]$ and $2[A_i] = [0]$, for $i = 1, \ldots, 2g+1$.

(ii) $\sum_{i=1}^{2g+1} [A_i] = [0]$.

(iii) Let $M \subseteq \mathrm{Cl}(F)$ be the subgroup of the divisor class group of F/K which is generated by $[A_1], [A_2], \ldots, [A_{2g+1}]$. Then $M \simeq (\mathbb{Z}/2\mathbb{Z})^{2g}$.

(iv) All divisor classes $[A] \in \mathrm{Cl}(F)$ with $2[A] = [0]$ (these are called 2-division classes of F) are in M. Hence the number of 2-division classes of F is 2^{2g}.

Remark. The previous exercise is a special case of the following much more general result. Let F/K be a function field of genus g over an algebraically closed field K. For $n \geq 1$ consider the group

$$\mathrm{Cl}(F)(n) := \{[A] \in \mathrm{Cl}(F) \,|\, n[A] = [0]\}\,.$$

If n is relatively prime to the characteristic of K, then $\mathrm{Cl}(F)(n) \simeq (\mathbb{Z}/n\mathbb{Z})^{2g}$.

6.14. Let E/K be an elliptic function field. For simplicity we assume that K is algebraically closed. We are going to study the structure of the automorphism group $\mathrm{Aut}\,(E/K)$.

Fix a place $P_0 \in \mathbb{P}_E$; then $\mathbb{P}_E$ carries the structure of an abelian group (see Proposition 6.1.7). The addition $\oplus$ on $\mathbb{P}_E$ is given by

$$P \oplus Q = R \iff P + Q \sim R + P_0$$

for $P, Q, R \in \mathbb{P}_E$ ($\sim$ means equivalence of divisors modulo principal divisors). For $P \in \mathbb{P}_E$ it follows from Riemann-Roch that $\ell(P + P_0) = 2$, hence there exists an element $x \in E$ whose pole divisor is $(x)_\infty = P + P_0$. We define the automorphism $\sigma_P \in \mathrm{Aut}\,(E/K)$ as

$$\sigma_P := \text{ the non-trivial automorphism of } E/K(x)\,.$$

We also define

$$\tau_P := \sigma_P \circ \sigma_{P_0}\,.$$

Observe that the definion of σ_P and τ_P depends on the choice of the place P_0. Prove:

(i) The automorphisms σ_P and τ_P are well-defined (i.e., they do not depend on the specific choice of the element x above).

(ii) For $P \neq Q$ we have $\sigma_P \neq \sigma_Q$ and $\tau_P \neq \tau_Q$. In particular, $\mathrm{Aut}\,(E/K)$ is an infinite group.

(iii) For all $P, Q \in \mathbb{P}_E$ holds $\sigma_P(Q) \oplus Q = P$ and $\tau_P(Q) = P \oplus Q$. Hence τ_P is called a translation automorphism.

(iv) The map $P \mapsto \tau_P$ is a group monomorphism from $\mathbb{P}_E$ into $\mathrm{Aut}\,(E/K)$. Its image $T := \{\tau_P \,|\, P \in \mathbb{P}_E\} \subseteq \mathrm{Aut}\,(E/K)$ is isomorphic to the group of divisor classes $\mathrm{Cl}^0(E)$ and hence an infinite abelian subgroup of $\mathrm{Aut}\,(E/K)$. It is called the translation group of E/K.

(v) The translation group T is independent of the choice of the place P_0 (which was used for the definition of the group structure on $\mathbb{P}_E$).

(vi) T is a normal subgroup of $\mathrm{Aut}\,(E/K)$, and the factor group $\mathrm{Aut}\,(E/K)/T$ is finite.

6.15. ($\mathrm{char}\, K = p > 0$) Consider the rational function field $K(x)/K$ and the element $y := x - x^{-p}$. Show:

(i) The extension $K(x)/K(y)$ is separable of degree $[K(x) : K(y)] = p + 1$.

(ii) The only place of $K(y)$ which is ramified in $K(x)/K(y)$, is the pole P_∞ of y. There are exactly 2 places of $K(x)$ lying over P_∞.

6.16. (char $K = p > 0$) Let F/K be a function field over a perfect constant field K of characteristic $p > 0$. Using the previous exercise, show that there exists an element $y \in F \setminus K$ with the following properties:

(i) The extension $F/K(y)$ is separable.

(ii) The pole of y is the only place of $K(y)$ which is ramified in $F/K(y)$.

Compare this result with Exercise 3.6 (ii).

7

Asymptotic Bounds for the Number of Rational Places

Let $F/\mathbb{F}_q$ be a function field over a finite field $\mathbb{F}_q$. We have seen in Chapter 5 that the number N of rational places of F over $\mathbb{F}_q$ satisfies the Hasse-Weil Bound $N \leq q + 1 + 2gq^{1/2}$, and that this upper bound can be attained only if $g \leq (q - q^{1/2})/2$. Here our aim is to investigate what happens if the genus is large with respect to q. The results of this chapter have interesting applications in coding theory, see Section 8.4.

In this chapter we consider function fields F over the finite field $\mathbb{F}_q$. The number of rational places of $F/\mathbb{F}_q$ is denoted by $N = N(F)$.

7.1 Ihara's Constant $A(q)$

In order to describe how many rational places a function field over $\mathbb{F}_q$ can have, we introduce the following notation.

Definition 7.1.1. *(a) For an integer $g \geq 0$ let*

$$N_q(g) := \max \{N(F) \mid F \textit{ is a function field over } \mathbb{F}_q \textit{ of genus } g\}.$$

(b) The real number

$$A(q) := \limsup_{g \to \infty} N_q(g)/g$$

is called Ihara's constant.

Remark 7.1.2. Note that $N_q(g) \leq q + 1 + g[2q^{1/2}]$ by Serre's Bound (Theorem 5.3.1). Therefore we have the trivial bound $0 \leq A(q) \leq [2q^{1/2}]$.

Our first aim is to improve the estimate $A(q) \leq [2q^{1/2}]$. The bound given in Theorem 7.1.3 below, is based on Serre's explicit formulas, cf. Proposition 5.3.4. For the convenience of the reader we recall this method briefly:

H. Stichtenoth, *Algebraic Function Fields and Codes*,
Graduate Texts in Mathematics 254,

Let $c_1, \ldots, c_m \geq 0$ be non-negative real numbers, not all of them equal to 0. We define the functions

$$\lambda_m(t) = \sum_{r=1}^{m} c_r t^r \quad and \quad f_m(t) = 1 + \lambda_m(t) + \lambda_m(t^{-1})$$

for $t \in \mathbb{C} \setminus \{0\}$, and we assume that

$$f_m(t) \geq 0 \text{ for all } t \in \mathbb{C} \text{ with } |t| = 1. \tag{7.1}$$

(Note that $f_m(t) \in \mathbb{R}$ holds for all $t \in \mathbb{C}$ with $|t| = 1$.) Then the number of rational places of each function field $F/\mathbb{F}_q$ of genus g is bounded by

$$N \leq \frac{g}{\lambda_m(q^{-1/2})} + \frac{\lambda_m(q^{1/2})}{\lambda_m(q^{-1/2})} + 1. \tag{7.2}$$

Theorem 7.1.3 (Drinfeld-Vladut Bound). *Ihara's constant $A(q)$ is bounded above by*

$$A(q) \leq q^{1/2} - 1\,.$$

Proof. With notation as above we set, for a fixed integer $m \geq 1$,

$$c_r := 1 - \frac{r}{m} \qquad for \quad r = 1, \ldots, m\,.$$

Then we have

$$\lambda_m(t) = \sum_{r=1}^{m} \left(1 - \frac{r}{m}\right) t^r\,.$$

In order to verify property (7.1) for the function $f_m(t) = 1 + \lambda_m(t) + \lambda_m(t^{-1})$, we consider the function

$$u(t) := \sum_{r=1}^{m} t^r = \frac{t^{m+1} - t}{t - 1}\,.$$

We have $u'(t) = \sum_{r=1}^{m} r t^{r-1}$ and hence

$$\frac{t \cdot u'(t)}{m} = \sum_{r=1}^{m} \frac{r}{m} t^r\,,$$

therefore

$$\begin{aligned} \lambda_m(t) &= \sum_{r=1}^{m} \left(1 - \frac{r}{m}\right) t^r = u(t) - \frac{t \cdot u'(t)}{m} \\ &= \frac{t}{t-1}(t^m - 1) - \frac{t}{m} \cdot \frac{(t-1)((m+1)t^m - 1) - (t^{m+1} - t)}{(t-1)^2} \\ &= \frac{t}{(t-1)^2}\left(\frac{t^m - 1}{m} + 1 - t\right). \end{aligned} \tag{7.3}$$

A straightforward calculation shows now that the function $f_m(t) = 1+\lambda_m(t)+\lambda_m(t^{-1})$ can be written as

$$f_m(t) = \frac{2-(t^m+t^{-m})}{m(t-1)(t^{-1}-1)}\,. \tag{7.4}$$

Since $t^{-1} = \bar{t}$ for $|t| = 1$, Equation (7.4) yields $f_m(t) \geq 0$ for all $t \in \mathbb{C}$ with $|t| = 1$, so the function $f_m(t)$ satisfies (7.1). Now we obtain from (7.2) the inequality

$$\frac{N}{g} \leq \frac{1}{\lambda_m(q^{-1/2})} + \frac{1}{g}\left(1+\frac{\lambda_m(q^{1/2})}{\lambda_m(q^{-1/2})}\right), \tag{7.5}$$

where N is the number of rational places of an arbitrary function field over $\mathbb{F}_q$ of genus g. Equation (7.3) implies, for $m \to \infty$,

$$\lambda_m(q^{-1/2}) \longrightarrow \frac{q^{-1/2}}{(q^{-1/2}-1)^2}(1-q^{-1/2}) = \frac{1}{q^{1/2}-1}\,.$$

Hence for each $\varepsilon > 0$ there exists $m_0 \in \mathbb{N}$ with

$$\lambda_{m_0}(q^{-1/2})^{-1} < q^{1/2} - 1 + \varepsilon/2\,.$$

We choose g_0 such that

$$\frac{1}{g_0}\left(1+\frac{\lambda_{m_0}(q^{1/2})}{\lambda_{m_0}(q^{-1/2})}\right) < \varepsilon/2\,,$$

and then, by using (7.5), we obtain the estimate

$$\frac{N}{g} < q^{1/2} - 1 + \varepsilon\,,$$

for all $g \geq g_0$ and all function fields F over $\mathbb{F}_q$ of genus g. This finishes the proof of Theorem 7.1.3. □

Remark 7.1.4. Here are some facts about Ihara's constant $A(q)$.

(a) $A(q) > 0$ for all prime powers $q = p^e$ (p a prime number, $e \geq 1$). More precisely, there exists a constant $c > 0$ such that $A(q) \geq c \cdot \log q$ for all q. This result is due to Serre [36], the proof uses class field theory and cannot be given within the scope of this book. For refinements of Serre's approach we refer to the book [31]. In Section 7.3 we will give a simple proof that $A(q) > 0$ for all $q = p^e$ with $e > 1$.

(b) If $q = \ell^2$ is a square then $A(q) = q^{1/2} - 1$; i.e., in this case the Drinfeld-Vladut Bound is attained. This equality was first proved by Ihara [22] and Tsfasman, Vladut and Zink [44], using the theory of modular curves. We will present a more elementary approach due to Garcia and Stichtenoth [12] in Section 7.4 below.

(c) If $q = \ell^3$ is a cube than $A(q) \geq 2(\ell^2 - 1)/(\ell + 2)$. For ℓ being a prime, this is a result of Zink [47]; for arbitrary ℓ this bound was first shown by Bezerra, Garcia and Stichtenoth [4]. In Theorem 7.4.17 we will give a simpler proof following Bassa, Garcia and Stichtenoth [2].

(d) The exact value of $A(q)$ is not known, for any non-square q.

7.2 Towers of Function Fields

Suppose that we have a sequence of function fields $F_i/\mathbb{F}_q$ $(i = 0, 1, 2, \ldots)$ with $g(F_i) \to \infty$ and $\lim_{i\to\infty} N(F_i)/g(F_i) > 0$ (where $N(F_i)$ denotes the number of rational places and $g(F_i)$ denotes the genus of F_i). Then it is clear that $A(q) \geq \lim_{i\to\infty} N(F_i)/g(F_i)$; so we obtain a non-trivial lower bound for $A(q)$. In this and the subsequent sections we will describe a systematic approach to study such sequences of function fields.

Definition 7.2.1. *A tower over $\mathbb{F}_q$ is an infinite sequence $\mathcal{F} = (F_0, F_1, F_2, \ldots)$ of function fields $F_i/\mathbb{F}_q$ such that the following hold:*

(i) $F_0 \subsetneqq F_1 \subsetneqq F_2 \subsetneqq \ldots \subsetneqq F_n \subsetneqq \ldots$;

(ii) each extension F_{i+1}/F_i is finite and separable;

(iii) the genera satisfy $g(F_i) \to \infty$ for $i \to \infty$.

Note that we always assume that $\mathbb{F}_q$ is the full constant field of F_i, for all $i \geq 0$.

Remark 7.2.2. The condition (iii) above follows from the conditions (i), (ii) and the following slightly weaker condition

(iii)* $g(F_j) \geq 2$ *for some* $j \geq 0$.

Proof. By Hurwitz Genus Formula one has

$$g(F_{i+1}) - 1 \geq [F_{i+1} : F_i](g(F_i) - 1) \quad \textit{for all } i.$$

Since $g(F_j) \geq 2$ and $[F_{i+1} : F_i] \geq 2$, it follows that

$$g(F_j) < g(F_{j+1}) < g(F_{j+2}) < \ldots ,$$

hence $g(F_i) \to \infty$ for $i \to \infty$. □

As we pointed out above, one is interested in the behavior of the quotient $N(F_i)/g(F_i)$ for $i \to \infty$. It is convenient to consider also the behavior of the number of rational places and the genus separately.

Lemma 7.2.3. *Let $\mathcal{F} = (F_0, F_1, F_2, \ldots)$ be a tower over $\mathbb{F}_q$. Then the following hold:*

(a) The sequence of rational numbers $(N(F_i)/[F_i : F_0])_{i\geq 0}$ *is monotonically decreasing and hence is convergent in* $\mathbb{R}^{\geq 0}$.

(b) The sequence of rational numbers $((g(F_i) - 1)/[F_i : F_0])_{i\geq 0}$ *is monotonically increasing and hence is convergent in* $\mathbb{R}^{\geq 0} \cup \{\infty\}$.

(c) Let $j \geq 0$ *such that* $g(F_j) \geq 2$. *Then the sequence* $(N(F_i)/(g(F_i) - 1))_{i\geq j}$ *is monotonically decreasing and hence is convergent in* $\mathbb{R}^{\geq 0}$.

Proof. (a) If Q is a rational place of F_{i+1}, then the restriction $P := Q \cap F_i$ of Q to F_i is a rational place of F_i. Conversely, at most $[F_{i+1} : F_i]$ rational places of F_{i+1} lie above a rational place of F_i (see Corollary 3.1.12). It follows that $N(F_{i+1}) \leq [F_{i+1} : F_i] \cdot N(F_i)$ and therefore

$$\frac{N(F_{i+1})}{[F_{i+1} : F_0]} \leq \frac{[F_{i+1} : F_i]}{[F_{i+1} : F_0]} \cdot N(F_i) = \frac{N(F_i)}{[F_i : F_0]}\,.$$

(b) Using Hurwitz Genus Formula for the field extension F_{i+1}/F_i we obtain $g(F_{i+1}) - 1 \geq [F_{i+1} : F_i]\,(g(F_i) - 1)$. Dividing by $[F_{i+1} : F_0]$ we get the desired inequality

$$\frac{g(F_i) - 1}{[F_i : F_0]} \leq \frac{g(F_{i+1}) - 1}{[F_{i+1} : F_0]}\,.$$

(c) The proof is similar to (a) and (b). □

Because of Lemma 7.2.3 the following definitions are meaningful.

Definition 7.2.4. *Let* $\mathcal{F} = (F_0, F_1, F_2, \ldots)$ *be a tower over* $\mathbb{F}_q$.

(a) The splitting rate $\nu(\mathcal{F}/F_0)$ *of* $\mathcal{F}$ *over* F_0 *is defined as*

$$\nu(\mathcal{F}/F_0) := \lim_{i\to\infty} N(F_i)/[F_i : F_0]\,.$$

(b) The genus $\gamma(\mathcal{F}/F_0)$ *of* $\mathcal{F}$ *over* F_0 *is defined as*

$$\gamma(\mathcal{F}/F_0) := \lim_{i\to\infty} g(F_i)/[F_i : F_0]\,.$$

(c) The limit $\lambda(\mathcal{F})$ *of the tower* $\mathcal{F}$ *is defined as*

$$\lambda(\mathcal{F}) := \lim_{i\to\infty} N(F_i)/g(F_i)\,.$$

It is clear from Lemma 7.2.3 and the definition of $A(q)$ that

$$\begin{aligned} 0 &\leq \nu(\mathcal{F}/F_0) < \infty\,, \\ 0 &< \gamma(\mathcal{F}/F_0) \leq \infty\,, \\ 0 &\leq \lambda(\mathcal{F}) \leq A(q)\,. \end{aligned}$$

Moreover we have the equation

$$\lambda(\mathcal{F}) = \nu(\mathcal{F}/F_0)/\gamma(\mathcal{F}/F_0). \tag{7.6}$$

This means in particular that $\lambda(\mathcal{F}) = 0$ if $\gamma(\mathcal{F}/F_0) = \infty$.

Definition 7.2.5. *A tower $\mathcal{F}$ over $\mathbb{F}_q$ is called*

(asymptotically) good,	*if* $\lambda(\mathcal{F}) > 0$,
(asymptotically) bad,	*if* $\lambda(\mathcal{F}) = 0$,
(asymptotically) optimal,	*if* $\lambda(\mathcal{F}) = A(q)$.

As an immediate consequence of Equation (7.6) we obtain the following characterization of good towers.

Proposition 7.2.6. *A tower $\mathcal{F} = (F_0, F_1, F_2, \ldots)$ over $\mathbb{F}_q$ is asymptotically good if and only if $\nu(\mathcal{F}/F_0) > 0$ and $\gamma(\mathcal{F}/F_0) < \infty$.*

It is a non-trivial task to find asymptotically good towers. If one attempts to construct good towers, it turns out in "most" cases that either $\nu(\mathcal{F}/F_0) = 0$ or that $\gamma(\mathcal{F}/F_0) = \infty$ (and hence the tower is bad). On the other hand, for applications in coding theory (and further applications in cryptography and other areas) it is of great importance to have some good towers explicitly. We will provide several examples in Sections 7.3 and 7.4.

Definition 7.2.7. *Let $\mathcal{F} = (F_0, F_1, F_2, \ldots)$ and $\mathcal{E} = (E_0, E_1, E_2, \ldots)$ be towers over $\mathbb{F}_q$. Then $\mathcal{E}$ is said to be a subtower of $\mathcal{F}$ if for each $i \geq 0$ there exists an index $j = j(i)$ and an embedding $\varphi_i : E_i \to F_j$ over $\mathbb{F}_q$.*

The following result is sometimes useful:

Proposition 7.2.8. *Let $\mathcal{E}$ be a subtower of $\mathcal{F}$. Then $\lambda(\mathcal{E}) \geq \lambda(\mathcal{F})$. In particular one has:*
(a) If $\mathcal{F}$ is asymptotically good then $\mathcal{E}$ is also asymptotically good.
(b) If $\mathcal{E}$ is asymptotically bad then $\mathcal{F}$ is also asymptotically bad.

Proof. Let $\varphi_i : E_i \to F_{j(i)}$ be an embedding of E_i into $F_{j(i)}$. Let H_i be the subfield of $F_{j(i)}$ which is uniquely determined by the following properties:

- $\varphi_i(E_i) \subseteq H_i \subseteq F_{j(i)}$.
- $H_i/\varphi_i(E_i)$ is separable.
- $F_{j(i)}/H_i$ is purely inseparable.

Then H_i is isomorphic to $F_{j(i)}$ by Proposition 3.10.2.(c) and therefore

$$\frac{N(F_{j(i)})}{g(F_{j(i)})-1} = \frac{N(H_i)}{g(H_i)-1} \le \frac{N(\varphi_i(E_i))}{g(\varphi_i(E_i))-1} = \frac{N(E_i)}{g(E_i)-1}.$$

The inequality above follows from Lemma 7.2.3(c). For $i \to \infty$ we obtain the desired result, $\lambda(\mathcal{F}) \le \lambda(\mathcal{E})$. □

In order to study the splitting rate $\nu(\mathcal{F}/F_0)$ and the genus $\gamma(\mathcal{F}/F_0)$ of a tower $\mathcal{F}$ over $\mathbb{F}_q$, we introduce the notions of splitting locus and ramification locus. Recall that, given a finite extension E/F of function fields, a place $P \in \mathbb{P}_F$ splits completely in E/F if there are exactly $n := [E : F]$ distinct places $Q_1, \ldots, Q_n \in \mathbb{P}_E$ with $Q_i|P$. Likewise, P is called ramified in E/F if there is at least one place $Q \in \mathbb{P}_E$ with $Q|P$ and $e(Q|P) > 1$.

Definition 7.2.9. *Let $\mathcal{F} = (F_0, F_1, F_2, \ldots)$ be a tower over $\mathbb{F}_q$.*
(a) The set

$$\mathrm{Split}(\mathcal{F}/F_0) := \{\, P \in \mathbb{P}_{F_0} \mid \deg P = 1 \textit{ and } P \textit{ splits completely in all extensions } F_n/F_0 \,\}$$

is called the splitting locus of $\mathcal{F}$ over F_0.
(b) The set

$$\mathrm{Ram}(\mathcal{F}/F_0) := \{\, P \in \mathbb{P}_{F_0} \,|\, P \textit{ is ramified in } F_n/F_0 \textit{ for some } n \ge 1 \,\}$$

is called the ramification locus of $\mathcal{F}$ over F_0.

Clearly the splitting locus $\mathrm{Split}(\mathcal{F}/F_0)$ is a finite set (which may be empty); the ramification locus may be finite or infinite.

Theorem 7.2.10. *Let $\mathcal{F} = (F_0, F_1, F_2, \ldots)$ be a tower over $\mathbb{F}_q$.*
(a) Let $s := |\mathrm{Split}(\mathcal{F}/F_0)|$. Then the splitting rate $\nu(\mathcal{F}/F_0)$ satisfies

$$\nu(\mathcal{F}/F_0) \ge s\,.$$

(b) Assume that the ramification locus $\mathrm{Ram}(\mathcal{F}/F_0)$ is finite and that for each place $P \in \mathrm{Ram}(\mathcal{F}/F_0)$ there is a constant $a_P \in \mathbb{R}$, such that for all $n \ge 0$ and for all places $Q \in \mathbb{P}_{F_n}$ lying above P, the different exponent $d(Q|P)$ is bounded by

$$d(Q|P) \le a_P \cdot e(Q|P)\,. \tag{7.7}$$

Then the genus $\gamma(\mathcal{F}/F_0)$ of the tower is finite, and we have the bound

$$\gamma(\mathcal{F}/F_0) \le g(F_0) - 1 + \frac{1}{2} \sum_{P \in \mathrm{Ram}(\mathcal{F}/F_0)} a_P \cdot \deg P\,.$$

(c) Now we assume that the splitting locus of $\mathcal{F}/F_0$ is non-empty and that $\mathcal{F}/F_0$ satisfies the conditions in (b). Then the tower $\mathcal{F}$ is asymptotically good, and its limit $\lambda(\mathcal{F})$ satisfies

$$\lambda(\mathcal{F}) \geq \frac{2s}{2g(F_0) - 2 + \sum_{P \in \mathrm{Ram}(\mathcal{F}/F_0)} a_P \cdot \deg P} > 0\,,$$

where $s = |\mathrm{Split}(\mathcal{F}/F_0)|$, and a_P is as in (7.7).

Proof. (a) Above each place $P \in \mathrm{Split}(\mathcal{F}/F_0)$ there are exactly $[F_n : F_0]$ places of F_n, and they are all rational. Hence $N(F_n) \geq [F_n : F_0] \cdot |\mathrm{Split}(\mathcal{F}/F_0)|$, and (a) follows immediately.

(b) For simplity we set $g_n := g(F_n)$ for all $n \geq 0$. The Hurwitz Genus Formula for F_n/F_0 gives

$$\begin{aligned}
2g_n - 2 &= [F_n : F_0](2g_0 - 2) + \sum_{P \in \mathrm{Ram}(\mathcal{F}/F_0)} \ \sum_{Q \in \mathbb{P}_{F_n},\, Q|P} d(Q|P) \cdot \deg Q \\
&\leq [F_n : F_0](2g_0 - 2) + \sum_{P \in \mathrm{Ram}(\mathcal{F}/F_0)} \ \sum_{Q \in \mathbb{P}_{F_n},\, Q|P} a_P \cdot e(Q|P) \cdot f(Q|P) \cdot \deg P \\
&= [F_n : F_0](2g_0 - 2) + \sum_{P \in \mathrm{Ram}(\mathcal{F}/F_0)} a_P \cdot \deg P \cdot \sum_{Q \in \mathbb{P}_{F_n},\, Q|P} e(Q|P) \cdot f(Q|P) \\
&= [F_n : F_0] \left(2g_0 - 2 + \sum_{P \in \mathrm{Ram}(\mathcal{F}/F_0)} a_P \cdot \deg P \right).
\end{aligned}$$

Here we have used (7.7) and the Fundamental Equality

$$\sum_{Q|P} e(Q|P) \cdot f(Q|P) = [F_n : F_0]\,,$$

see Theorem 3.1.11. Dividing the inequality above by $2[F_n : F_0]$ and letting $n \to \infty$ we obtain the inequality

$$\gamma(\mathcal{F}/F_0) \leq g(F_0) - 1 + \frac{1}{2} \sum_{P \in \mathrm{Ram}(\mathcal{F}/F_0)} a_P \cdot \deg P\,.$$

(c) follows immediately from (a) and (b), since $\lambda(\mathcal{F}) = \nu(\mathcal{F}/F_0)/\gamma(\mathcal{F}/F_0)$ (see Equation (7.6)). □

The assumption (7.7) about different exponents holds in particular in the case of tame ramification. We call the tower $\mathcal{F} = (F_0, F_1, F_2, \ldots)$ *tame* if all ramification indices $e(Q|P)$ in all extensions F_n/F_0 are relatively prime to the characteristic of $\mathbb{F}_q$; otherwise we say that the tower $\mathcal{F}/F_0$ is *wild.* Then we obtain the following corollary to Theorem 7.2.10.

Corollary 7.2.11. *Let $\mathcal{F} = (F_0, F_1, F_2, \ldots)$ be a tame tower. Assume that the splitting locus* $\mathrm{Split}(\mathcal{F}/F_0)$ *is non-empty and that the ramification locus* $\mathrm{Ram}(\mathcal{F}/F_0)$ *is finite. Then $\mathcal{F}$ is asymptotically good. More precisely, setting*

$$s := |\mathrm{Split}(\mathcal{F}/F_0)| \quad \text{and} \quad r := \sum_{P \in \mathrm{Ram}(\mathcal{F}/F_0)} \deg P\,,$$

we obtain

$$\lambda(\mathcal{F}) \geq \frac{2s}{2g(F_0) - 2 + r}\,.$$

Proof. For a tamely ramified place $Q|P$ we have $d(Q|P) = e(Q|P) - 1 \leq e(Q|P)$, by Dedekind's Different Theorem. Therefore we can choose $a_P := 1$ in (7.7), and Theorem 7.2.10(c) gives the desired result. □

The towers that we will construct in Sections 7.3 and 7.4 are often given in a recursive manner. Let us give a precise definition of what this means. Recall that a rational function $f(T)$ over $\mathbb{F}_q$ is a quotient of two polynomials $f(T) = f_1(T)/f_2(T)$ with $f_1(T), f_2(T) \in \mathbb{F}_q[T]$ and $f_2(T) \neq 0$. We can assume that $f_1(T), f_2(T)$ are relatively prime, and then we put $\deg f(T) := \max\{\deg f_1(T), \deg f_2(T)\}$, and call it the degree of $f(T)$. If $f(T) \in \mathbb{F}_q(T) \backslash \mathbb{F}_q$ then $f(T)$ is said to be non-constant; this is equivalent to the condition $\deg f(T) \geq 1$. Note that every polynomial $f(T) \in \mathbb{F}_q[T]$ can also be regarded as a rational function.

Definition 7.2.12. *Let $f(Y) \in \mathbb{F}_q(Y)$ and $h(X) \in \mathbb{F}_q(X)$ be non-constant rational functions, and let $\mathcal{F} = (F_0, F_1, F_2, \ldots)$ be a sequence of function fields. Suppose that there exist elements $x_i \in F_i$ $(i = 0, 1, 2, \ldots)$ such that*

(i) x_0 is transcendental over $\mathbb{F}_q$ and $F_0 = \mathbb{F}_q(x_0)$; i.e., F_0 is a rational function field.

(ii) $F_i = \mathbb{F}_q(x_0, x_1, \ldots, x_i)$ for all $i \geq 0$.

(iii) For all $i \geq 0$ the elements x_i, x_{i+1} satisfy $f(x_{i+1}) = h(x_i)$.

(iv) $[F_1 : F_0] = \deg f(Y)$.

Then we say that the sequence $\mathcal{F}$ is recursively defined over $\mathbb{F}_q$ by the equation

$$f(Y) = h(X)\,.$$

Remark 7.2.13. With the notation of Definition 7.2.12 it follows that

$$[F_{i+1} : F_i] \leq \deg f(Y) \quad \textit{for all } i \geq 0\,.$$

Proof. We write $f(Y) = f_1(Y)/f_2(Y)$ with relatively prime polynomials $f_1(Y)$, $f_2(Y) \in \mathbb{F}_q[Y]$. Then $F_{i+1} = F_i(x_{i+1})$, and x_{i+1} is a zero of the polynomial

$$\varphi_i(Y) := f_1(Y) - h(x_i) f_2(Y) \in F_i[Y]\,,$$

which has degree $\deg \varphi_i(Y) = \deg f(Y)$. □

Remark 7.2.14. Condition (iv) in Definition 7.2.12 is not really needed in most proofs on recursive towers. However this assumption holds in all examples of recursive towers that will be considered in this book, and it will later make some statements smoother. Clearly (iv) is equivalent to the condition that the polynomial

$$\varphi(Y) := f_1(Y) - h(x_0)f_2(Y) \in \mathbb{F}_q(x_0)[Y]$$

is irreducible over the field $\mathbb{F}_q(x_0)$. We remark that the degree $[F_{i+1} : F_i]$ may be less than $\deg f(Y)$ for some $i \geq 1$.

Here are two typical examples of recursively defined sequences of function fields. The first one is given by the equation

$$Y^m = a(X+b)^m + c \quad , \tag{7.8}$$

where $a, b, c \in \mathbb{F}_q^\times$, $m > 1$ and $\gcd(m, q) = 1$. The second one is defined by the equation

$$Y^\ell - Y = \frac{X^\ell}{1 - X^{\ell-1}} \quad , \tag{7.9}$$

where $q = \ell^2$ is a square. We will study these two sequences in detail in Sections 7.3 and 7.4.

Now we consider a sequence of fields $\mathcal{F} = (F_0, F_1, F_2, \ldots)$ which is recursively defined by an equation $f(Y) = h(X)$ with some non-constant rational functions $f(Y) \in \mathbb{F}_q(Y)$ and $h(X) \in \mathbb{F}_q(X)$. In order to decide whether the sequence $\mathcal{F}$ is a *tower* over $\mathbb{F}_q$, one has first to answer the following questions:

- Is $F_n \subsetneqq F_{n+1}$ for all $n \geq 0$?
- Is $\mathbb{F}_q$ the full constant field of F_n for all $n \geq 0$?

Our next proposition (which does not only apply to *recursively* defined sequences of fields) gives sufficient conditions for an affirmative answer to these questions.

Proposition 7.2.15. *Consider a sequence of fields $F_0 \subseteq F_1 \subseteq F_2 \subseteq \ldots$, where F_0 is a function field with the exact constant field $\mathbb{F}_q$ and $[F_{n+1} : F_n] < \infty$ for all $n \geq 0$. Suppose that for all n there exist places $P_n \in \mathbb{P}_{F_n}$ and $Q_n \in \mathbb{P}_{F_{n+1}}$ with $Q_n|P_n$ and ramification index $e(Q_n|P_n) > 1$. Then it follows that $F_n \subsetneqq F_{n+1}$.*
If we assume furthermore that $e(Q_n|P_n) = [F_{n+1} : F_n]$ for all n, then $\mathbb{F}_q$ is the full constant field of F_n for all $n \geq 0$.

Proof. By the Fundamental Equality we have $[F_{n+1} : F_n] \geq e(Q_n|P_n)$ and therefore $F_n \subsetneqq F_{n+1}$. If we assume the equality $e(Q_n|P_n) = [F_{n+1} : F_n]$, then F_n and F_{n+1} have the same constant field (since constant field extensions are unramified by Theorem 3.6.3). □

We illustrate the use of Proposition 7.2.15 with a simple example.

Example 7.2.16. Let q be a power of an *odd* prime number. We claim that the sequence $\mathcal{F} = (F_0, F_1, F_2, \ldots)$ which is recursively defined over $\mathbb{F}_q$ by the equation

$$Y^2 = \frac{X^2+1}{2X} \tag{7.10}$$

is a tower over $\mathbb{F}_q$. Hence we need to prove the following:

(i) $F_n \subsetneqq F_{n+1}$, and F_{n+1}/F_n is separable for all $n \geq 0$.

(ii) $\mathbb{F}_q$ is the full constant field of all F_n.

(iii) $g(F_j) \geq 2$ for some j.

First we observe that $F_{n+1} = F_n(x_{n+1})$ and $x_{n+1}^2 = (x_n^2+1)/2x_n$, hence $[F_{n+1} : F_n] \leq 2$. Since $q \equiv 1 \bmod 2$, it follows that F_{n+1}/F_n is separable. Our next aim is to find places $P_n \in \mathbb{P}_{F_n}$ and $Q_n \in \mathbb{P}_{F_{n+1}}$ with $Q_n|P_n$ and $e(Q_n|P_n) = 2$. Having found such places, items (i) and (ii) will follow immediately from Proposition 7.2.15. We proceed as follows: Let $P_0 \in \mathbb{P}_{F_0}$ be the unique pole of x_0 in the rational function field $F_0 = \mathbb{F}_q(x_0)$ and choose some place $Q_0 \in \mathbb{P}_{F_1}$ lying above P_0. From the equation $x_1^2 = (x_0^2+1)/2x_0$ we conclude that

$$2v_{Q_0}(x_1) = v_{Q_0}(x_1^2) = e(Q_0|P_0) \cdot v_{P_0}\left(\frac{x_0^2+1}{2x_0}\right) = e(Q_0|P_0) \cdot (-1),$$

therefore $e(Q_0|P_0) = 2$ and $v_{Q_0}(x_1) = -1$ (observe here that $e(Q_0|P_0) \leq [F_1 : F_0] \leq 2$).

In the next step we take P_1 as Q_0, and choose Q_1 as a place of F_2 lying above P_1. Again we see from the equation $x_2^2 = (x_1^2+1)/2x_1$ that $e(Q_1|P_1) = 2$ and $v_{Q_1}(x_2) = -1$. By iterating this process we obtain the desired places $P_n \in \mathbb{P}_{F_n}$ and $Q_n \in \mathbb{P}_{F_{n+1}}$ with $Q_n|P_n$ and $e(Q_n|P_n) = 2$, for all $n \geq 0$.

It remains to prove (iii). From the equation $x_1^2 = (x_0^2+1)/2x_0$ we see that exactly the following places of F_0 are ramified in F_1/F_0:

- the zero and the pole of x_0;
- the two zeros of x_0^2+1, if x_0^2+1 splits into linear factors in $\mathbb{F}_q[x_0]$, or the place of degree 2 corresponding to x_0^2+1, if x_0^2+1 is irreducible.

In each case the different degree of F_1/F_0 is $\deg \operatorname{Diff}(F_1/F_0) = 4$ by Dedekind's Different Theorem, and then the Hurwitz Genus Formula for F_1/F_0 gives $g(F_1) = 1$. In the extension F_2/F_1 there is at least one ramified place (namely the place $Q_1|P_1$ as constructed above) and hence $\deg \operatorname{Diff}(F_2/F_1) \geq 1$. Again by Hurwitz Genus Formula we obtain that $g(F_2) \geq 2$. This proves (iii).

Remark 7.2.17. In the following, we will study towers of function fields over $\mathbb{F}_q$ which are recursively defined by an equation $f(Y) = h(X)$. Since each step in a tower is by definition a separable extension, this implies that the rational function $f(Y)$ must be separable (i.e., $f(Y) \notin \mathbb{F}_q(Y^p)$ where $p = \operatorname{char} \mathbb{F}_q$).

Our main goal is to construct *asymptotically good* towers $\mathcal{F}$ over $\mathbb{F}_q$. According to Theorem 7.2.10 we are therefore interested in formulating criteria which ensure a *non-empty* splitting locus $\mathrm{Split}(\mathcal{F}/F_0)$ and a *finite* ramification locus $\mathrm{Ram}(\mathcal{F}/F_0)$. In order to tackle these problems for recursively defined towers, we introduce the notion of the *basic function field.*

Definition 7.2.18. *Let $\mathcal{F} = (F_0, F_1, F_2, \ldots)$ be a sequence of function fields which is recursively defined over $\mathbb{F}_q$ by the equation $f(Y) = h(X)$, with non-constant rational functions $f(Y), h(X)$ over $\mathbb{F}_q$. We define the basic function field F corresponding to the tower $\mathcal{F}$ by*

$$F := \mathbb{F}_q(x, y) \quad \textit{with the relation } f(y) = h(x)\,.$$

Observe that the extension $F/\mathbb{F}_q(x)$ is separable by Remark 7.2.17. As a consequence of condition (iv) in Definition 7.2.12 we have

$$[F : \mathbb{F}_q(x)] = \deg f(Y) \ \textit{ and } \ [F : \mathbb{F}_q(y)] = \deg h(X)\,.$$

Moreover, all subfields of F_n of the form $\mathbb{F}_q(x_i, x_{i+1})$ (with $0 \le i \le n-1$) are $\mathbb{F}_q$ - isomorphic to the basic function field $F = \mathbb{F}_q(x, y)$ under the map $x \mapsto x_i,\ y \mapsto x_{i+1}$.

Before giving a first application of the basic function field in Proposition 7.2.20, we introduce a convenient notation for the rational places of a rational function field.

Definition 7.2.19. *Let $K(z)$ be a rational function field over an arbitrary field K. Then we denote for $\alpha \in K$ by $(z = \alpha)$ the unique place of $K(z)$ which is a zero of $z - \alpha$. Likewise we denote by $(z = \infty)$ the unique pole of z in $K(z)$.*

Observe that this notation differs from the notation for rational places of $K(z)$ as used in Section 1.2.

Proposition 7.2.20. *Let $\mathcal{F} = (F_0, F_1, F_2, \ldots)$ be a tower over $\mathbb{F}_q$ which is recursively defined by the equation $f(Y) = h(X)$, and let $F = \mathbb{F}_q(x, y)$ be the corresponding basic function field with the relation $f(y) = h(x)$. Assume that $\Sigma \subseteq \mathbb{F}_q \cup \{\infty\}$ is a non-empty set which satisfies the following two conditions:*

(1) For all $\alpha \in \Sigma$, the place $(x = \alpha)$ of $\mathbb{F}_q(x)$ splits completely in the extension $F/\mathbb{F}_q(x)$.

(2) If $\alpha \in \Sigma$ and Q is a place of F above the place $(x = \alpha)$, then $y(Q) \in \Sigma$.

Then for all $\alpha \in \Sigma$, the place $(x_0 = \alpha)$ of $F_0 = \mathbb{F}_q(x_0)$ splits completely in $\mathcal{F}/F_0$. In particular, the splitting locus of $\mathcal{F}/F_0$ satisfies

$$|\mathrm{Split}(\mathcal{F}/F_0)| \ge |\Sigma|\,,$$

and therefore we obtain a lower bound for the splitting rate $\nu(\mathcal{F}/F_0)$ of the tower $\mathcal{F}$,

$$\nu(\mathcal{F}/F_0) \ge |\Sigma|\,.$$

Proof. Let $\alpha \in \Sigma$. We show by induction that the place $(x_0 = \alpha)$ splits completely in F_n/F_0 for all $n \geq 0$. This is trivial for $n = 0$, and we assume now that the assertion holds for some n. We have to show that every place $Q \in \mathbb{P}_{F_n}$ lying above $(x_0 = \alpha)$ splits completely in F_{n+1}/F_n. By condition (2) we know that $x_n(Q) =: \beta \in \Sigma$, and then it follows from condition (1) that the place $(x_n = \beta)$ splits completely in the extension $\mathbb{F}_q(x_n, x_{n+1})/\mathbb{F}_q(x_n)$. Since F_{n+1} is the composite field of F_n and $\mathbb{F}_q(x_n, x_{n+1})$, the place Q splits completely in F_{n+1}/F_n by Proposition 3.9.6. The inequality $\nu(\mathcal{F}/F_0) \geq |\Sigma|$ follows now from Theorem 7.2.10(a). □

Corollary 7.2.21. *Let $\mathcal{F} = (F_0, F_1, F_2, \ldots)$ be a tower over $\mathbb{F}_q$ which is recursively defined by the equation $f(Y) = h(X)$. Let $m := \deg f(Y)$. Assume that $\Sigma \subseteq \mathbb{F}_q$ is a non-empty set such that the following condition holds:*

> *For all $\alpha \in \Sigma$ we have $h(\alpha) \in \mathbb{F}_q$ (i.e., α is not a pole of the rational function $h(X)$), and the equation $f(t) = h(\alpha)$ has m distinct roots $t = \beta$ in Σ.*

Then all places $(x_0 = \alpha)$ with $\alpha \in \Sigma$ are in the splitting locus of $\mathcal{F}$ over F_0, and the splitting rate satisfies $\nu(\mathcal{F}/F_0) \geq |\Sigma|$.

Proof. Consider the basic function field $F = \mathbb{F}_q(x, y)$ with defining equation $f(y) = h(x)$. Let $P = (x = \alpha)$ with $\alpha \in \Sigma$ be the place of $\mathbb{F}_q(x)$ which is the zero of $x - \alpha$, and $\mathcal{O}_P \subseteq \mathbb{F}_q(x)$ be the corresponding valuation ring. We write $f(Y) = f_1(Y)/f_2(Y)$ with relatively prime polynomials $f_1(Y), f_2(Y) \in \mathbb{F}_q[Y]$ and $\max\{\deg f_1(Y), \deg f_2(Y)\} = m$, say $f_1(Y) = a_m Y^m + \ldots + a_0$ and $f_2(Y) = b_m Y^m + \ldots + b_0$. Then $y \in F$ is a root of the polynomial

$$\varphi(Y) = f_1(Y) - h(x) f_2(Y) \in \mathbb{F}_q(x)[Y].$$

By our assumption, the equation $f_1(t) - h(\alpha) f_2(t) = 0$ has m distinct roots $\beta_1, \ldots, \beta_m \in \Sigma$, so the leading coefficient $a_m - b_m h(\alpha)$ is nonzero. This means that the function $a_m - b_m h(x) \in \mathbb{F}_q(x)$ is a unit in $\mathcal{O}_P$. Dividing $\varphi(Y)$ by $a_m - b_m h(x)$ gives now an integral equation for y over $\mathcal{O}_P$, whose reduction modulo P has the roots $\beta_1, \ldots, \beta_m$. By Kummer's Theorem 3.3.7 there exist m distinct places $Q_1, \ldots, Q_m \in \mathbb{P}_F$ lying above P, with $y(Q_i) = \beta_i$ for $i = 1, \ldots, m$. We have thus verified the conditions (1), (2) in Proposition 7.2.20, and hence the result follows. □

Example 7.2.16(cont.). We return to the tower $\mathcal{F}$ in Example 7.2.16; i.e., $\mathcal{F}$ is recursively given by the equation $f(Y) = h(X)$ with $f(Y) = Y^2$ and $h(X) = (X^2 + 1)/2X$ over a field $\mathbb{F}_q$ of odd characteristic. One can show that for all squares $q = \ell^2$ (ℓ being a power of an odd prime) the splitting locus of $\mathcal{F}/F_0$ is non-empty. Since the proof requires tools, which are not covered in this book, we shall be content with proving the case $q = 9$. The field $\mathbb{F}_9$ can be represented as $\mathbb{F}_9 = \mathbb{F}_3(\delta)$ with $\delta^2 = -1$, so we have $\mathbb{F}_9 = \{0, \pm1, \pm\delta, \pm(\delta+1), \pm(\delta-1)\}$. We claim that the set $\Sigma := \{\pm(\delta+1), \pm(\delta-1)\}$ satisfies the condition of Corollary 7.2.21. In fact one finds by straightforward calculation that

$$h\,(\delta+1) = h(\delta-1) = \delta = f(\delta-1) = f(-\delta+1)\,,$$
$$h\,(-\delta-1) = h(-\delta+1) = -\delta = f(\delta+1) = f(-\delta-1)\,.$$

We conclude from Corollary 7.2.21 that the splitting rate of $\mathcal{F}$ over F_0 satisfies $\nu(\mathcal{F}/F_0) \geq 4$.

Next we turn our attention to ramification in towers.

Remark 7.2.22. Let $\mathcal{F} = (F_0, F_1, F_2, \ldots)$ be a tower over $\mathbb{F}_q$ and let $L \supseteq \mathbb{F}_q$ be an algebraic field extension of $\mathbb{F}_q$. Then we can consider the *constant field extension* $\mathcal{F}' := \mathcal{F}L$ of $\mathcal{F}$ by L, which is defined as the sequence of fields

$$\mathcal{F}' = (F_0', F_1', F_2', \ldots) \qquad \textit{with } F_i' := F_i L\,.$$

It follows that $[F_{i+1}' : F_i'] = [F_{i+1} : F_i]$, and L is the full constant field of F_i', for all $i \geq 0$. Moreover we have $g(F_i') = g(F_i)$. A place $P \in \mathbb{P}_{F_i}$ is ramified in F_{i+1}/F_i if and only if the places $P' \in \mathbb{P}_{F_i'}$ above P are ramified in F_{i+1}'/F_i', therefore

$$\mathrm{Ram}(\mathcal{F}'/F_0') = \{P' \in \mathbb{P}_{F_0'} \mid P' \cap F_0 \in \mathrm{Ram}(\mathcal{F}/F_0)\}$$

and

$$\gamma(\mathcal{F}'/F_0') = \gamma(\mathcal{F}/F_0)\,.$$

In case of a finite ramification locus we define the *ramification divisor* of $\mathcal{F}/F_0$ by

$$R(\mathcal{F}/F_0) := \sum_{P \in \mathrm{Ram}(\mathcal{F}/F_0)} P\,;$$

then we also have

$$R(\mathcal{F}'/F_0') = \mathrm{Con}_{F_0'/F_0} R(\mathcal{F}/F_0)$$

and

$$\deg R(\mathcal{F}'/F_0') = \deg R(\mathcal{F}/F_0)\,.$$

All statements above follow immediately from Section 3.6.

We can now prove a useful criterion for finiteness of the ramification locus of recursive towers.

Proposition 7.2.23. *Let $\mathcal{F} = (F_0, F_1, F_2, \ldots)$ be a recursive tower over $\mathbb{F}_q$, defined by the equation $f(Y) = h(X)$, and let $\mathcal{F}' = \mathcal{F}L = (F_0', F_1', F_2', \ldots)$ be the constant field extension of $\mathcal{F}$ by an algebraic extension field $L \supseteq \mathbb{F}_q$. We denote by F (resp. F') the basic function field of $\mathcal{F}$ (resp. $\mathcal{F}'$), so $F = \mathbb{F}_q(x, y)$*

and $F' = FL = L(x,y)$ with the relation $f(y) = h(x)$. We assume that all places of $L(x)$, which ramify in the extension $F'/L(x)$, are rational and hence the set

$$\Lambda_0 := \{x(P) \,|\, P \in \mathbb{P}_{L(x)} \text{ is ramified in } F'/L(x)\}$$

is contained in $L \cup \{\infty\}$. Suppose that Λ is a finite subset of $L \cup \{\infty\}$ such that the following two conditions hold:

(1) $\Lambda_0 \subseteq \Lambda$.

(2) If $\beta \in \Lambda$ and $\alpha \in \bar{\mathbb{F}}_q \cup \{\infty\}$ satisfy the equation $f(\beta) = h(\alpha)$, then $\alpha \in \Lambda$.

Then the ramification locus $\mathrm{Ram}(\mathcal{F}'/F_0')$ *(and hence also* $\mathrm{Ram}(\mathcal{F}/F_0)$*) is finite, and we have*

$$\mathrm{Ram}(\mathcal{F}'/F_0') \subseteq \{P \in \mathbb{P}_{F_0'} \,|\, x_0(P) \in \Lambda\}\,.$$

Proof. By definition, the field F_0' is the rational function field $F_0' = L(x_0)$ over L. Let $P \in \mathrm{Ram}(\mathcal{F}'/F_0')$. There is some $n \geq 0$ and some place Q of F_n' above P such that Q is ramified in the extension F_{n+1}'/F_n'. Setting $R := Q \cap L(x_n)$, we have the situation as shown in Figure 7.1 below.

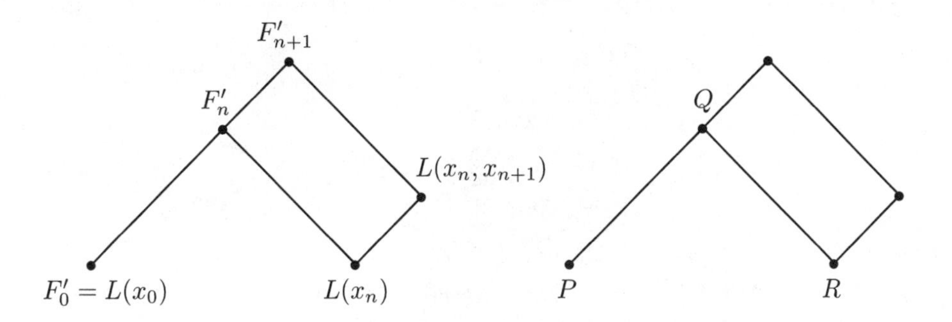

Fig. 7.1.

Since Q is ramified in F_{n+1}'/F_n', it follows by Abhyankar's Lemma that the place R is ramified in the extension $L(x_n, x_{n+1})/L(x_n)$. So we have $\beta_n := x_n(Q) \in \Lambda_0$. Setting $\beta_i := x_i(Q)$ we obtain $f(\beta_{i+1}) = h(\beta_i)$ for $i = 0, \ldots, n-1$, and from Condition (2) we conclude that $\beta_0 = x_0(Q) = x_0(P)$ is in Λ. □

Example 7.2.16(cont.). We apply Proposition 7.2.23 to the recursive tower $\mathcal{F}$ over $\mathbb{F}_q$, which was considered in Example 7.2.16. Recall that $\mathcal{F} = (F_0, F_1, F_2, \ldots)$ is recursively defined by the equation

$$Y^2 = (X^2 + 1)/2X$$

over a field $\mathbb{F}_q$ of odd characteristic. We fix a finite field $L \supseteq \mathbb{F}_q$ which contains an element δ with $\delta^2 = -1$. The set Λ_0 in Proposition 7.2.23 is given here by

$$\Lambda_0 = \{\infty, 0, \pm\delta\},$$

as a direct consequence of Proposition 3.7.3. We consider now the set

$$\Lambda := \{0, \infty, \pm 1, \pm\delta\} \subseteq L \cup \{\infty\},$$

and we claim that Λ satisfies Condition (2) of Proposition 7.2.23. So we have to show that for all $\beta \in \Lambda$, all solutions $\alpha \in \bar{\mathbb{F}}_q \cup \{\infty\}$ of the equation $(\alpha^2+1)/2\alpha = \beta^2$ are in Λ. This is easily checked as follows:

$$\begin{array}{ll} \textit{if } \beta = \infty & \textit{then } \alpha = 0 \textit{ or } \alpha = \infty\,; \\ \textit{if } \beta = 0 & \textit{then } \alpha = \pm\delta\,; \\ \textit{if } \beta = \pm 1 & \textit{then } \alpha = 1\,; \\ \textit{if } \beta = \pm\delta & \textit{then } \alpha = -1\,. \end{array}$$

Proposition 7.2.23 implies now that the ramification locus of $\mathcal{F}' = \mathcal{F}L$ over $F_0' = L(x_0)$ is contained in the set

$$\{(x_0 = 0), (x_0 = \infty), (x_0 = 1), (x_0 = -1), (x_0 = \delta), (x_0 = -\delta)\} \subseteq \mathbb{P}_{L(x_0)}\,.$$

The tower $\mathcal{F}$ is tame, as $\operatorname{char}(\mathbb{F}_q) \neq 2$ and all extensions F_{i+1}/F_i are of degree $[F_{i+1} : F_i] = 2$. So we get the bound

$$\gamma(\mathcal{F}/F_0) \le -1 + \frac{1}{2} \sum_{P \in \operatorname{Ram}(\mathcal{F}/F_0)} \deg P \le -1 + \frac{6}{2} = 2$$

for genus of the tower, by Theorem 7.2.10(b)
In the special case $q = 9$ we have proved (just before Remark 7.2.22) the inequality $\nu(\mathcal{F}/F_0) \ge 4$ for the splitting rate $\nu(\mathcal{F}/F_0)$, hence it follows that $\lambda(\mathcal{F}) = \nu(\mathcal{F}/F_0)/\gamma(\mathcal{F}/F_0) \ge 4/2 = 2$. Note that Ihara's constant $A(9)$ satisfies the inequality $A(9) \le \sqrt{9} - 1 = 2$, so we have $2 \le \lambda(\mathcal{F}) \le A(9) \le 2$.

We summarize the results of Example 7.2.16 as follows:

- The tower $\mathcal{F}$ over $\mathbb{F}_9$, which is recursively defined by the equation $Y^2 = (X^2+1)/2X$, is asymptotically optimal.
- $A(9) = 2$.

Observe that the second assertion is a special case of the equality $A(q^2) = q-1$ that will be proved in Section 7.4.

7.3 Some Tame Towers

In this section we present two examples of asymptotically good tame towers. Both of them are special cases of the next theorem. As usual, $\bar{\mathbb{F}}_q$ denotes the algebraic closure of $\mathbb{F}_q$.

Theorem 7.3.1. *Let $m \geq 2$ be an integer with $q \equiv 1 \bmod m$. Assume that the polynomial $h(X) \in \mathbb{F}_q[X]$ has the following properties:*

(1) $\deg h(X) = m$, and the leading coefficient of $h(X)$ is an m-th power of some element $c \in \mathbb{F}_q^\times$.

(2) $h(0) = 0$ and $h'(0) \neq 0$; i.e., 0 is a simple zero of $h(X)$.

(3) There exists a subset $\Lambda \subseteq \mathbb{F}_q$ such that for all $\beta, \gamma \in \bar{\mathbb{F}}_q$,
  (a) $h(\gamma) = 0 \implies \gamma \in \Lambda$, and
  (b) $\alpha \in \Lambda$ and $\beta^m = h(\alpha) \implies \beta \in \Lambda$.

Then the equation

$$Y^m = h(X) \tag{7.11}$$

defines an asymptotically good tower $\mathcal{F}$ over $\mathbb{F}_q$ with limit

$$\lambda(\mathcal{F}) \geq \frac{2}{|\Lambda| - 2}.$$

Proof. We consider the sequence $\mathcal{F} = (F_0, F_1, F_2, ...)$, where $F_0 = \mathbb{F}_q(x_0)$ is a rational function field and for all $n \geq 0$,

$$F_{n+1} = F_n(x_{n+1}) \quad \textit{with} \quad x_{n+1}^m = h(x_n) \ . \tag{7.12}$$

Clearly F_{n+1}/F_n is a separable extension of degree $\leq m$. We show now that at each step F_{n+1}/F_n, there is a place with ramification index $e = m$ (which implies that $[F_{n+1} : F_n] = m$). Denote by $P_0 = (x_0 = 0)$, the zero of x_0 in F_0. For $n \geq 0$ we choose recursively a place P_{n+1} of F_{n+1} with $P_{n+1}|P_n$, and we claim that

$$v_{P_n}(x_n) = 1 \quad \textit{and} \quad e(P_{n+1}|P_n) = m \tag{7.13}$$

holds for all $n \geq 0$. For $n = 0$ we have $v_{P_0}(x_0) = 1$. Condition (2) implies that P_0 is a simple zero of the function $h(x_0)$, therefore $v_{P_0}(h(x_0)) = 1$. From (7.12) we get

$$m \cdot v_{P_1}(x_1) = v_{P_1}(x_1^m) = v_{P_1}(h(x_0)) = e(P_1|P_0) \leq [F_1 : F_0] \leq m \ ,$$

hence (7.13) holds for $n = 0$. Using the same argument, one shows by induction that (7.13) holds for all $n \geq 0$. We conclude that $[F_{n+1} : F_n] = m$ and that $\mathbb{F}_q$ is the full constant field of F_n for all $n \geq 0$, cf. Proposition 7.2.15.

We claim that the pole $(x_0 = \infty)$ of x_0 splits completely in all extensions F_n/F_0. In order to prove this we consider the basic function field

$$F = \mathbb{F}_q(x, y) \quad \textit{with} \quad y^m = h(x) \ .$$

According to Proposition 7.2.20, it is sufficient to show:

(i) the place $P_\infty := (x = \infty)$ splits completely in $F/\mathbb{F}_q(x)$, and
(ii) if $Q \in \mathbb{P}_F$ is a place with $Q|P_\infty$, then $y(Q) = \infty$.

By condition (1) we have $y^m = c^m x^m + \dots$, so the element $z := y/x$ satisfies an equation

$$z^m = c^m + r(1/x) \;\; with \;\; v_{P_\infty}(r(1/x)) > 0 \;\; and \;\; c \in \mathbb{F}_q^\times .$$

Hence z is integral over the valuation ring $\mathcal{O}_{P_\infty}$. The equation $Z^m = c^m$ has m distinct roots in $\mathbb{F}_q$ (here we use the assumption $q \equiv 1 \bmod m$), and it follows from Kummer's Theorem 3.3.7 that P_∞ splits completely in $F/\mathbb{F}_q(x)$. This is assertion (i). Assertion (ii) follows immediately from the equation $y^m = h(x)$.

We have thus shown that the splitting locus of $\mathcal{F}/F_0$ is non-empty. As a consequence, the number $N(F_n)$ of rational places of $F_n/\mathbb{F}_q$ satisfies the inequality $N(F_n) \ge m^n$ and therefore the genus $g(F_n)$ tends to infinity as $n \to \infty$. So $\mathcal{F}$ is in fact a tower over $\mathbb{F}_q$.

The only places of $\mathbb{F}_q(x)$ which ramify in the extension $F/\mathbb{F}_q(x)$ are zeros of $h(x)$, by Theorem 3.7.3. Hence Proposition 7.2.23 implies that the ramification locus of $\mathcal{F}/F_0$ is finite and

$$|\mathrm{Ram}(\mathcal{F}/F_0)| \le |\Lambda| .$$

(Note that the assumptions of Proposition 7.2.23 follow from the conditions 3(a),(b).) Now Corollary 7.2.11 gives the desired estimate

$$\lambda(\mathcal{F}) \ge 2/(|\Lambda| - 2)$$

for the limit of the tower $\mathcal{F}$. □

We remark that the cardinality of the set Λ above is certainly larger than 2 since a tame extension of a rational function field with at most 2 ramified places (of degree 1) is rational by the Hurwitz Genus Formula.

Proposition 7.3.2. *Let $q = \ell^2$ be a square, $\ell > 2$. Then the equation*

$$Y^{\ell-1} = 1 - (X+1)^{\ell-1}$$

defines an asymptotically good tower $\mathcal{F}$ over $\mathbb{F}_q$ with limit

$$\lambda(\mathcal{F}) \ge 2/(\ell - 1).$$

For $\ell = 3$ this tower is optimal over the field $\mathbb{F}_9$.

Proof. We set $h(X) = 1 - (X+1)^{\ell-1}$ and $\Lambda = \mathbb{F}_\ell$. We need to check that the assumptions of Theorem 7.3.1 are satisfied:
(1) The leading coefficient of $h(X)$ is -1, which is a square in $\mathbb{F}_q$ since $q = \ell^2$.

(2) The condition $h(0) = 0$ is clear, and $h'(0) \neq 0$ follows from the equation $h'(X) = (X+1)^{\ell-2}$.
(3)(a) Let $h(\gamma) = 0$. Then $(\gamma+1)^{\ell-1} = 1$, hence $\gamma + 1 \in \mathbb{F}_\ell^\times$ and $\gamma \in \mathbb{F}_\ell = \Lambda$.
(3)(b) Let $\alpha \in \mathbb{F}_\ell$ and $\beta^{\ell-1} = h(\alpha) = 1 - (\alpha+1)^{\ell-1}$. Then $h(\alpha) = 0$ (if $\alpha \neq -1$) or $h(\alpha) = 1$ (if $\alpha = -1$), and therefore $\beta \in \mathbb{F}_\ell = \Lambda$.
The use of Proposition 7.3.1 now completes the proof. □

Next we give examples of asymptotically good towers over all non prime fields.

Proposition 7.3.3. *Let $q = \ell^e$ with $e \geq 2$. Then the equation*

$$Y^m = 1 - (X+1)^m$$

with $m := (q-1)/(\ell-1)$ defines an asymptotically good tower $\mathcal{F}$ over $\mathbb{F}_q$ with limit

$$\lambda(\mathcal{F}) \geq 2/(q-2).$$

For $\ell = e = 2$ this is an optimal tower over $\mathbb{F}_4$.

Proof. In this case we have $h(X) = 1-(X+1)^m$ and we set $\Lambda := \mathbb{F}_q$. Observe that the map $\gamma \mapsto \gamma^m$ is the norm map from $\mathbb{F}_q$ to $\mathbb{F}_\ell$ and hence is surjective. Moreover, every element $\beta \in \bar{\mathbb{F}}_q$ with $\beta^m \in \mathbb{F}_\ell$ is in $\mathbb{F}_q$. Using these facts, the proof of this proposition is essentially the same as that of Proposition 7.3.2. □

Recall that Ihara's constant $A(q)$ is positive for all prime powers $q = p^e$, see Remark 7.1.4(a). Proposition 7.3.3 provides a simple proof of this fact in the case $e > 1$. The lower bound $A(q) \geq 2/(q-2)$ is however rather weak for $q \neq 4$.

7.4 Some Wild Towers

Ihara's constant $A(q)$ attains the Drinfeld-Vladut Bound $A(q) = q^{1/2} - 1$ when $q = \ell^2$ is a square, cf. Remark 7.1.4. In this section we will prove this result, by providing a recursive tower $\mathcal{G} = (G_0, G_1, G_2, \ldots)$ over $\mathbb{F}_q$ with limit $\lambda(\mathcal{G}) = \ell - 1$, when $q = \ell^2$. We also present a recursive tower $\mathcal{H} = (H_0, H_1, H_2, \ldots)$ over a cubic field $\mathbb{F}_q$ with $q = \ell^3$, whose limit satisfies $\lambda(\mathcal{H}) \geq 2(\ell^2-1)/(\ell+2)$, thus obtaining the bound $A(\ell^3) \geq 2(\ell^2-1)/(\ell+2)$. Both towers $\mathcal{G}$ and $\mathcal{H}$ are wild towers; i.e., for some $i \geq 1$ there are places which are wildly ramified in the extensions G_i/G_0 (resp. H_i/H_0).

We begin with the tower $\mathcal{G}$ over a quadratic field $\mathbb{F}_q$.

Definition 7.4.1. *Let $q = \ell^2$ where ℓ is a power of some prime number p. We define the tower $\mathcal{G} = (G_0, G_1, G_2, \ldots)$ of function fields G_i over $\mathbb{F}_q$ recursively by the equation*

$$Y^\ell - Y = \frac{X^\ell}{1 - X^{\ell-1}}\,; \tag{7.14}$$

i.e., $G_0 = \mathbb{F}_q(x_0)$ is the rational function field, and for all $i \geq 0$ we have $G_{i+1} = G_i(x_{i+1})$ with

$$x_{i+1}^\ell - x_{i+1} = \frac{x_i^\ell}{1 - x_i^{\ell-1}}\,. \tag{7.15}$$

We will show that Equation (7.14) actually defines a tower (see Lemma 7.4.3) and that this tower attains the Drinfeld-Vladut Bound over $\mathbb{F}_q$.

Remark 7.4.2. With the notations of Definition 7.4.1 we set $\tilde{x}_i := \zeta x_i$ for $i = 0, 1, 2, \ldots$, where $\zeta \in \mathbb{F}_q$ satisfies the equation $\zeta^{\ell-1} = -1$. Then it follows that

$$\begin{aligned}\tilde{x}_{i+1}^\ell + \tilde{x}_{i+1} &= \zeta^\ell x_{i+1}^\ell + \zeta x_{i+1} = -\zeta(x_{i+1}^\ell - x_{i+1}) \\ &= -\zeta \cdot \frac{x_i^\ell}{1 - x_i^{\ell-1}} = \frac{\zeta^\ell x_i^\ell}{1 + \zeta^{\ell-1} x_i^{\ell-1}} = \frac{\tilde{x}_i^\ell}{\tilde{x}_i^{\ell-1} + 1}\,.\end{aligned}$$

This shows that the tower $\mathcal{G}$ can also be defined by the equation

$$Y^\ell + Y = \frac{X^\ell}{X^{\ell-1} + 1}\,. \tag{7.16}$$

In fact, the tower $\mathcal{G}$ was first introduced by Garcia-Stichtenoth [12] with Equation (7.16) as its defining equation. The reason why we prefer Equation (7.14) here, is that then the analogy between the towers $\mathcal{H}$, considered below, and $\mathcal{G}$ becomes more obvious.

Lemma 7.4.3. *Equation* (7.14) *defines a recursive tower $\mathcal{G} = (G_0, G_1, G_2, \ldots)$ over $\mathbb{F}_q$. All extensions G_{i+1}/G_i are Galois of degree $[G_{i+1} : G_i] = \ell$, and the place $(x_0 = \infty)$ of G_0 is totally ramified in all extensions G_n/G_0.*

Proof. It is clear that the equation $Y^\ell - Y = X^\ell/(1 - X^{\ell-1})$ is separable and hence all extensions G_{i+1}/G_i are separable of degree $[G_{i+1} : G_i] \leq \ell$. Let $P_0 := (x_0 = \infty)$ be the pole of x_0 in the rational function field $G_0 = \mathbb{F}_q(x_0)$. For all $i \geq 0$ we choose recursively a place $P_{i+1} \in \mathbb{P}_{G_{i+1}}$ with $P_{i+1}|P_i$, and we claim that $e(P_{i+1}|P_i) = \ell$. From the equation $x_1^\ell - x_1 = x_0^\ell/(1 - x_0^{\ell-1})$ we see that

$$v_{P_1}(x_1^\ell - x_1) = e(P_1|P_0) \cdot (-1) < 0\,,$$

so P_1 is a pole of x_1, and then $v_{P_1}(x_1^\ell - x_1) = \ell \cdot v_{P_1}(x_1)$ by the Strict Triangle Inequality. It follows that

$$-\ell \cdot v_{P_1}(x_1) = e(P_1|P_0) \le [G_1 : G_0] \le \ell,$$

hence $e(P_1|P_0) = \ell$ and $v_{P_1}(x_1) = -1$. By induction we obtain $e(P_{i+1}|P_i) = \ell$ and $v_{P_{i+1}}(x_{i+1}) = -1$ for all $i \ge 0$. We conclude that $[G_{i+1} : G_i] = \ell$, and $\mathbb{F}_q$ is the full constant field of G_i for all $i \ge 0$. Since the equation of x_{i+1} over G_i is an Artin-Schreier equation $y^\ell - y = x_i^\ell/(1 - x_i^{\ell-1})$, it also follows that G_{i+1}/G_i is Galois.
It remains to show that $g(G_n) \ge 2$ for some n. Since

$$G_1 = G_0(x_1) \quad \textit{with } x_1^\ell - x_1 = x_0^\ell/(1 - x_0^{\ell-1})$$

and the right hand side of this equation has ℓ simple poles in G_0 (namely the places $(x_0 = \infty)$ and $(x_0 = \alpha)$ for $\alpha^{\ell-1} = 1$), the genus of G_1 is

$$g(G_1) = \frac{\ell - 1}{2}(-2 + 2\ell) = (\ell - 1)^2$$

by Proposition 3.7.8. For all $\ell \ne 2$ we have thus $g(G_1) \ge 2$ as desired. For $\ell = 2$ we have $g(G_1) = 1$. In the extension G_2/G_1 at least one place (namely the pole of x_0) ramifies, and then the Hurwitz Genus Formula for G_2/G_1 shows that $g(G_2) \ge 2$. □

Our next step is to estimate the splitting rate $\nu(\mathcal{G}/G_0)$ over the field $\mathbb{F}_q$. We are going to apply Corollary 7.2.21.

Lemma 7.4.4. *Let $\mathcal{G}$ be the tower over $\mathbb{F}_q$ (with $q = \ell^2$), which is recursively defined by Equation (7.14). Then the splitting locus of $\mathcal{G}/G_0$ satisfies*

$$\mathrm{Split}(\mathcal{G}/G_0) \supseteq \{(x_0 = \alpha) \,|\, \alpha \in \mathbb{F}_q \setminus \mathbb{F}_\ell\},$$

and the splitting rate $\nu(\mathcal{G}/G_0)$ satisfies

$$\nu(\mathcal{G}/G_0) \ge \ell^2 - \ell.$$

Proof. We want to show that the set $\Sigma := \mathbb{F}_q \setminus \mathbb{F}_\ell$ satisfies the condition of Corollary 7.2.21. So let $\alpha \in \Sigma$; then

$$\frac{\alpha^\ell}{1 - \alpha^{\ell-1}} \in \mathbb{F}_q \quad \textit{since } \alpha^{\ell-1} \ne 1.$$

Consider an element $\beta \in \bar{\mathbb{F}}_q$ (the algebraic closure of $\mathbb{F}_q$) with

$$\beta^\ell - \beta = \frac{\alpha^\ell}{1 - \alpha^{\ell-1}}. \tag{7.17}$$

Then

$$\beta^{\ell^2} - \beta^\ell = \frac{\alpha^{\ell^2}}{(1 - \alpha^{\ell-1})^\ell} = \frac{\alpha}{(1 - \alpha^{\ell-1})^\ell}. \tag{7.18}$$

Adding (7.17) and (7.18) we get

$$\begin{aligned}\beta^{\ell^2} - \beta &= \frac{\alpha^\ell}{1-\alpha^{\ell-1}} + \frac{\alpha}{(1-\alpha^{\ell-1})^\ell} = \frac{(\alpha^\ell - \alpha^{\ell^2}) + (\alpha - \alpha^\ell)}{(1-\alpha^{\ell-1})^{\ell+1}} \\ &= \frac{\alpha - \alpha^{\ell^2}}{(1-\alpha^{\ell-1})^{\ell+1}} = 0\,,\end{aligned}$$

since $\alpha \in \mathbb{F}_q$ and $q = \ell^2$. Hence $\beta^{\ell^2} = \beta$; i.e., $\beta \in \mathbb{F}_q$. Since $\beta^\ell - \beta = \alpha^\ell/(1-\alpha^{\ell-1}) \neq 0$, it follows that $\beta \notin \mathbb{F}_\ell$ and hence $\beta \in \Sigma$. It is also clear that Equation (7.17) has ℓ distinct roots β. Thus we have verified the condition of Corollary 7.2.21. This finishes the proof. □

Using Proposition 7.2.23, the ramification locus $\mathrm{Ram}(\mathcal{G}/G_0)$ can be determined as follows. We denote by

$$G := \mathbb{F}_q(x,y) \text{ with the relation } y^\ell - y = x^\ell/(1-x^{\ell-1})$$

the basic function field of the tower $\mathcal{G}$. By the theory of Artin-Schreier extensions, exactly the places $(x=\infty)$ and $(x=\gamma)$ with $\gamma^{\ell-1}=1$ (i.e., $\gamma \in \mathbb{F}_\ell^\times$) are ramified in $G/\mathbb{F}_q(x)$. So we have

$$\Lambda_0 := \{x(P) \,|\, P \in \mathbb{P}_{\mathbb{F}_q(x)} \text{ is ramified in } G/\mathbb{F}_q(x)\} = \mathbb{F}_\ell^\times \cup \{\infty\}\,.$$

We set $\Lambda := \mathbb{F}_\ell \cup \{\infty\}$. In order to verify condition (2) of Proposition 7.2.23, we must show the following: if $\beta \in \Lambda$ and $\alpha \in \bar{\mathbb{F}}_q \cup \{\infty\}$ satisfy the relation

$$\beta^\ell - \beta = \frac{\alpha^\ell}{1-\alpha^{\ell-1}}\,, \tag{7.19}$$

then $\alpha \in \Lambda$. We distinguish two cases:

Case 1. $\beta \in \mathbb{F}_\ell$. Then $\beta^\ell - \beta = 0$, and from (7.19) follows $\alpha = 0 \in \Lambda$.

Case 2. $\beta = \infty$. Then it follows from (7.19) that $\alpha = \infty$ or $\alpha^{\ell-1} = 1$, so we have again $\alpha \in \Lambda$.

Thus we can apply Proposition 7.2.23 and obtain:

Lemma 7.4.5. *The tower $\mathcal{G}$ over $\mathbb{F}_q$ (with $q = \ell^2$) which is recursively defined by Equation* (7.14) *has a finite ramification locus. More precisely one has*

$$\mathrm{Ram}(\mathcal{G}/G_0) \subseteq \{(x_0 = \beta) \,|\, \beta \in \mathbb{F}_\ell \cup \{\infty\}\}\,.$$

It is easy to show that the tower $\mathcal{G}$ is a wild tower. As we have seen in Lemma 7.4.3, in each step G_{i+1}/G_i there are places which are totally (and hence wildly) ramified. So we do *not* have the estimate $d(Q|P) \le e(Q|P)$ for all $P \in \mathrm{Ram}(\mathcal{G}/G_0)$ and Q lying above P in some extension $G_n \supseteq G_0$. However, the following weaker assertion holds.

Lemma 7.4.6. *Let $\mathcal{G} = (G_0, G_1, G_2, \ldots)$ be the tower over $\mathbb{F}_q$ which is defined by Equation (7.14). Let $P \in \mathbb{P}_{G_0}$ be a place in the ramification locus of $\mathcal{G}/G_0$ and let $Q \in \mathbb{P}_{G_n}$ be a place lying above P. Then*

$$d(Q|P) = 2e(Q|P) - 2 .$$

We postpone the proof of this lemma, and we first draw an important conclusion:

Theorem 7.4.7 (Garcia-Stichtenoth). *Let $q = \ell^2$. Then the equation*

$$Y^\ell - Y = X^\ell/(1 - X^{\ell-1})$$

defines a recursive tower $\mathcal{G} = (G_0, G_1, G_2, \ldots)$ over $\mathbb{F}_q$, whose limit is

$$\lambda(\mathcal{G}) = \ell - 1 = q^{1/2} - 1 .$$

Therefore the tower $\mathcal{G}$ is optimal.

In fact, the first example of a tower over $\mathbb{F}_q$ with $q = \ell^2$, attaining the limit $\ell - 1$, was provided by Y. Ihara. He used modular curves for his construction.

Corollary 7.4.8 (Ihara). $A(q) = q^{1/2} - 1$, *if q is a square.*

Proof of Theorem 7.4.7. We will apply Theorem 7.2.10. The splitting locus $\mathrm{Split}(\mathcal{G}/G_0)$ has cardinality

$$|\mathrm{Split}(\mathcal{G}/G_0)| =: s \geq \ell^2 - \ell$$

by Lemma 7.4.4. The ramification locus $\mathrm{Ram}(\mathcal{G}/G_0)$ has cardinality

$$|\mathrm{Ram}(\mathcal{G}/G_0)| \leq \ell + 1$$

by Lemma 7.4.5. All places $P \in \mathrm{Ram}(\mathcal{G}/G_0)$ have degree one, and for all $n \geq 0$ and all $Q \in \mathbb{P}_{G_n}$ above P the different exponent is bounded by

$$d(Q|P) \leq 2e(Q|P)$$

by Lemma 7.4.6. Now the formula in Theorem 7.2.10(c) gives

$$\lambda(\mathcal{G}) \geq \frac{2(\ell^2 - \ell)}{-2 + 2(\ell + 1)} = \ell - 1 .$$

By the Drinfeld-Vladut Bound we have also the opposite inequality $\lambda(\mathcal{G}) \leq \ell - 1$, therefore the equality $\lambda(\mathcal{G}) = \ell - 1$ holds. □

Remark 7.4.9. One can easily show that in Lemma 7.4.4 and Lemma 7.4.5 equality holds; i.e.,

$$\begin{aligned} \mathrm{Split}(\mathcal{G}/G_0) &= \{(x_0 = \alpha) \mid \alpha \in \mathbb{F}_q \setminus \mathbb{F}_\ell\} , \text{ and} \\ \mathrm{Ram}(\mathcal{G}/G_0) &= \{(x_0 = \beta) \mid \beta \in \mathbb{F}_\ell \text{ or } \beta = \infty\} . \end{aligned}$$

Remark 7.4.10. One can determine exactly the genus and the number of rational places for each field G_n in the tower $\mathcal{G}$. However, this requires lengthy and very technical calculations.

We still have to prove Lemma 7.4.6. To this end we introduce some notation which will be useful also in connection with other wild towers. We recall a property of different exponents in a Galois extension E/F of function fields of degree $[E:F] = p = \operatorname{char} F$. If $P \in \mathbb{P}_F$ and $Q \in \mathbb{P}_E$ are places with $Q|P$, then

$$d(Q|P) = k \cdot (e(Q|P) - 1) \quad \textit{for some integer } k \geq 2\,. \tag{7.20}$$

This follows immediately from Hilbert's Different Formula. Note that (7.20) holds also if $e(Q|P) = 1$. In the case of $e(Q|P) = p$, the integer k in (7.20) is determined by the higher ramification groups of $Q|P$ as follows:

$$\operatorname{ord} G_i(Q|P) = p \textit{ for } 0 \leq i < k\,, \textit{ and } \operatorname{ord} G_k(Q|P) = 1\,.$$

If we have $k = 2$ in (7.20), we say that $Q|P$ is *weakly ramified.* We want to generalize this notion to certain extensions of degree $p^m \geq p$.

Remark 7.4.11. Let E/F be an extension of function fields of degree $[E:F] = p^m$ with $p = \operatorname{char} F$. Assume that there exists a chain of intermediate fields $F = E_0 \subseteq E_1 \subseteq \ldots \subseteq E_n = E$ with the property

$$E_{i+1}/E_i \textit{ is Galois for all } 0 \leq i < n\,.$$

Let $P \in \mathbb{P}_F$ and $Q \in \mathbb{P}_E$ with $Q|P$, and denote the restriction of Q to E_i by $Q_i := Q \cap E_i$. Then the following conditions are equivalent:

(1) $d(Q|P) = 2(e(Q|P) - 1)$.

(2) $d(Q_{i+1}|Q_i) = 2(e(Q_{i+1}|Q_i) - 1)$ for all $i = 0, \ldots, n-1$.

Proof. (2) $\Rightarrow$ (1): We assume (2) and show by induction that

$$d(Q_{i+1}|P) = 2(e(Q_{i+1}|P) - 1) \tag{7.21}$$

holds for $0 \leq i \leq n-1$. The case $i = 0$ is trivial since $P = Q_0$. Assume (7.21) for some i with $0 \leq i \leq n-2$. Then we obtain, by transitivity of different exponents,

$$\begin{aligned} d(Q_{i+2}|P) &= e(Q_{i+2}|Q_{i+1}) \cdot d(Q_{i+1}|P) + d(Q_{i+2}|Q_{i+1}) \\ &= e(Q_{i+2}|Q_{i+1}) \cdot 2(e(Q_{i+1}|P) - 1) + 2(e(Q_{i+2}|Q_{i+1}) - 1) \\ &= 2(e(Q_{i+2}|P) - 1)\,. \end{aligned}$$

Thus we have established the induction step. Setting $i := n-1$ in (7.21), we get $d(Q|P) = 2(e(Q|P) - 1)$.

(1) $\Rightarrow$ (2): Now we assume that (1) holds. Since the extensions E_{i+1}/E_i are Galois extensions of degree p^{n_i} (for some $n_i \geq 0$), Hilbert's Different Formula shows that $d(Q_{i+1}|Q_i) \geq 2(e(Q_{i+1}|Q_i) - 1)$ holds for all i. If this inequality is strict for some $i \in \{0, \ldots, n-1\}$, then the transitivity of different exponents yields $d(Q|P) > 2(e(Q|P) - 1)$, as in the proof of (2) $\Rightarrow$ (1). This contradicts the assumption (1), so we have $d(Q_{i+1}|Q_i) = 2(e(Q_{i+1}|Q_i) - 1)$ for all i. □

Definition 7.4.12. *Let F be a function field with* $\operatorname{char} F = p > 0$. *A finite extension E/F is said to be weakly ramified, if the following conditions hold:*

(1) There exist intermediate fields $F = E_0 \subseteq E_1 \subseteq \ldots \subseteq E_n = E$ such that all extensions E_{i+1}/E_i are Galois p-extensions (i.e., $[E_{i+1} : E_i]$ is a power of p), for $i = 0, 1, \ldots, n-1$.

(2) For all places $P \in \mathbb{P}_F$ and $Q \in \mathbb{P}_E$ with $Q|P$, the different exponent is given by $d(Q|P) = 2(e(Q|P) - 1)$.

The following proposition is crucial for the proof of Lemma 7.4.6 (and also for proving that some other wild towers are asymptotically good).

Proposition 7.4.13. *Let E/F be a finite extension of function fields and let M, N be intermediate fields of $E \supseteq F$ such that $E = MN$ is the compositum of M and N. Assume that both extensions M/F and N/F are weakly ramified. Then E/F is weakly ramified.*

Proof. The special case $[M : F] = [N : F] = p$ has been considered in Proposition 3.9.4. The idea of proof here is to reduce the general case to this special case. There is a sequence of intermediate fields

$$F = M_0 \subseteq M_1 \subseteq \ldots \subseteq M_k = M \tag{7.22}$$

such that all extensions M_{i+1}/M_i are weakly ramified Galois p-extensions. It is a well-known fact from group theory that every finite p-group G contains a chain of subgroups $\{1\} = G_0 \subseteq G_1 \subseteq \ldots \subseteq G_s = G$, where G_j is a normal subgroup of G_{j+1} of index $(G_{j+1} : G_j) = p$ for $j = 0, \ldots, s-1$. By Galois theory we can therefore refine the extensions $M_i \subseteq M_{i+1}$, to obtain Galois steps of degree p; i.e.,

$$M_i = M_i^{(0)} \subseteq M_i^{(1)} \subseteq \ldots \subseteq M_i^{(k_i)} = M_{i+1}$$

with weakly ramified Galois extensions $M_i^{(j+1)}/M_i^{(j)}$ of degree p. Therefore we can assume *a priori* that the extensions M_{i+1}/M_i in the chain (7.22) are all Galois of degree p.

In the same way we split the extension N/F into weakly ramified Galois steps of degree p. By induction over the degree $[N : F]$, the proof of Proposition 7.4.13 is thus reduced to the case where N/F is Galois of degree $[N : F] = p$. So we have the following situation: $F = M_0 \subseteq M_1 \subseteq \ldots \subseteq M_k = M$ with weakly ramified Galois extensions M_{i+1}/M_i of degree p, and $E = MN$ where N/F is a weakly ramified Galois extension of degree p. The extension E/M is then Galois of degree 1 or p and we have to show that E/M is also weakly ramified. If $[E : M] = 1$ there is nothing to prove, so we consider now the case $[E : M] = p$. Setting $N_i := M_i N$ for $i = 0, \ldots, k$ we obtain a chain of Galois extensions $N = N_0 \subseteq N_1 \subseteq \ldots \subseteq N_k = E$ with $[N_{i+1} : N_i] = p$, and also the extensions N_i/M_i are Galois of degree p. Then it follows from Proposition 3.9.4 by induction over i that all extensions N_i/M_i are weakly ramified, and hence E/M is weakly ramified. □

Now we are in a position to prove Lemma 7.4.6, and thereby to finish the proof of Theorem 7.4.7.

Proof of Lemma 7.4.6. We consider again the recursive tower $\mathcal{G} = (G_0, G_1, \ldots)$ over $\mathbb{F}_q$ with $q = \ell^2$, which is defined recursively by the equation $Y^\ell - Y = X^\ell/(1 - X^{\ell-1})$. The assertion of Lemma 7.4.6 is that all extensions G_n/G_0 are weakly ramified. We prove this by induction over n. The fields G_0 and G_1 are given by $G_0 = \mathbb{F}_q(x_0)$ and $G_1 = \mathbb{F}_q(x_0, x_1)$ with

$$x_1^\ell - x_1 = x_0^\ell/(1 - x_0^{\ell-1}). \tag{7.23}$$

As follows from Proposition 3.7.8, exactly the places $(x_0 = \infty)$ and $(x_0 = \beta)$ with $\beta \in \mathbb{F}_\ell^\times$ are ramified in the extensions G_1/G_0, with ramification index $e = \ell$ and different exponent $d = 2(\ell - 1)$. So G_1/G_0 is weakly ramified. For the induction step we assume that G_n/G_0 is weakly ramified, and we have to show that also the extension G_{n+1}/G_n is weakly ramified. We set $L_i := \mathbb{F}_q(x_1, \ldots, x_i)$ for $1 \le i \le n+1$, see Figure 7.2 below. The field L_{n+1} is $\mathbb{F}_q$-isomorphic to G_n via the isomorphism $\varphi : x_j \to x_{j+1}$ for $0 \le j \le n$; therefore L_{n+1}/L_1 is weakly ramified by induction hypothesis.

Now we observe that Equation (7.23) can be rewritten as

$$\left(\frac{1}{x_0}\right)^\ell - \frac{1}{x_0} = \frac{1}{x_1^\ell - x_1}.$$

This shows that also the extension G_1/L_1 is an Artin-Schreier extension of degree ℓ. In the same way as for the extension G_1/G_0, it follows from Proposition 3.7.10 that G_1/L_1 is weakly ramified (the ramified places are just the places $(x_1 = \alpha)$ with $\alpha \in \mathbb{F}_\ell$). Since G_{n+1} is the compositum of G_1 and L_{n+1} over L_1, Proposition 7.4.13 implies that G_{n+1}/L_1 is weakly ramified. Consequently the extension G_{n+1}/G_n is weakly ramified, by Remark 7.4.11. □

Although each step G_{i+1}/G_i in the tower $\mathcal{G}$ is Galois (by Lemma 7.4.3), the extensions G_i/G_0 are *not* Galois for $i \ge 2$. This follows from the fact that the place $(x_0 = 0) \in \mathbb{P}_{G_0}$ has unramified as well as ramified extensions in G_i. We therefore ask if it is possible to "extend" the tower $\mathcal{G} = (G_0, G_1, G_2, \ldots)$

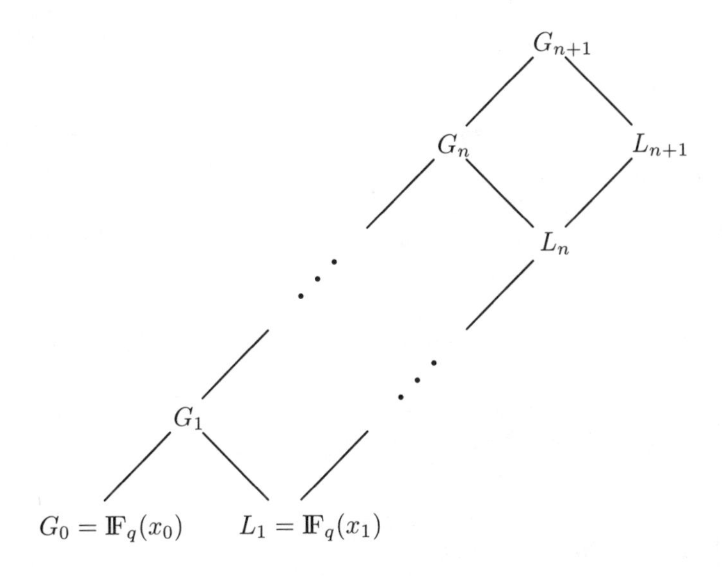

Fig. 7.2.

to a tower $\mathcal{G}^* = (G_0^*, G_1^*, G_2^*, \ldots)$ of function fields G_i^* over $\mathbb{F}_q$ having the following properties:

(1) $\mathcal{G}$ is a subtower of $\mathcal{G}^*$,

(2) G_i^*/G_0^* is Galois for all $i \geq 0$,

(3) the tower $\mathcal{G}^*$ is optimal over $\mathbb{F}_q$; i.e., $\lambda(\mathcal{G}^*) = q^{1/2} - 1$.

An obvious choice would be to take G_i^* as the Galois closure of G_i over G_0. With regard to an application in coding theory in Chapter 8, we proceed in a slightly different manner. We will need the following lemma.

Lemma 7.4.14. *Let $K(x)$ be the rational function field over a field $K \supseteq \mathbb{F}_\ell$. Consider the subfields $K(u) \subseteq K(t) \subseteq K(x)$ with*

$$t := x^\ell - x \quad \text{and} \quad u := (x^\ell - x)^{\ell-1} + 1 = t^{\ell-1} + 1\,. \tag{7.24}$$

Then the following hold:

(a) The extensions $K(x)/K(u)$, $K(x)/K(t)$ and $K(t)/K(u)$ are Galois of degree

$$[K(x) : K(u)] = \ell(\ell-1)\,,\ [K(x) : K(t)] = \ell\,, \ \text{ and } \ [K(t) : K(u)] = \ell - 1\,.$$

(b) The place $(u = \infty)$ of $K(u)$ is totally ramified in the extension $K(x)/K(u)$. The place of $K(x)$ above $(u = \infty)$ is the pole $(x = \infty)$ of x, and the place of $K(t)$ above $(u = \infty)$ is the pole $(t = \infty)$ of t. The ramification indices and different exponents are

$$\begin{aligned} e((t=\infty)|(u=\infty)) &= \ell-1 \quad \textit{and} \quad d((t=\infty)|(u=\infty)) = \ell-2\,, \\ e((x=\infty)|(t=\infty)) &= \ell \quad \textit{and} \quad d((x=\infty)|(t=\infty)) = 2(\ell-1)\,, \\ e((x=\infty)|(u=\infty)) &= \ell(\ell-1) \quad \textit{and} \quad d((x=\infty)|(u=\infty)) = \ell^2-2\,. \end{aligned}$$

Hence $(x=\infty)|(t=\infty)$ is weakly ramified.

(c) The place $(u=1)$ is totally ramified in $K(t)/K(u)$, the place in $K(t)$ above $(u=1)$ is the place $(t=0)$, with ramification index $e((t=0)|(u=1))=\ell-1$ and different exponent $d((t=0)|(u=1))=\ell-2$. In $K(x)/K(t)$ the place $(t=0)$ splits completely. The places of $K(x)$ above $(t=0)$ are exactly the places $(x=\beta)$ with $\beta\in\mathbb{F}_\ell$.

(d) No other places of $K(u)$, except $(u=\infty)$ and $(u=1)$, are ramified in $K(x)/K(u)$.

(e) If $\mathbb{F}_{\ell^2}\subseteq K$ then the place $(u=0)$ of $K(u)$ splits completely in $K(x)/K(u)$; the places of $K(x)$ above $(u=0)$ are exactly the places $(x=\alpha)$ with $\alpha\in\mathbb{F}_{\ell^2}\setminus\mathbb{F}_\ell$.

Proof. (a) We consider the two subgroups U_0,U_1 of the automorphism group of $K(x)/K$ which are defined by

$$\begin{aligned} U_0 &:= \{\sigma : x\mapsto ax+b \mid a\in\mathbb{F}_\ell^\times\,,\, b\in\mathbb{F}_\ell\}, \\ U_1 &:= \{\sigma : x\mapsto x+b \mid b\in\mathbb{F}_\ell\} \subseteq U_0\,. \end{aligned}$$

It is clear that $\operatorname{ord} U_0=\ell(\ell-1)$ and $\operatorname{ord} U_1=\ell$, and U_1 is a normal subgroup of U_0. One checks that u is invariant under all $\sigma\in U_0$, and t is invariant under all $\sigma\in U_1$. Since $[K(x):K(u)]=\ell(\ell-1)$ and $[K(x):K(t)]=\ell$ by (7.24), it follows that $K(x)/K(u)$ is Galois with Galois group U_0, and $K(x)/K(t)$ is Galois with Galois group U_1.

(b) The assertions concerning ramification indices are obvious, since u and t are polynomials in x. The only non-trivial statements of (b) are

$$d((x=\infty)|(t=\infty)) = 2(\ell-1) \quad \textit{and} \quad d((x=\infty)|(u=\infty)) = \ell^2-2\,. \tag{7.25}$$

In order to prove this, we note that the extension $K(x)/K(t)$ is Galois of order ℓ and $e((x=\infty)|(t=\infty))=\ell$. Therefore

$$d((x=\infty)|(t=\infty)) \geq 2(\ell-1)$$

by Hilbert's Different Formula. On the other hand, Hurwitz Genus Formula for $K(x)/K(t)$ yields

$$-2 = -2\ell + \deg \operatorname{Diff}(K(x)/K(t))$$

and hence

$$d((x=\infty)|(t=\infty)) = \deg \operatorname{Diff}(K(x)/K(t)) = 2(\ell-1)\,.$$

So we have $d((x=\infty)|(t=\infty))=2(\ell-1)$. By using the transitivity of the different exponent in the extensions $K(x)\supseteq K(t)\supseteq K(u)$, we obtain easily that $d((x=\infty)|(u=\infty))=\ell^2-2$.

(c) Since $u-1=t^{\ell-1}$, it is clear that $(t=0)$ is the only place of $K(t)$ above the place $(u=1)$ of $K(u)$, with ramification index $e((t=0)|(u=1))=\ell-1$ and different exponent $d((t=0)|(u=1))=\ell-2$. The equation $x^\ell-x=t$ shows that $(t=0)$ splits completely in $K(x)/K(t)$, and the places of $K(x)$ above $(t=0)$ are exactly the places $(x=\beta)$ with $\beta^\ell-\beta=0$.

(d) This follows by Hurwitz Genus Formula for the extension $K(x)/K(u)$ and parts (b) and (c).

(e) Note that

$$u=(x^\ell-x)^{\ell-1}+1=(x^{\ell^2}-x)/(x^\ell-x)=\prod_{\alpha\in\mathbb{F}_{\ell^2}\setminus\mathbb{F}_\ell}(x-\alpha)\,,$$

so that assertion (e) follows immediately. □

We return to the construction of the tower $\mathcal{G}^*$ over $\mathbb{F}_q$ with $q=\ell^2$. We start with the recursive tower $\mathcal{G}=(G_0,G_1,G_2,\ldots)$ with the defining equation $Y^\ell-Y=X^\ell/(1-X^{\ell-1})$, so we have

$$G_0=\mathbb{F}_q(x_0)\ \ and\ \ G_{i+1}=G_i(x_{i+1})\ \ with\ \ x_{i+1}^\ell-x_{i+1}=\frac{x_i^\ell}{1-x_i^{\ell-1}}$$

for all $i\geq 0$. We set

$$t_0:=x_0^\ell-x_0\ \ and\ \ u_0:=t_0^{\ell-1}+1=(x_0^\ell-x_0)^{\ell-1}+1\,. \tag{7.26}$$

Then we define $G_0^*:=\mathbb{F}_q(u_0)$ and

$$G_i^*:=\ Galois\ closure\ of\ G_i\ over\ \mathbb{F}_q(u_0)\,,\ for\ i\geq 1\,. \tag{7.27}$$

Thus we obtain a sequence of fields

$$G_0^*=\mathbb{F}_q(u_0)\subseteq\mathbb{F}_q(t_0)\subseteq\mathbb{F}_q(x_0)\subseteq G_1^*\subseteq G_2^*\subseteq\ldots\,,$$

where $\mathbb{F}_q(t_0)/\mathbb{F}_q(u_0)$ is a Galois extension of degree $\ell-1$ and $\mathbb{F}_q(x_0)/\mathbb{F}_q(t_0)$ is Galois of degree ℓ. The extension $\mathbb{F}_q(x_0)/\mathbb{F}_q(u_0)$ is also Galois, by Lemma 7.4.14.

Theorem 7.4.15. *Let $q=\ell^2$ be a square. With the above notation, the following hold:*

(a) $\mathbb{F}_q$ is the full constant field of G_i^* for all $i\geq 0$. Therefore the sequence

$$\mathcal{G}^*:=(G_0^*,G_1^*,G_2^*,\ldots)$$

is a tower over $\mathbb{F}_q$, where $G_0^*=\mathbb{F}_q(u_0)$ is rational and all extensions G_i^*/G_0^* are Galois.

(b) The place $(u_0=0)$ of $\mathbb{F}_q(u_0)$ splits completely in all extensions $G_i^*/\mathbb{F}_q(u_0)$.

(c) The places $(u_0 = \infty)$ *and* $(u_0 = 1)$ *are the only places of* $\mathbb{F}_q(u_0)$ *which are ramified in the tower* $\mathcal{G}^*$ *over* $\mathbb{F}_q(u_0)$. *Both of them are totally ramified in the extension* $\mathbb{F}_q(t_0)/\mathbb{F}_q(u_0)$ *of degree* $\ell - 1$.

(d) The extensions $G_i^*/\mathbb{F}_q(t_0)$ *are weakly ramified Galois p-extensions (with* $p = \operatorname{char} \mathbb{F}_q$*).*

(e) The tower $\mathcal{G}^*$ *attains the Drinfeld-Vladut Bound*

$$\lambda(\mathcal{G}^*) = \ell - 1 = q^{1/2} - 1 .$$

Proof. We consider the field extensions

$$\mathbb{F}_q(u_0) \subseteq \mathbb{F}_q(t_0) \subseteq \mathbb{F}_q(x_0) = G_0 \subseteq G_1 \subseteq G_2 \subseteq \ldots . \tag{7.28}$$

The first step $\mathbb{F}_q(t_0)/\mathbb{F}_q(u_0)$ is Galois of degree $\ell - 1$, and all other steps in (7.28) are weakly ramified Galois extensions of degree ℓ (by Lemma 7.4.14 and Lemma 7.4.6). The place $(u_0 = 0)$ of $\mathbb{F}_q(u_0)$ splits completely in $\mathbb{F}_q(x_0)/\mathbb{F}_q(u_0)$. The places of $\mathbb{F}_q(x_0)$ above $(u_0 = 0)$ are exactly the places $(x_0 = \alpha)$ with $\alpha \in \mathbb{F}_q \setminus \mathbb{F}_\ell$ (Lemma 7.4.14(e)). These places $(x_0 = \alpha)$ are in the splitting locus of the tower $\mathcal{G}$ over $G_0 = \mathbb{F}_q(x_0)$ (Lemma 7.4.4), and we conclude that the place $(u_0 = 0)$ splits completely in all extensions $G_i/\mathbb{F}_q(u_0)$. It follows from Corollary 3.9.7, that $(u_0 = 0)$ splits completely in the Galois closure G_i^* of G_i over $\mathbb{F}_q(u_0)$ and that $\mathbb{F}_q$ is the full constant field of G_i^*. We have thus proved parts (a) and (b).

(c) Let P be a place of $\mathbb{F}_q(u_0)$ which is different from $(u_0 = 1)$ and $(u_0 = \infty)$. Then P is unramified in $G_0 = \mathbb{F}_q(x_0)$, and the places of G_0 lying above P are different from the places $(x_0 = \beta)$ with $\beta \in \mathbb{F}_\ell \cup \{\infty\}$ (Lemma 7.4.14). By Lemma 7.4.5, the ramification locus of $\mathcal{G}$ over G_0 is contained in the set $\{(x_0 = \beta) \mid \beta \in \mathbb{F}_\ell \cup \{\infty\}\}$. It follows that P is unramified in all extensions $G_i/\mathbb{F}_q(u_0)$. Thus P is unramified in the Galois closure G_i^* of G_i over $\mathbb{F}_q(u_0)$, by Corollary 3.9.3. The rest of part (c) follow immediately from Lemma 7.4.14.

(d) We denote the Galois group of $G_i^*/\mathbb{F}_q(u_0)$ by Γ_i. Every $\tau \in \Gamma_i$ maps the field $\mathbb{F}_q(t_0)$ to itself since $\mathbb{F}_q(t_0)/\mathbb{F}_q(u_0)$ is Galois. By Lemma 7.4.14(b) and Lemma 7.4.6, the extension $G_i/\mathbb{F}_q(t_0)$ is weakly ramified, hence $\tau(G_i)/\mathbb{F}_q(t_0)$ is weakly ramified for all $\tau \in \Gamma_i$. The field G_i^* is the compositum of the fields $\tau(G_i)$ with $\tau \in \Gamma_i$, and we conclude from Proposition 7.4.13 that also the extension $G_i^*/\mathbb{F}_q(t_0)$ is weakly ramified.

(e) We will apply Theorem 7.2.10(c) to the tower $\mathcal{G}^*$ over $\mathbb{F}_q(u_0)$; to this end we need an estimate for the different exponents of ramified places in $G_n^*/\mathbb{F}_q(u_0)$ as in (7.7). So we consider a place P of $\mathbb{F}_q(u_0)$ which ramifies in G_n^* (we have $P = (u_0 = \infty)$ or $P = (u_0 = 1)$ by (c)). Let Q^* be a place of G_i^* lying above P and set $Q := Q^* \cap \mathbb{F}_q(t_0)$. Then $Q|P$ is tamely ramified with ramification index $e(Q|P) = \ell - 1$ by (c), and $Q^*|Q$ is weakly ramified by (d). Using transitivity of different exponents we obtain

$$\begin{aligned} d(Q^*|P) &= d(Q^*|Q) + e(Q^*|Q) \cdot d(Q|P) \\ &= 2(e(Q^*|Q) - 1) + (\ell - 2) \cdot e(Q^*|Q) \\ &< \ell \cdot e(Q^*|Q) = \frac{\ell}{\ell - 1} \cdot e(Q^*|P)\,. \end{aligned} \tag{7.29}$$

So we have the inequality

$$d(Q^*|P) \leq a_P \cdot e(Q^*|P) \;\; with \;\; a_P := \frac{\ell}{\ell - 1}\,.$$

As we have shown in part (b), the place $(u_0 = 0)$ of $\mathbb{F}_q(u_0)$ splits completely in the tower $\mathcal{G}^*$. Now Theorem 7.2.10 gives the estimate

$$\lambda(\mathcal{G}^*) \geq \frac{2}{-2 + 2 \cdot \frac{\ell}{\ell - 1}} = \ell - 1\,.$$

The inequality $\lambda(\mathcal{G}^*) \leq \ell - 1$ follows from the Drinfeld-Vladut Bound, hence we have $\lambda(\mathcal{G}^*) = \ell - 1$. □

In connection with an application in coding theory (see Section 8.4) we note some specific properties of the tower $\mathcal{G}^*$:

Corollary 7.4.16. *With the notations of Theorem 7.4.15, we set*

$$n_i := [G_i^* : \mathbb{F}_q(u_0)] \;=\; (\ell - 1) \cdot m_i \tag{7.30}$$

for every $i \geq 0$, so $m_i = [G_i^* : \mathbb{F}_q(t_0)]$ is a power of $p = \operatorname{char} \mathbb{F}_q$. Denote by

$e_i^{(0)}$, *the ramification index of the place $(t_0 = 0)$ in $G_i^*/\mathbb{F}_q(t_0)$,*
$e_i^{(\infty)}$, *the ramification index of the place $(t_0 = \infty)$ in $G_i^*/\mathbb{F}_q(t_0)$.*

Thus the principal divisor of t_0 in G_i^* is

$$(t_0)^{G_i^*} \;=\; e_i^{(0)} A_i - e_i^{(\infty)} B_i \tag{7.31}$$

with positive divisors $A_i, B_i \in \operatorname{Div}(G_i^*)$. Then the following hold:
(a) The function field G_i^* has the genus

$$g(G_i^*) \;=\; 1 + m_i \left(1 - \frac{1}{e_i^{(0)}} - \frac{1}{e_i^{(\infty)}} \right).$$

(b) The zero divisor of u_0 in G_i^* has the form

$$D_i \;=\; \sum_{j=1}^{n_i} P_j^{(i)}\,, \tag{7.32}$$

with n_i rational places $P_j^{(i)} \in \mathbb{P}_{G_i^*}$.

(c) The differential $\eta^{(i)} := du_0/u_0$ of G_i^* has the divisor

$$(\eta^{(i)}) = (\ell e_i^{(0)} - 2)A_i + (e_i^{(\infty)} - 2)B_i - D_i\,,$$

where the divisors A_i, B_i, D_i are defined by (7.31) and (7.32). At all places $P = P_j^{(i)} \le D_i$, the residue of $\eta^{(i)}$ is

$$\mathrm{res}_P(\eta^{(i)}) = 1\,.$$

Proof. The places $(t_0 = 0)$ and $(t_0 = \infty)$ are the only places of $\mathbb{F}_q(t_0)$ which ramify in the extension $G_i^*/\mathbb{F}_q(t_0)$. Since they are weakly ramified (Theorem 7.4.15(c), (d)), the different of $G_i^*/\mathbb{F}_q(t_0)$ is given by

$$\mathrm{Diff}(G_i^*/\mathbb{F}_q(t_0)) = (2e_i^{(0)} - 2)A_i + (2e_i^{(\infty)} - 2)B_i\ ,$$

(with the divisors A_i, B_i as in (7.31)). Then the differential dt_0 (as a differential of the field G_i^*) has the divisor

$$\begin{aligned}(dt_0) &= -2e_i^{(\infty)}B_i + \mathrm{Diff}(G_i^*/\mathbb{F}_q(t_0))\\ &= (2e_i^{(0)} - 2)A_i - 2B_i\,,\end{aligned} \tag{7.33}$$

by Remark 4.3.7. Since $\deg A_i = m_i/e_i^{(0)}$ and $\deg B_i = m_i/e_i^{(\infty)}$, we obtain

$$2g(G_i^*) - 2 = \deg(dt_0) = (2e_i^{(0)} - 2)\cdot\frac{m_i}{e_i^{(0)}} - 2\cdot\frac{m_i}{e_i^{(\infty)}}\,.$$

This proves (a). Part (b) is clear from Theorem 7.4.15(b). In order to show part (c) we note that $u_0 = 1 + t_0^{\ell-1}$ by (7.26), hence $du_0 = -t_0^{\ell-2}dt_0$ and

$$\begin{aligned}(du_0) &= (\ell-2)\cdot(e_i^{(0)}A_i - e_i^{(\infty)}B_i) + (dt_0)\\ &= (\ell e_i^{(0)} - 2)A_i - ((\ell-2)e_i^{(\infty)} + 2)B_i\,.\end{aligned}$$

The principal divisor of u_0 in G_i^* is

$$(u_0)^{G_i^*} = D_i - (\ell-1)e_i^{(\infty)}B_i\,,$$

so the differential $\eta^{(i)} := du_0/u_0$ has the divisor (in G_i^*)

$$(\eta^{(i)}) = (\ell e_i^{(0)} - 2)A_i + (e_i^{(\infty)} - 2)B_i - D_i\,.$$

At all places P in the support of D_i, the element u_0 is a prime element, consequently we have $\mathrm{res}_P(du_0/u_0) = 1$ by the definition of the residue of a differential. □

Now we turn to finite fields $\mathbb{F}_q$ where $q = \ell^3$ is a cube. We investigate the recursive tower $\mathcal{H}$ over $\mathbb{F}_q$, which is given by the equation

$$(Y^\ell - Y)^{\ell-1} + 1 = \frac{-X^{\ell(\ell-1)}}{(X^{\ell-1} - 1)^{\ell-1}}\,. \tag{7.34}$$

Our main result is

Theorem 7.4.17. *Equation* (7.34) *defines a recursive tower* $\mathcal{H}$ *over the field* $\mathbb{F}_q$ *with* $q = \ell^3$. *Its limit* $\lambda(\mathcal{H})$ *satisfies*

$$\lambda(\mathcal{H}) \geq \frac{2(\ell^2 - 1)}{\ell + 2} .$$

The following lower bound for $A(\ell^3)$ is due to T. Zink in the case of a prime number ℓ; for arbitrary ℓ it was first shown by Bezerra, Garcia and Stichtenoth.

Corollary 7.4.18. *For* $q = \ell^3$, *Ihara's constant* $A(q)$ *is bounded below by*

$$A(q) \geq \frac{2(\ell^2 - 1)}{\ell + 2} .$$

We will prove Theorem 7.4.17 in several steps. It is convenient to consider $\mathcal{H}$ not only over the field $\mathbb{F}_q$ but also over any field K which contains $\mathbb{F}_\ell$. So we have $\mathcal{H} = (H_0, H_1, H_2, \ldots)$ where $H_0 = K(y_0)$ is a rational function field over K, and

$$H_{i+1} = H_i(y_{i+1}) \text{ with } (y_{i+1}^\ell - y_{i+1})^{\ell-1} + 1 = \frac{-y_i^{\ell(\ell-1)}}{(y_i^{\ell-1} - 1)^{\ell-1}} \tag{7.35}$$

for all $i \geq 0$. We denote the generators of $\mathcal{H}$ by $y_0, y_1, y_2, \ldots$ in order to avoid confusion with the generators $x_0, x_1, x_2, \ldots$ of the tower $\mathcal{G}$.

The best way to understand the tower $\mathcal{H}$ is to study its *second basic function field* $H := K(x, y, z)$ which is defined by the equations

$$(y^\ell - y)^{\ell-1} + 1 = \frac{-x^{\ell(\ell-1)}}{(x^{\ell-1} - 1)^{\ell-1}} =: u \tag{7.36}$$

and

$$(z^\ell - z)^{\ell-1} + 1 = \frac{-y^{\ell(\ell-1)}}{(y^{\ell-1} - 1)^{\ell-1}} =: v . \tag{7.37}$$

First we determine the degree of H over some of its subfields as shown in Figure 7.3 below.

Lemma 7.4.19. *Assume that* $\mathbb{F}_\ell \subseteq K$. *With the above notation we have:*

(a) The extensions $K(x)/K(u)$, $K(y)/K(u)$, $K(y)/K(v)$ *and* $K(z)/K(v)$ *are Galois of degree* $\ell(\ell - 1)$.

(b) Also the extensions at the next level, $K(x, y)/K(x)$, $K(x, y)/K(y)$, $K(y, z)/K(y)$ *and* $K(y, z)/K(z)$ *are Galois of degree* $\ell(\ell - 1)$.

(c) The extensions $K(x, y, z)/K(x, y)$ *and* $K(x, y, z)/K(y, z)$ *are Galois of degree* ℓ.

(d) K *is the full constant field of* $H = K(x, y, z)$.

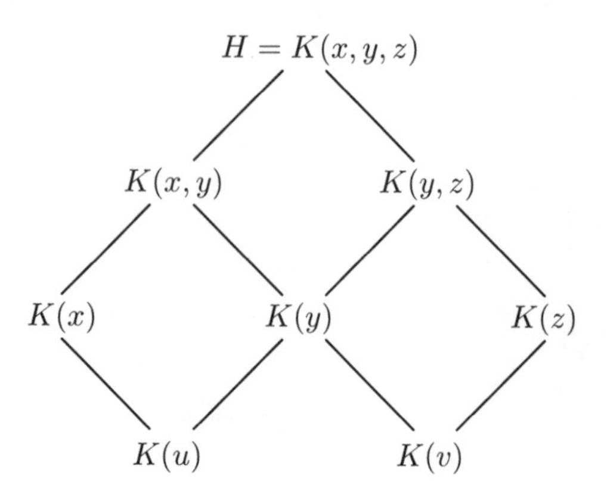

Fig. 7.3. The second basic function field H.

Proof. (a) By Lemma 7.4.14(a), the extension $K(y)/K(u)$ is Galois of degree $[K(y):K(u)] = \ell(\ell-1)$, since $u = (y^\ell - y)^{\ell-1} + 1$. Now we observe that the equation $u = -x^{\ell(\ell-1)}/(x^{\ell-1}-1)^{\ell-1}$ is equivalent to

$$\left(\left(\frac{1}{x}\right)^\ell - \frac{1}{x}\right)^{\ell-1} + 1 = \frac{u-1}{u}\,. \tag{7.38}$$

As $K(x) = K(1/x)$ and $K(u) = K((u-1)/u)$, we conclude again from Lemma 7.4.14(a) that $K(x)/K(u)$ is a Galois extension of degree $\ell(\ell-1)$. The assertions for the extensions $K(y)/K(v)$ and $K(z)/K(v)$ follow analogously.

(b) We consider the extension $K(x,y)/K(x)$. Since $K(y)/K(u)$ is Galois, it follows from Galois theory that $K(x,y)/K(x)$ is Galois, and its Galois group $\mathrm{Gal}(K(x,y)/K(x))$ is isomorphic to a subgroup of $\mathrm{Gal}(K(y)/K(u))$. It remains to show that $[K(x,y):K(x)] \geq \ell(\ell-1)$. To this end we consider ramification in some subextensions of the field H as in Figure 7.4.

For a place $R \in \mathbb{P}_H$ we denote its restrictions to the corresponding subfields of H according to Figure 7.4 ; this means for instance that $P = R \cap K(x)$ and $\tilde{Q} = R \cap K(v)$. Specifically we choose a place Q of $K(x,y)$ which is a zero of the element u. We then have $P^* = (u=0)$. From Equation (7.36) we see that $P = (x=0)$ and $e(P|P^*) = \ell(\ell-1)$. On the other hand it follows from Lemma 7.4.14(d) that $e(Q^*|P^*) = 1$ and hence $e(Q|Q^*) = \ell(\ell-1)$. We conclude that $[K(x,y):K(x)] = [K(x,y):K(y)] \geq \ell(\ell-1)$.

Thus we have shown that $K(x,y)/K(x)$ is a Galois extension of degree $\ell(\ell-1)$. The corresponding claims for the other extensions $K(x,y)/K(y)$, $K(y,z)/K(y)$ and $K(y,z)/K(z)$ follow immediately.

(c) Next we choose a place R of $K(x,y,z)$ which is a pole of x. Thus $P = (x=\infty)$, and we obtain from Equations (7.36) and (7.37) that the places P^*,

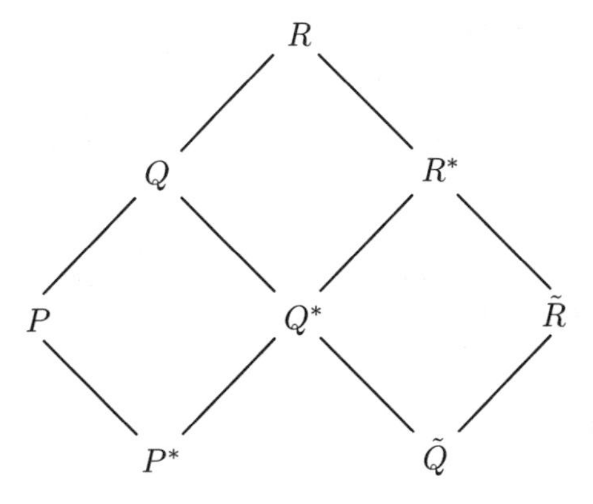

Fig. 7.4. Places of subfields of H.

Q^*, $\tilde{Q}$ and $\tilde{R}$ are as follows:

$$\begin{aligned} P^* &= (u=\infty) \;\; \textit{with} \;\; e(P|P^*) = \ell - 1\,, \\ Q^* &= (y=\infty) \;\; \textit{with} \;\; e(Q^*|P^*) = \ell(\ell - 1)\,, \\ \tilde{Q} &= (v=\infty) \;\; \textit{with} \;\; e(Q^*|\tilde{Q}) = \ell - 1\,, \\ \tilde{R} &= (z=\infty) \;\; \textit{with} \;\; e(\tilde{R}|\tilde{Q}) = \ell(\ell - 1)\,. \end{aligned}$$

Now Abhyankar's Lemma yields $e(Q|Q^*) = 1$, $e(R^*|Q^*) = \ell$ and $e(R|Q) = \ell$ (observe the notation in Figure 7.4), and it follows that

$$[K(x,y,z) : K(x,y)] \geq e(R\,|\,Q) = \ell\,. \tag{7.39}$$

On the other hand we get from (7.36) and (7.37), that

$$\begin{aligned} (z^\ell - z)^{\ell-1} &= \frac{-y^{\ell(\ell-1)}}{(y^{\ell-1}-1)^{\ell-1}} - 1 \\ &= \frac{-((y^\ell - y)^{\ell-1} + 1)}{(y^{\ell-1}-1)^{\ell-1}} \\ &= \frac{x^{\ell(\ell-1)}}{(x^{\ell-1}-1)^{\ell-1}\,(y^{\ell-1}-1)^{\ell-1}}\,. \end{aligned}$$

Therefore we have

$$z^\ell - z = \mu \cdot \frac{x^\ell}{(x^{\ell-1}-1)(y^{\ell-1}-1)} \tag{7.40}$$

for some $\mu \in \mathbb{F}_\ell^\times$. This is an equation for z over $K(x,y)$ of degree ℓ, so $[K(x,y,z) : K(x,y)] \leq \ell$. With (7.39) we conclude that the extension

$K(x,y,z)/K(x,y)$ has degree ℓ. From Equation (7.40) it also follows that $K(x,y,z)/K(x,y)$ is Galois.

(d) We have seen in the proof of (b) and (c) that in both extensions $K(x,y)/K(x)$ and $K(x,y,z)/K(x,y)$ there are totally ramified places. Theorem 3.6.3(a) implies that K is the full constant field of $K(x,y)$ and of $K(x,y,z)$. □

Next we determine all places of the field $H = K(x,y,z)$ which are ramified over the field $K(u)$ (where u is given by Equation (7.36)). For simplicity we can assume that K contains the field $\mathbb{F}_{\ell^2}$, since the ramification behavior (i.e., ramification index and different degree) does not change under constant field extensions. The following lemma is of fundamental importance for the investigation of the tower $\mathcal{H}$.

Lemma 7.4.20. *We maintain the notation of Equations* (7.36), (7.37) *and Figures* 7.3, 7.4, *and also assume that* $\mathbb{F}_{\ell^2} \subseteq K$. *Let R be a place of H which is ramified in the extension $H/K(u)$. Then the restriction $P^* = R \cap K(u)$ of R to $K(u)$ is one of the places $(u=0)$, $(u=1)$ or $(u=\infty)$. More precisely the following hold:*

Case 1. $P^* = (u=0)$. *Then* $P = (x=0)$,

$$e(Q|P) = e(R|Q) = 1 \quad \text{and} \quad e(R|R^*) = \ell\,.$$

Case 2. $P^* = (u=1)$. *Then* $P = (x=\beta)$ *with* $\beta \in \mathbb{F}_{\ell^2} \setminus \mathbb{F}_\ell$,

$$e(Q|P) = \ell - 1\,, \quad e(R|Q) \in \{1, \ell\} \quad \text{and} \quad e(R|R^*) = 1\,.$$

Case 3. $P^* = (u=\infty)$. *Then* $P = (x=\alpha)$ *with some* $\alpha \in \mathbb{F}_\ell^\times \cup \{\infty\}$, $Q^* = (y=\infty)$,

$$e(Q|P) = e(R|Q) = \ell \quad \text{and} \quad e(Q|Q^*) = e(R|R^*) = e(R^*|\tilde{R}) = 1\,.$$

In all cases above where the ramification index is $e = \ell$, the corresponding different exponent is $d = 2(\ell - 1)$.

Proof. Since $R|P^*$ is ramified, at least one of the places $P|P^*$ or $R^*|P^*$ is ramified, see Figure 7.4. We distinguish several cases:

(i) Assume that $P|P^*$ is ramified. Since $K(x) = K(1/x)$ and

$$\left(\left(\frac{1}{x}\right)^\ell - \frac{1}{x}\right)^{\ell-1} + 1 = \frac{u-1}{u} =: u'$$

(see Equation (7.38)), it follows from Lemma 7.4.14 that $P^* = (u' = \infty)$ or $P^* = (u' = 1)$. Since $u' = \infty \Leftrightarrow u = 0$ and $u' = 1 \Leftrightarrow u = \infty$, we conclude that $P^* = (u=0)$ or $P^* = (u=\infty)$.

(ii) Now we assume that $R^*|P^*$ is ramified. Then one of the places $Q^*|P^*$ or $R^*|Q^*$ is ramified.

(ii$_1$) If $Q^*|P^*$ is ramified then $P^* = (u = \infty)$ or $P^* = (u = 1)$ by Lemma 7.4.14.

(ii$_2$) If $R^*|Q^*$ is ramified then $\tilde{R}|\tilde{Q}$ is ramified (see Figure 7.4), hence $\tilde{Q} = (v = \infty)$ or $\tilde{Q} = (v = 1)$ by Lemma 7.4.14. First we discuss the case $\tilde{Q} = (v = \infty)$. By Equation (7.37) we have then $Q^* = (y = \infty)$ or $Q^* = (y = \gamma)$ with $\gamma \in \mathbb{F}_\ell^\times$. If $Q^* = (y = \infty)$ then $P^* = (u = \infty)$, and if $Q^* = (y = \gamma)$ then $P^* = (u = 1)$ by Equation (7.36).

It remains to consider the case $\tilde{Q} = (v = 1) = (v' = 0)$ where $v' := (v - 1)/v$. From the equation

$$\left(\left(\frac{1}{y}\right)^\ell - \frac{1}{y}\right)^{\ell-1} + 1 = v'$$

(see Equation (7.38)) and Lemma 7.4.14(e) we obtain $Q^* = (1/y = \alpha)$ with $\alpha \in \mathbb{F}_{\ell^2} \setminus \mathbb{F}_\ell$, hence $Q^* = (y = \delta)$ with $\delta = \alpha^{-1} \in \mathbb{F}_{\ell^2} \setminus \mathbb{F}_\ell$. Again it follows from Lemma 7.4.14(e) that $P^* = (u = 0)$.

So far we have shown that if $R|P^*$ is ramified then $P^* = (u = 0)$ or $(u = 1)$ or $(u = \infty)$. Now we have to discuss these three cases.

Case 1. $P^* = (u = 0)$. Using Equations (7.36), (7.37) and Lemma 7.4.14 in the same way as above we find that

$$\begin{aligned} P &= (x = 0) \ \textit{and} \ Q^* = (y = \gamma) \ \textit{with} \ \gamma \in \mathbb{F}_{\ell^2} \setminus \mathbb{F}_\ell\,, \\ \tilde{Q} &= (v = 1) \ \textit{and} \ \tilde{R} = (z = \beta) \ \textit{with} \ \beta \in \mathbb{F}_\ell\,, \end{aligned}$$

with ramification indices (resp. different exponents)

$$\begin{aligned} e(P\,|\,P^*) &= \ell(\ell - 1) \ \textit{and} \ d(P\,|\,P^*) = \ell^2 - 2\,, \\ e(Q^*\,|\,P^*) &= e(Q^*\,|\,\tilde{Q}) = 1\,, \\ e(\tilde{R}\,|\,\tilde{Q}) &= \ell - 1\,. \end{aligned}$$

Going up one level in Figure 7.4 it follows immediately that

$$\begin{aligned} e(Q\,|\,P) &= 1 \ \textit{and} \ e(R^*\,|\,Q^*) = \ell - 1\,, \\ e(Q\,|\,Q^*) &= \ell(\ell - 1) \ \textit{and} \ d(Q\,|\,Q^*) = \ell^2 - 2\,. \end{aligned}$$

We go up another level in Figure 7.4, apply Abhyankar's Lemma and get

$$\begin{aligned} e(R\,|\,Q) &= 1\,, \\ e(R\,|\,R^*) &= \ell \ \textit{and} \ d(R\,|\,R^*) = 2(\ell - 1)\,. \end{aligned}$$

This finishes the proof of Lemma 7.4.20 in Case 1. The other two cases are similar. □

Corollary 7.4.21. *Both extensions $K(x,y,z)/K(x,y)$ and $K(x,y,z)/K(y,z)$ are weakly ramified Galois extensions of degree ℓ.*

We are now able to prove that Equation (7.34) actually defines a tower.

Proposition 7.4.22. *Let K be a field with $\mathbb{F}_\ell \subseteq K$, and consider the sequence $\mathcal{H} = (H_0, H_1, H_2, \ldots)$ where $H_0 = K(y_0)$ is a rational function field, and $H_{i+1} = H_i(y_{i+1})$ with*

$$(y_{i+1}^{\ell} - y_{i+1})^{\ell-1} + 1 = \frac{-y_i^{\ell(\ell-1)}}{(y_i^{\ell-1} - 1)^{\ell-1}}$$

for all $i \geq 0$. Then $\mathcal{H}$ is a tower over K. The extension H_1/H_0 is Galois of degree $[H_1 : H_0] = \ell(\ell-1)$, and for all $i \geq 1$, the extension H_{i+1}/H_i is Galois of degree $[H_{i+1} : H_i] = \ell$.

Proof. The field $H_2 = K(y_0, y_1, y_2)$ is isomorphic to the function field $H = K(x, y, z)$ that we studied in Lemmas 7.4.19 and 7.4.20. Therefore we know already that H_1/H_0 is Galois of degree $\ell(\ell-1)$, the extension H_2/H_1 is Galois of degree ℓ and K is the full constant field of H_2. Now let $i \geq 2$. Since $H_{i+1} = H_i(y_{i+1})$ and y_{i+1} satisfies the equation

$$y_{i+1}^{\ell} - y_{i+1} = \mu \cdot \frac{y_{i-1}^{\ell}}{(y_{i-1}^{\ell-1} - 1)(y_i^{\ell-1} - 1)}$$

with some element $\mu \in \mathbb{F}_\ell^{\times}$ (cf. Equation (7.40)), the extension H_{i+1}/H_i is Galois of degree $[H_{i+1} : H_i] \leq \ell$. We show now by induction the following claim which readily implies the remaining assertions of Proposition 7.4.22. We fix a place $P_1 \in \mathbb{P}_{H_1}$ which is a pole of the element y_0 in H_1.

Claim. Let $i \geq 2$ and let P_i be a place of H_i lying above P_1. Then P_i is also a pole of y_i, and we have

$$e(P_i \,|\, P_{i-1}) = \ell \ \textit{and}\ e(P_i \,|\, (y_i = \infty)) = 1\,.$$

Proof of the Claim. The case $i = 2$ follows from Case 3 of Lemma 7.4.20. So we assume now that the claim holds for some $i \geq 2$. Choose a place P_{i+1} of H_{i+1} which lies above P_i. The field H_{i+1} is the compositum of the fields H_i and $K(y_i, y_{i+1})$ over $K(y_i)$. By induction hypothesis we have that $P_i \cap K(y_i) = (y_i = \infty)$ and $e(P_i|(y_i = \infty)) = 1$. We set $P_{i+1}^* := P_{i+1} \cap K(y_i, y_{i+1})$ and $\tilde{P}_{i+1} := P_{i+1} \cap K(y_{i+1})$. It follows from Case 3 of Lemma 7.4.20 that $\tilde{P}_{i+1}$ is the pole of y_{i+1} in $K(y_{i+1})$ and that $e(P_{i+1}^*|(y_i = \infty)) = \ell$ and $e(P_{i+1}^*|\tilde{P}_{i+1}) = 1$. The situation is shown below in Figure 7.5.

From this picture we see that $e(P_{i+1}|P_i) = \ell$ and $e(P_{i+1}|(y_{i+1} = \infty)) = 1$. This completes the proof of Proposition 7.4.22. □

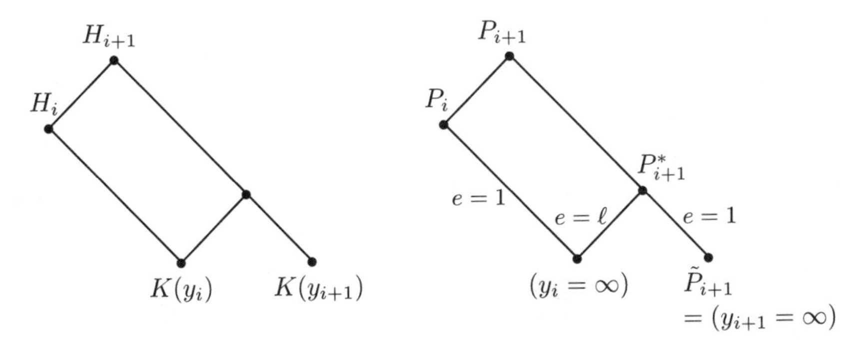

Fig. 7.5.

Thus we have proved that Equation (7.34) defines a tower $\mathcal{H}=(H_0, H_1, H_2, \ldots)$ over every constant field $K \supseteq \mathbb{F}_\ell$, and that all steps in the tower are Galois, of degree $[H_1 : H_0] = \ell(\ell-1)$ and $[H_{i+1} : H_i] = \ell$ for $i \geq 1$.

Remark 7.4.23. One of the defining properties of a tower is that it must contain a function field of genus $g \geq 2$ (see Definition 7.2.1 and Remark 7.2.2). We have not verified this property for the tower $\mathcal{H}$ yet. One way is to show directly (using Lemma 7.4.20) that the second basic function field $H = K(x, y, z)$ has genus $g \geq 2$. Another way is to show that the number of rational places $N(H_i)$ tends to infinity as $i \to \infty$. We will prove this fact in Lemma 7.4.26 below.

Next we consider the ramification locus of the tower $\mathcal{H}$ over H_0. The following lemma is analogous to Lemma 7.4.5.

Lemma 7.4.24. *Assume that* $\mathbb{F}_{\ell^2} \subseteq K$. *Then the ramification locus of* $\mathcal{H}$ *over* H_0 *satisfies*

$$\mathrm{Ram}(\mathcal{H}/H_0) \subseteq \{(y_0 = \beta) \,|\, \beta \in \mathbb{F}_{\ell^2} \cup \{\infty\}\}\,.$$

Before giving the proof of this lemma, we observe that the defining equation of the tower $\mathcal{H}$ can be written in a slightly different way as follows. We set

$$f(T) := (T^\ell - T)^{\ell-1} + 1 = (T^{\ell^2} - T)/(T^\ell - T)\,. \tag{7.41}$$

Then Equation (7.34) is equivalent to the equation

$$f(Y) = \frac{1}{1 - f(1/X)}\,. \tag{7.42}$$

From (7.41) it is obvious that the zeros of the polynomial $f(T)$ in $\bar{K}$ are exactly the elements $\gamma \in \mathbb{F}_{\ell^2} \setminus \mathbb{F}_\ell$.

Proof of Lemma 7.4.24. We want to apply Proposition 7.2.23. We set

$$\Lambda_0 := \{y_0(P) \mid P \in \mathbb{P}_{H_0} \text{ is ramified in } H_1/H_0\}.$$

Then $\Lambda_0 = \mathbb{F}_{\ell^2}^\times \cup \{\infty\}$ by Lemma 7.4.20, Cases 2 and 3. We claim that the set

$$\Lambda := \Lambda_0 \cup \{0\} = \mathbb{F}_{\ell^2} \cup \{\infty\}$$

satisfies condition (2) of Proposition 7.2.23. So we consider $\beta \in \Lambda$ and $\alpha \in \bar{K} \cup \{\infty\}$ where

$$(\beta^\ell - \beta)^{\ell-1} + 1 = \frac{-\alpha^{\ell(\ell-1)}}{(\alpha^{\ell-1} - 1)^{\ell-1}} = \frac{1}{1 - f(1/\alpha)} \tag{7.43}$$

(note Equation (7.42)). We must show that $\alpha \in \Lambda$.

Case 1. $\beta \in \mathbb{F}_\ell$. Then we have

$$1 = \frac{1}{1 - f(1/\alpha)}$$

and hence $f(1/\alpha) = 0$. It follows that $1/\alpha \in \mathbb{F}_{\ell^2} \setminus \mathbb{F}_\ell$ and therefore $\alpha \in \mathbb{F}_{\ell^2} \setminus \mathbb{F}_\ell \subseteq \Lambda$.

Case 2. $\beta = \infty$. It follows from Equation (7.43) that $\alpha = \infty$ or $\alpha \in \mathbb{F}_\ell^\times$, hence $\alpha \in \Lambda$.

Case 3. $\beta \in \mathbb{F}_{\ell^2} \setminus \mathbb{F}_\ell$. Then $(\beta^\ell - \beta)^{\ell-1} + 1 = 0$ and therefore $\alpha = 0 \in \Lambda$. This completes the proof, by Proposition 7.2.23. □

Corollary 7.4.25. *The genus $\gamma(\mathcal{H}/H_0)$ of the tower $\mathcal{H}$ is finite; it is bounded by*

$$\gamma(\mathcal{H}/H_0) \le \frac{\ell^2 + 2\ell}{2}.$$

Proof. Recall that both extensions $H_2/K(y_0, y_1)$ and $H_2/K(y_1, y_2)$ are weakly ramified Galois extensions of degree ℓ by Corollary 7.4.21. For all $n \ge 2$ we consider the field H_n as the compositum of the fields H_{n-1} and $K(y_{n-2}, y_{n-1}, y_n)$ over $K(y_{n-2}, y_{n-1})$, and then it follows by induction from Proposition 7.4.13 that H_n/H_1 is weakly ramified of degree $[H_n : H_1] = \ell^{n-1}$.

Now we estimate the different exponent $d(Q|P)$ where $P \in \mathrm{Ram}(\mathcal{H}/H_0)$ and Q is a place of H_n lying above P.

Case 1. $P = (y_0 = \alpha)$ with $\alpha \in \mathbb{F}_\ell \cup \{\infty\}$. It follows from Lemma 7.4.20 that P is weakly ramified in H_2/H_0, hence $Q|P$ is weakly ramified in H_n/H_0 and we obtain in Case 1, that

$$d(Q \mid P) = 2(e(Q \mid P) - 1) \le 2e(Q \mid P).$$

Case 2. $P = (y_0 = \beta)$ with $\beta \in \mathbb{F}_{\ell^2} \setminus \mathbb{F}_\ell$. Let $P_1 := Q \cap H_1$, then $e(P_1|P) = \ell - 1$ and $Q|P_1$ is weakly ramified by Lemma 7.4.20. It follows that

$$\begin{aligned} d(Q\,|\,P) &= e(Q\,|\,P_1)\cdot d(P_1\,|\,P) + d(Q\,|\,P_1) \\ &= e(Q\,|\,P_1)\cdot(\ell-2) + 2(e(Q\,|\,P_1)-1) \\ &= \frac{\ell}{\ell-1}\cdot e(Q\,|\,P) - 2 \le \frac{\ell}{\ell-1}\cdot e(Q\,|\,P)\,. \end{aligned}$$

There are $(l+1)$ places in Case 1 and $(\ell^2-\ell)$ places in Case 2, and we obtain from Theorem 7.2.10(b) the estimate

$$\gamma(\mathcal{H}/H_0) \le -1 + \frac{1}{2}\cdot(\ell+1)\cdot 2 + \frac{1}{2}\cdot(\ell^2-\ell)\cdot\frac{\ell}{\ell-1} = \frac{\ell^2+2\ell}{2}\,.$$

□

Observe that Corollary 7.4.25 holds for every constant field $K \supseteq \mathbb{F}_\ell$. One does not need the assumption that $K \supseteq \mathbb{F}_{\ell^2}$, since the genus $\gamma(\mathcal{H}/H_0)$ does not change under a constant field extension.

Eventually we investigate completely splitting places in the tower $\mathcal{H}$. We show that the splitting behavior of the tower $\mathcal{H}$ over the constant field $\mathbb{F}_{\ell^3}$ is very similar to the splitting behavior of the tower $\mathcal{G}$ over $\mathbb{F}_{\ell^2}$ (see Lemma 7.4.4).

Lemma 7.4.26. *Consider the tower $\mathcal{H} = (H_0, H_1, H_2, \ldots)$, which is defined by Equation (7.34) over the cubic field $K = \mathbb{F}_{\ell^3}$. Then all places $(y_0 = \alpha)$ with $\alpha \in \mathbb{F}_{\ell^3} \setminus \mathbb{F}_\ell$ split completely in $\mathcal{H}/H_0$; i.e.,*

$$\mathrm{Split}(\mathcal{H}/H_0) \supseteq \{(y_0 = \alpha)\,|\,\alpha \in \mathbb{F}_{\ell^3} \setminus \mathbb{F}_\ell\}\,.$$

Therefore the splitting rate $\nu(\mathcal{H}/H_0)$ satisfies

$$\nu(\mathcal{H}/H_0) \ge \ell^3 - \ell\,.$$

Proof. We will use Corollary 7.2.21. Accordingly we set

$$f(Y) := (Y^\ell - Y)^{\ell-1} + 1\,, \tag{7.44}$$

$$h(X) := \frac{-X^{\ell(\ell-1)}}{(X^{\ell-1}-1)^{\ell-1}} = \frac{1}{1-f(1/X)}\,, \tag{7.45}$$

$$\Sigma := \mathbb{F}_{\ell^3} \setminus \mathbb{F}_\ell\,,$$

and we must show that the following condition holds:

Claim: For all $\alpha \in \Sigma$ we have $h(\alpha) \ne \infty$, and the equation $f(t) = h(\alpha)$ has $\ell(\ell-1)$ distinct roots $t = \gamma \in \Sigma$.

All assertions of Lemma 7.4.26 follow then immediately by Corollary 7.2.21. In order to prove the claim we introduce the polynomial

$$g(Y) := (Y^{\ell^3} - Y)/(Y^\ell - Y) \in K[Y]\,,$$

whose roots are exactly the elements of Σ. We have the polynomial identity

$$f(Y)^\ell(f(Y)-1)+1 = g(Y)\,, \tag{7.46}$$

since

$$\begin{aligned} f(Y)^\ell(f(Y)-1)+1 &= \frac{(Y^{\ell^2}-Y)^\ell}{(Y^\ell-Y)^\ell}\left(\frac{Y^{\ell^2}-Y}{Y^\ell-Y}-1\right)+1 \\ &= \frac{Y^{\ell^3}-Y^\ell}{(Y^\ell-Y)^{\ell+1}}\left(Y^{\ell^2}-Y-(Y^\ell-Y)\right)+1 \\ &= \frac{Y^{\ell^3}-Y^\ell}{Y^\ell-Y}+1 = \frac{Y^{\ell^3}-Y}{Y^\ell-Y} = g(Y)\,. \end{aligned}$$

Now let $\alpha \in \Sigma = \mathbb{F}_{\ell^3} \setminus \mathbb{F}_\ell$, then $h(\alpha) \neq \infty$ by (7.45). It also follows that $1/\alpha \in \Sigma$ and hence $g(1/\alpha) = 0$. From (7.46) we conclude that

$$f(1/\alpha)^\ell = \frac{1}{1-f(1/\alpha)}\,. \tag{7.47}$$

Let γ be an element in the algebraic closure $\bar{K}$ with $f(\gamma) = h(\alpha)$, so

$$f(\gamma) = \frac{1}{1-f(1/\alpha)} = f(1/\alpha)^\ell \tag{7.48}$$

by (7.45) and (7.47). Then we obtain

$$\begin{aligned} g(\gamma) &= f(\gamma)^\ell(f(\gamma)-1)+1 \\ &= \left(\frac{1}{1-f(1/\alpha)}\right)^\ell (f(1/\alpha)^\ell - 1) + 1 = -1+1 = 0\,, \end{aligned}$$

using (7.46) and (7.48). We conclude that $\gamma \in \Sigma$, since the zeros of the polynomial $g(Y)$ are exactly the elements of Σ. The polynomal $f(t)-h(\alpha) \in K[t]$ has no multiple roots since the derivative $f'(t) = (t^\ell - t)^{\ell-2}$ does not have a common root with $f(t)-h(\alpha) = (t^\ell-t)^{\ell-1}+1-h(\alpha)$. This completes the proof of the claim and hence of Lemma 7.4.26. □

Putting together the results of Corollary 7.4.25 and Lemma 7.4.26, we get the estimate

$$\lambda(\mathcal{H}) = \frac{\nu(\mathcal{H}/H_0)}{\gamma(\mathcal{H}/H_0)} \geq \frac{\ell^3-\ell}{(\ell^2+2\ell)/2} = \frac{2(\ell^2-1)}{\ell+2}$$

for the limit of the tower $\mathcal{H}$. This completes the proof of Theorem 7.4.17.

7.5 Exercises

7.1. Suppose that $\mathcal{F} = (F_0, F_1, F_2, \ldots)$ and $\mathcal{G} = (G_0, G_1, G_2, \ldots)$ are towers over $\mathbb{F}_q$. Call $\mathcal{F}$ and $\mathcal{G}$ equivalent if for all $i \geq 0$ there exist $j, k \geq 0$ such that $F_i \subseteq E_j \subseteq F_k$. Show:

(i) This is an equivalence relation.

(ii) Suppose that $\mathcal{F}$ and $\mathcal{G}$ are equivalent. Then the following hold:

(a) $\lambda(\mathcal{F}) = \lambda(\mathcal{G})$.

(b) $\gamma(\mathcal{F}/F_0) < \infty \quad \Leftrightarrow \quad \gamma(\mathcal{G}/G_0) < \infty$.

(c) $\nu(\mathcal{F}/F_0) > 0 \quad \Leftrightarrow \quad \nu(\mathcal{G}/G_0) > 0$.

7.2. Let $\mathcal{F} = (F_0, F_1, F_2, \ldots)$ be a tame tower over $\mathbb{F}_q$. Assume that its ramification locus $\mathrm{Ram}(\mathcal{F}/F_0)$ is finite and that its splitting locus $\mathrm{Split}(\mathcal{F}/F_0)$ is non-empty. Define $\mathcal{F}^* = (F_0^*, F_1^*, F_2^*, \ldots)$ where F_i^* is the Galois closure of F_i/F_0. Show that $\mathcal{F}^* = (F_0^*, F_1^*, F_2^*, \ldots)$ is an asymptotically good tower over $\mathbb{F}_q$.

7.3. Assume that $\mathcal{F} = (F_0, F_1, F_2, \ldots)$ is an asymptotically good Galois tower (i.e., all extensions F_i/F_0 are Galois). Prove:

(i) The ramification locus $\mathrm{Ram}(\mathcal{F}/F_0)$ is finite.

(ii) For some $n \geq 0$ there exists a rational place $P \in \mathbb{P}_{F_n}$ which splits completely in all extensions F_m/F_n with $m \geq n$.

7.4. Let $\mathcal{F} = (F_0, F_1, F_2, \ldots)$ be a tower of function fields over $\mathbb{F}_q$. For $n \geq 1$ we set $D_n := \deg(\mathrm{Diff}(F_n/F_{n-1}))$.

(i) Assume that there is a real number ϵ with $0 \leq \epsilon < 1$ and an integer $m \geq 1$ such that $D_{n+1} \leq \epsilon \cdot [F_{n+1} : F_n] \cdot D_n$ holds for all $n \geq m$. Show that the genus of $\mathcal{F}/F_0$ satisfies $\gamma(\mathcal{F}/F_0) < \infty$.

(ii) Now assume that $D_m \neq 0$ for some m, and $D_{n+1} \geq [F_{n+1} : F_n] \cdot D_n$ for all $n \geq m$. Show that the genus of $\mathcal{F}/F_0$ is $\gamma(\mathcal{F}/F_0) = \infty$. In particular, the tower is asymptotically bad.

7.5. Suppose that the tower $\mathcal{F} = (F_0, F_1, F_2, \ldots)$ is recursively defined by the equation $f(Y) = h(X)$ with $f(Y) \in \mathbb{F}_q(Y)$ and $h(X) \in \mathbb{F}_q(x)$; i.e.,

(1) $F_0 = \mathbb{F}_q(x_0)$ and $F_{n+1} = F_n(x_{n+1})$ for all $n \geq 0$,

(2) $f(x_{n+1}) = h(x_n)$ for all $n \geq 0$.

Assume moreover that

(3) both rational functions $f(Y)$ and $h(X)$ are separable,

(4) The equation $f(Y) = h(x_n)$ is absolutely irreducible over F_n, for all $n \geq 0$,

(5) $\deg f(Y) \neq \deg h(X)$.

Show that $\mathcal{F}$ is asymptotically bad.

7.6. (char $\mathbb{F}_q \neq 2$) Suppose that $f_1(X), f_2(X) \in \mathbb{F}_q[X]$ are polynomials with $\deg f_1(X) = 1 + \deg f_2(X)$. Show that the equation

$$Y^2 = f_1(X)/f_2(X)$$

defines a recursive tower over $\mathbb{F}_q$. If moreover $\deg f_1 \geq 3$, this tower is asymptotically bad.

7.7. (q is a square and char $\mathbb{F}_q \neq 2$) Consider the tower $\mathcal{F} = (F_0, F_1, F_2, \ldots)$ which is recursively defined over $\mathbb{F}_q$ by the equation

$$Y^2 = X(1-X)/(X+1)\,.$$

(Note that this equation defines a tower over $\mathbb{F}_q$, by the previous exercise.) Show that the ramification locus of $\mathcal{F}/F_0$ is the set of places $P \in \mathbb{P}_{F_0}$ with

$$x_0(P) \in \{0, 1, -1, \infty\}$$

or

$$x_0(P) \in \{\alpha \in \mathbb{F}_q \,|\, (\alpha^2+1)(\alpha^2-2\alpha-1)(\alpha^2+2\alpha-1) = 0\}\,.$$

Give an upper bound for $\gamma(\mathcal{F}/F_0)$.

7.8. Consider the tower of the previous exercise over the field $\mathbb{F}_{81}$. Show that at least 8 rational places of F_0 are completely splitting in the tower, and conclude that the limit of the tower satisfies $\lambda(\mathcal{F}) \geq 2$.

7.9. Prove Remark 7.4.9.

7.10. Let $c \in \mathbb{F}_8$ be an element with $c^3 + c + 1 = 0$.

(i) Show that the equation

$$Y^2 + Y = 1/(X^2 + cX)$$

defines a recursive tower $\mathcal{F} = (F_0, F_1, F_2, \ldots)$ over $\mathbb{F}_8$.

(ii) Show that the ramification locus of $\mathcal{F}/F_0$ is finite.

(iii) Prove that $\mathcal{F}/F_0$ is weakly ramified, and give an upper bound for the genus $\gamma(\mathcal{F}/F_0)$.

(iv) Is the tower asymptotically good?

7.11. Let $F_0 = \mathbb{F}_2(x)$ be the rational function field over $\mathbb{F}_2$. Consider for all $n \geq 1$ the field $F_n = F_0(y_1, \ldots, y_n)$, where

$$\begin{aligned}
y_1^2 + y_1 &= x(x^2+x)\,,\\
y_2^2 + y_2 &= x^3(x^2+x)\,,\\
&\cdots\\
y_n^2 + y_n &= x^{2n-1}(x^2+x)\,.
\end{aligned}$$

(i) Show that F_n/F_0 is Galois of degree $[F_n : F_0] = 2^n$, and that the pole of x is totally ramified in F_n/F_0.

(ii) Determine the genus and the number of rational places of $F_n/\mathbb{F}_2$.

(iii) Conclude that the tower $\mathcal{F} = (F_0, F_1, F_2, \ldots)$ over $\mathbb{F}_2$ is asymptotically bad.

(iv) Generalize this example to arbitrary finite fields $\mathbb{F}_q$.

Remark. Part (iii) of the above exercise is in fact a special case of the following general result: If a tower $\mathcal{F} = (F_0, F_1, F_2, \ldots)$ over $\mathbb{F}_q$ has the property that all extensions F_n/F_0 are Galois with abelian Galois groups, then the genus of $\mathcal{F}/F_0$ is infinite, and hence the tower is asymptotically bad.

7.12. Let $\mathcal{F} = (F_0, F_1, F_2, \ldots)$ be an asymptotically good tower of function fields over $\mathbb{F}_q$. Consider a finite extension field $E \supseteq F_0$ having the same constant field $\mathbb{F}_q$, and define the sequence $\mathscr{E} := (E_0, E_1, E_2, \ldots)$ by setting $E_j := EF_j$ for all $j \geq 0$. Assume that there is at least one rational place of $E/\mathbb{F}_q$ whose restriction to F_0 is in the splitting locus of $\mathscr{F}/F_0$. Show that $\mathscr{E}$ is an asymptotically good tower over $\mathbb{F}_q$.

7.13. ($q = p^a$ with $a \geq 2$) Let $F/\mathbb{F}_q$ be a function field having at least one rational place. Show that there exists an asymptotically good tower $\mathcal{F} = (F_0, F_1, F_2, \ldots)$ over $\mathbb{F}_q$ with $F_0 = F$.

8

More about Algebraic Geometry Codes

In Chapter 2 we studied algebraic geometry (AG) codes associated with divisors of an algebraic function field over $\mathbb{F}_q$. Here we continue their investigation. Let us fix some notation for the whole of Chapter 8.

> *$F/\mathbb{F}_q$ is an algebraic function field of genus g and $\mathbb{F}_q$ is the full constant field of F.*
>
> *$P_1, \ldots, P_n \in \mathbb{P}_F$ are pairwise distinct places of degree one.*
>
> *$D = P_1 + \ldots + P_n$.*
>
> *G is a divisor of F with* $\operatorname{supp} G \cap \operatorname{supp} D = \emptyset$.
>
> *$C_{\mathscr{L}}(D,G) = \{(x(P_1), \ldots, x(P_n)) \in \mathbb{F}_q^n \,|\, x \in \mathscr{L}(G)\}$ is the algebraic geometry code associated with D and G.*
>
> *$C_\Omega(D,G) = \{(\omega_{P_1}(1), \ldots, \omega_{P_n}(1)) \,|\, \omega \in \Omega_F(G-D)\}$ is the dual code of $C_{\mathscr{L}}(D,G)$.*

8.1 The Residue Representation of $C_\Omega(D,G)$

Let $P \in \mathbb{P}_F$ be a place of degree one and let $\omega \in \Omega_F$ be a Weil differential. In Chapter 4 we identified Ω_F with the differential module Δ_F (cf. Remark 4.3.7(a)). Via this identification the local component of ω at the place P can be evaluated by means of the residue of ω at P, namely $\omega_P(u) = \operatorname{res}_P(u\omega)$ for all $u \in F$ (Theorem 4.3.2(d)). In particular we have $\omega_P(1) = \operatorname{res}_P(\omega)$. Hence we have the following alternative description of the code $C_\Omega(D,G)$.

Proposition 8.1.1.

$$C_\Omega(D,G) = \{(\operatorname{res}_{P_1}(\omega), \ldots, \operatorname{res}_{P_n}(\omega)) \,|\, \omega \in \Omega_F(G-D)\}\,.$$

It is this representation that is most commonly used in the literature to define the code $C_\Omega(D,G)$.

H. Stichtenoth, *Algebraic Function Fields and Codes*,
Graduate Texts in Mathematics 254,

By Proposition 2.2.10 the code $C_\Omega(D,G)$ can also be written as $C_{\mathscr{L}}(D,H)$ where $H = D-G+(\eta)$ and η is a differential with $v_{P_i}(\eta) = -1$ and $\eta_{P_i}(1) = 1$ for $i = 1,\ldots,n$. Using results from Chapter 4 one can easily construct such a differential η.

Proposition 8.1.2. *Let t be an element of F such that $v_{P_i}(t) = 1$ for $i = 1,\ldots,n$. Then the following hold:*

(a) The differential $\eta := dt/t$ satisfies $v_{P_i}(\eta) = -1$ and $\mathrm{res}_{P_i}(\eta) = 1$ for $i = 1,\ldots,n$.

(b) $C_\Omega(D,G) = C_{\mathscr{L}}\big(D, D-G+(dt)-(t)\big)$.

Proof. (a) Since t is a prime element of $P := P_i$, the P-adic power series of $\eta = dt/t$ with respect to t is

$$\eta = \frac{1}{t}\,dt\,.$$

Hence $v_P(\eta) = -1$ and $\mathrm{res}_P(\eta) = 1$.

(b) Follows immediately from (a) and Proposition 2.2.10. □

Corollary 8.1.3. *Suppose that $t \in F$ is a prime element for all places $P_1, ..., P_n$.*

(a) If $2G - D \leq (dt/t)$ then the code $C_{\mathscr{L}}(D,G)$ is self-orthogonal; i.e.,

$$C_{\mathscr{L}}(D,G) \subseteq C_{\mathscr{L}}(D,G)^\perp\,.$$

(b) If $2G - D = (dt/t)$ then $C_{\mathscr{L}}(D,G)$ is self-dual.

Proof. This is an immediate consequence of Corollary 2.2.11. □

8.2 Automorphisms of AG Codes

The symmetric group $\mathcal{S}_n$ (whose elements are the permutations of the set $\{1,\ldots,n\}$) acts on the vector space $\mathbb{F}_q^n$ via

$$\pi(c_1,\ldots,c_n) := (c_{\pi(1)},\ldots,c_{\pi(n)})$$

for $\pi \in \mathcal{S}_n$ and $c = (c_1,\ldots,c_n) \in \mathbb{F}_q^n$.

Definition 8.2.1. *The automorphism group of a code $C \subseteq \mathbb{F}_q^n$ is defined by*

$$\mathrm{Aut}(C) := \{\pi \in \mathcal{S}_n \mid \pi(C) = C\}\,.$$

Obviously $\mathrm{Aut}(C)$ is a subgroup of $\mathcal{S}_n$. Many interesting codes have a non-trivial automorphism group. In this section we study automorphisms of algebraic geometry codes that are induced by automorphisms of the corresponding function field.

Let $F/\mathbb{F}_q$ be a function field and let $\mathrm{Aut}(F/\mathbb{F}_q)$ be the group of automorphisms of F over $\mathbb{F}_q$ (i.e., $\sigma(a) = a$ for $\sigma \in \mathrm{Aut}(F/\mathbb{F}_q)$ and $a \in \mathbb{F}_q$). The group $\mathrm{Aut}(F/\mathbb{F}_q)$ acts on $\mathbb{P}_F$ by setting $\sigma(P) := \{\sigma(x) \mid x \in P\}$, cf. Lemma 3.5.2. The corresponding valuations v_P and $v_{\sigma(P)}$ are related as follows:

$$v_{\sigma(P)}(y) = v_P(\sigma^{-1}(y)) \quad \textit{for all} \quad y \in F\,. \tag{8.1}$$

Moreover, $\deg \sigma(P) = \deg P$ since σ induces an isomorphism of the residue class fields of P and $\sigma(P)$ given by $\sigma(z(P)) := \sigma(z)(\sigma(P))$. The action of $\mathrm{Aut}(F/\mathbb{F}_q)$ on $\mathbb{P}_F$ extends to an action on the divisor group by setting

$$\sigma\left(\sum n_P P\right) := \sum n_P \sigma(P)\,.$$

As before we consider divisors $D = P_1 + \ldots + P_n$ and G of $F/\mathbb{F}_q$ where $P_1, \ldots, P_n$ are distinct places of degree one and $\mathrm{supp}\, G \cap \mathrm{supp}\, D = \emptyset$.

Definition 8.2.2. *We define*

$$\mathrm{Aut}_{D,G}(F/\mathbb{F}_q) := \{\sigma \in \mathrm{Aut}(F/\mathbb{F}_q) \mid \sigma(D) = D \textit{ and } \sigma(G) = G\}\,.$$

Observe that an automorphism $\sigma \in \mathrm{Aut}_{D,G}(F/\mathbb{F}_q)$ need not fix the places $P_1, \ldots, P_n$, but it yields a permutation of $P_1, \ldots, P_n$. From (8.1) it follows easily that

$$\sigma(\mathscr{L}(G)) = \mathscr{L}(G) \tag{8.2}$$

for $\sigma \in \mathrm{Aut}_{D,G}(F/\mathbb{F}_q)$, because $\sigma(G) = G$. Now we show that every automorphism $\sigma \in \mathrm{Aut}_{D,G}(F/\mathbb{F}_q)$ induces an automorphism of the corresponding code $C_{\mathscr{L}}(D, G)$.

Proposition 8.2.3. *(a)* $\mathrm{Aut}_{D,G}(F/\mathbb{F}_q)$ *acts on the code* $C_{\mathscr{L}}(D,G)$ *by*

$$\sigma((x(P_1), \ldots, x(P_n))) := (x(\sigma(P_1)), \ldots, x(\sigma(P_n)))$$

(for $x \in \mathscr{L}(G)$*). This yields a homomorphism from* $\mathrm{Aut}_{D,G}(F/\mathbb{F}_q)$ *into* $\mathrm{Aut}(C_{\mathscr{L}}(D,G))$.

(b) If $n > 2g+2$, *the above homomorphism is injective. Hence* $\mathrm{Aut}_{D,G}(F/\mathbb{F}_q)$ *can be regarded as a subgroup of* $\mathrm{Aut}(C_{\mathscr{L}}(D,G))$.

Proof. (a) We begin with the following assertion: given a place P of degree one and an element $y \in F$ with $v_P(y) \geq 0$, we have

$$\sigma(y)(\sigma(P)) = y(P)\,. \tag{8.3}$$

In fact, setting $a := y(P) \in \mathbb{F}_q$, we obtain $y - a \in P$. Hence $\sigma(y) - a = \sigma(y - a) \in \sigma(P)$, and (8.3) follows.

For the proof of (a) we have to show that for every $x \in \mathscr{L}(G)$ and $\sigma \in \mathrm{Aut}_{D,G}(F/\mathbb{F}_q)$ the vector $(x(\sigma(P_1)), \ldots, x(\sigma(P_n)))$ is in $C_{\mathscr{L}}(D, G)$. As $\mathscr{L}(G) = \sigma(\mathscr{L}(G))$ by (8.2), we can write $x = \sigma(y)$ with $y \in \mathscr{L}(G)$, so

$$(x(\sigma(P_1)), \ldots, x(\sigma(P_n))) = (y(P_1), \ldots, y(P_n)) \in C_{\mathscr{L}}(D, G),$$

by (8.3).

(b) It is sufficient to prove that the only automorphism of $F/\mathbb{F}_q$ fixing more than $2g + 2$ places of degree one is the identity. So we assume that $\sigma(Q) = Q$ and $\sigma(Q_i) = Q_i$ for $i = 1, \ldots, 2g + 2$, where $\sigma \in \mathrm{Aut}(F/\mathbb{F}_q)$ and $Q, Q_1, \ldots, Q_{2g+2}$ are distinct places of degree one. Choose $x, z \in F$ such that $(x)_\infty = 2gQ$ and $(z)_\infty = (2g + 1)Q$ (this is possible by the Riemann-Roch Theorem). Then $\mathbb{F}_q(x, z) = F$ since the degrees $[F : \mathbb{F}_q(x)] = 2g$ and $[F : \mathbb{F}_q(z)] = 2g + 1$ are relatively prime. The elements $x - \sigma(x)$ and $z - \sigma(z)$ have at least $2g + 2$ zeros (namely $Q_1, \ldots, Q_{2g+2}$) but their pole divisor has degree $\leq 2g + 1$ because Q is their only pole. We conclude $\sigma(x) = x$ and $\sigma(z) = z$, hence σ is the identity. □

Example 8.2.4. As an example we consider a BCH code C of length n over $\mathbb{F}_q$. As shown in Section 2.3, C can be realized as a subfield subcode of a rational AG code as follows: let $n \mid (q^m - 1)$ and let $\beta \in \mathbb{F}_{q^m}$ be a primitive n-th root of unity. Consider the rational function field $F = \mathbb{F}_{q^m}(z)$. For $i = 1, \ldots, n$ let P_i be the zero of $z - \beta^{i-1}$, and set $D_\beta := P_1 + \ldots + P_n$. Denote by P_0 resp. P_∞ the zero resp. the pole of z in F. Then

$$C = C_{\mathscr{L}}(D_\beta, rP_0 + sP_\infty)\,|_{\mathbb{F}_q}$$

with $r, s \in \mathbb{Z}$ (see Proposition 2.3.9). The automorphism $\sigma \in \mathrm{Aut}(F/\mathbb{F}_{q^m})$ given by $\sigma(z) = \beta^{-1}z$ leaves the places P_0 and P_∞ invariant, and we have

$$\sigma(P_i) = P_{i+1} \quad (i = 1, \ldots, n-1) \quad \textit{and } \sigma(P_n) = P_1 .$$

Hence, by Proposition 8.2.3, σ induces the following automorphism of the code $C_{\mathscr{L}}(D_\beta, rP_0 + sP_\infty)$:

$$\sigma(c_1, \ldots, c_n) = (c_2, \ldots, c_n, c_1) . \tag{8.4}$$

This means (in the usual terminology of coding theory) that BCH codes are *cyclic* codes.

8.3 Hermitian Codes

In Chapter 6 we discussed several examples of algebraic function fields. One can use all these examples for the explicit construction of algebraic geometry codes. In this section we investigate some codes which are constructed by means of the Hermitian function field. This class of codes provides interesting and non-trivial examples of AG codes. These codes are codes over $\mathbb{F}_{q^2}$, they are not too short compared with the size of the alphabet, and their parameters k and d are fairly good.

First we recall some properties of the Hermitian function field H (cf. Lemma 6.4.4). H is a function field over $\mathbb{F}_{q^2}$; it can be represented as

$$H = \mathbb{F}_{q^2}(x,y) \quad \textit{with} \quad y^q + y = x^{q+1}\,. \tag{8.5}$$

The genus of H is $g = q(q-1)/2$, and H has $N = 1 + q^3$ places of degree one, namely

- the unique common pole Q_∞ of x and y, and
- for each pair $(\alpha,\beta) \in \mathbb{F}_{q^2} \times \mathbb{F}_{q^2}$ with $\beta^q + \beta = \alpha^{q+1}$ there is a unique place $P_{\alpha,\beta} \in \mathbb{P}_H$ of degree one such that $x(P_{\alpha,\beta}) = \alpha$ and $y(P_{\alpha,\beta}) = \beta$.

Observe that for all $\alpha \in \mathbb{F}_{q^2}$ there exist q distinct elements $\beta \in \mathbb{F}_{q^2}$ with $\beta^q + \beta = \alpha^{q+1}$, hence the number of places $P_{\alpha,\beta}$ is q^3.

Definition 8.3.1. *For $r \in \mathbb{Z}$ we define the code*

$$C_r := C_{\mathscr{L}}(D, rQ_\infty)\,, \tag{8.6}$$

where

$$D := \sum_{\beta^q+\beta=\alpha^{q+1}} P_{\alpha,\beta} \tag{8.7}$$

is the sum of all places of degree one (except Q_∞) of the Hermitian function field $H/\mathbb{F}_{q^2}$. The codes C_r are called Hermitian codes.

Hermitian codes are codes of length $n = q^3$ over the field $\mathbb{F}_{q^2}$. For $r \le s$ we obviously have $C_r \subseteq C_s$. Let us first discuss some trivial cases. For $r < 0$, $\mathscr{L}(rQ_\infty) = 0$ and therefore $C_r = 0$. For $r > q^3 + q^2 - q - 2 = q^3 + (2g-2)$, Theorem 2.2.2 and the Riemann-Roch Theorem yield

$$\begin{aligned} \dim C_r &= \ell(rQ_\infty) - \ell(rQ_\infty - D) \\ &= (r+1-g) - (r - q^3 + 1 - g) = q^3 = n\,. \end{aligned}$$

Hence $C_r = \mathbb{F}_{q^2}^n$ in this case, and it remains to study Hermitian codes with $0 \le r \le q^3 + q^2 - q - 2$.

Proposition 8.3.2. *The dual code of C_r is*

$$C_r^\perp = C_{q^3+q^2-q-2-r}\,.$$

Hence C_r is self-orthogonal if $2r \le q^3 + q^2 - q - 2$, and C_r is self-dual for $r = (q^3 + q^2 - q - 2)/2$.

Proof. Consider the element

$$t := \prod_{\alpha \in \mathbb{F}_{q^2}} (x - \alpha) = x^{q^2} - x \,.$$

t is a prime element for all places $P_{\alpha,\beta} \le D$, and its principal divisor is $(t) = D - q^3 Q_\infty$. Since $dt = d(x^{q^2} - x) = -dx$, the differential dt has the divisor $(dt) = (dx) = (q^2 - q - 2)Q_\infty$ (Lemma 6.4.4). Now Theorem 2.2.8 and Proposition 8.1.2 imply

$$\begin{aligned} C_r^\perp &= C_\Omega(D, rQ_\infty) = C_\mathscr{L}\big(D, D - rQ_\infty + (dt) - (t)\big) \\ &= C_\mathscr{L}\big(D, (q^3 + q^2 - q - 2 - r)Q_\infty\big) = C_{q^3+q^2-q-2-r} \,. \end{aligned}$$

□

Our next aim is to determine the parameters of C_r. We consider the set I of pole numbers of Q_∞ (cf. Definition 1.6.7); i.e.,

$$I = \{\, n \ge 0 \,|\, \textit{there is an element } z \in H \textit{ with } (z)_\infty = nQ_\infty \,\} \,.$$

For $s \ge 0$ let

$$I(s) := \{n \in I \,|\, n \le s\} \,. \tag{8.8}$$

Then $|I(s)| = \ell(sQ_\infty)$, and the Riemann-Roch Theorem gives

$$|I(s)| = s + 1 - q(q-1)/2 \;\; \textit{for} \;\; s \ge 2g - 1 = q(q-1) - 1 \,.$$

From Lemma 6.4.4 we obtain the following description of $I(s)$:

$$I(s) = \{n \le s \,|\, n = iq + j(q+1) \;\; \textit{with } i \ge 0 \textit{ and } 0 \le j \le q - 1\} \,,$$

hence

$$|I(s)| = \big|\{\, (i,j) \in \mathbb{N}_0 \times \mathbb{N}_0 \,;\, j \le q - 1 \;\; \textit{and} \;\; iq + j(q+1) \le s \,\}\big| \,.$$

Proposition 8.3.3. *Suppose that $0 \le r \le q^3 + q^2 - q - 2$. Then the following hold:*

(a) The dimension of C_r is given by

$$\dim C_r = \begin{cases} |I(r)| & \textit{for } \; 0 \le r < q^3 \,, \\ q^3 - |I(s)| & \textit{for } \; q^3 \le r \le q^3 + q^2 - q - 2 \,, \end{cases}$$

where $s := q^3 + q^2 - q - 2 - r$ and $I(r)$ is defined by (8.8).

(b) For $q^2 - q - 2 < r < q^3$ we have

$$\dim C_r = r + 1 - q(q-1)/2 \,.$$

(c) The minimum distance d of C_r satisfies

$$d \geq q^3 - r \, .$$

If $0 \leq r < q^3$ and both numbers r and $q^3 - r$ are pole numbers of Q_∞, then

$$d = q^3 - r \, .$$

Proof. (a) For $0 \leq r < q^3$ Corollary 2.2.3 gives

$$\dim C_r = \dim \mathscr{L}(rQ_\infty) = |I(r)| \, .$$

For $q^3 \leq r \leq q^3 + q^2 - q - 2$ we set $s := q^3 + q^2 - q - 2 - r$. Then $0 \leq s \leq q^2 - q - 2 < q^3$. By Proposition 8.3.2 we obtain

$$\dim C_r = q^3 - \dim C_s = q^3 - |I(s)| \, .$$

(b) For $q^2 - q - 2 = 2g - 2 < r < q^3$, Corollary 2.2.3 gives

$$\dim C_r = r + 1 - g = r + 1 - q(q-1)/2 \, .$$

(c) The inequality $d \geq q^3 - r$ follows from Theorem 2.2.2. Now let $0 \leq r < q^3$ and assume that both numbers r and $q^3 - r$ are pole numbers of Q_∞. In order to prove the equality $d = q^3 - r$ we distinguish three cases.

Case 1: $r = q^3 - q^2$. Choose $i := q^2 - q$ distinct elements $\alpha_1, \ldots, \alpha_i \in \mathbb{F}_{q^2}$. Then the element

$$z := \prod_{\nu=1}^{i} (x - \alpha_\nu) \in \mathscr{L}(rQ_\infty)$$

has exactly $qi = r$ distinct zeros $P_{\alpha,\beta}$ of degree one, and the weight of the corresponding codeword $\mathrm{ev}_D(z) \in C_r$ is $q^3 - r$. Hence $d = q^3 - r$.

Case 2: $r < q^3 - q^2$. We write $r = iq + j(q+1)$ with $i \geq 0$ and $0 \leq j \leq q - 1$, so $i \leq q^2 - q - 1$. Fix an element $0 \neq \gamma \in \mathbb{F}_q$ and consider the set $A := \{\alpha \in \mathbb{F}_{q^2} \mid \alpha^{q+1} \neq \gamma\}$. Then $|A| = q^2 - (q+1) \geq i$, and we can choose distinct elements $\alpha_1, \ldots, \alpha_i \in A$. The element

$$z_1 := \prod_{\nu=1}^{i} (x - \alpha_\nu)$$

has iq distinct zeros $P_{\alpha,\beta} \leq D$. Next we choose j distinct elements $\beta_1, \ldots, \beta_j \in \mathbb{F}_{q^2}$ with $\beta_\mu^q + \beta_\mu = \gamma$ and set

$$z_2 := \prod_{\mu=1}^{j} (y - \beta_\mu) \, .$$

z_2 has $j(q+1)$ zeros $P_{\alpha,\beta} \le D$, and all of them are distinct from the zeros of z_1 because $\beta_\mu^q + \beta_\mu = \gamma \neq \alpha_\nu^{q+1}$ for $\mu = 1, \ldots, j$ and $\nu = 1, \ldots, i$. Hence

$$z := z_1 z_2 \in \mathscr{L}\big((iq + j(q+1))Q_\infty\big) = \mathscr{L}(rQ_\infty)$$

has r distinct zeros $P_{\alpha,\beta} \le D$. The corresponding codeword $\mathrm{ev}_D(z) \in C_r$ has weight $q^3 - r$.

Case 3: $q^3 - q^2 < r < q^3$. By assumption, $s := q^3 - r$ is a pole number and $0 < s < q^2 \le q^3 - q^2$. By case 2 there exists an element $z \in H$ with principal divisor $(z) = D' - sQ_\infty$ where $0 \le D' \le D$ and $\deg D' = s$. The element $u := x^{q^2} - x \in H$ has the divisor $(u) = D - q^3 Q_\infty$, hence

$$(z^{-1}u) = (D - D') - (q^3 - s)Q_\infty = (D - D') - rQ_\infty\,.$$

The codeword $\mathrm{ev}_D(z^{-1}u) \in C_r$ has weight $q^3 - r$. □

We mention that the minimum distance of C_r is known also in the remaining cases (where $r \ge q^3$, or one of the numbers r or $q^3 - r$ is a gap of Q_∞).

One can easily specify a generator matrix for the Hermitian codes C_r. We fix an ordering of the set $T := \{(\alpha, \beta) \in \mathbb{F}_{q^2} \times \mathbb{F}_{q^2} \mid \beta^q + \beta = \alpha^{q+1}\}$. For $s = iq + j(q+1)$ (where $i \ge 0$ and $0 \le j \le q-1$) we define the vector

$$u_s := \big(\alpha^i \beta^j\big)_{(\alpha,\beta)\in T} \in (\mathbb{F}_{q^2})^{q^3}\,.$$

Then we have:

Corollary 8.3.4. *Suppose that* $0 \le r < q^3$. *Let* $0 = s_1 < s_2 < \ldots < s_k \le r$ *be all pole numbers* $\le r$ *of* Q_∞. *Then the* $k \times q^3$ *matrix* M_r *whose rows are* $u_{s_1}, \ldots, u_{s_k}$, *is a generator matrix of* C_r.

Proof. Corollary 2.2.3. □

In the same manner we obtain a parity check matrix for C_r (for $r > q^2 - q - 2$), since the dual of C_r is the code C_s with $s = q^3 + q^2 - q - 2 - r$.

Finally we study automorphisms of Hermitian codes. Let $H = \mathbb{F}_{q^2}(x, y)$ as before, cf. (8.5). Let

$$\varepsilon \in \mathbb{F}_{q^2} \setminus \{0\}\,, \quad \delta \in \mathbb{F}_{q^2} \ \textit{and} \ \mu^q + \mu = \delta^{q+1}\,. \tag{8.9}$$

Then $\mu \in \mathbb{F}_{q^2}$, and there exists an automorphism $\sigma \in \mathrm{Aut}(H/\mathbb{F}_{q^2})$ with

$$\sigma(x) = \varepsilon x + \delta \ \textit{and} \ \sigma(y) = \varepsilon^{q+1} y + \varepsilon \delta^q x + \mu\,. \tag{8.10}$$

(The existence of an automorphism σ satisfying (8.10) follows from the fact that $\sigma(y)$ and $\sigma(x)$ satisfy the equation $\sigma(y)^q + \sigma(y) = \sigma(x)^{q+1}$, which is a consequence of (8.9).) The set of all automorphisms (8.10) of $H/\mathbb{F}_{q^2}$ constitutes a group $\Gamma \subseteq \mathrm{Aut}(H/\mathbb{F}_{q^2})$ of order $q^3(q^2-1)$ (as $\varepsilon \neq 0$ and δ are arbitrary, and for each δ there are q possible values of μ). Clearly $\sigma(Q_\infty) = Q_\infty$ for all $\sigma \in \Gamma$, and σ permutes the places $P_{\alpha,\beta}$ of H since they are the only places of H of degree one other than Q_∞. By Proposition 6.3.3, Γ acts as a group of automorphisms on the Hermitian codes C_r. We have proved:

Proposition 8.3.5. *The automorphism group* $\mathrm{Aut}(C_r)$ *of the Hermitian code* C_r *contains a subgroup of order* $q^3(q^2-1)$.

Remark 8.3.6. It is easily seen that Γ acts *transitively* on the places $P_{\alpha,\beta}$; i.e., given $P_{\alpha,\beta}$ and $P_{\alpha',\beta'}$ then there exists some $\sigma \in \Gamma$ with $\sigma(P_{\alpha,\beta}) = P_{\alpha',\beta'}$.

8.4 The Tsfasman-Vladut-Zink Theorem

It is well-known in coding theory that large block lengths (hence large dimension and large minimum distance) are required to achieve reliable transmission of information. We introduce some notation that will simplify discussion of asymptotic performance of codes.

Definition 8.4.1. *(a) Given an* $[n, k, d]$ *code* C *over* $\mathbb{F}_q$*, we define its information rate*

$$R = R(C) := k/n$$

and its relative minimum distance

$$\delta = \delta(C) := d/n\,.$$

(b) Let $V_q := \{(\delta(C), R(C)) \in [0,1]^2 \,|\, C \text{ is a code over } \mathbb{F}_q\}$ *and* $U_q \subseteq [0,1]^2$ *be the set of limit points of* V_q.

This means: a point $(\delta, R) \in \mathbb{R}^2$ is in U_q if and only if there are codes C over $\mathbb{F}_q$ of arbitrary large length such that the point $(\delta(C), R(C))$ is arbitrarily close to (δ, R).

Proposition 8.4.2. *There is a continuous function* $\alpha_q : [0,1] \to [0,1]$ *such that*

$$U_q = \{(\delta, R) \,|\, 0 \le \delta \le 1 \text{ and } 0 \le R \le \alpha_q(\delta)\}\,.$$

Moreover the following hold: $\alpha_q(0) = 1$, $\alpha_q(\delta) = 0$ *for* $1 - q^{-1} \le \delta \le 1$, *and* α_q *is decreasing in the interval* $0 \le \delta \le 1 - q^{-1}$.

The proof of this proposition requires only elementary techniques of coding theory; we refer to [29].

For $0 < \delta < 1 - q^{-1}$ the exact value of $\alpha_q(\delta)$ is unknown. However, several upper and lower bounds are available. In the following propositions we state some of these bounds. Proofs can be found in most books on coding theory, e.g. in [28]. The *q-ary entropy function* $H_q : [0, 1-q^{-1}] \to \mathbb{R}$ is defined by $H_q(0) := 0$ and

$$H_q(x) := x \log_q(q-1) - x \log_q(x) - (1-x) \log_q(1-x)$$

for $0 < x \leq 1 - q^{-1}$.

Proposition 8.4.3. *The following upper bounds for $\alpha_q(\delta)$ hold:*
(a) (Plotkin Bound) For $0 \leq \delta \leq 1 - q^{-1}$,

$$\alpha_q(\delta) \leq 1 - \frac{q}{q-1} \cdot \delta .$$

(b) (Hamming Bound) For $0 \leq \delta \leq 1$,

$$\alpha_q(\delta) \leq 1 - H_q(\delta/2) .$$

(c) (Bassalygo-Elias Bound) For $0 \leq \delta \leq \theta := 1 - q^{-1}$,

$$\alpha_q(\delta) \leq 1 - H_q\big(\theta - \sqrt{\theta(\theta - \delta)}\big) .$$

Out of the upper bounds in Proposition 8.4.3, the Bassalygo-Elias Bound is always the best, see Figure 8.1 below; an even better upper bound (which is more complicated to state and more difficult to prove) is the *McEliece-Rodemich-Rumsey-Welch Bound*, see [28],[32].
Perhaps more important than *upper* bounds are *lower* bounds for $\alpha_q(\delta)$, because every non-trivial lower bound for $\alpha_q(\delta)$ guarantees the existence of arbitrary long codes with good parameters $(\delta(C), R(C))$.

Proposition 8.4.4 (Gilbert-Varshamov Bound). *For $0 \leq \delta \leq 1 - q^{-1}$,*

$$\alpha_q(\delta) \geq 1 - H_q(\delta) .$$

The Gilbert-Varshamov bound is the best lower bound for $\alpha_q(\delta)$ which is known from elementary coding theory. However, its proof is not constructive (i.e., it does not provide a simple algebraic algorithm for the construction of good long codes).

Our aim is to construct algebraic geometry codes of large length in order to improve the Gilbert-Varshamov Bound. Given an algebraic function field $F/\mathbb{F}_q$ with $N = N(F)$ places of degree one, the length of any AG code $C_{\mathscr{L}}(D,G)$ (resp. $C_\Omega(D,G)$) associated with divisors D and G of F is bounded by N, since D is a sum of places of degree one. In fact, this is the only restriction on the length of an AG code which can be constructed by means of the function field F.

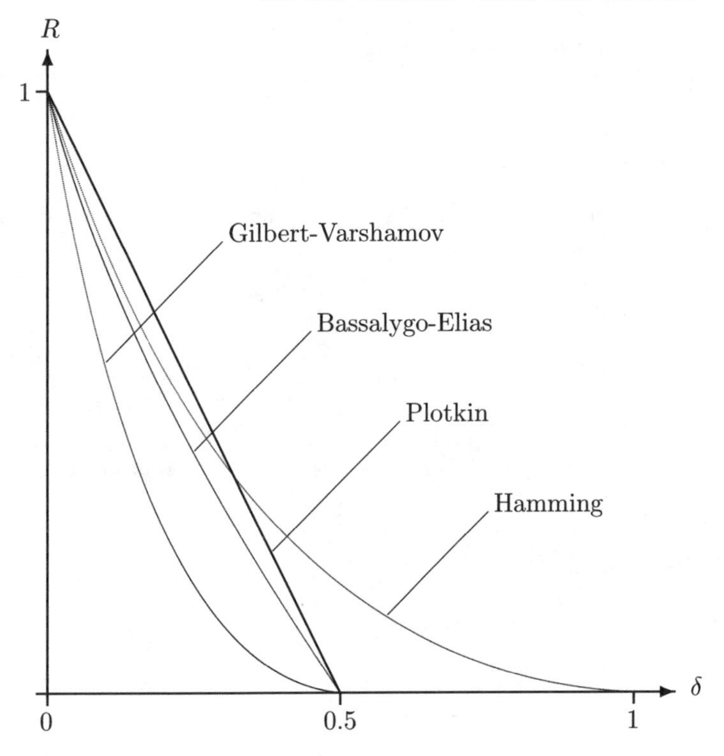

Fig. 8.1. Bounds for $q = 2$.

Lemma 8.4.5. *Suppose that* $P_1, \ldots, P_n$ *are distinct places of* $F/\mathbb{F}_q$ *of degree one. Then there exists, for each* $r \geq 0$*, a divisor* G *such that* $\deg G = r$ *and* $P_i \notin \operatorname{supp} G$ *(for* $i = 1, \ldots, n$*).*

Proof. The lemma is trivial if there is another place Q of degree one, different from $P_1, \ldots, P_n$. In this case we set $G := rQ$. If $P_1, \ldots, P_n$ are all the places of $F/\mathbb{F}_q$ of degree one, we choose a divisor $G \sim rP_1$ (i.e., G is equivalent to rP_1) such that $v_{P_i}(G) = 0$ for $i = 1, \ldots, n$. This is possible by the Approximation Theorem. □

According to Lemma 8.4.5 one needs function fields over $\mathbb{F}_q$ having many rational places in order to construct long AG codes. We recall the definition of Ihara's constant $A(q)$ given in Chapter 7. For $g \geq 0$ let

$$N_q(g) := \max\{N(F) \mid F \text{ is a function field over } \mathbb{F}_q \text{ of genus } g\},$$

where $N(F)$ denotes the number of places of $F/\mathbb{F}_q$ of degree one. Then $A(q)$ is defined as

$$A(q) = \limsup_{g \to \infty} \frac{N_q(g)}{g}.$$

Proposition 8.4.6. *Suppose that $A(q) > 1$. Then*

$$\alpha_q(\delta) \geq (1 - A(q)^{-1}) - \delta$$

in the interval $0 \leq \delta \leq 1 - A(q)^{-1}$.

Proof. Let $\delta \in [0, 1 - A(q)^{-1}]$. Choose a sequence of function fields $F_i/\mathbb{F}_q$ of genus g_i such that

$$g_i \to \infty \quad \textit{and} \quad n_i/g_i \to A(q)\,, \tag{8.11}$$

where $n_i := N(F_i)$. Choose $r_i > 0$ such that

$$r_i/n_i \to 1 - \delta\,. \tag{8.12}$$

This is possible as $n_i \to \infty$ for $i \to \infty$. Let D_i be the sum of all places of $F_i/\mathbb{F}_q$ of degree one, thus $\deg D_i = n_i$. By Lemma 8.4.5 there exists a divisor G_i of $F_i/\mathbb{F}_q$ such that $\deg G_i = r_i$ and $\operatorname{supp} G_i \cap \operatorname{supp} D_i = \emptyset$. Consider the code $C_i := C_{\mathscr{L}}(D_i, G_i)$; this is an $[n_i, k_i, d_i]$ code whose parameters k_i and d_i satisfy the inequalities

$$k_i \geq \deg G_i + 1 - g_i = r_i + 1 - g_i \quad \textit{and} \quad d_i \geq n_i - \deg G_i = n_i - r_i$$

(cf. Corollary 2.2.3). Hence

$$R_i := R(C_i) \geq \frac{r_i + 1}{n_i} - \frac{g_i}{n_i} \quad \textit{and} \quad \delta_i := \delta(C_i) \geq 1 - \frac{r_i}{n_i}\,. \tag{8.13}$$

W.l.o.g. we can assume that the sequences $(R_i)_{i \geq 1}$ and $(\delta_i)_{i \geq 1}$ are convergent (otherwise we choose an appropriate subsequence), say $R_i \to R$ and $\delta_i \to \tilde{\delta}$. From (8.11), (8.12) and (8.13) it follows that $R \geq 1 - \delta - A(q)^{-1}$ and $\tilde{\delta} \geq \delta$. So $\alpha_q(\tilde{\delta}) \geq R \geq 1 - \delta - A(q)^{-1}$. Since α_q is non-increasing, this implies

$$\alpha_q(\delta) \geq \alpha_q(\tilde{\delta}) \geq 1 - \delta - A(q)^{-1}\,.$$

□

Now we can easily prove the main result of this section.

Theorem 8.4.7 (Tsfasman-Vladut-Zink Bound). *Let $q = \ell^2$ be a square. Then we have for all δ with $0 \leq \delta \leq 1 - (q^{1/2} - 1)^{-1}$,*

$$\alpha_q(\delta) \geq \left(1 - \frac{1}{q^{1/2} - 1}\right) - \delta\,.$$

Proof. By Corollary 7.4.8 we have $A(q) = q^{1/2} - 1$ if q is a square. Now the assertion follows immediately from Proposition 8.4.6. □

For all $q \geq 49$ the Tsfasman-Vladut-Zink Bound improves the Gilbert-Varshamov Bound in a certain interval, see Figure 8.2 .

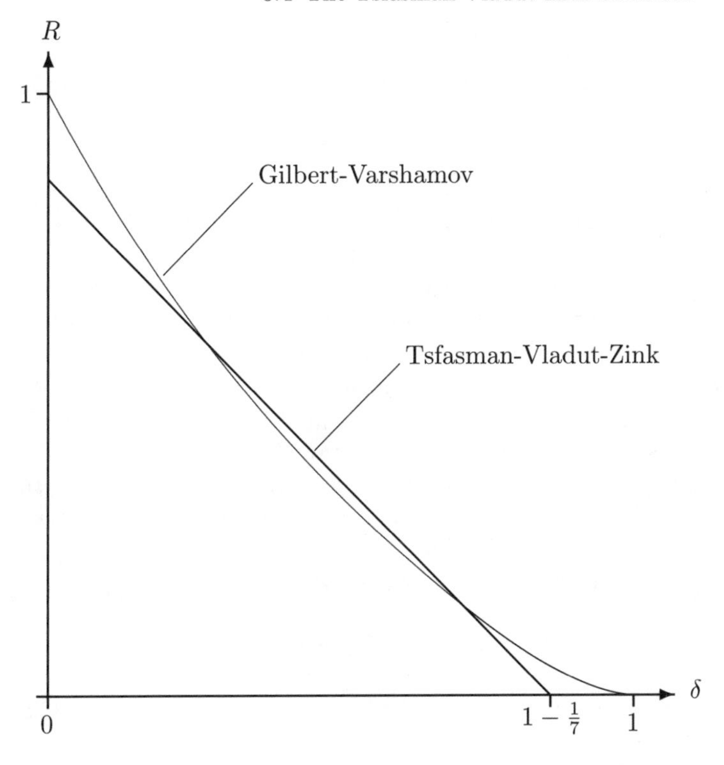

Fig. 8.2. Bounds for $q = 64$.

Remark 8.4.8. Also if q is not a square one can obtain an improvement of the Gilbert-Varshamov Bound if a good lower bound for Ihara's constant $A(q)$ is available. For instance, let $q = \ell^3$ be a cube. Then we have for all δ with $0 \le \delta \le 1 - (\ell+2)/(2(\ell^2-1))$ the following lower bound for $\alpha_q(\delta)$:

$$\alpha_q(\delta) \ge \left(1 - \frac{\ell+2}{2(\ell^2-1)}\right) - \delta \, . \tag{8.14}$$

The proof of this bound is exactly the same as in Theorem 8.4.7; one just uses the bound for $A(\ell^3)$ given in Corollary 7.4.18. We note that (8.14) improves the Gilbert-Varshamov Bound for all cubes $q \ge 7^3$.

In the proof of the Tsfasman-Vladut-Zink Theorem we have only used that there *exists* a sequence of function fields $F_i/\mathbb{F}_q$ (for $q = \ell^2$) with $\lim_{n\to\infty} N(F_i)/g(F_i) = \ell - 1$. If the function fields F_i have additional nice properties, one can hope that the corresponding AG codes also have nice properties. As an example for this idea we shall prove the existence of long self-dual codes whose parameters attain the Tsfasman-Vladut-Zink Bound.

Theorem 8.4.9. *Let $q = \ell^2$ be a square. Then there exists a sequence of self-dual codes $(C_i)_{i\geq 0}$ over $\mathbb{F}_q$ with parameters $[n_i, k_i, d_i]$ such that $n_i \to \infty$ and*

$$\liminf_{i\to\infty} \frac{d_i}{n_i} \geq \frac{1}{2} - \frac{1}{\ell - 1} . \tag{8.15}$$

Note that Inequality (8.6) just says that the sequence $(C_i)_{i\geq 0}$ attains the Tsfasman-Vladut-Zink Bound, because the information rate of a self-dual code C is $R(C) = 1/2$. As a consequence we obtain that there are self-dual codes over $\mathbb{F}_q$ (with $q = \ell^2 \geq 49$) of arbitrary large length whose performance is better than the Gilbert-Varshamov bound.

Proof of Theorem 8.4.9. For simplicity we will assume that q is even; i.e., $\operatorname{char} \mathbb{F}_q = 2$ (the assertion is also true in the case of odd characteristic, but the proof is then a bit more complicated). We will use the Galois tower $\mathcal{G}^* = (G_0^*, G_1^*, G_2^*, ...)$ of function fields over $\mathbb{F}_q$ that was studied in Section 7.4, see Theorem 7.4.15 and Corollary 7.4.16. We recall briefly the properties of this tower that will be needed below.

The field $G_0^* = \mathbb{F}_q(u_0)$ is a rational function field. For $i \geq 1$ we have

$$n_i = [G_i^* : G_0^*] = (\ell - 1)m_i ,$$

where $m_i \geq \ell$ is a power of $p = \operatorname{char} \mathbb{F}_q$. The zero divisor of u_0 in G_i^* has the form

$$(u_0)_0^{G_i^*} = D_i = \sum_{j=1}^{n_i} P_j^{(i)} \tag{8.16}$$

with pairwise distinct places $P_j^{(i)}$ of degree one, and the divisor of the differential $\eta^{(i)} = du_0/u_0$ in the function field G_i^* is given by

$$(\eta^{(i)}) = (\ell e_i^{(0)} - 2)A_i + (e_i^{(\infty)} - 2)B_i - D_i \tag{8.17}$$

with positive divisors A_i, B_i and

$$\big(\operatorname{supp} A_i \cup \operatorname{supp} B_i\big) \cap \operatorname{supp} D_i = \emptyset .$$

Moreover the degrees of the divisors A_i, B_i satisfy

$$e_i^{(0)} \cdot \deg A_i = e_i^{(\infty)} \cdot \deg B_i = n_i/(\ell - 1) . \tag{8.18}$$

with certain integers $e_i^{(0)}, e_i^{(\infty)}$. Now we define the divisor $H_i \in \operatorname{Div}(G_i^*)$ as

$$H_i := (\frac{\ell e_i^{(0)} - 2}{2})A_i + (\frac{e_i^{(\infty)} - 2}{2})B_i .$$

At this point we have used the assumption that q (and hence ℓ) is even. Since $2H_i - D_i = (\eta^{(i)})$ by (8.17), it follows from Corollary 8.1.3 that the code

$$C_i := C_{\mathscr{L}}(D_i, H_i) \subseteq \mathbb{F}_q^{n_i}$$

is self-dual. By Theorem 2.2.2 its minimum distance $d_i := d(C_i)$ can be estimated by

$$\begin{aligned}
d_i &\geq \deg(D_i - H_i) = \deg D_i - \deg H_i \\
&= n_i - \Big(\frac{\ell e_i^{(0)} - 2}{2}\Big)\deg A_i - \Big(\frac{e_i^{(\infty)} - 2}{2}\Big)\deg B_i \\
&\geq n_i - \frac{1}{2}\Big(\ell e_i^{(0)} \deg A_i + e_i^{(\infty)} \deg B_i\Big) \\
&= n_i - \frac{1}{2}\Big(\ell \frac{n_i}{\ell - 1} + \frac{n_i}{\ell - 1}\Big) \\
&= n_i\Big(\frac{1}{2} - \frac{1}{\ell - 1}\Big)
\end{aligned}$$

(here we have used Equations (8.16) and (8.18)). Therefore we obtain

$$\delta_i := \delta(C_i) = \frac{d_i}{n_i} \geq \frac{1}{2} - \frac{1}{\ell - 1}\,.$$

□

8.5 Decoding AG Codes

For a code to have practical use, it is essential that one has an effective decoding algorithm. Let us briefly explain what this means. We consider an $[n, k, d]$ code $C \subseteq \mathbb{F}_q^n$. Then C is t-error correcting for all $t \leq (d-1)/2$, cf. Section 2.1. Suppose $a \in \mathbb{F}_q^n$ is an n-tuple such that

$$a = c + e\,, \tag{8.19}$$

where $c \in C$ is a codeword and $e \in \mathbb{F}_q^n$ has weight

$$\mathrm{wt}(e) \leq (d-1)/2\,. \tag{8.20}$$

Then c is uniquely determined by a and the conditions (8.19) and (8.20); it is the unique codeword whose distance to a is minimal. The vector e in (8.19) is called the *error vector* of a with respect to C. A *decoding algorithm* is an algorithm which calculates for every element $a \in \mathbb{F}_q^n$ satisfying (8.19) and (8.20) the corresponding codeword c (or, equivalently, the corresponding error vector e).

For algebraic geometry codes a very general decoding algorithm is available. We consider the code

$$C_\Omega := C_\Omega(D, G) \tag{8.21}$$

with $D = P_1 + \ldots + P_n$ and $\operatorname{supp} D \cap \operatorname{supp} G = \emptyset$, as always in this chapter. For $b = (b_1, \ldots, b_n) \in \mathbb{F}_q^n$ and $f \in \mathscr{L}(G)$ we define the *syndrome*

$$[b, f] := \sum_{\nu=1}^{n} b_\nu \cdot f(P_\nu) . \tag{8.22}$$

The symbol $[\ ,\ ]$ is obviously bilinear. As C_Ω is the dual of $C_\mathscr{L}(D, G)$ (Theorem 2.2.8), we have

$$C_\Omega = \{ b \in \mathbb{F}_q^n \mid [b, f] = 0 \ \textit{for all} \ f \in \mathscr{L}(G) \} . \tag{8.23}$$

Let $t \geq 0$ be an integer and let G_1 be a divisor of $F/\mathbb{F}_q$ satisfying the following conditions:

$$\begin{aligned} &\operatorname{supp} G_1 \cap \operatorname{supp} D = \emptyset , \\ &\deg G_1 < \deg G - (2g - 2) - t , \\ &\ell(G_1) > t . \end{aligned} \tag{8.24}$$

We will show that under these assumptions all errors of weight $\leq t$ can be corrected by means of a simple algorithm.

Remark 8.5.1. Let $d^* = \deg G - (2g - 2)$ be the designed distance of C_Ω, cf. Section 2.2. We know that $d^* \leq d(C_\Omega)$ where $d(C_\Omega)$ denotes the minimum distance of C_Ω (Theorem 2.2.7). The following assertions hold:

(a) If G_1 and t satisfy (8.24) then $t \leq (d^* - 1)/2$.

(b) If $0 \leq t \leq (d^* - 1 - g)/2$ then there exists a divisor G_1 such that (8.24) is satisfied.

Proof. (a) By (8.24) and Inequality (1.21) of Chapter 1 we have $t \leq \ell(G_1) - 1 \leq \deg G_1$ and $t \leq \deg G - \deg G_1 - 2g + 1$. Adding these inequalities yields $2t \leq \deg G + 1 - 2g = d^* - 1$, hence $t \leq (d^* - 1)/2$.

(b) Now we assume that $t \leq (d^* - 1 - g)/2$. We choose a divisor G_1 such that

$$\deg G_1 = g + t \quad \textit{and} \quad \operatorname{supp} G_1 \cap \operatorname{supp} D = \emptyset . \tag{8.25}$$

This is possible by the Approximation Theorem. By the Riemann-Roch Theorem,

$$\ell(G_1) \geq \deg G_1 + 1 - g = t + 1 > t .$$

The assumption $t \leq (d^* - 1 - g)/2$ is equivalent to $d^* - 2t - g \geq 1$. So

$$\deg G - (2g - 2) - t - \deg G_1 = d^* - 2t - g > 0 .$$

This proves (8.24). □

From now on we assume (8.24). Suppose that

$$a = c + e \ \textit{with} \ c \in C_\Omega \ \textit{and} \ \mathrm{wt}(e) \le t\,. \tag{8.26}$$

Let

$$e = (e_1, \dots, e_n) \ \textit{and} \ I := \{\, \nu \mid 1 \le \nu \le n \ \textit{and} \ e_\nu \ne 0 \,\} \tag{8.27}$$

be the set of *error positions* (so $|I| = \mathrm{wt}(e) \le t$). In the first step of the decoding algorithm we shall construct an *error locator function*; i.e., an element $0 \ne f \in \mathscr{L}(G_1)$ with the property $f(P_\nu) = 0$ for all $\nu \in I$. That means that the error positions are contained in the set

$$N(f) := \{\, \nu \mid 1 \le \nu \le n \ \textit{and} \ f(P_\nu) = 0 \,\}\,. \tag{8.28}$$

In the second step we determine the *error values* e_ν for all $\nu \in N(f)$. As $e_\nu = 0$ for $\nu \notin N(f)$, this yields the error vector e. We shall see that each of these steps requires the solution of a certain system of linear equations. We specify bases

$$\begin{aligned} &\{f_1, \dots, f_l\} \ \textit{of} \ \mathscr{L}(G_1)\,, \\ &\{g_1, \dots, g_k\} \ \textit{of} \ \mathscr{L}(G - G_1)\,, \\ &\{h_1, \dots, h_m\} \ \textit{of} \ \mathscr{L}(G)\,. \end{aligned} \tag{8.29}$$

Note that the choice of these bases does not depend on the vector a which is to be decoded. It is obvious that $f_\lambda g_\rho \in \mathscr{L}(G)$ for $1 \le \lambda \le l$ and $1 \le \rho \le k$.

Consider the following system of linear equations which plays an essential role in the decoding algorithm.

$$\sum_{\lambda=1}^{l} [a, f_\lambda g_\rho] \cdot x_\lambda = 0\,, \quad \textit{for} \ \rho = 1, \dots, k\,. \tag{8.30}$$

Proposition 8.5.2. *With the above notations and assumptions (in particular* (8.24) *and* (8.26) - (8.29)*), the system* (8.30) *has a non-trivial solution. If* $(\alpha_1, \dots, \alpha_l)$ *is a non-trivial solution of* (8.30) *we set*

$$f := \sum_{\lambda=1}^{l} \alpha_\lambda f_\lambda \in \mathscr{L}(G_1)\,. \tag{8.31}$$

Then $f(P_\nu) = 0$ *for all error positions* $\nu \in I$*; i.e.,* f *is an error locator function.*

Proof. $I \subseteq \{1, \dots, n\}$ is the set of error positions, see (8.27). As $|I| \le t$ by (8.26) and $\ell(G_1) > t$ by (8.24), we have $\ell(G_1 - \sum_{\nu \in I} P_\nu) > 0$, cf. Lemma 1.4.8. Choose $0 \ne z \in \mathscr{L}(G_1 - \sum_{\nu \in I} P_\nu)$ and write

$$z = \sum_{\lambda=1}^{l} \gamma_\lambda f_\lambda \quad \textit{with} \quad \gamma_\lambda \in \mathbb{F}_q\,.$$

Then $zg_\rho \in \mathscr{L}(G)$ for $1 \le \rho \le k$, and we obtain

$$[a, zg_\rho] = \sum_{\lambda=1}^{l} [a, f_\lambda g_\rho] \cdot \gamma_\lambda \,. \tag{8.32}$$

On the other hand, since $c \in C_\Omega$ and $zg_\rho \in \mathscr{L}(G)$, we have $[c, zg_\rho] = 0$ by (8.23). Hence

$$[a, zg_\rho] = [c + e, zg_\rho] = [e, zg_\rho] = \sum_{\nu=1}^{n} e_\nu \cdot z(P_\nu) \cdot g_\rho(P_\nu) = 0\,. \tag{8.33}$$

(Observe that $e_\nu = 0$ for $\nu \notin I$ and $z(P_\nu) = 0$ for $\nu \in I$, because z is an element of $\mathscr{L}(G_1 - \sum_{\nu\in I} P_\nu)$.) Equations (8.32) and (8.33) show that $(\gamma_1, \ldots, \gamma_l)$ is a non-trivial solution of (8.30).

Now we take an arbitrary solution $(\alpha_1, \ldots, \alpha_l)$ of the system (8.30) and set $f := \sum_{\lambda=1}^{l} \alpha_\lambda f_\lambda$. Suppose there is an error position $\nu_0 \in I$ such that $f(P_{\nu_0}) \neq 0$. By (8.24),

$$\deg\Big(G - G_1 - \sum_{\nu\in I} P_\nu\Big) \ge \deg G - \deg G_1 - t > 2g - 2\,.$$

This implies

$$\mathscr{L}\Big(G - G_1 - \sum_{\nu\in I} P_\nu\Big) \subsetneqq \mathscr{L}\Big(G - G_1 - \sum_{\nu\in I\setminus\{\nu_0\}} P_\nu\Big)\,.$$

So we find an element $h \in \mathscr{L}(G - G_1)$ with $h(P_{\nu_0}) \neq 0$ and $h(P_\nu) = 0$ for all $\nu \in I \setminus \{\nu_0\}$. We obtain

$$[a, fh] = [e, fh] = \sum_{\nu=1}^{n} e_\nu \cdot f(P_\nu) \cdot h(P_\nu) = e_{\nu_0} \cdot f(P_{\nu_0}) \cdot h(P_{\nu_0}) \neq 0\,. \tag{8.34}$$

However, h is a linear combination of $g_1, \ldots, g_k$ and

$$[a, fg_\rho] = \sum_{\lambda=1}^{l} [a, f_\lambda g_\rho] \cdot \alpha_\lambda = 0\,,$$

since $(\alpha_1, \ldots, \alpha_l)$ is a solution of (8.30). This contradicts (8.34). □

The cardinality of $N(f) = \{\nu \,|\, 1 \le \nu \le n \text{ and } f(P_\nu) = 0\}$ does not exceed the degree of G_1, because $f \in \mathscr{L}\big(G_1 - \sum_{\nu\in N(f)} P_\nu\big)$ implies

$$\deg G_1 - |N(f)| \ge 0\,. \tag{8.35}$$

Observe that in general not all $\nu \in N(f)$ are actually error positions.

In order to determine the error values e_ν, we consider another system of linear equations:

$$\sum_{\nu\in N(f)} h_\mu(P_\nu)\cdot z_\nu = [a,h_\mu]\,, \qquad \textit{for } \mu=1,\ldots,m\,. \tag{8.36}$$

(Recall that $\{h_1,\ldots,h_m\}$ is a basis of $\mathscr{L}(G)$, see (8.29).)

Proposition 8.5.3. *Under the above hypotheses the system* (8.36) *has a unique solution, namely the vector* $(e_\nu)_{\nu\in N(f)}$.

Proof. As $h_\mu\in\mathscr{L}(G)$, we have

$$[a,h_\mu]=[c+e,h_\mu]=[e,h_\mu]=\sum_{\nu=1}^{n} e_\nu\cdot h_\mu(P_\nu)=\sum_{\nu\in N(f)} h_\mu(P_\nu)\cdot e_\nu\,.$$

(Note that $e_\nu=0$ for $\nu\notin N(f)$ by Proposition 8.5.2.) Hence $(e_\nu)_{\nu\in N(f)}$ is a solution of the system (8.36).

Suppose that $(b_\nu)_{\nu\in N(f)}$ is another solution of (8.36). Define the vector $b:=(b_1,\ldots,b_n)\in\mathbb{F}_q^n$ by setting $b_\nu:=0$ for $\nu\notin N(f)$. Then

$$[b,h_\mu]=\sum_{\nu\in N(f)} h_\mu(P_\nu)\cdot b_\nu=[a,h_\mu]=[e,h_\mu]$$

for $\mu=1,\ldots,m$. As $\{h_1,\ldots,h_m\}$ is a basis of $\mathscr{L}(G)$, this implies $b-e\in C_\Omega$, by (8.23). The weight of $b-e$ can be estimated as follows:

$$\mathrm{wt}(b-e)\le|N(f)|\le\deg G_1<\deg G-(2g-2)=d^*\,.$$

(We have used (8.35) and (8.24) once again.) Since the minimum distance of C_Ω is $\ge d^*$ we conclude that $b=e$. □

Proposition 8.5.2 and 8.5.3 are summarized in the following decoding algorithm for the code C_Ω. We maintain all previous notation.

Decoding Algorithm 8.5.4. *Let an element $a\in\mathbb{F}_q^n$ be given.*

(1) Find a non-trivial solution $(\alpha_1,\ldots,\alpha_l)$ of the system (8.30), *and set $f:=\sum_{\lambda=1}^{l}\alpha_\lambda f_\lambda$. (If* (8.30) *has only the trivial solution, we cannot decode a.)*

(2) Determine $N(f)=\{\nu\,|\,1\le\nu\le n$ and $f(P_\nu)=0\}$. (This can be done by evaluating $f(P_\nu)=\sum_{\lambda=1}^{l}\alpha_\lambda f_\lambda(P_\nu)$ for $\nu=1,\ldots,n$.)

(3) If the system (8.36) *has a unique solution $(e_\nu)_{\nu\in N(f)}$, we set $e:=(e_1,\ldots,e_n)$ with $e_\nu=0$ for $\nu\notin N(f)$. (If* (8.36) *is not uniquely solvable, we cannot decode a.)*

(4) Check whether $c:=a-e$ is an element of C_Ω (by calculating the syndromes $[c,h_\mu]$ for $\mu=1,\ldots,m$) and whether $\mathrm{wt}(e)\le t$. If the answer is yes, we decode a to the codeword c. If the answer is no, we cannot decode a.

Theorem 8.5.5 (Skorobogatov-Vladut). *(a) Provided G_1 and t satisfy* (8.24), *the algorithm* 8.5.4 *decodes all errors of weight* $\leq t$.

(b) One can choose the divisor G_1 in such a way that the algorithm 8.5.4 *decodes all errors e of weight*

$$\operatorname{wt}(e) \leq (d^* - 1 - g)/2\,,$$

where $d^* = \deg G - (2g-2)$ *is the designed distance of* C_Ω.

Proof. (a) is obvious from Proposition 8.5.2 and 8.5.3, and (b) follows from Remark 8.5.1(b). □

Remark 8.5.6. (a) It can happen that steps (1) - (3) of the decoding algorithm work but that $a - e \notin C_\Omega$ or $\operatorname{wt}(e) > t$. In either case there is no codeword $c \in C_\Omega$ with the property $\operatorname{wt}(c) \leq t$.

(b) The decoding algorithm 8.5.4 is due to A.N.Skorobogatov and S.G.Vladut, following an idea of J.Justesen et al. One deficiency of the algorithm is that it decodes (in general) only errors of weight $\leq (d^*-1-g)/2$ but not all errors of weight $\leq (d^*-1)/2$. There are several variants of the algorithm which correct, under additional hypotheses, more than $(d^* - 1 - g)/2$ errors.

(c) There are some completely different approaches to decode AG codes: we mention the method due to G. L. Feng and T. R. N. Rao (see Chapter 10 in [32]), and the decoding algorithm of M. Sudan (see [17]); this algorithm is known as *list decoding.*

8.6 Exercises

8.1. Let F be a function field over a finite field $\mathbb{F}_q$. We consider divisors G and $D = P_1 + \ldots + P_n$, where $P_1, \ldots, P_n$ are distinct places of degree one and $\operatorname{supp} G \cap \operatorname{supp} D = \emptyset$ as usual. Show that $C_\mathscr{L}(D,G) = C_\mathscr{L}(\sigma(D), \sigma(G))$ and $C_\Omega(D,G) = C_\Omega(\sigma(D), \sigma(G))$, for every automorphism σ of $F/\mathbb{F}_q$.

8.2. *(i)* Show that the Tsfasman-Vladut-Zink Bound improves the Gilbert-Varshamov Bound for all $q = \ell^2 \geq 49$, but not for $q \leq 25$.

(ii) Let $q = \ell^3$ be a cube. Show that the bound

$$\alpha_q(\delta) \geq \left(1 - \frac{\ell+2}{2(\ell^2-1)}\right) - \delta$$

improves the Gilbert-Varshamov bound for all $q = \ell^3 \geq 343$ (see Remark 8.4.8).

8.3. (This exercise will be useful for the following exercises.) Consider the tower $\mathcal{G}^* = (G_0^*, G_1^*, G_2^*, \ldots)$ over the field $\mathbb{F}_q$ with $q = \ell^2$ as in Theorem 7.4.15. It has amongst others the following properties:

(1) $G_0^* = \mathbb{F}_q(u_0)$ is rational,
(2) all extensions G_i^*/G_0^* are Galois,
(3) the only places of $\mathbb{F}_q(u_0)$ which are ramified in some extension G_n^*/G_0^*, are the places $(u_0 = 1)$ and $(u_0 = \infty)$,
(4) the place $(u_0 = 0)$ splits completely in all extensions G_n^*/G_0^*.

Show that the ramification indices of the places $(u_0 = 1)$ and $(u_0 = \infty)$ in G_n^*/G_0^* tend to infinity, as $n \to \infty$.

8.4. A code $C \subseteq \mathbb{F}_q^n$ is said to be transitive, if its automorphism group Aut (C) is a transitive subgroup of the symmetric group $\mathcal{S}_n$ (i.e., for any two indices $i, j \in \{1, 2, \ldots, n\}$ there is an automorphism $\pi \in \text{Aut}\,(C)$ such that $\pi(i) = j$). This is obviously a generalization of the notion of cyclic codes.
Show that the class of transitive codes over $\mathbb{F}_q$, where $q = \ell^2$ is a square, attains the Tsfasman-Vladut-Zink Bound. More precisely, let $R, \delta \geq 0$ be real numbers with $R = 1 - \delta - 1/(\ell - 1)$. Show that there exists a sequence $(C_j)_{j \geq 0}$ of linear codes over $\mathbb{F}_q$ with parameters $[n_j, k_j, d_j]$ having the following properties:

(1) all C_j are transitive,
(2) $n_j \to \infty$ as $j \to \infty$,
(3) $\lim_{j \to \infty} k_j/n_j \geq R$ and $\lim_{j \to \infty} d_j/n_j \geq \delta$.

8.5. Assume that $q = \ell^2$ is a square. Let $0 \leq R \leq 1/2$ and $\delta \geq 0$ with $R = 1 - \delta - 1/(\ell - 1)$. Show that there is a sequence $(C_j)_{j \geq 0}$ of linear codes over $\mathbb{F}_q$ with parameters $[n_j, k_j, d_j]$ having the following properties:

(1) all C_j are self-orthogonal; i.e., $C_j \subseteq C_j^{\perp}$,
(2) $n_j \to \infty$ as $j \to \infty$,
(3) $\lim_{j \to \infty} k_j/n_j \geq R$ and $\lim_{j \to \infty} d_j/n_j \geq \delta$,
(4) also the dual codes $(C_j^{\perp})_{j \geq 0}$ attain the Tsfasman-Vladut-Zink Bound.

8.6. Let $q = \ell^2$ be a square. Show that there exists a sequence of self-dual codes $(C_i)_{i \geq 0}$ over $\mathbb{F}_q$ with parameters $[n_i, k_i, d_i]$ such that $n_i \to \infty$ and

$$\liminf_{i \to \infty} \frac{d_i}{n_i} \geq \frac{1}{2} - \frac{1}{\ell - 1}\,.$$

Hint. This exercise is just the assertion of Theorem 8.4.9 which we proved only in the case where q is even. So you can assume now that q is odd. Again use the tower $(G_0^*, G_1^*, G_2^*, \ldots)$ as in Theorem 7.4.15. Consider the elements u_0 and t_0 as defined in Equation (7.26); they satisfy the equation $u_0 = t_0^{\ell-1} + 1$. Choose $\epsilon \in \mathbb{F}_q$ with $\epsilon^{\ell-1} = -1$ and consider the differential $\omega := \epsilon \cdot dt_0/u_0$ in G_n^*. Show that its divisor has the form $(\omega) = 2H - D$ where $H \geq 0$ and D is the sum of rational places of G_n^*. Calculate the residues of ω at these rational places and show that they are squares in $\mathbb{F}_q^{\times}$. Conclude that the code $C_{\mathscr{L}}(D, H)$ is equivalent to a self-dual code (see Definition 2.2.13).

9

Subfield Subcodes and Trace Codes

A very useful method of constructing codes over $\mathbb{F}_q$ is to restrict codes which are defined over an extension field $\mathbb{F}_{q^m}$. This means that, given a code $C \subseteq (\mathbb{F}_{q^m})^n$, one considers the subfield subcode $C|_{\mathbb{F}_q} = C \cap \mathbb{F}_q^n$. Many well-known codes can be defined in this way, for instance BCH codes, Goppa codes and, more generally, alternant codes (cf. Section 2.3).

There is yet another method of defining a code over $\mathbb{F}_q$ if a code over $\mathbb{F}_{q^m}$ is given. This construction uses the trace mapping $\mathrm{Tr} : \mathbb{F}_{q^m} \to \mathbb{F}_q$. An important class of codes which can be represented as trace codes in a natural manner is the class of cyclic codes (see Example 9.2.4 below). The subfield subcode construction and the trace construction are closely related by Delsarte's Theorem 9.1.2.

In this chapter we present some results on subfield subcodes and trace codes. It is a surprising fact that the study of trace codes leads to a second non-trivial relation between coding theory and the theory of algebraic function fields, see Section 9.2.

9.1 On the Dimension of Subfield Subcodes and Trace Codes

We consider the field extension $\mathbb{F}_{q^m}/\mathbb{F}_q$; this is a Galois extension of degree $[\mathbb{F}_{q^m} : \mathbb{F}_q] = m$. Let

$$\mathrm{Tr} : \mathbb{F}_{q^m} \longrightarrow \mathbb{F}_q$$

denote the trace mapping (cf. Appendix A). For $a = (a_1, \ldots, a_n) \in (\mathbb{F}_{q^m})^n$ we define

$$\mathrm{Tr}(a) := \big(\mathrm{Tr}(a_1), \ldots, \mathrm{Tr}(a_n)\big) \in \mathbb{F}_q^n .$$

In this manner we obtain an $\mathbb{F}_q$-linear map $\mathrm{Tr} : (\mathbb{F}_{q^m})^n \to \mathbb{F}_q^n$.

H. Stichtenoth, *Algebraic Function Fields and Codes*,
Graduate Texts in Mathematics 254,

Definition 9.1.1. *Let $C \subseteq (\mathbb{F}_{q^m})^n$ be a code over $\mathbb{F}_{q^m}$.*
(a) $C|_{\mathbb{F}_q} := C \cap \mathbb{F}_q^n$ is called the subfield subcode (or the restriction of C to $\mathbb{F}_q$).
(b) $\mathrm{Tr}(C) := \{\mathrm{Tr}(c) \,|\, c \in C\} \subseteq \mathbb{F}_q^n$ is called the trace code of C.

Note that the subfield subcode and the trace code of a code $C \subseteq (\mathbb{F}_{q^m})^n$ are codes over $\mathbb{F}_q$ of length n.

Theorem 9.1.2 (Delsarte). *For a code C over $\mathbb{F}_{q^m}$,*

$$(C|_{\mathbb{F}_q})^\perp = \mathrm{Tr}(C^\perp)\,.$$

Proof. Recall that we denote by $\langle\ ,\ \rangle$ the canonical inner product on $\mathbb{F}_q^n$ (resp. on $(\mathbb{F}_{q^m})^n$). In order to prove $(C|_{\mathbb{F}_q})^\perp \supseteq \mathrm{Tr}(C^\perp)$ we have to show that

$$\langle c, \mathrm{Tr}(a)\rangle = 0 \quad \textit{for all } c \in C|_{\mathbb{F}_q} \textit{ and } a \in C^\perp\,. \tag{9.1}$$

Write $c = (c_1, \ldots, c_n)$ and $a = (a_1, \ldots, a_n)$; then

$$\langle c, \mathrm{Tr}(a)\rangle = \sum_{i=1}^{n} c_i \cdot \mathrm{Tr}(a_i) = \mathrm{Tr}\big(\sum c_i a_i\big) = \mathrm{Tr}(\langle a, c\rangle) = \mathrm{Tr}(0) = 0\,.$$

We have used here the $\mathbb{F}_q$-linearity of the trace and the fact that $\langle a, c\rangle = 0$ (which follows from $c \in C$ and $a \in C^\perp$). This proves (9.1).

Next we show that $(C|_{\mathbb{F}_q})^\perp \subseteq \mathrm{Tr}(C^\perp)$. This assertion is equivalent to

$$\mathrm{Tr}(C^\perp)^\perp \subseteq C|_{\mathbb{F}_q}\,. \tag{9.2}$$

Suppose that (9.2) does not hold. Then there exists some $u \in \mathrm{Tr}(C^\perp)^\perp \setminus C$, hence an element $v \in C^\perp$ with $\langle u, v\rangle \neq 0$. As $\mathrm{Tr} : \mathbb{F}_{q^m} \to \mathbb{F}_q$ is not the zero map, there is an element $\gamma \in \mathbb{F}_{q^m}$ such that $\mathrm{Tr}(\gamma \cdot \langle u, v\rangle) \neq 0$. Hence

$$\langle u, \mathrm{Tr}(\gamma v)\rangle = \mathrm{Tr}(\langle u, \gamma v\rangle) = \mathrm{Tr}(\gamma \cdot \langle u, v\rangle) \neq 0\,.$$

But on the other hand we have $\langle u, \mathrm{Tr}(\gamma v)\rangle = 0$ because $u \in \mathrm{Tr}(C^\perp)^\perp$ and $\gamma v \in C^\perp$. This contradiction proves (9.2). □

There are obvious upper bounds for the dimension of subfield subcodes and trace codes, namely

$$\dim C|_{\mathbb{F}_q} \leq \dim C \tag{9.3}$$

and

$$\dim \mathrm{Tr}(C) \leq m \cdot \dim C\,. \tag{9.4}$$

(9.3) follows from the fact that a basis of $C|_{\mathbb{F}_q}$ over $\mathbb{F}_q$ is also linearly independent over $\mathbb{F}_{q^m}$, and (9.4) follows since $\mathrm{Tr} : C \to \mathrm{Tr}(C)$ is a surjective

$\mathbb{F}_q$-linear mapping and the dimension of C, regarded as a vector space over the field $\mathbb{F}_q$, is $m \cdot \dim C$.

Using Delsarte's Theorem we also obtain lower bounds for the dimensions of subfield subcodes and trace codes:

Lemma 9.1.3. *Let C be a code of length n over $\mathbb{F}_{q^m}$. Then*

$$\dim C \leq \dim \mathrm{Tr}(C) \leq m \cdot \dim C \tag{9.5}$$

and

$$\dim C - (m-1)(n - \dim C) \leq \dim C|_{\mathbb{F}_q} \leq \dim C\,. \tag{9.6}$$

Proof. By Delsarte's Theorem and (9.3) we have

$$\dim \mathrm{Tr}(C) = \dim(C^\perp|_{\mathbb{F}_q})^\perp = n - \dim C^\perp|_{\mathbb{F}_q} \geq n - \dim C^\perp = \dim C\,.$$

This proves (9.5). The lower estimate in (9.6) is proved in an analogous manner. □

The bounds given in Lemma 9.1.3 are general bounds, valid for each code over $\mathbb{F}_{q^m}$. Our aim is to improve these estimates in specific cases, in particular for certain algebraic geometry codes.

In the following, a *subcode* of a code $C \subseteq (\mathbb{F}_{q^m})^n$ means an $\mathbb{F}_{q^m}$-subspace $U \subseteq C$. By U^q we denote the set

$$U^q := \{(a_1^q, \ldots, a_n^q) \mid (a_1, \ldots, a_n) \in U\}\,.$$

It is obvious that U^q is also an $\mathbb{F}_{q^m}$-subspace of $(\mathbb{F}_{q^m})^n$.

Proposition 9.1.4. *Let C be a code over $\mathbb{F}_{q^m}$ and let $U \subseteq C$ be a subcode with the additional property $U^q \subseteq C$. Then*

$$\dim \mathrm{Tr}(C) \leq m \cdot (\dim C - \dim U) + \dim U|_{\mathbb{F}_q}\,.$$

Proof. We consider the $\mathbb{F}_q$-linear map $\phi : U \to C$, given by $\phi(u) := u^q - u$. The kernel of ϕ is easily seen to be

$$\mathrm{Ker}(\phi) = U|_{\mathbb{F}_q}\,. \tag{9.7}$$

Since $\mathrm{Tr}(a^q) = \mathrm{Tr}(a)$ for $a \in \mathbb{F}_{q^m}$, the image of ϕ is contained in the kernel of the trace map $\mathrm{Tr} : C \to \mathrm{Tr}(C)$; i.e.,

$$\mathrm{Im}(\phi) \subseteq \mathrm{Ker}(\mathrm{Tr})\,. \tag{9.8}$$

From (9.7) and (9.8) we obtain

$$\begin{aligned}\dim \mathrm{Tr}(C) &= \dim_{\mathbb{F}_q} C - \dim \mathrm{Ker}(\mathrm{Tr}) \leq m \cdot \dim C - \dim \mathrm{Im}(\phi) \\ &= m \cdot \dim C - (\dim_{\mathbb{F}_q} U - \dim \mathrm{Ker}(\phi)) \\ &= m \cdot (\dim C - \dim U) + \dim U|_{\mathbb{F}_q}\,.\end{aligned}$$

□

Note that Proposition 9.1.4 improves the upper bound (9.5) for the dimension of trace codes. We also obtain an improvement of the lower bound (9.6) for the dimension of subfield subcodes:

Corollary 9.1.5. *Let C be a code of length n over $\mathbb{F}_{q^m}$ and let $V \subseteq C^\perp$ be a subcode of $C^\perp$ satisfying $V^q \subseteq C^\perp$. Then*

$$\begin{aligned}\dim C|_{\mathbb{F}_q} &\geq \dim C - (m-1)(n - \dim C) + m \cdot \dim V - \dim V|_{\mathbb{F}_q} \\ &\geq \dim C - (m-1)(n - \dim C - \dim V).\end{aligned}$$

Proof. We use Proposition 9.1.4 and Delsarte's Theorem:

$$\begin{aligned}\dim C|_{\mathbb{F}_q} &= \dim \mathrm{Tr}(C^\perp)^\perp = n - \dim \mathrm{Tr}(C^\perp) \\ &\geq n - \big(m \cdot (\dim C^\perp - \dim V) + \dim V|_{\mathbb{F}_q}\big) \\ &= \dim C - (m-1)(n - \dim C) + m \cdot \dim V - \dim V|_{\mathbb{F}_q}.\end{aligned}$$

□

Now we apply the above results to AG codes.

Theorem 9.1.6. *Let F be an algebraic function field of genus g over the constant field $\mathbb{F}_{q^m}$. Consider the AG codes*

$$C_{\mathscr{L}} := C_{\mathscr{L}}(D,G) \quad \text{and} \quad C_\Omega := C_\Omega(D,G), \tag{9.9}$$

where $D = P_1 + \ldots + P_n$ (with pairwise distinct places $P_1, \ldots, P_n$ of degree one), $\mathrm{supp}\, D \cap \mathrm{supp}\, G = \emptyset$ and $\deg G < n$. Suppose that G_1 is a divisor of F satisfying

$$G_1 \leq G \quad \text{and} \quad q \cdot G_1 \leq G. \tag{9.10}$$

Then

$$\dim \mathrm{Tr}(C_{\mathscr{L}}) \leq \begin{cases} m(\ell(G) - \ell(G_1)) + 1 & \text{if } G_1 \geq 0, \\ m(\ell(G) - \ell(G_1)) & \text{if } G_1 \ngeq 0, \end{cases} \tag{9.11}$$

and

$$\dim C_\Omega|_{\mathbb{F}_q} \geq \begin{cases} n - 1 - m(\ell(G) - \ell(G_1)) & \text{if } G_1 \geq 0, \\ n - m(\ell(G) - \ell(G_1)) & \text{if } G_1 \ngeq 0. \end{cases} \tag{9.12}$$

Proof. Let $U := C_{\mathscr{L}}(D, G_1)$. It follows from (9.10) that $U^q \subseteq C_{\mathscr{L}}$. We can apply Proposition 9.1.4 and obtain

$$\dim \mathrm{Tr}(C_{\mathscr{L}}) \leq m(\ell(G) - \ell(G_1)) + \dim U|_{\mathbb{F}_q}. \tag{9.13}$$

So we have to determine the subfield subcode $U|_{\mathbb{F}_q} = C_{\mathscr{L}}(D, G_1)|_{\mathbb{F}_q}$. Consider an element $x \in \mathscr{L}(G_1)$ such that $x(P_i) \in \mathbb{F}_q$ for $i = 1, \ldots, n$. Then $x^q - x \in$

$\mathscr{L}(G)$ and $(x^q - x)(P_i) = 0$, hence $x^q - x \in \mathscr{L}(G - D)$. Since we assumed that $\deg(G - D) < 0$, it follows that $x^q - x = 0$; i.e., $x \in \mathbb{F}_q$. Consequently

$$\dim U|_{\mathbb{F}_q} = \begin{cases} 1 & \textit{if } G_1 \geq 0\,, \\ 0 & \textit{if } G_1 \ngeq 0\,. \end{cases}$$

Substituting this into (9.13) we get the desired estimate (9.11) for the dimension of $\mathrm{Tr}(C_{\mathscr{L}})$.

The corresponding estimate (9.12) for the dimension of $C_\Omega|_{\mathbb{F}_q}$ follows from Corollary 9.1.5. □

Remark 9.1.7. In addition to the hypotheses of Theorem 9.1.6 assume that $\deg G_1 > 2g - 2$. Then we can replace the terms $\ell(G)$ and $\ell(G_1)$ in (9.11) and (9.12) by $\deg G$ and $\deg G_1$. This follows immediately from the Riemann-Roch Theorem.

Example 9.1.8. As an illustration of Theorem 9.1.6 we consider a Goppa code

$$\Gamma(L, g(z)) = C_\Omega(D_L, G_0 - P_\infty)|_{\mathbb{F}_q}$$

(notation as in Definition 2.3.10 and Proposition 2.3.11). Let $g_1(z) \in \mathbb{F}_{q^m}[z]$ be the polynomial of maximal degree such that $g_1(z)^q$ divides $g(z)$. We set $G_1 := (g_1(z))_0 - P_\infty$ where $(g_1(z))_0$ is the zero divisor of $g_1(z)$, and obtain from (9.12) the estimate

$$\dim \Gamma(L, g(z)) \geq n - m(\deg g(z) - \deg g_1(z))\,. \tag{9.14}$$

In many cases, equality holds in (9.14). This will be shown in Proposition 9.2.13.

9.2 Weights of Trace Codes

In this section we investigate some specific trace codes. The main idea is to relate the weights of their codewords to the number of rational places in certain algebraic function fields. The Hasse-Weil-Serre Bound then yields estimates for the weights and the minimum distance of these codes.

First we introduce the codes to be considered.

Definition 9.2.1. *Let F be an algebraic function field over the constant field $\mathbb{F}_{q^m}$ and let $V \subseteq F$ be a finite-dimensional $\mathbb{F}_{q^m}$-subspace of F. Let $P_1, \ldots, P_n \in \mathbb{P}_F$ be n distinct places of degree one such that $v_{P_i}(f) \geq 0$ for all $f \in V$ and $i = 1, \ldots, n$. Set $D := P_1 + \ldots + P_n$. Then we define*

$$C(D, V) := \{(f(P_1), \ldots, f(P_n)) \mid f \in V\} \subseteq (\mathbb{F}_{q^m})^n$$

and

$$\mathrm{Tr}_D(V) := \mathrm{Tr}(C(D,V)) \subseteq \mathbb{F}_q^n\,;$$

i.e., $\mathrm{Tr}_D(V)$ *is the trace code of* $C(D,V)$ *with respect to the extension* $\mathbb{F}_{q^m}/\mathbb{F}_q$.

Note that $C(D,V)$ is a code over $\mathbb{F}_{q^m}$, whereas $\mathrm{Tr}_D(V)$ is a code over $\mathbb{F}_q$. Our main objective in this section is to study the codes $\mathrm{Tr}_D(V)$. Let us first give some examples of such codes.

Example 9.2.2. The codes $C(D,V)$ are a generalization of algebraic geometry codes. Choosing $V := \mathscr{L}(G)$ where G is a divisor with $\operatorname{supp} G \cap \operatorname{supp} D = \emptyset$ as usual, we obtain $C(D,V) = C_{\mathscr{L}}(D,G)$.

Example 9.2.3. Every code $C \subseteq \mathbb{F}_q^n$ over $\mathbb{F}_q$ can be represented as $C = \mathrm{Tr}_D(V)$ for a suitable choice of V and D. This can be seen as follows. Choose $m \in \mathbb{N}$ sufficiently large such that $q^m \geq n$. Let $F := \mathbb{F}_{q^m}(z)$ be the rational function field over $\mathbb{F}_{q^m}$. Choose n distinct elements $\alpha_1, \ldots, \alpha_n \in \mathbb{F}_{q^m}$ and denote by $P_i \in \mathbb{P}_F$ the zero of $z - \alpha_i$. Choose a basis $\{a^{(1)}, \ldots, a^{(k)}\}$ of C over $\mathbb{F}_q$. Write $a^{(j)} = (a_1^{(j)}, \ldots, a_n^{(j)})$. For $j = 1, \ldots, k$ choose a polynomial $f_j = f_j(z) \in \mathbb{F}_{q^m}[z]$ satisfying $f_j(\alpha_i) = a_i^{(j)}$ for $i = 1, \ldots, n$. Let $V \subseteq F$ be the $\mathbb{F}_{q^m}$-vector space generated by $f_1, \ldots, f_k$. Then it is easily verified that $C = \mathrm{Tr}_D(V)$.

More interesting than the previous example is the fact that specific classes of codes over $\mathbb{F}_q$ can be represented as trace codes in a natural manner. In the following we give such a representation for cyclic codes.

A code C over $\mathbb{F}_q$ of length n is said to be *cyclic* if its automorphism group $\mathrm{Aut}(C)$ contains the cyclic shift; i.e.,

$$(c_0, c_1, \ldots, c_{n-1}) \in C \Longrightarrow (c_1, \ldots, c_{n-1}, c_0) \in C\,.$$

As is common in coding theory, we identify $\mathbb{F}_q^n$ with the vector space of polynomials of degree $\leq n-1$ over $\mathbb{F}_q$ via

$$c = (c_o, \ldots, c_{n-1}) \longleftrightarrow c(x) = c_o + c_1 x + \ldots + c_{n-1}x^{n-1} \in \mathbb{F}_q[x]\,. \tag{9.15}$$

We shall always assume that

$$\gcd(n,q) = 1\,. \tag{9.16}$$

Let m be the least integer ≥ 1 satisfying $q^m \equiv 1 \bmod n$. Then the polynomial $x^n - 1$ factors over the field $\mathbb{F}_{q^m} \supseteq \mathbb{F}_q$ as

$$x^n - 1 = \prod_{\nu=0}^{n-1} (x - \beta^\nu)\,, \tag{9.17}$$

where $\beta \in \mathbb{F}_{q^m}$ is a primitive n-th root of unity. All linear factors in (9.17) are distinct.

Let us briefly recall some basic facts about cyclic codes, cf. [28]. Given a cyclic code $C \neq \{0\}$ of length n over $\mathbb{F}_q$, there exists a unique monic polynomial $g(x) \in C$ of minimal degree; it is called the *generator polynomial* of C. The generator polynomial divides $x^n - 1$, so

$$g(x) = \prod_{\nu \in I} (x - \beta^\nu), \tag{9.18}$$

where β is a primitive n-th root of unity as in (9.17) and I is a certain subset of $\{0, \ldots, n-1\}$. The elements β^ν with $\nu \in I$ are called the *zeros* of C because one has the following description of C: for $c(x) \in \mathbb{F}_q[x]$ with $\deg c(x) \leq n-1$,

$$c(x) \in C \iff c(\beta^\nu) = 0 \quad \textit{for all} \quad \nu \in I. \tag{9.19}$$

The conditions on the right-hand side of (9.19) can be weakened. To this end we define the *cyclotomic coset* $\mathcal{C}(i)$ of an integer $i \in \mathbb{Z}$, $0 \leq i \leq n-1$, by

$$\mathcal{C}(i) := \{j \in \mathbb{Z} \,|\, 0 \leq j \leq n-1 \ \textit{and} \ j \equiv q^l i \bmod n \ \textit{for some} \ l \geq 0\}.$$

It is easily checked that either $\mathcal{C}(i) = \mathcal{C}(i')$ or $\mathcal{C}(i) \cap \mathcal{C}(i') = \emptyset$. Hence the set $\{0, 1, \ldots, n-1\}$ is partitioned into pairwise disjoint cyclotomic cosets; i.e., $\{0, 1, \ldots, n-1\} = \bigcup_{\mu=1}^{s} \mathcal{C}_\mu$ (with $s \leq n$ and $\mathcal{C}_\mu = \mathcal{C}(i_\mu)$ for some i_μ, $0 \leq i_\mu \leq n-1$).

For $\nu \in \mathbb{Z}$ denote by $\tilde{\nu}$ the unique integer in $\{0, 1, \ldots, n-1\}$ with $\nu \equiv \tilde{\nu} \bmod n$. Let $\emptyset \neq M \subseteq \{0, 1, \ldots, n-1\}$. A subset $M_0 \subseteq \mathbb{Z}$ is called a *complete set of cyclotomic coset representatives* of M, if for each $\nu \in M$ there exists a unique $\nu_0 \in M_0$ such that $\tilde{\nu}_0 \in \mathcal{C}(\nu)$. It is evident that one can always find a complete set of cyclotomic coset representatives of M which is contained in $\{0, 1, \ldots, n-1\}$.

Now consider the set $I \subseteq \{0, 1, \ldots, n-1\}$ given by (9.18); i.e., $\{\beta^\nu \,|\, \nu \in I\}$ is the set of zeros of the cyclic code C. Let I_0 be a complete set of cyclotomic coset representatives of I. Since for $c(x) \in \mathbb{F}_q[x]$,

$$c(\beta^\nu) = 0 \iff c(\beta^{q^l \nu}) = 0,$$

we can replace (9.19) by the following condition:

$$c(x) \in C \iff c(\beta^\nu) = 0 \quad \textit{for all} \ \nu \in I_0, \tag{9.20}$$

where $c(x) \in \mathbb{F}_q[x]$ and $\deg c(x) \leq n-1$.

The dual code $C^\perp$ of a cyclic code C is cyclic as well. Let $g(x) \in \mathbb{F}_q[x]$ be the generator polynomial of C and

$$h(x) := (x^n - 1)/g(x) \in \mathbb{F}_q[x]. \tag{9.21}$$

The polynomial $h(x)$ is called the *check polynomial* of C. The reciprocal polynomial $h^{\perp}(x)$ of $h(x)$,

$$h^{\perp}(x) := h(0)^{-1} \cdot x^{\deg h(x)} \cdot h(x^{-1}), \tag{9.22}$$

is the generator polynomial of $C^{\perp}$. Write

$$h^{\perp}(x) = \prod_{\rho \in J} (x - \beta^{\rho}) \quad \textit{with } J \subseteq \{0, 1, \ldots, n-1\}. \tag{9.23}$$

It follows from (9.21), (9.22) and (9.23) that

$$\begin{aligned} \rho \in J &\iff h^{\perp}(\beta^{\rho}) = 0 \iff h(\beta^{-\rho}) = 0 \\ &\iff g(\beta^{-\rho}) \neq 0 \iff \rho \not\equiv -\nu \bmod n \quad \textit{for all} \quad \nu \in I. \end{aligned}$$

Let J_0 be a complete set of cyclotomic coset representatives of J. From (9.20) we obtain for $a(x) = a_0 + a_1 x + \ldots + a_{n-1} x^{n-1} \in \mathbb{F}_q[x]$,

$$a(x) \in C^{\perp} \iff a(\beta^{\rho}) = 0 \quad \textit{for all } \rho \in J_0. \tag{9.24}$$

Using the canonical inner product $\langle\ ,\ \rangle$ on $(\mathbb{F}_{q^m})^n$, the equation $a(\beta^{\rho}) = 0$ can also be written as

$$\big\langle (a_0, a_1, \ldots, a_{n-1}), (1, \beta^{\rho}, \beta^{2\rho}, \ldots, \beta^{(n-1)\rho}) \big\rangle = 0. \tag{9.25}$$

Now we are in a position to represent the given cyclic code C as a trace code. We consider the rational function field $F = \mathbb{F}_{q^m}(z)$ and the vector space $V \subseteq F$ generated by the set $\{z^{\rho} \mid \rho \in J_0\}$. Denote by $P_i \in \mathbb{P}_F$ the zero of $z - \beta^{i-1}$ and set $D = P_1 + \ldots + P_n$. By definition, the code $C(D, V)$ is generated – as a vector space over $\mathbb{F}_{q^m}$ – by the vectors $(1, \beta^{\rho}, \beta^{2\rho}, \ldots, \beta^{(n-1)\rho})$ with $\rho \in J_0$. We conclude from (9.24) and (9.25) that

$$C^{\perp} = C(D, V)^{\perp}|_{\mathbb{F}_q}.$$

Applying Delsarte's theorem we finally obtain

$$C = \big(C(D, V)^{\perp}|_{\mathbb{F}_q}\big)^{\perp} = \mathrm{Tr}(C(D, V)) = \mathrm{Tr}_D(V).$$

We summarize:

Proposition 9.2.4. *Let C be a cyclic code of length n over $\mathbb{F}_q$ with generator polynomial $g(x)$, and let $\mathbb{F}_{q^m} = \mathbb{F}_q(\beta)$ where β is a primitive n-th root of unity. Let $J = \{0 \le \rho \le n-1 \mid g(\beta^{-\rho}) \neq 0\}$ and let J_0 be a complete set of cyclotomic coset representatives of J. Consider the rational function field $F = \mathbb{F}_{q^m}(z)$ over $\mathbb{F}_{q^m}$ and the $\mathbb{F}_{q^m}$-vector space $V \subseteq F$ which is generated by $\{z^{\rho} \mid \rho \in J_0\}$. Then $C = \mathrm{Tr}_D(V)$ where $D = P_1 + \ldots + P_n$ and $P_i \in \mathbb{P}_F$ is the zero divisor of $z - \beta^{i-1}$ $(i = 1, \ldots, n)$.*

We return to the general situation as described at the beginning of this section. Given a finite-dimensional vector space $\{0\} \neq V \subseteq F$, there exists a unique effective divisor A of smallest degree such that $(f)_\infty \leq A$ for all $0 \neq f \in V$. One can describe A as follows: choose a basis $\{f_1, \ldots, f_k\}$ of V, then

$$v_P(A) = \max\{v_P((f_i)_\infty) \mid 1 \leq i \leq k\} \qquad (9.26)$$

for all $P \in \mathbb{P}_F$. For instance, if $V = \mathscr{L}(G)$ and $G = G_+ - G_-$ where $G_+ \geq 0$ and $G_- \geq 0$, then $A \leq G_+$. We associate with the divisor A a second divisor A^0 defined by

$$A^0 := \sum_{P \in \operatorname{supp} A} P. \qquad (9.27)$$

To avoid complications, we restrict ourselves to the case where $q = p$ is a prime number and F is the rational function field over $\mathbb{F}_{p^m}$.

Theorem 9.2.5. *Let $F = \mathbb{F}_{p^m}(z)$ be the rational function field over $\mathbb{F}_{p^m}$ (where p is a prime number), $D = P_1 + \ldots + P_n$ with pairwise distinct places $P_i \in \mathbb{P}_F$ of degree one, and $s := p^m + 1 - n$. Let $V \subseteq F$ be a finite-dimensional $\mathbb{F}_{p^m}$-subspace of F such that $v_{P_i}(f) \geq 0$ for all $f \in V$ and $i = 1, \ldots, n$. Then the weight w of a codeword in $\mathrm{Tr}_D(V)$ satisfies $w = 0$, $w = n$ or*

$$\left| w - \frac{p-1}{p} n \right| \leq \frac{p-1}{2p}(-2 + \deg A + \deg A^0) \cdot [2p^{m/2}] + \frac{p-1}{p} s,$$

where the divisors A and A^0 are defined by (9.26) *and* (9.27).

The proof of this theorem requires some preparation. Let us first introduce some notation. For $f \in V$ let

$$\mathrm{Tr}_D(f) := \big(\mathrm{Tr}(f(P_1)), \ldots, \mathrm{Tr}(f(P_n))\big) \in \mathbb{F}_p^n,$$

so that $\mathrm{Tr}_D(V) = \{\mathrm{Tr}_D(f) \mid f \in V\}$.

Definition 9.2.6. *An element $f \in F$ is said to be degenerate if f can be written as $f = \gamma + (h^p - h)$, with $\gamma \in \mathbb{F}_{p^m}$ and $h \in F$. Otherwise f is called non-degenerate.*

Lemma 9.2.7. *Suppose $f \in V$ is degenerate. Then*

$$\mathrm{Tr}_D(f) = (\alpha, \alpha, \ldots, \alpha) \quad \textit{with } \alpha \in \mathbb{F}_p.$$

Hence the weight of $\mathrm{Tr}_D(f)$ is 0 or n.

Proof. Write $f = \gamma + (h^p - h)$ with $\gamma \in \mathbb{F}_{p^m}$ and $h \in F$. Since $v_{P_i}(f) \geq 0$, the Triangle Inequality yields $v_{P_i}(h) \geq 0$ for $1 \leq i \leq n$. Setting $\gamma_i := h(P_i) \in \mathbb{F}_{p^m}$ we obtain

$$\mathrm{Tr}(f(P_i)) = \mathrm{Tr}(\gamma) + \mathrm{Tr}(\gamma_i^p - \gamma_i) = \mathrm{Tr}(\gamma),$$

independent of i (note that Tr is the trace mapping to $\mathbb{F}_p$, so $\mathrm{Tr}(\gamma_i^p) = \mathrm{Tr}(\gamma_i)$). □

The non-degenerate case is clearly more interesting. In the following proposition we retain notation of Theorem 9.2.5.

Proposition 9.2.8. *Suppose $f \in V$ is non-degenerate. Then the polynomial $\varphi(Y) := Y^p - Y - f \in F[Y]$ is irreducible over F. Let $E_f := F(y)$ where y is a root of $\varphi(Y)$; i.e.,*

$$E_f = F(y) \quad \textit{with} \quad y^p - y = f\,.$$

The field extension E_f/F is cyclic of degree p, and $\mathbb{F}_{p^m}$ is the full constant field of E_f. Let $S := \{P \in \mathbb{P}_F \mid \deg P = 1$ and $P \notin \operatorname{supp} D\}$, so $|S| = s = p^m + 1 - n$. Let $\bar{S} := \{Q \in \mathbb{P}_{E_f} \mid \deg Q = 1$ and $Q \cap F \in S\}$, and denote by $\bar{s}$ the cardinality of $\bar{S}$. Then

$$0 \le \bar{s} \le ps\,, \tag{9.28}$$

and the weight $w_f := w(\mathrm{Tr}_D(f))$ is given by the formula

$$w_f = n - \frac{N(E_f) - \bar{s}}{p}\,. \tag{9.29}$$

(As usual, $N(E_f)$ denotes the number of all places of degree one of $E_f/\mathbb{F}_{p^m}$).

Proof. We use some facts about Artin-Schreier extensions, cf. Appendix A. The polynomial $\varphi(Y) = Y^p - Y - f$ is either irreducible in $F[Y]$ or it has a root in F; i.e., $f = h^p - h$ with $h \in F$. Since f is assumed to be non-degenerate, the irreducibility of $\varphi(Y)$ follows, and the field $E_f = F(y)$ defined by $y^p - y = f$ is a cyclic extension of F of degree p.

Suppose that the constant field L of E_f is strictly larger than $\mathbb{F}_{p^m}$, the constant field of F. Then $L/\mathbb{F}_{p^m}$ is cyclic of degree p, hence $L = \mathbb{F}_{p^m}(\delta)$ with $\delta^p - \delta = \epsilon \in \mathbb{F}_{p^m}$ for some $\delta \in L$. As $\delta \notin F$, it is an Artin-Schreier generator of E_f/F as well, so $f = \lambda\epsilon + (h^p - h)$ with $0 \ne \lambda \in \mathbb{F}_p$ and $h \in F$ (see Appendix A). This is a contradiction as f is non-degenerate, and proves that $\mathbb{F}_{p^m}$ is the full constant field of E_f.

The support of D consists of two disjoint subsets $\{P_1, \ldots, P_n\} = N \cup Z$ where

$$N := \{P_i \in \operatorname{supp} D \mid \mathrm{Tr}(f(P_i)) \ne 0\}$$

and

$$Z := \{P_i \in \operatorname{supp} D \mid \mathrm{Tr}(f(P_i)) = 0\}\,.$$

We determine the decomposition behavior of the places $P_i \in N$ (resp. Z) in the extension E_f/E.

Hilbert's Theorem 90 (Appendix A) states that for $\gamma \in \mathbb{F}_{p^m}$,

$$\mathrm{Tr}(\gamma) = 0 \Longleftrightarrow \gamma = \beta^p - \beta \quad \textit{for some } \beta \in \mathbb{F}_{p^m}\,. \tag{9.30}$$

Let $P_i \in N$ and $\gamma_i := f(P_i)$. The Artin-Schreier polynomial $Y^p - Y - \gamma_i$ has no root in $\mathbb{F}_{p^m}$ by (9.30), hence it is irreducible over $\mathbb{F}_{p^m}$. Now Kummer's

Theorem shows that P_i has a unique extension $Q \in \mathbb{P}_{E_f}$, with relative degree $f(Q|P_i) = p$. Consequently there are no places of E_f of degree one lying over a place $P_i \in N$.

Next we consider a place $P_i \in Z$. By (9.30), $f(P_i)$ can be written as $f(P_i) =: \gamma_i = \beta_i^p - \beta_i$ with $\beta_i \in \mathbb{F}_{p^m}$, hence the polynomial $Y^p - Y - \gamma_i$ factors over $\mathbb{F}_{p^m}$ into p distinct linear factors. In this case Kummer's Theorem implies that P_i decomposes into p distinct places of E_f, all of degree one.

The above considerations imply that

$$N(E_f) = |\bar{S}| + p \cdot |Z| = \bar{s} + p(n - |N|) . \tag{9.31}$$

Since $w_f = w(\mathrm{Tr}_D(f)) = |N|$, (9.29) is an immediate consequence of (9.31). The estimate (9.28) is trivial. □

The next aim is to determine the genera $g(E_f)$ of the function fields E_f (where $f \in V$ is non-degenerate). Using Proposition 3.7.8 one can sometimes calculate $g(E_f)$ precisely. We shall be satisfied with an upper bound on $g(E_f)$.

Lemma 9.2.9. *Suppose $f \in V$ is non-degenerate. Then*

$$g(E_f) \le \frac{p-1}{2}(-2 + \deg A + \deg A^0) ,$$

where A and A^0 are defined by (9.26) *and* (9.27).

Proof. This lemma is an easy application of Proposition 3.7.8(d). Observe that all places $P \notin \mathrm{supp}\, A$ are unramified in E_f/F, and for $P \in \mathrm{supp}\, A$ the integer m_P (as defined in Proposition 3.7.8) is obviously bounded by $v_P(A)$. □

Proof of Theorem 9.2.5. By Lemma 9.2.7 we can assume that $w = w_f$ is the weight of a codeword $\mathrm{Tr}_D(f)$ where $f \in V$ is non-degenerate. We use the notation of Proposition 9.2.8. Equation (9.29) yields

$$N(E_f) = p(n - w_f) + \bar{s} .$$

We subtract $p^m + 1$, apply the Serre Bound (Theorem 5.3.1) and obtain

$$\left| p(n - w_f) + \bar{s} - (p^m + 1) \right| \le g(E_f) \cdot [2p^{m/2}] . \tag{9.32}$$

Since $p^m + 1 = s + n$, we have

$$p(n - w_f) + \bar{s} - (p^m + 1) = (p-1)n - pw_f + (\bar{s} - s) .$$

We substitute this into (9.32), divide by p and estimate $g(E_f)$ by Lemma 9.2.9. The result is

$$\left|w_f - \frac{p-1}{p}n - \frac{\bar{s}-s}{p}\right| \leq \frac{p-1}{2p}(-2+\deg A + \deg A^0) \cdot [2p^{m/2}]. \qquad (9.33)$$

Finally (9.28) yields that $-s \leq \bar{s} - s \leq (p-1)s$, and therefore

$$\left|\frac{\bar{s}-s}{p}\right| \leq \frac{p-1}{p} \cdot s.$$

This finishes the proof of Theorem 9.2.5. □

Remark 9.2.10. (a) In some specific cases the number $\bar{s}$ can be determined more precisely. In such cases, (9.33) provides a better estimate than does Theorem 9.2.5, cf. Example 9.2.12 below.

(b) Clearly Theorem 9.2.5 gives non-trivial bounds for the weights of codewords in $\mathrm{Tr}_D(V)$ only if the length of the code is large in comparison with m and the degree of A. It turns out that – under this restriction – the estimates given in Theorem 9.2.5 are often fairly good.

Corollary 9.2.11. *Notation as in Theorem* 9.2.5. *If* $V \neq \{0\}$ *and* $V \neq \mathbb{F}_{p^m}$, *the minimum distance* d *of* $\mathrm{Tr}_D(V)$ *is bounded from below by*

$$d \geq \frac{p-1}{p}n - \frac{s}{p} - \frac{p-1}{2p}(-2+\deg A + \deg A^0) \cdot [2p^{m/2}]. \qquad (9.34)$$

Proof. The assumption $V \neq \{0\}$ and $V \neq \mathbb{F}_{p^m}$ implies that $\deg A > 0$, hence the right-hand side of (9.34) is $\leq n$. Therefore it is sufficient to estimate the weight $w_f = w(\mathrm{Tr}_D(f))$ for a non-degenerate element $f \in V$. By (9.33),

$$w_f \geq \frac{p-1}{p}n + \frac{\bar{s}-s}{p} - \frac{p-1}{2p}(-2+\deg A + \deg A^0) \cdot [2p^{m/2}].$$

Since $\bar{s} \geq 0$, the corollary follows. □

Example 9.2.12. We consider the dual $C^\perp$ of the BCH code C over $\mathbb{F}_p$ of length $n = p^m - 1$ and designed distance $\delta = 2t+1 > 1$. Thus

$$C = \left\{(c_0, c_1, \ldots, c_{n-1}) \in \mathbb{F}_p^n \;\middle|\; \sum_{i=0}^{n-1} c_i\beta^{i\lambda} = 0 \quad \textit{for} \quad \lambda = 1, \ldots, \delta - 1\right\},$$

where $\beta \in \mathbb{F}_{p^m}$ is a primitive (p^m-1)-th root of unity, cf. Definition 2.3.8. Then the weight w of a codeword in the dual code $C^\perp$ satisfies $w = 0, w = n$ or

$$\left|w - p^m(1-\frac{1}{p})\right| \leq \frac{(p-1)(2t-1)}{2p} \cdot [2p^{m/2}]. \qquad (9.35)$$

In the case $p = 2$ this can be improved to

$$\left|w - 2^{m-1}\right| \le \frac{t-1}{2} \cdot \left[2^{\frac{m}{2}+1}\right]. \tag{9.36}$$

This is the so-called *Carlitz-Uchiyama Bound*, cf. [28]. Note that the bounds (9.35) and (9.36) are not covered by Theorem 9.2.5.

We show now how (9.35) and (9.36) follow from our previous results. Let $F = \mathbb{F}_{p^m}(z)$. For $1 \le i \le n$ let $P_i \in \mathbb{P}_F$ be the zero of $z - \beta^{i-1}$. Set $D = P_1 + \ldots + P_n$ and $(z) = P_0 - P_\infty$. By Proposition 2.3.9 and Delsarte's Theorem,

$$C^\perp = \left(C_{\mathscr{L}}(D, aP_0 + bP_\infty)^\perp\big|_{\mathbb{F}_p}\right)^\perp = \mathrm{Tr}_D(\mathscr{L}(aP_0 + bP_\infty)),$$

where $a = -1$ and $\delta = a + b + 2 = b + 1$. Since $\delta = 2t + 1$ we find that

$$C^\perp = \mathrm{Tr}_D(\mathscr{L}(-P_0 + 2tP_\infty)).$$

Let $f \in \mathscr{L}(-P_0 + 2tP_\infty)$ be non-degenerate. Consider the corresponding field extension $E_f = F(y)$, defined by $y^p - y = f$. Since $f(P_0) = 0$, the place P_0 has p extensions of degree one in E_f. The pole P_∞ of z ramifies in E_f/F. With notation as in Proposition 9.2.8 we obtain $\bar{s} = p + 1$. Hence

$$\left|w_f - \left(1 - \frac{1}{p}\right) \cdot p^m\right| \le \frac{g(E_f)}{p} \cdot [2p^{m/2}],$$

by (9.32). Since $f \in \mathscr{L}(-P_0 + 2tP_\infty)$, Proposition 3.7.8 yields

$$g(E_f) \le \frac{p-1}{2} \cdot (-2 + (2t+1)) = \frac{(p-1)(2t-1)}{2}.$$

This proves (9.35). In case $p = 2$ the genus is $\le (2t-2)/2$ because the integer m_{P_∞} in Proposition 3.7.8 is relatively prime to the characteristic and therefore $m_{P_\infty} \le 2t - 1$ for $p = 2$, which gives (9.36).

The method of proof of Theorem 9.2.5 can also be used to calculate precisely the dimension of certain trace codes and of their duals. We illustrate this idea by an example which also shows that the estimate for the dimension of trace codes given in Theorem 9.1.6 is often tight.

Proposition 9.2.13. *Consider a Goppa code $\Gamma(L, g(z))$ over $\mathbb{F}_p$, where $L = \mathbb{F}_{p^m}$ and $g(z) \in \mathbb{F}_{p^m}[z]$ is a monic polynomial without zeros in $\mathbb{F}_{p^m}$ (see Definition 2.3.10). Write*

$$g(z) = \prod_{j=1}^{l} h_j(z)^{a_j}$$

with pairwise distinct irreducible monic polynomials $h_j(z) \in \mathbb{F}_{p^m}[z]$ and exponents $a_j > 0$. Let $a_j = pb_j + c_j$ with $0 \le c_j \le p-1$. Set

$$g_1(z) := \prod_{j=1}^{l} h_j(z)^{b_j} \quad \text{and} \quad g_2(z) := \prod_{j=1}^{l} h_j(z)\,.$$

Suppose that

$$2(p^m + 1) > (-2 + \deg g(z) + \deg g_2(z)) \cdot [2p^{m/2}] \tag{9.37}$$

holds. Then the dimension of $\Gamma(L, g(z))$ *is*

$$\dim \Gamma(L, g(z)) = p^m - m(\deg g(z) - \deg g_1(z))\,. \tag{9.38}$$

Proof. By Proposition 2.3.11 and Delsarte's Theorem we can represent the dual code $\Gamma(L, g(z))^\perp$ as follows:

$$\Gamma(L, g(z))^\perp = \left(C_{\mathscr{L}}(D, G_0 - P_\infty)^\perp\big|_{\mathbb{F}_p}\right)^\perp = \mathrm{Tr}_D(\mathscr{L}(G_0 - P_\infty)) \tag{9.39}$$

with the usual notation: P_∞ is the pole divisor of z in the rational function field $F = \mathbb{F}_{p^m}(z)$, the divisor G_0 is the zero divisor of the Goppa polynomial $g(z)$, and D is the sum of all places of $F/\mathbb{F}_{p^m}$ of degree one except P_∞ (note that $L = \mathbb{F}_{p^m}$). The $\mathbb{F}_p$-linear map $\mathrm{Tr}_D : \mathscr{L}(G_0 - P_\infty) \to \mathrm{Tr}_D(\mathscr{L}(G_0 - P_\infty))$ is surjective, and we wish to determine its kernel.

Claim. If $f \in \mathscr{L}(G_0 - P_\infty)$ and $\mathrm{Tr}_D(f) = 0$, then f is degenerate.

Proof of the Claim. Suppose that $f \in \mathscr{L}(G_0 - P_\infty)$ is non-degenerate and $\mathrm{Tr}_D(f) = 0$. We consider the corresponding field extension E_f/F of degree p (see Proposition 9.2.8). As $f(P_\infty) = 0$, the place P_∞ decomposes into p distinct places of E_f of degree one (by Kummer's Theorem). So (9.29) yields $0 = w_f = p^m - p^{-1}(N(E_f) - p)$; i.e.,

$$N(E_f) = p(p^m + 1)\,. \tag{9.40}$$

The genus $g(E_f)$ is bounded by

$$g(E_f) \le \frac{p-1}{2}(-2 + \deg g(z) + \deg g_2(z))\,, \tag{9.41}$$

by Proposition 3.7.8(d). Combining (9.40), (9.41) and the Serre Bound we obtain

$$p(p^m + 1) \le p^m + 1 + \frac{p-1}{2}(-2 + \deg g(z) + \deg g_2(z)) \cdot [2p^{m/2}]\,,$$

which contradicts (9.37). This proves the claim.

Let $G_1 := (g_1(z))_0$ denote the zero divisor of the polynomial $g_1(z)$. We have an $\mathbb{F}_p$-linear map

$$\phi : \begin{cases} \mathscr{L}(G_1 - P_\infty) & \longrightarrow \ \mathrm{Ker}(\mathrm{Tr}_D) \subseteq \mathscr{L}(G_0 - P_\infty)\,, \\ h & \longmapsto \ h^p - h\,. \end{cases}$$

Since $h^p - h = 0$ if and only if $h \in \mathbb{F}_p$, the kernel of ϕ is $\mathbb{F}_p \cap \mathscr{L}(G_1 - P_\infty) = \{0\}$. Thus ϕ is injective.

ϕ is also surjective. In order to see this, let $f \in \mathrm{Ker}(\mathrm{Tr}_D)$. By the claim, $f = h_1^p - h_1 + \gamma$ with $h_1 \in F, \gamma \in \mathbb{F}_{p^m}$. From $\mathrm{Tr}_D(f) = 0$ follows that $\mathrm{Tr}(\gamma) = 0$. This implies by Hilbert's Theorem 90 that $\gamma = \alpha^p - \alpha$ with $\alpha \in \mathbb{F}_{p^m}$. Hence $f = h^p - h$ with $h := h_1 + \alpha \in F$. The Triangle Inequality shows that $h \in \mathscr{L}(G_1)$ (because $f = h^p - h \in \mathscr{L}(G_0)$). Moreover it follows from

$$\prod_{\mu \in \mathbb{F}_p} (h - \mu) = h^p - h = f \in \mathscr{L}(G_0 - P_\infty),$$

that P_∞ is a zero of one of the factors $h - \mu$. This factor $h - \mu$ lies in the space $\mathscr{L}(G_1 - P_\infty)$, hence $f = (h-\mu)^p - (h - \mu) = \phi(h - \mu)$ is in the image of ϕ.

In what follows, we denote by $\dim V$ the dimension of a vector space V over $\mathbb{F}_p$. We conclude that

$$\dim \mathrm{Ker}(\mathrm{Tr}_D) = \dim \mathscr{L}(G_1 - P_\infty) = m \cdot \deg g_1(z),$$

hence

$$\begin{aligned} \dim\ & \Gamma(L, g(z))^\perp = \dim \mathrm{Tr}_D(\mathscr{L}(G_0 - P_\infty)) \\ &= \dim \mathscr{L}(G_0 - P_\infty) - \dim \mathrm{Ker}(\mathrm{Tr}_D) = m(\deg g(z) - \deg g_1(z)). \end{aligned}$$

The dimension of the dual code is therefore

$$\dim \Gamma(L, g(z)) = p^m - m(\deg g(z) - \deg g_1(z)).$$

□

9.3 Exercises

9.1. Find a larger class of trace codes for which the method of Proposition 9.2.13 applies.

9.2. Consider the extension $\mathbb{F}_{q^m}/\mathbb{F}_q$ of degree $m \geq 1$ and an arbitrary $\mathbb{F}_q$-linear map $\lambda : \mathbb{F}_{q^m} \to \mathbb{F}_q$ which is not identically zero. Extend this to a map $\lambda : (\mathbb{F}_{q^m})^n \to \mathbb{F}_q^n$ by setting $\lambda(c_1, \ldots, c_n) = (\lambda(c_1), \ldots, \lambda(c_n))$. For a code C of length n over $\mathbb{F}_{q^m}$ define

$$\lambda(C) := \{\lambda(c) \,|\, c \in C\},$$

which is obviously a code over $\mathbb{F}_q$ of length n. Show that $\lambda(C) = \mathrm{Tr}\,(C)$.

9.3. Let C be a code of length n over $\mathbb{F}_{q^m}$. Assume there is an $r \times n$ matrix M of rank s with coefficients in $\mathbb{F}_q$ such that $M \cdot c^t = 0$ for all $c \in C$. Show that

$$\dim C|_{\mathbb{F}_q} \geq \dim C - (m-1)(n - s - \dim C).$$

Give examples (with $m > 1$) of codes where the above estimate is sharp.

9.4. Let $n = p^m - 1$ where p is prime and $m > 1$. Consider a cyclic code C over $\mathbb{F}_p$ of length n such that $(1, 1, \ldots, 1) \notin C$.

(i) Show that there exists a complete set of coset representatives for $C^\perp$ of the form $J_0 = \{\rho_1, \ldots, \rho_s\} \subseteq \{1, \ldots, n-1\}$ such that $(p, \rho_i) = 1$ for $i = 1, \ldots, s$.

(ii) Show that the weights w of nonzero codewords $c \in C$ satisfy

$$\left| w - p^m(1 - p^{-1}) \right| \leq \frac{(p-1)(\rho-1)}{2p} \cdot \lfloor 2p^{m/2} \rfloor$$

with $\rho := \max\{\rho_1, \ldots, \rho_s\}$.

Appendix A. Field Theory

We put together some facts from field theory that we frequently call upon. Proofs can be found in all standard textbooks on algebra, e.g. [23]

A.1 Algebraic Field Extensions

Let L be a field that contains K as a subfield. Then L/K is called a *field extension*. Considering L as a vector space over K, its dimension is called the *degree* of L/K and denoted by $[L : K]$.

L/K is said to be a *finite extension* if $[L : K] = n < \infty$. Then there exists a *basis* $\{\alpha_1, \ldots, \alpha_n\}$ of L/K; i.e., every $\gamma \in L$ has a unique representation $\gamma = \sum_{i=1}^{n} c_i \alpha_i$ with $c_i \in K$. If L/K and M/L are finite extensions, then M/K is finite as well, and the degree is $[M : K] = [M : L] \cdot [L : K]$.

An element $\alpha \in L$ is *algebraic over* K if there is a non-zero polynomial $f(X) \in K[X]$ (the polynomial ring over K) such that $f(\alpha) = 0$. Among all such polynomials there is a unique polynomial of smallest degree that is monic (i.e., its leading coefficient is 1); this is called the *minimal polynomial* of α over K. The minimal polynomial is irreducible in the ring $K[X]$, hence it is often called the *irreducible polynomial* of α over K.

The field extension L/K is called an *algebraic extension* if all elements $\alpha \in L$ are algebraic over K.

Let $\gamma_1, \ldots, \gamma_r \in L$. The smallest subfield of L that contains K and all elements $\gamma_1, \ldots, \gamma_r$ is denoted by $K(\gamma_1, \ldots, \gamma_r)$. The extension $K(\gamma_1, \ldots, \gamma_r)/K$ is finite if and only if all γ_i are algebraic over K.

In particular, $\alpha \in L$ is algebraic over K if and only if $[K(\alpha) : K] < \infty$. Let $p(X) \in K[X]$ be the minimal polynomial of α over K and $r = \deg p(X)$. Then $[K(\alpha) : K] = r$, and the elements $1, \alpha, \alpha^2, \ldots, \alpha^{r-1}$ form a basis of $K(\alpha)/K$.

A.2 Embeddings and K-Isomorphisms

Consider field extensions L_1/K and L_2/K. A field homomorphism $\sigma : L_1 \to L_2$ is called an *embedding* of L_1 into L_2 over K, if $\sigma(a) = a$ for all $a \in K$. It follows that σ is injective and yields an isomorphism of L_1 onto the subfield $\sigma(L_1) \subseteq L_2$. A surjective (hence bijective) embedding of L_1 into L_2 over K is a *K-isomorphism.*

A.3 Adjoining Roots of Polynomials

Given a field K and a non-constant polynomial $f(X) \in K[X]$, there exists an algebraic extension field $L = K(\alpha)$ with $f(\alpha) = 0$. If $f(X)$ is irreducible, this extension field is unique up to K-isomorphism. This means: if $L' = K(\alpha')$ is another extension field with $f(\alpha') = 0$, then there exists a K-isomorphism $\sigma : L \to L'$ with $\sigma(\alpha) = \alpha'$. We say that $L = K(\alpha)$ is obtained by *adjoining a root* of $p(X)$ to K.

If $f_1(X), \ldots, f_r(X) \in K[X]$ are monic polynomials of degree $d_i \geq 1$, there exists an extension field $Z \supseteq K$ such that all $f_i(X)$ split into linear factors $f_i(X) = \prod_{j=1}^{d_i}(X - \alpha_{ij})$ with $\alpha_{ij} \in Z$, and $Z = K(\{\alpha_{ij} \mid 1 \leq i \leq r \textit{ and } 1 \leq j \leq d_i\})$. The field Z is unique up to K-isomorphism; it is called the *splitting field* of $f_1, \ldots, f_r$ over K.

A.4 Algebraic Closure

A field M is called *algebraically closed* if every polynomial $f(X) \in M[X]$ of degree ≥ 1 has a root in M.

For every field K there exists an algebraic extension $\bar{K}/K$ with an algebraically closed field $\bar{K}$. The field $\bar{K}$ is unique up to K-isomorphism; it is called the *algebraic closure* of K.

Given an algebraic field extension L/K, there exists an embedding $\sigma : L \to \bar{K}$ over K. If $[L : K] < \infty$, the number of distinct embeddings of L to $\bar{K}$ over K is at most $[L : K]$.

A.5 The Characteristic of a Field

Let K be a field and let $1 \in K$ be the neutral element with respect to multiplication. For each integer $m > 0$, let $\bar{m} = 1 + 1 + \ldots + 1 \in K$ (m summands). If $\bar{m} \neq 0$ (the zero element of K) for all $m > 0$, we say that K has characteristic zero. Otherwise there exists a unique prime number $p \in \mathbb{N}$ such that $\bar{p} = 0$, and K is said to have characteristic p. We use the abbreviation $\operatorname{char} K$. It is convenient to identify an integer $m \in \mathbb{Z}$ with the element $\bar{m} \in K$; i.e., we simply write $m = \bar{m} \in K$.

If $\operatorname{char} K = 0$, then K contains the field $\mathbb{Q}$ of rational numbers (up to isomorphism). In case $\operatorname{char} K = p > 0$, K contains the field $\mathbb{F}_p = \mathbb{Z}/p\mathbb{Z}$.

In a field of characteristic $p > 0$ we have $(a+b)^q = a^q + b^q$ for all $a, b \in K$ and $q = p^j$, $j \geq 0$.

A.6 Separable Polynomials

Let $f(X) \in K[X]$ be a monic polynomial of degree $d \geq 1$. Over some extension field $L \supseteq K$, $f(X)$ splits into linear factors $f(X) = \prod_{i=1}^{d}(X - \alpha_i)$. The polynomial $f(X)$ is called *separable* if $\alpha_i \neq \alpha_j$ for all $i \neq j$; otherwise, f is an *inseparable* polynomial.

If $\operatorname{char} K = 0$, all irreducible polynomials are separable. In case $\operatorname{char} K = p > 0$, an irreducible polynomial $f(X) = \sum a_i X^i \in K[X]$ is separable if and only if $a_i \neq 0$ for some $i \not\equiv 0 \bmod p$.

The *derivative* of $f(X) = \sum a_i X^i \in K[X]$ is defined in the usual manner by $f'(X) = \sum i a_i X^{i-1}$ (where $i \in \mathbb{N}$ is considered as an element of K as in A.5). An irreducible polynomial $f(X) \in K[X]$ is separable if and only if $f'(X) \neq 0$.

A.7 Separable Field Extensions

Let L/K be an algebraic field extension. An element $\alpha \in L$ is called *separable* over K if its minimal polynomial $p(X) \in K[X]$ is a separable polynomial. L/K is a *separable extension* if all $\alpha \in L$ are separable over K. If $\operatorname{char} K = 0$, then all algebraic extensions L/K are separable.

Let Φ be an algebraically closed field, $\Phi \supseteq K$, and suppose that L/K is a finite extension of degree $[L : K] = n$. Then L/K is separable if and only if there exist n distinct embeddings $\sigma_1, \ldots, \sigma_n : L \to \Phi$ over K (cf. A.4). In this case an element $\gamma \in L$ is in K if and only if $\sigma_i(\gamma) = \gamma$ for $i = 1, \ldots, n$.

Given a tower $M \supseteq L \supseteq K$ of algebraic field extensions, the extension M/K is separable if and only if both extensions M/L and L/K are separable.

A.8 Purely Inseparable Extensions

Consider an algebraic extension L/K where $\operatorname{char} K = p > 0$. An element $\gamma \in L$ is called *purely inseparable* over K if $\gamma^{p^r} \in K$ for some $r \geq 0$. In this case the minimal polynomial of γ over K has the form $f(X) = X^{p^e} - c$ with

$c \in K$ (and $e \leq r$). The extension L/K is *purely inseparable* if all elements $\gamma \in L$ are purely inseparable over K.

Given an arbitrary algebraic extension L/K, there exists a unique intermediate field S, $K \subseteq S \subseteq L$, such that S/K is separable and L/S is purely inseparable.

A.9 Perfect Fields

A field K is called *perfect* if all algebraic extensions L/K are separable. Fields of characteristic 0 are always perfect. A field K of characteristic $p > 0$ is perfect if and only if every $\alpha \in K$ can be written as $\alpha = \beta^p$, for some $\beta \in K$. All finite fields are perfect (cf. A.15).

A.10 Simple Algebraic Extensions

An algebraic extension L/K is called *simple* if $L = K(\alpha)$ for some $\alpha \in L$. The element α is called a *primitive element* for L/K. Every finite separable algebraic field extension is simple.

Suppose that $L = K(\alpha_1, \ldots, \alpha_r)$ is a finite separable extension and $K_0 \subseteq K$ is an infinite subset of K. Then there exists a primitive element α of the form $\alpha = \sum_{i=1}^{r} c_i \alpha_i$ with $c_i \in K_0$.

A.11 Galois Extensions

For a field extension L/K we denote the group of automorphisms of L over K by $\mathrm{Aut}(L/K)$. That is, an element $\sigma \in \mathrm{Aut}(L/K)$ is a K-isomorphism of L onto L. If $[L : K] < \infty$, the order of $\mathrm{Aut}(L/K)$ is always $\leq [L : K]$. The extension L/K is said to be *Galois* if the order of $\mathrm{Aut}(L/K)$ is $[L : K]$. In this case we call $\mathrm{Gal}(L/K) := \mathrm{Aut}(L/K)$ the *Galois group* of L/K. The following conditions are equivalent, for a field extension L/K of finite degree:

(1) L/K is Galois.

(2) L is the splitting field of separable polynomials $f_1(X), \ldots, f_r(X) \in K[X]$ over K.

(3) L/K is separable, and every irreducible polynomial $p(X) \in K[X]$ that has a root in L, splits into linear factors in $L[X]$.

Given a finite separable extension L/K and an algebraically closed field $\Phi \supseteq L$, there exists a unique field M, $L \subseteq M \subseteq \Phi$, with the following properties:

(a) M/K is Galois, and

(b) if $L \subseteq N \subseteq \Phi$ and N/K is Galois, then $M \subseteq N$.

This field M is called the *Galois closure* of L/K. Another characterization of M is that it is the compositum of the fields $\sigma(L)$ where σ runs over all embeddings of L into Φ over K.

A.12 Galois Theory

We consider a Galois extension L/K with Galois group $G = \mathrm{Gal}(L/K)$. Let

$$\mathcal{U} := \{U \subseteq G \,|\, U \text{ is a subgroup of } G\}$$

and

$$\mathcal{F} := \{N \,|\, N \text{ is an intermediate field of } L/K\}.$$

For each intermediate field N of L/K the extension L/N is Galois, thus we have a mapping

$$\begin{aligned} \mathcal{F} &\longrightarrow \mathcal{U}, \\ N &\longmapsto \mathrm{Gal}(L/N). \end{aligned} \tag{$*$}$$

On the other hand, given a subgroup $U \subseteq G$ we define the *fixed field* of U by

$$L^U := \{c \in L \,|\, \sigma(c) = c \text{ for all } \sigma \in U\}.$$

In this manner we obtain a mapping

$$\begin{aligned} \mathcal{U} &\longrightarrow \mathcal{F}, \\ U &\longmapsto L^U. \end{aligned} \tag{$**$}$$

Now we can formulate the main results of Galois theory:

(1) The mappings $(*)$ and $(**)$ are inverse to each other. They yield a 1–1 correspondence between $\mathcal{U}$ and $\mathcal{F}$ (*Galois correspondence*).

(2) For $U \in \mathcal{U}$ we have

$$[L : L^U] = \mathrm{ord}\, U \quad \text{and} \quad [L^U : K] = (G : U).$$

(3) If $U \subseteq G$ is a subgroup, then $U = \mathrm{Gal}(L/L^U)$.

(4) If N is an intermediate field of L/K, then $N = L^U$ with $U = \mathrm{Gal}(L/N)$.

(5) For subgroups $U, V \subseteq G$ we have $\quad U \subseteq V \iff L^U \supseteq L^V$.

(6) Let N_1 and N_2 be intermediate fields of L/K and $N = N_1N_2$ be the compositum of N_1 and N_2. Then $\mathrm{Gal}(L/N) = \mathrm{Gal}(L/N_1) \cap \mathrm{Gal}(L/N_2)$.

(7) If N_1 and N_2 are intermediate fields of L/K, then the Galois group of $L/(N_1 \cap N_2)$ is the subgroup of G that is generated by $\mathrm{Gal}(L/N_1)$ and $\mathrm{Gal}(L/N_2)$.

(8) A subgroup $U \subseteq G$ is a normal subgroup of G if and only if the extension L^U/K is Galois. In this case, $\mathrm{Gal}(L^U/K)$ is isomorphic to the factor group G/U.

A.13 Cyclic Extensions

A Galois extension L/K is said to be *cyclic* if $\mathrm{Gal}(L/K)$ is a cyclic group. Two special cases are of particular interest: Kummer extensions and Artin-Schreier extensions.

(1) (*Kummer Extensions.*) Suppose that L/K is a cyclic extension of degree $[L:K]=n$, where n is relatively prime to the characteristic of K, and K contains all n-th roots of unity (i.e., the polynomial X^n-1 splits into linear factors in $K[X]$). Then there exists an element $\gamma \in L$ such that $L=K(\gamma)$ with

$$\gamma^n = c \in K\,, \quad \textit{and } c \neq w^d \textit{ for all } w \in K \textit{ and } d \mid n,\, d>1\,. \qquad (\circ)$$

Such a field extension is called a *Kummer Extension*. The automorphisms $\sigma \in \mathrm{Gal}(L/K)$ are given by $\sigma(\gamma)=\zeta\cdot\gamma$, where $\zeta \in K$ is an n-th root of unity.

Conversely, if K contains all n-th roots of unity (n relatively prime to $\mathrm{char}\,K$), and $L=K(\gamma)$ where γ satisfies the conditions $(\circ)$, then L/K is a cyclic extension of degree n.

(2) (*Artin-Schreier Extensions.*) Let L/K be a cyclic extension of degree $[L:K]=p=\mathrm{char}\,K$. Then there exists an element $\gamma \in L$ such that $L=K(\gamma)$,

$$\gamma^p - \gamma = c \in K\,, \quad \textit{and } c \neq \alpha^p - \alpha \textit{ for all } \alpha \in K\,. \qquad (\circ\circ)$$

Such a field extension is called an *Artin-Schreier Extension* of degree p. The automorphisms of L/K are given by $\sigma(\gamma)=\gamma+\nu$ with $\nu \in \mathbb{Z}/p\mathbb{Z} \subseteq K$.

An element $\gamma_1 \in L$ such that $L=K(\gamma_1)$ and $\gamma_1^p-\gamma_1 \in K$, is called an *Artin-Schreier generator* for L/K. Any two Artin-Schreier generators γ and γ_1 of L/K are related as follows: $\gamma_1=\mu\cdot\gamma+(b^p-b)$ with $0\neq\mu\in\mathbb{Z}/p\mathbb{Z}$ and $b\in K$.

Conversely, if we have a field extension $L=K(\gamma)$ where γ satisfies $(\circ\circ)$ (and $\mathrm{char}\,K=p$), then L/K is cyclic of degree p.

A.14 Norm and Trace

Let L/K be a field extension of degree $[L:K]=n<\infty$. Each element $\alpha \in L$ yields a K-linear map $\mu_\alpha : L \to L$, defined by $\mu_\alpha(z) := \alpha\cdot z$ for $z \in L$. We define the *norm* (resp. the *trace*) of α with respect to the extension L/K by

$$\mathrm{N}_{L/K}(\alpha) := \det(\mu_\alpha) \quad \textit{resp.} \quad \mathrm{Tr}_{L/K}(\alpha) := \mathrm{Trace}(\mu_\alpha)\,.$$

This means: if $\{\alpha_1,\ldots,\alpha_n\}$ is a basis of L/K and

$$\alpha\cdot\alpha_i = \sum_{j=1}^{n} a_{ij}\alpha_j \quad \textit{with} \quad a_{ij} \in K\,,$$

then

$$\mathrm{N}_{L/K}(\alpha) = \det\big(a_{ij}\big)_{1\le i,j\le n} \quad \textit{and} \quad \mathrm{Tr}_{L/K}(\alpha) = \sum_{i=1}^{n} a_{ii}\,.$$

We note some properties of norms and traces.

(1) The norm map is multiplicative; i.e., $\mathrm{N}_{L/K}(\alpha\cdot\beta) = \mathrm{N}_{L/K}(\alpha)\cdot\mathrm{N}_{L/K}(\beta)$ for all $\alpha, \beta \in L$. Moreover, $\mathrm{N}_{L/K}(\alpha) = 0 \iff \alpha = 0$, and for $a \in K$ we have $\mathrm{N}_{L/K}(a) = a^n$, with $n = [L : K]$.

(2) For $\alpha, \beta \in L$ and $a \in K$ the following hold:

$$\begin{aligned}
&\mathrm{Tr}_{L/K}(\alpha+\beta) = \mathrm{Tr}_{L/K}(\alpha) + \mathrm{Tr}_{L/K}(\beta)\,,\\
&\mathrm{Tr}_{L/K}(a\cdot\alpha) = a\cdot \mathrm{Tr}_{L/K}(\alpha)\,, \quad \textit{and}\\
&\mathrm{Tr}_{L/K}(a) = n\cdot a\,, \quad \textit{with } n = [L:K]\,.
\end{aligned}$$

In particular $\mathrm{Tr}_{L/K}$ is a K-linear map.

(3) If L/K and M/L are finite extensions, then

$$\begin{aligned}
&\mathrm{Tr}_{M/K}(\alpha) = \mathrm{Tr}_{L/K}(\mathrm{Tr}_{M/L}(\alpha)) \quad \textit{and}\\
&\mathrm{N}_{M/K}(\alpha) = \mathrm{N}_{L/K}(\mathrm{N}_{M/L}(\alpha))
\end{aligned}$$

for all $\alpha \in M$.

(4) A finite field extension L/K is separable if and only if there exists an element $\gamma \in L$ such that $\mathrm{Tr}_{L/K}(\gamma) \neq 0$ (since the trace map is K-linear, it follows then that $\mathrm{Tr}_{L/K} : L \to K$ is surjective).

(5) Let $f(X) = X^r + a_{r-1}X^{r-1} + \ldots + a_0 \in K[X]$ be the minimal polynomial of α over K, and $[L : K] = n = rs$ (with $s = [L : K(\alpha)]$). Then

$$\mathrm{N}_{L/K}(\alpha) = (-1)^n a_0^s \quad \textit{and} \quad \mathrm{Tr}_{L/K}(\alpha) = -s a_{r-1}\,.$$

(6) Suppose now that L/K is separable of degree n. Consider the n distinct embeddings $\sigma_1, \ldots, \sigma_n : L \to \Phi$ of L over K into an algebraically closed field $\Phi \supseteq K$. Then,

$$\mathrm{N}_{L/K}(\alpha) = \prod_{i=1}^{n} \sigma_i(\alpha) \quad \textit{and} \quad \mathrm{Tr}_{L/K}(\alpha) = \sum_{i=1}^{n} \sigma_i(\alpha)$$

for $\alpha \in L$.

(7) In particular, if L/K is Galois with Galois group $G = \mathrm{Gal}(L/K)$, then

$$\mathrm{N}_{L/K}(\alpha) = \prod_{\sigma\in G} \sigma(\alpha) \quad \textit{and} \quad \mathrm{Tr}_{L/K}(\alpha) = \sum_{\sigma\in G} \sigma(\alpha)$$

for $\alpha \in L$.

A.15 Finite Fields

Let $p > 0$ be a prime number and let $q = p^n$ be a power of p. Then there exists a finite field $\mathbb{F}_q$ with $|\mathbb{F}_q| = q$, and $\mathbb{F}_q$ is unique up to isomorphism; it is the splitting field of the polynomial $X^q - X$ over the field $\mathbb{F}_p := \mathbb{Z}/p\mathbb{Z}$. In this manner, we obtain *all* finite fields of characteristic p.

The multiplicative group $\mathbb{F}_q^\times$ of $\mathbb{F}_q$ is a cyclic group of order $q-1$; i.e.,

$$\mathbb{F}_q = \{0, \beta, \beta^2, \ldots, \beta^{q-1} = 1\},$$

where β is a generator of $\mathbb{F}_q^\times$.

Let $m \geq 1$. Then $\mathbb{F}_q \subseteq \mathbb{F}_{q^m}$, and the extension $\mathbb{F}_{q^m}/\mathbb{F}_q$ is a Galois extension of degree m. The Galois group $\mathrm{Gal}(\mathbb{F}_{q^m}/\mathbb{F}_q)$ is cyclic; it is generated by the *Frobenius automorphism*

$$\varphi : \begin{cases} \mathbb{F}_{q^m} & \longrightarrow \mathbb{F}_{q^m}, \\ \alpha & \longmapsto \alpha^q. \end{cases}$$

In particular, all finite fields are perfect.

The norm and trace map from $\mathbb{F}_{q^m}$ to $\mathbb{F}_q$ are given by

$$\begin{aligned} \mathrm{N}_{\mathbb{F}_{q^m}/\mathbb{F}_q}(\alpha) &= \alpha^{1+q+q^2+\ldots+q^{m-1}}, \\ \mathrm{Tr}_{\mathbb{F}_{q^m}/\mathbb{F}_q}(\alpha) &= \alpha + \alpha^q + \ldots + \alpha^{q^{m-1}}. \end{aligned}$$

Hilbert's Theorem 90 states, for $\alpha \in \mathbb{F}_{q^m}$,

$$\mathrm{Tr}_{\mathbb{F}_{q^m}/\mathbb{F}_q}(\alpha) = 0 \iff \alpha = \beta^q - \beta \quad \textit{for some } \beta \in \mathbb{F}_{q^m}.$$

A.16 Transcendental Extensions

Let L/K be a field extension. An element $x \in L$ that is not algebraic over K is called *transcendental* over K. A finite subset $\{x_1, \ldots, x_n\} \subseteq L$ is *algebraically independent* over K if there does not exist a non-zero polynomial $f(X_1, \ldots, X_n) \in K[X_1, \ldots, X_n]$ with $f(x_1, \ldots, x_n) = 0$. An arbitrary subset $S \subseteq L$ is algebraically independent over K if all finite subsets of S are algebraically independent over K.

A *transcendence basis* of L/K is a maximal algebraically independent subset of L. Any two transcendence bases of L/K have the same cardinality, the *transcendence degree* of L/K.

If L/K has finite transcendence degree n and $\{x_1, \ldots, x_n\}$ is a transcendence basis if L/K, the field $K(x_1, \ldots, x_n) \subseteq L$ is K-isomorphic to $K(X_1, \ldots, X_n)$, the quotient field of the polynomial ring $K[X_1, \ldots, X_n]$ in n variables over K. The extension $L/K(x_1, \ldots, x_n)$ is algebraic.

Appendix B. Algebraic Curves and Function Fields

This appendix contains a brief survey of the relations between algebraic curves and algebraic function fields. For details and proofs we refer to the literature on algebraic geometry, for instance [11],[18],[37],[38].

We assume that K is an algebraically closed field.

B.1 Affine Varieties

The n-dimensional *affine space* $\mathbf{A}^n = \mathbf{A}^n(K)$ is the set of all n-tuples of elements of K. An element $P = (a_1, \ldots, a_n) \in \mathbf{A}^n$ is a *point*, and $a_1, \ldots, a_n$ are the *coordinates* of P.

Let $K[X_1, \ldots, X_n]$ be the ring of polynomials in n variables over K. A subset $V \subseteq \mathbf{A}^n$ is an *algebraic set* if there exists a set $M \subseteq K[X_1, \ldots, X_n]$ such that

$$V = \{P \in \mathbf{A}^n \mid F(P) = 0 \text{ for all } F \in M\}.$$

Given an algebraic set $V \subseteq \mathbf{A}^n$, the set of polynomials

$$I(V) = \{F \in K[X_1, \ldots, X_n] \mid F(P) = 0 \text{ for all } P \in V\}$$

is called the *ideal* of V. $I(V)$ is obviously an ideal in $K[X_1, \ldots, X_n]$, and it can be generated by finitely many polynomials $F_1, \ldots, F_r \in K[X_1, \ldots, X_n]$. Thus we have

$$V = \{P \in \mathbf{A}^n \mid F_1(P) = \ldots = F_r(P) = 0\}.$$

An algebraic set $V \subseteq \mathbf{A}^n$ is called *irreducible* if it cannot be written as $V = V_1 \cup V_2$, where V_1 and V_2 are proper algebraic subsets of V. Equivalently, V is irreducible if and only if the corresponding ideal $I(V)$ is a prime ideal. An *affine variety* is an irreducible algebraic set $V \subseteq \mathbf{A}^n$.

The *coordinate ring* of an affine variety V is the residue class ring $\Gamma(V) = K[X_1, \ldots, X_n]/I(V)$. As $I(V)$ is a prime ideal, $\Gamma(V)$ is an integral domain. Every $f = F + I(V) \in \Gamma(V)$ induces a function $f : V \to K$ by setting $f(P) := F(P)$ for $P \in V$. The quotient field

$$K(V) = \mathrm{Quot}(\Gamma(V))$$

is called the *field of rational functions* (or the *function field*) of V. It contains K as a subfield. The *dimension* of V is the transcendence degree of $K(V)/K$.

For a point $P \in V$ let

$$\mathcal{O}_P(V) = \{f \in K(V) \mid f = g/h \text{ with } g, h \in \Gamma(V) \text{ and } h(P) \neq 0\}.$$

This is a local ring with quotient field $K(V)$, its unique maximal ideal is

$$M_P(V) = \{f \in K(V) \mid f = g/h \text{ with } g, h \in \Gamma(V), h(P) \neq 0 \text{ and } g(P) = 0\}.$$

$\mathcal{O}_P(V)$ is called the *local ring* of V at P. For $f = g/h \in \mathcal{O}_P(V)$ with $h(P) \neq 0$, the *value* of f at P is defined to be $f(P) := g(P)/h(P)$.

B.2 Projective Varieties

On the set $\mathbf{A}^{n+1} \setminus \{(0, \ldots, 0)\}$ an equivalence relation $\sim$ is given by

$$(a_0, a_1, \ldots, a_n) \sim (b_0, b_1, \ldots, b_n) :\Longleftrightarrow$$
$$\text{there is an element } 0 \neq \lambda \in K \text{ such that } b_i = \lambda a_i \text{ for } 0 \leq i \leq n.$$

The equivalence class of $(a_0, a_1, \cdots, a_n)$ with respect to $\sim$ is denoted by $(a_0 : a_1 : \ldots : a_n)$. The n-dimensional *projective space* $\mathbf{P}^n = \mathbf{P}^n(K)$ is the set of all equivalence classes

$$\mathbf{P}^n = \{(a_0 : \ldots : a_n) \mid a_i \in K, \text{ not all } a_i = 0\}.$$

An element $P = (a_0 : \ldots : a_n) \in \mathbf{P}^n$ is a *point*, and $a_0, \ldots, a_n$ are called *homogeneous coordinates* of P.

A *monomial* of degree d is a polynomial $G \in K[X_0, \ldots, X_n]$ of the form

$$G = a \cdot \prod_{i=0}^n X_i^{d_i} \text{ with } 0 \neq a \in K \text{ and } \sum_{i=0}^n d_i = d.$$

A polynomial F is a *homogeneous polynomial* if F is the sum of monomials of the same degree. An ideal $I \subseteq K[X_0, \ldots, X_n]$ which is generated by homogeneous polynomials is called a *homogeneous ideal.*

Let $P = (a_0 : \ldots : a_n) \in \mathbf{P}^n$ and let $F \in K[X_0, \ldots, X_n]$ be a homogeneous polynomial. We say that $F(P) = 0$ if $F(a_0, \ldots, a_n) = 0$. This makes sense: since $F(\lambda a_0, \ldots, \lambda a_n) = \lambda^d \cdot F(a_0, \ldots, a_n)$ (with $d = \deg F$), one has $F(a_0, \ldots, a_n) = 0 \Longleftrightarrow F(\lambda a_0, \ldots, \lambda a_n) = 0$.

A subset $V \subseteq \mathbf{P}^n$ is a *projective algebraic set* if there exists a set of homogeneous polynomials $M \subseteq K[X_0, \ldots, X_n]$ such that

$$V = \{P \in \mathbf{P}^n \mid F(P) = 0 \textit{ for all } F \in M\}.$$

The ideal $I(V) \subseteq K[X_0, \ldots, X_n]$, which is generated by all homogeneous polynomials F with $F(P) = 0$ for all $P \in V$, is called the *ideal* of V. It is a homogeneous ideal. *Irreducibility* of projective algebraic sets is defined as in the affine case. Again, $V \subseteq \mathbf{P}^n$ is irreducible if and only if $I(V)$ is a homogeneous prime ideal in $K[X_0, \ldots, X_n]$. A *projective variety* is an irreducible projective algebraic set.

Given a non-empty variety $V \subseteq \mathbf{P}^n$, we define its *homogeneous coordinate ring* by

$$\Gamma_h(V) = K[X_0, \ldots, X_n]/I(V);$$

this is an integral domain containing K. An element $f \in \Gamma_h(V)$ is said to be a *form* of degree d if $f = F + I(V)$ for some homogeneous polynomial $F \in K[X_0, \ldots, X_n]$ with $deg\, F = d$. The *function field* of V is defined by

$$K(V) := \left\{ \frac{g}{h} \,\middle|\, g, h \in \Gamma_h(V) \textit{ are forms of the same degree and } h \neq 0 \right\},$$

which is a subfield of $\mathrm{Quot}(\Gamma_h(V))$, the quotient field of $\Gamma_h(V)$.

The *dimension* of V is the transcendence degree of $K(V)$ over K.

Let $P = (a_0 : \ldots : a_n) \in V$ and $f \in K(V)$. Write $f = g/h$ where $g = G + I(V)$, $h = H + I(V) \in \Gamma_h(V)$ and G, H are homogeneous polynomials of degree d. Since

$$\frac{G(\lambda a_0, \ldots, \lambda a_n)}{H(\lambda a_0, \ldots, \lambda a_n)} = \frac{\lambda^d \cdot G(a_0, \ldots, a_n)}{\lambda^d \cdot H(a_0, \ldots, a_n)} = \frac{G(a_0, \ldots, a_n)}{H(a_0, \ldots, a_n)},$$

we can set $f(P) := G(a_0, \ldots, a_n)/H(a_0, \ldots, a_n) \in K$, if $H(P) \neq 0$. Then we say that f is *defined* at P and call $f(P)$ the *value* of f at P. The ring

$$\mathcal{O}_P(V) = \{f \in K(V) \mid f \textit{ is defined at } P\} \subseteq K(V)$$

is a local ring with maximal ideal

$$M_P(V) = \{f \in \mathcal{O}_P(V) \mid f(P) = 0\}.$$

B.3 Covering Projective Varieties by Affine Varieties

For $0 \leq i \leq n$ we consider the mapping $\varphi_i : \mathbf{A}^n \to \mathbf{P}^n$ given by

$$\varphi_i(a_0, \ldots, a_{n-1}) = (a_0 : \ldots : a_{i-1} : 1 : a_i : \ldots : a_{n-1}).$$

φ_i is a bijection from $\mathbf{A}^n$ onto the set

$$U_i = \{(c_0 : \ldots : c_n) \in \mathbf{P}^n \,|\, c_i \neq 0\},$$

and $\mathbf{P}^n = \bigcup_{i=0}^n U_i$. So $\mathbf{P}^n$ is covered by $n+1$ copies of the affine space $\mathbf{A}^n$ (this is not a disjoint union).

Let $V \subseteq \mathbf{P}^n$ be a projective variety, then $V = \bigcup_{i=0}^n (V \cap U_i)$. Suppose that $V \cap U_i \neq \emptyset$. Then

$$V_i := \varphi_i^{-1}(V \cap U_i) \subseteq \mathbf{A}^n$$

is an affine variety, and the ideal $I(V_i)$ (in the sense of B.1) is given by

$$I(V_i) = \{F(X_0, \ldots, X_{i-1}, 1, X_{i+1}, \ldots X_n) \,|\, F \in I(V)\}.$$

For convenience we restrict ourselves in the following to the case $i = n$ (and $V \cap U_n \neq \emptyset$). The complement $H_n = \mathbf{P}^n \setminus U_n = \{(a_0 : \ldots : a_n) \in \mathbf{P}^n \,|\, a_n = 0\}$ is called the *hyperplane at infinity*, and the points $P \in V \cap H_n$ are the *points* of V *at infinity*.

There is a natural K-isomorphism α from $K(V)$ (the function field of the projective variety V) onto $K(V_n)$ (the function field of the affine variety $V_n = \varphi_n^{-1}(V \cap U_n)$). This isomorphism is defined as follows: Let $f = g/h \in K(V)$ where $g, h \in \Gamma_h(V)$ are forms of the same degree and $h \neq 0$. Choose homogeneous polynomials $G, H \in K[X_0, \ldots, X_n]$ which represent g resp. h. Let $G_* = G(X_0, \ldots, X_{n-1}, 1)$ and $H_* = H(X_0, \ldots, X_{n-1}, 1) \in K[X_0, \ldots, X_{n-1}]$. Their residue classes in $\Gamma(V_n) = K[X_0, \ldots, X_{n-1}]/I(V_n)$ are g_* resp. h_*. Then $\alpha(f) = g_*/h_*$. Under this isomorphism, the local ring of a point $P \in V \cap U_n$ is mapped onto the local ring of $\varphi_n^{-1}(P) \in V_n$, hence these local rings are isomorphic.

B.4 The Projective Closure of an Affine Variety

For a polynomial $F = F(X_0, \ldots, X_{n-1}) \in K[X_0, \ldots, X_{n-1}]$ of degree d we set

$$F^* = X_n^d \cdot F(X_0/X_n, \ldots, X_{n-1}/X_n) \in K[X_0, \ldots, X_n].$$

F^* is a homogeneous polynomial of degree d in $n+1$ variables.

Consider now an affine variety $V \in \mathbf{A}^n$ and the corresponding ideal $I(V) \subseteq K[X_0, \ldots, X_{n-1}]$. Define the projective variety $\bar{V} \subseteq \mathbf{P}^n$ as follows:

$$\bar{V} = \{P \in \mathbf{P}^n \,|\, F^*(P) = 0 \textit{ for all } F \in I(V)\}.$$

This variety $\bar{V}$ is called the *projective closure* of V. One can recover V from $\bar{V}$ by the process described in B.3, namely

$$V = \varphi_n^{-1}(\bar{V} \cap U_n) = (\bar{V})_n.$$

Consequently the function fields of V and $\bar{V}$ are naturally isomorphic, and V and $\bar{V}$ have the same dimension.

B.5 Rational Maps and Morphisms

Let $V \subseteq \mathbf{P}^m$ and $W \subseteq \mathbf{P}^n$ be projective varieties. Suppose that $F_0, \ldots, F_n \in K[X_0, \ldots, X_m]$ are homogeneous polynomials with the following properties:

(a) $F_0, \ldots, F_n$ have the same degree;

(b) not all F_i are in $I(V)$;

(c) for all $H \in I(W)$ holds $H(F_0, \ldots, F_n) \in I(V)$.

Let $Q \in V$ and assume that $F_i(Q) \neq 0$ for at least one $i \in \{0, \ldots, n\}$ (by (b) such a point exists). Then the point $(F_0(Q) : \ldots : F_n(Q)) \in \mathbf{P}^n$ lies in W, by (c). Let $(G_0, \ldots, G_n)$ be another n-tuple of homogeneous polynomials satisfying (a), (b) and (c). We say that $(F_0, \ldots, F_n)$ and $(G_0, \ldots, G_n)$ are equivalent if

(d) $F_i G_j \equiv F_j G_i \mod I(V)$ for $0 \leq i, j \leq n$.

The equivalence class of $(F_0, \ldots, F_n)$ with respect to this equivalence relation is denoted by

$$\phi = (F_0 : \ldots : F_n),$$

and ϕ is called a *rational map* from V to W.

A rational map $\phi = (F_0 : \ldots : F_n)$ is *regular* (or *defined*) at the point $P \in V$ if there exist homogeneous polynomials $G_0, \ldots, G_n \in K[X_0, \ldots, X_m]$ such that $\phi = (G_0 : \ldots : G_n)$ and $G_i(P) \neq 0$ for at least one i. Then we set

$$\phi(P) = (G_0(P) : \ldots : G_n(P)) \in W,$$

which is well-defined by (a) and (d).

Two varieties V_1, V_2 are *birationally equivalent* if there are rational maps $\phi_1 : V_1 \to V_2$ and $\phi_2 : V_2 \to V_1$ such that $\phi_1 \circ \phi_2$ and $\phi_2 \circ \phi_1$ are the identity maps on V_2 and V_1, respectively. V_1 and V_2 are birationally equivalent if and only if their function fields $K(V_1)$ and $K(V_2)$ are K-isomorphic.

A rational map $\phi : V \to W$ which is regular at all points $P \in V$ is called a *morphism*. It is called an *isomorphism* if there is a morphism $\psi : W \to V$ such that $\phi \circ \psi$ and $\psi \circ \phi$ are the identity maps on W and V, respectively. In this case V and W are said to be *isomorphic*. Clearly, isomorphy implies birational equivalence, but the converse is not true in general.

B.6 Algebraic Curves

A *projective (affine) algebraic curve* V is a projective (affine) variety of dimension one. This means that the field $K(V)$ of rational functions on V is an algebraic function field of one variable.

A point $P \in V$ is *non-singular* (or *simple*) if the local ring $\mathcal{O}_P(V)$ is a discrete valuation ring (i.e., $\mathcal{O}_P(V)$ is a principal ideal domain with exactly one maximal ideal $\neq \{0\}$). There exist only finitely many singular points on a curve. The curve V is called *non-singular* (or *smooth*) if all points $P \in V$ are non-singular.

A *plane affine curve* is an affine curve $V \subseteq \mathbf{A}^2$. Its ideal $I(V) \subseteq K[X_0, X_1]$ is generated by an irreducible polynomial $G \in K[X_0, X_1]$ (which is unique up to a constant factor). Conversely, given an irreducible polynomial $G \in K[X_0, X_1]$, the set $V = \{P \in \mathbf{A}^2 \,|\, G(P) = 0\}$ is a plane affine curve, and G generates the corresponding ideal $I(V)$. A point $P \in V$ is non-singular if and only if

$$G_{X_0}(P) \neq 0 \;\; or \;\; G_{X_1}(P) \neq 0 \quad (or\ both),$$

where $G_{X_i} \in K[X_0, X_1]$ is the partial derivative of G with respect to X_i (*Jacobi-Criterion*).

Accordingly, the ideal of a *plane projective curve* $V \subseteq \mathbf{P}^2$ is generated by an irreducible homogeneous polynomial $H \in K[X_0, X_1, X_2]$. A point $P \in V$ is non-singular if and only if $H_{X_i}(P) \neq 0$ for at least one $i \in \{0, 1, 2\}$.

If $V = \{P \in \mathbf{A}^2 \,|\, G(P) = 0\}$ is a plane affine curve (with an irreducible polynomial $G \in K[X_0, X_1]$ of degree d), the projective closure $\bar{V} \subseteq \mathbf{P}^2$ is the set of zeros of the homogeneous polynomial $G^* = X_2^d \cdot G(X_0/X_2, X_1/X_2)$.

B.7 Maps Between Curves

We consider rational maps $\phi : V \to W$ where V and W are projective curves. The following hold:

(a) ϕ is defined at all non-singular points $P \in V$. Hence, if V is a non-singular curve, then ϕ is a morphism.

(b) If V is non-singular and ϕ is non-constant then ϕ is surjective.

B.8 The Non-Singular Model of a Curve

Let V be a projective curve. Then there exists a non-singular projective curve V' and a birational morphism $\phi' : V' \to V$. The pair (V', ϕ') is unique in the following sense: given another non-singular curve V'' and a birational morphism $\phi'' : V'' \to V$, there exists a unique isomorphism $\phi : V' \to V''$ such that $\phi' = \phi'' \circ \phi$. Hence V' (more precisely: the pair (V', ϕ')) is called *the non-singular model* of V.

If $\phi' : V' \to V$ is the non-singular model of V and $P \in V$ is non-singular, there exists a unique $P' \in V'$ with $\phi'(P') = P$; for a singular point $P \in V$ the number of $P' \in V'$ with $\phi'(P') = P$ is finite (it may be one).

B.9 The Curve Associated with an Algebraic Function Field

Starting from an algebraic function field of one variable F/K, there exists a non-singular projective curve V (unique up to isomorphism) whose function field $K(V)$ is (K-isomorphic to) F. One can construct V as follows: choose $x, y \in F$ such that $F = K(x, y)$ (this is possible by Proposition 3.10.2). Let $G(X, Y) \in K[X, Y]$ be the irreducible polynomial with $G(x, y) = 0$. Let $W = \{P \in \mathbf{A}^2 \mid G(P) = 0\}$ and $\bar{W} \subseteq \mathbf{P}^2$ be the projective closure of W. Let V be the non-singular model of $\bar{W}$; then $K(V) \simeq F$.

B.10 Non-Singular Curves and Algebraic Function Fields

Let V be a non-singular projective curve and let $F = K(V)$ be its function field. There is a 1-1 correspondence between the points $P \in V$ and the places of F/K, given by

$$P \longmapsto M_P(V),$$

the maximal ideal of the local ring $\mathcal{O}_P(V)$. This correspondence makes it possible to translate definitions and results from algebraic function fields to algebraic curves (and vice versa). We give some examples:

– The *genus* of the curve V is the genus of the function field $K(V)$.

– A *divisor* of V is a formal sum $D = \sum_{P \in V} n_P P$ where $n_P \in \mathbb{Z}$ and almost all $n_P = 0$. The *degree* of D is $\deg D = \sum_{P \in V} n_P$. The divisors of V form an additive group $\mathrm{Div}(V)$, the *divisor group* of V.

– The *order of a rational function* at a point $P \in V$ is defined to be $v_P(f)$, where v_P is the discrete valuation of $K(V)$ corresponding to the valuation ring $\mathcal{O}_P(V)$.

– The *principal divisor* (f) of a rational function $0 \neq f \in K(V)$ is $(f) = \sum_{P \in V} v_P(f) P$. The degree of a principal divisor is 0.

– The principal divisors form a subgroup $\mathrm{Princ}(V)$ of the divisor group $\mathrm{Div}(V)$. The factor group $\mathrm{Jac}(V) = \mathrm{Div}^0(V)/\mathrm{Princ}(V)$, where $\mathrm{Div}^0(V)$ is the group of divisors of degree 0, is called the *Jacobian* of V.

– For $D \in \mathrm{Div}(V)$ the space $\mathscr{L}(D)$ is defined as in the function field case. It is a finite-dimensional vector space over K, its dimension is given by the Riemann-Roch Theorem.

B.11 Varieties over a Non-Algebraically Closed Field

Thus far it was assumed that K is an algebraically closed field. Now we drop this assumption and suppose only that K is a *perfect* field. Let $\bar{K} \supseteq K$ be the algebraic closure of K.

An affine variety $V \subseteq \mathbf{A}^n(\bar{K})$ is said to be *defined over* K if its ideal $I(V) \subseteq \bar{K}[X_1, \ldots, X_n]$ can be generated by polynomials $F_1, \ldots, F_r \in K[X_1, \ldots, X_n]$. If V is defined over K, the set

$$V(K) = V \cap \mathbf{A}^n(K) = \{P = (a_1, \ldots, a_n) \in V \mid \textit{all } a_i \in K\}$$

is called the set of *K-rational points* of V.

Similarly, a projective variety $V \subseteq \mathbf{P}^n(\bar{K})$ is *defined over* K if $I(V)$ is generated by homogeneous polynomials $F_1, \ldots, F_r \in K[X_0, \ldots, X_n]$. A point $P \in V$ is called *K-rational* if there exist homogeneous coordinates $a_0, \ldots, a_n$ of P which are in K, and we set

$$V(K) = \{P \in V \mid P \textit{ is } K\textit{-rational}\}.$$

Let $V \subseteq \mathbf{A}^n(\bar{K})$ be an affine variety defined over K. Define the ideal

$$I(V/K) = I(V) \cap K[X_1, \ldots, X_n]$$

and the residue class ring

$$\Gamma(V/K) = K[X_1, \ldots, X_n]/I(V/K).$$

The quotient field

$$K(V) = \mathrm{Quot}(\Gamma(V/K)) \subseteq \bar{K}(V)$$

is the field of *K-rational functions* of V. The field extension $K(V)/K$ is finetely generated, its transcendence degree is the dimension of V. In the same manner, the field of K-rational functions of a projective variety can be defined.

Consider two varieties $V \subseteq \mathbf{P}^m(\bar{K})$ and $W \subseteq \mathbf{P}^n(\bar{K})$. A rational map $\phi : V \to W$ is *defined over* K if there exist homogeneous polynomials $F_0, \ldots, F_n \in K[X_0, \ldots, X_m]$ satisfying the conditions (a), (b) and (c) of B.5, such that $\phi = (F_0 : \ldots : F_n)$.

Another way to describe K-rational points, K-rational functions etc. on a variety which is defined over K is the following: Let $\mathcal{G}_{\bar{K}/K}$ be the Galois group of $\bar{K}/K$. The action of $\mathcal{G}_{\bar{K}/K}$ on $\bar{K}$ extends naturally to an action on the sets $\mathbf{A}^n(\bar{K})$, $\mathbf{P}^n(\bar{K})$, $\bar{K}[X_1, \ldots, X_n]$, V, $\Gamma(V)$, $\bar{K}(V)$ etc. For instance, consider a projective variety $V \subseteq \mathbf{P}^n(\bar{K})$ (defined over K), a point $P = (a_0 : \ldots : a_n) \in V$ and an automorphism $\sigma \in \mathcal{G}_{\bar{K}/K}$; then $P^\sigma = (a_0^\sigma : \ldots : a_n^\sigma)$. It is easily seen that

$$V(K) = \{P \in V \mid P^\sigma = P \textit{ for all } \sigma \in \mathcal{G}_{\bar{K}/K}\},$$

$$K(V) = \{f \in \bar{K}(V) \mid f^\sigma = f \textit{ for all } \sigma \in \mathcal{G}_{\bar{K}/K}\},$$

and so on.

B.12 Curves over a Non-Algebraically Closed Field

Consider a projective curve $V \subseteq \mathbf{P}^n(\bar{K})$ which is defined over K (where K is perfect and $\bar{K}$ is the algebraic closure of K as in B.11). Then the field $K(V)$ of K-rational functions on V is an algebraic function field of one variable over K, and $\bar{K}(V)$ is the constant field extension of $K(V)$ with $\bar{K}$.

A divisor $D = \sum_{P \in V} n_P P \in \mathrm{Div}(V)$ is *defined over* K if $D^\sigma = D$ for all $\sigma \in \mathcal{G}_{\bar{K}/K}$ (this means that $n_{P^\sigma} = n_P$ for all $P \in V$). The divisors of V defined over K form a subgroup $\mathrm{Div}(V/K) \subseteq \mathrm{Div}(V)$. For $D \in \mathrm{Div}(V/K)$ the space $\mathscr{L}_K(D)$ is given by

$$\mathscr{L}_K(D) = K(V) \cap \mathscr{L}(D).$$

It is a finite-dimensional K-vector space, and its dimension (over K) equals the dimension of $\mathscr{L}(D)$ (over $\bar{K}$), by Theorem 3.6.3(d).

A divisor $Q \in \mathrm{Div}(V/K)$ with $Q > 0$ is called a *prime divisor* of V/K if Q cannot be written as $Q = Q_1 + Q_2$ with effective divisors $Q_1, Q_2 \in \mathrm{Div}(V/K)$. It is easily seen that the divisor group $\mathrm{Div}(V/K)$ is the free abelian group generated by the prime divisors. Prime divisors of V/K correspond to the places of the function field $K(V)/K$; under this correspondence, prime divisors of degree one (i.e., K-rational points) of V correspond to the places of $K(V)/K$ of degree one.

B.13 An Example

Let K be a perfect field of characteristic $p \geq 0$ and let

$$G(X,Y) = aX^n + bY^n + c \ \ with\ \ a,b,c \in K \setminus \{0\},\ \ n \geq 1\ \ and\ \ p \nmid n.$$

(This is Example 6.3.4.) The polynomial $G(X,Y)$ is irreducible (which follows easily from Eisenstein's Criterion, cf. Proposition 3.1.15). The affine curve $V = \{P \in \mathbf{A}^2(\bar{K}) \mid G(P) = 0\}$ is non-singular since for $P = (\alpha, \beta) \in V$,

$$G_X(\alpha,\beta) = na\alpha^{n-1} \neq 0 \ \ or \ \ G_Y(\alpha,\beta) = nb\beta^{n-1} \neq 0.$$

Let $G^*(X,Y,Z) = aX^n + bY^n + cZ^n$; then the projective closure of V is the curve

$$\bar{V} = \{(\alpha : \beta : \gamma) \in \mathbf{P}^2(\bar{K}) \mid G^*(\alpha,\beta,\gamma) = 0\}.$$

We consider the points $P \in \bar{V}$ at infinity; i.e., $P = (\alpha : \beta : 0)$ with $(\alpha, \beta) \neq (0,0)$ and $G^*(\alpha, \beta, 0) = 0$. From the equation $0 = G^*(\alpha,\beta,0) = a\alpha^n + b\beta^n$ follows that $\beta \neq 0$, so we can set $\beta = 1$; i.e., $P = (\alpha : 1 : 0)$. The equation $a\alpha^n + b = 0$ has n distinct roots $\alpha \in K$, so there are n distinct points of $\bar{V}$ at infinity. All of them are non-singular since $G_Y^*(\alpha, 1, 0) = nb \neq 0$.

In the special case $K = \mathbb{F}_{q^2}$ and $G(X,Y) = X^{q+1} + Y^{q+1} - 1$ (the *Hermitian* curve, cf. Example 6.3.6) we want to determine the K-rational points $P = (\alpha : \beta : \gamma) \in \bar{V}(K)$. First let $\gamma \neq 0$; i.e., $P = (\alpha : \beta : 1)$. For all $\alpha \in K$ with $\alpha^{q+1} \neq 1$ there are $q+1$ distinct elements $\beta \in K$ with $G^*(\alpha, \beta, 1) = 0$. If $\alpha^{q+1} = 1$, $\beta = 0$ is the only root of $G^*(\alpha, \beta, 1) = 0$. Finally, if $\gamma = 0$, there are $q+1$ points $P = (\alpha : 1 : 0) \in \bar{V}(K)$. We have thus constructed all K-rational points on the Hermitian curve over $\mathbb{F}_{q^2}$. Their number is $|(\bar{V}(\mathbb{F}_{q^2})| = q^3 + 1$, in accordance with Lemma 6.4.4.

List of Notations

$\tilde{K}$	field of constants of F/K, 1
$\mathbb{P}_F$	set of places of F/K, 4
$\mathcal{O}_P$	valuation ring of the place P, 4
v_P	valuation corresponding to the place P, 5
F_P	residue class field of the place P, 6
$x(P)$	residue class of $x \in \mathcal{O}_P$ in F_P, 6
$\deg P$	degree of the place P, 6
$P_{p(x)}$	the place of $K(x)$ corresponding to an irreducible polynomial $p(x)$, 9
P_∞	the infinite place of $K(x)$, 9
$\mathbf{P}^1(K)$	projective line, 11
$\mathrm{Div}(F)$	divisor group of F, 15
$\mathrm{supp}\, D$	support of D, 15
$v_Q(D)$	16
$\deg D$	degree of the divisor D, 16
$(x)_0,\ (x)_\infty,\ (x)$	zero, pole, principal divisor of x, 16
$\mathrm{Princ}(F)$	group of principal divisors of F, 17
$\mathrm{Cl}(F)$	divisor class group of F, 17
$[D]$	divisor class of the divisor D, 17
$D \sim D'$	equivalence of divisors, 17
$\mathscr{L}(A)$	Riemann-Roch space of the divisor A, 17
$\ell(A)$	dimension of the Riemann-Roch space $\mathscr{L}(A)$, 19
g	the genus of F, 22
$i(A)$	index of specialty of A, 24
$\mathcal{A}_F$	adele space of F, 24
$\mathcal{A}_F(A)$	24
Ω_F	space of Weil differentials of F, 27
$\Omega_F(A)$	27
(ω)	divisor of the Weil differential ω, 29
$v_P(\omega)$	29
$\iota_P(x)$	38
ω_P	local component of a Weil differential, 38
$d(a,b)$	Hamming distance, 45

$\mathrm{wt}(a)$	weight of a vector $a \in \mathbb{F}_q^n$, 45
$d(C)$	minimum distance of the code C, 46
$[n,k,d]$ code	code with parameters n, k, d, 46
$\langle a, b \rangle$	the canonical inner product on $\mathbb{F}_q^n$, 46
$C^\perp$	the dual code of C, 46
$C_{\mathscr{L}}(D,G)$	the algebraic geometry code associated to the divisors D and G, 48
$\mathrm{ev}_D(x)$	49
$C_\Omega(D,G)$	differential code associated with divisors D and G, 51
$\mathrm{GRS}_k(\alpha, v)$	Generalized Reed-Solomon code, 57
$C\vert_{\mathbb{F}_q}$	subfield subcode, 59
$P'\vert P$	the place P' lies over P, 69
$e(P'\vert P)$	ramification index of $P'\vert P$, 71
$f(P'\vert P)$	relative degree of $P'\vert P$, 71
$\mathrm{Con}_{F'/F}(P)$	conorm of P in the extension F'/F, 72
$\mathcal{O}_S$	holomorphy ring corresponding to the set $S \subseteq \mathbb{P}_F$, 77
$\mathrm{ic}_F(R)$	integral closure of R in F, 78
$\mathcal{O}'_P$	integral closure of $\mathcal{O}_P$ in F', 85
$\mathcal{C}_P$	complementary module over $\mathcal{O}_P$, 91
$d(P'\vert P)$	different exponent of $P'\vert P$, 92
$\mathrm{Diff}(F'/F)$	different of F'/F, 92
$G_Z(P'\vert P)$	decomposition group of $P'\vert P$, 130
$G_T(P'\vert P)$	inertia group of $P'\vert P$, 130
$G_i(P'\vert P)$	i-th ramification group of $P'\vert P$, 133
δ_x	derivation with respect to x, 158
Der_F	module of derivations of F/K, 158
$u\,dx$	differential of F/K, 159
$\Delta_F,\ (\Delta_F, d)$	differential module of F/K, 159
$(\hat{T}, \hat{v})$	completion of the valued field (T, v), 162
$\hat{F}_P$	P-adic completion of F, 164
$\mathrm{res}_{P,t}(z)$	residue of z with respect to P and t, 167
$\mathrm{res}_P(\omega)$	residue of ω at the place P, 169
$\delta(x)$	the Weil differential associated with x, 170
$\mathrm{Div}^0(F)$	group of divisors of degree 0, 186
$\mathrm{Cl}^0(F)$	group of divisor classes of degree 0, 186
$h,\ h_F$	class number of $F/\mathbb{F}_q$, 186
∂	187
$Z(t),\ Z_F(t)$	Zeta function of $F/\mathbb{F}_q$, 188
$F_r = F\mathbb{F}_{q^r}$	the constant field extension of $F/\mathbb{F}_q$ of degree r, 190
$L(t),\ L_F(t)$	the L-polynomial of $F/\mathbb{F}_q$, 193
N_r	number of rational places of F_r, 195
B_r	number of places of degree r, 206
$N_q(g)$	243
$A(q)$	Ihara's constant, 243
$\lambda_m(t)$	244
$f_m(t)$	244
$\mathcal{F} = (F_0, F_1, F_2, \ldots)$	a tower of function fields, 246
$\nu(\mathcal{F}/F_0)$	splitting rate of the tower $\mathcal{F}/F_0$, 247
$\gamma(\mathcal{F}/F_0)$	genus of the tower $\mathcal{F}/F_0$, 247

References

1. E. Artin, *Algebraic numbers and algebraic functions*, Gordon and Breach, New York, 1967.
2. A.Bassa, A.Garcia and H.Stichtenoth, *A new tower over cubic finite fields*, Moscow Math. J., to appear.
3. E. Berlekamp, *Algebraic coding theory*, McGraw-Hill, New York, 1968.
4. J. Bezerra, A. Garcia and H. Stichtenoth, *An explicit tower of function fields over cubic finite fields and Zink's lower bound*, J. Reine Angew. Math. **589**, 2005, pp. 159-199.
5. E. Bombieri, *Counting points on curves over finite fields*, Sem. Bourbaki, No. 430, Lecture Notes Math. **383**, Springer-Verlag, Berlin-Heidelberg-New York, 1974, pp. 234-241.
6. C. Chevalley, *Introduction to the theory of algebraic functions of one variable*, AMS Math. Surveys No. **6**, 1951.
7. M. Deuring, *Lectures on the theory of algebraic functions of one variable*, Lecture Notes in Math. **314**, Springer-Verlag, Berlin-Heidelberg-New York, 1973.
8. V. G. Drinfeld and S. G. Vladut, *The number of points of an algebraic curve*, Func. Anal. **17**, 1983, pp. 53-54.
9. M. Eichler, *Introduction to the theory of algebraic numbers and functions*, Academic Press, New York, 1966.
10. H. M. Farkas and I. Kra, *Riemann surfaces*, 2nd edition, Graduate Texts in Math. **71**, Springer-Verlag, New York, 1991.
11. W. Fulton, *Algebraic curves*, Benjamin, New York, 1969.
12. A. Garcia and H. Stichtenoth, *On the asymptotic behavior of some towers of functions fields over finite fields*, J. Number Theory **61**, 1996, pp. 248-273.
13. A. Garcia and H. Stichtenoth (eds.), *Topics in geometry, coding theory and cryptography*, Algebr. Appl. **6**, Springer-Verlag, Dordrecht, 2007.
14. D. M. Goldschmidt, *Algebraic functions and projective curves*, Graduate Texts in Math. **215**, Springer-Verlag, New York, 2002.
15. V. D. Goppa, *Codes on algebraic curves*, Soviet Math. Dokl. **24**, No. 1, 1981, pp. 170-172.
16. V. D. Goppa, *Geometry and codes*, Kluwer Academic Publ., Dordrecht, 1988.
17. V. Guruswami and M. Sudan, *Improved decoding of Reed-Solomon and algebraic geometry codes*, IEEE Trans. Inform. Th. **45**, 1999, pp. 1757-1768.

18. R. Hartshorne, *Algebraic geometry*, Graduate Texts in Math. **52**, Springer-Verlag, New York-Heidelberg-Berlin, 1977.
19. H. Hasse, *Theorie der relativ-zyklischen algebraischen Funktionenkörper, insbesondere bei endlichem Konstantenkörper*, J. Reine Angew. Math. **172**, 1934, pp. 37-54.
20. K. Hensel and G. Landsberg, *Theorie der algebraischen Funktionen einer Variablen*, Leipzig, 1902. Reprinted by Chelsea Publ. Comp., New York, 1965.
21. J. W. P. Hirschfeld, G. Korchmaros and F. Torres, *Algebraic curves over a finite field*, Princeton Ser. in Applied Math., Princeton Univ. Press, Princeton and Oxford, 2008.
22. Y. Ihara, *Some remarks on the number of rational points of algebraic curves over finite fields*, J. Fac. Sci. Univ. Tokyo Sect. IA Math. **28**, 1981, pp. 721-724.
23. S. Lang, *Algebra*, 3rd edition, Graduate Texts in Math. **211**, Springer-Verlag, New York, 2002.
24. R. Lidl and H. Niederreiter, *Finite fields*, 2nd edition, Cambridge Univ. Press, Cambridge, 1997.
25. J. H. van Lint, *Introduction to coding theory*, 2nd edition, Graduate Texts in Math. **86**, Springer-Verlag, Berlin-Heidelberg-New York, 1992.
26. J. H. van Lint and G. van der Geer, *Introduction to coding theory and algebraic geometry*, DMV Seminar **12**, Birkhäuser-Verlag, Basel-Boston-Berlin, 1988.
27. W. Lütkebohmert, *Codierungstheorie*, Vieweg-Verlag, Braunschweig, 2003.
28. F. J. MacWilliams and N. J. A. Sloane, *The theory of error-correcting codes*, North-Holland, Amsterdam, 1977.
29. Y. I. Manin, *What is the maximum number of points on a curve over* $\mathbb{F}_2$*?*, J. Fac. Sci. Univ. Tokyo Sect. IA Math. **28**, 1981, pp. 715-720.
30. C. Moreno, *Algebraic curves over finite fields*, Cambridge Tracts in Math. **97**, Cambridge Univ. Press, Cambridge, 1991.
31. H. Niederreiter and C. P. Xing, *Rational points on curves over finite fields*, London Math. Soc. Lecture Notes Ser. **285**, Cambridge Univ. Press, Cambridge, 2001.
32. V. Pless and W. C. Huffmann (eds.), *Handbook of coding theory*, Vol. I and II, Elsevier, Amsterdam, 1998.
33. O. Pretzel, *Codes and algebraic curves*, Oxford Lecture Ser. in Math. and its Applications **8**, Oxford Univ. Press, Oxford, 1998.
34. H.-G. Rück and H. Stichtenoth, *A characterization of Hermitian function fields over finite fields*, J. Reine Angew. Math. **457**, 1994, pp. 185-188.
35. J.-P. Serre, *Local fields*, Springer-Verlag, New York-Berlin, 1979.
36. J.-P. Serre, *Sur le nombre des points rationnels d'une courbe algébrique sur un corps fini*, C. R. Acad. Sci. Paris **296**, 1983, pp. 397-402.
37. I. R. Shafarevic, *Basic algebraic geometry*, Grundlehren der math. Wissensch. **213**, Springer-Verlag, Berlin-Heidelberg-New York, 1977.
38. J. H. Silverman, *The arithmetic of elliptic curves*, Graduate Texts in Math. **106**, Springer-Verlag, Berlin-Heidelberg-New York, 1986.
39. A. N. Skorobogatov and S. G. Vladut, *On the decoding of algebraic-geometric codes*, IEEE Trans. Inform. Th. **36**, 1990, pp. 1461-1463.
40. S. A. Stepanov, *Arithmetic of algebraic curves*, Monographs in Contemp. Math., Plenum Publ. Corp., New York, 1994.
41. H. Stichtenoth, *Über die Automorphismengruppe eines algebraischen Funktionenkörpers von Primzahlcharakteristik*, Arch. Math. **24**, 1973, pp. 527-544 and 615-631.

42. M. A. Tsfasman and S. G. Vladut, *Algebraic-geometric codes*, Kluwer Academic Publ., Dordrecht-Boston-London, 1991.
43. M. A. Tsfasman, S. G. Vladut and D. Nogin, *Algebraic-geometric codes: basic notions*, AMS Math. Surveys and Monographs **139**, 2007.
44. M. A. Tsfasman, S. G. Vladut and T. Zink, *Modular curves, Shimura curves, and Goppa codes, better than the Varshamov-Gilbert bound*, Math. Nachr. **109**, 1982, pp. 21-28.
45. G. D. Villa Salvador, *Topics in the theory of algebraic function fields*, Birkhäuser-Verlag, Boston, 2006.
46. B. L. van der Waerden, *Algebra*, Teil 1 und 2, 7th edition, Springer-Verlag, Berlin-Heidelberg-New York, 1966.
47. T. Zink, *Degeneration of Shimura surfaces and a problem in coding theory*, in Fundamentals of Computation Theory (L. Budach, ed.), Lecture Notes in Computer Science **199**, Springer-Verlag, Berlin, 1985, pp. 503-511.

A more comprehensive list of references to recent work on function fields and algebraic geometry codes can be found in [13],[21],[31],[42],[43] and [45].

Index

Graduate Texts in Mathematics

(continued from page ii)

126 BOREL. Linear Algebraic Groups. 2nd ed.
127 MASSEY. A Basic Course in Algebraic Topology.
128 RAUCH. Partial Differential Equations.
129 FULTON/HARRIS. Representation Theory: A First Course.
Readings in Mathematics
130 DODSON/POSTON. Tensor Geometry.
131 LAM. A First Course in Noncommutative Rings. 2nd ed.
132 BEARDON. Iteration of Rational Functions.
133 HARRIS. Algebraic Geometry: A First Course.
134 ROMAN. Coding and Information Theory.
135 ROMAN. Advanced Linear Algebra. 2nd ed.
136 ADKINS/WEINTRAUB. Algebra: An Approach via Module Theory.
137 AXLER/BOURDON/RAMEY. Harmonic Function Theory. 2nd ed.
138 COHEN. A Course in Computational Algebraic Number Theory.
139 BREDON. Topology and Geometry.
140 AUBIN. Optima and Equilibria. An Introduction to Nonlinear Analysis.
141 BECKER/WEISPFENNING/KREDEL. Gröbner Bases. A Computational Approach to Commutative Algebra.
142 LANG. Real and Functional Analysis. 3rd ed.
143 DOOB. Measure Theory.
144 DENNIS/FARB. Noncommutative Algebra.
145 VICK. Homology Theory. An Introduction to Algebraic Topology. 2nd ed.
146 BRIDGES. Computability: A Mathematical Sketchbook.
147 ROSENBERG. Algebraic K-Theory and Its Applications.
148 ROTMAN. An Introduction to the Theory of Groups. 4th ed.
149 RATCLIFFE. Foundations of Hyperbolic Manifolds.
150 EISENBUD. Commutative Algebra with a View Toward Algebraic Geometry.
151 SILVERMAN. Advanced Topics in the Arithmetic of Elliptic Curves.
152 ZIEGLER. Lectures on Polytopes.
153 FULTON. Algebraic Topology: A First Course.
154 BROWN/PEARCY. An Introduction to Analysis.
155 KASSEL. Quantum Groups.
156 KECHRIS. Classical Descriptive Set Theory.
157 MALLIAVIN. Integration and Probability.
158 ROMAN. Field Theory.
159 CONWAY. Functions of One Complex Variable II.
160 LANG. Differential and Riemannian Manifolds.
161 BORWEIN/ERDÉLYI. Polynomials and Polynomial Inequalities.
162 ALPERIN/BELL. Groups and Representations.
163 DIXON/MORTIMER. Permutation Groups.
164 NATHANSON. Additive Number Theory: The Classical Bases.
165 NATHANSON. Additive Number Theory: Inverse Problems and the Geometry of Sumsets.
166 SHARPE. Differential Geometry: Cartan's Generalization of Klein's Erlangen Program.
167 MORANDI. Field and Galois Theory.
168 EWALD. Combinatorial Convexity and Algebraic Geometry.
169 BHATIA. Matrix Analysis.
170 BREDON. Sheaf Theory. 2nd ed.
171 PETERSEN. Riemannian Geometry.
172 REMMERT. Classical Topics in Complex Function Theory.
173 DIESTEL. Graph Theory. 3rd ed.
174 BRIDGES. Foundations of Real and Abstract Analysis.
175 LICKORISH. An Introduction to Knot Theory.
176 LEE. Riemannian Manifolds.
177 NEWMAN. Analytic Number Theory.
178 CLARKE/LEDYAEV/STERN/WOLENSKI. Nonsmooth Analysis and Control Theory.
179 DOUGLAS. Banach Algebra Techniques in Operator Theory. 2nd ed.
180 SRIVASTAVA. A Course on Borel Sets.
181 KRESS. Numerical Analysis.
182 WALTER. Ordinary Differential Equations.
183 MEGGINSON. An Introduction to Banach Space Theory.
184 BOLLOBAS. Modern Graph Theory.
185 COX/LITTLE/O'SHEA. Using Algebraic Geometry. 2nd ed.
186 RAMAKRISHNAN/VALENZA. Fourier Analysis on Number Fields.
187 HARRIS/MORRISON. Moduli of Curves.
188 GOLDBLATT. Lectures on the Hyperreals: An Introduction to Nonstandard Analysis.
189 LAM. Lectures on Modules and Rings.

190 ESMONDE/MURTY. Problems in Algebraic Number Theory. 2nd ed.
191 LANG. Fundamentals of Differential Geometry.
192 HIRSCH/LACOMBE. Elements of Functional Analysis.
193 COHEN. Advanced Topics in Computational Number Theory.
194 ENGEL/NAGEL. One-Parameter Semigroups for Linear Evolution Equations.
195 NATHANSON. Elementary Methods in Number Theory.
196 OSBORNE. Basic Homological Algebra.
197 EISENBUD/HARRIS. The Geometry of Schemes.
198 ROBERT. A Course in p-adic Analysis.
199 HEDENMALM/KORENBLUM/ZHU. Theory of Bergman Spaces.
200 BAO/CHERN/SHEN. An Introduction to Riemann–Finsler Geometry.
201 HINDRY/SILVERMAN. Diophantine Geometry: An Introduction.
202 LEE. Introduction to Topological Manifolds.
203 SAGAN. The Symmetric Group: Representations, Combinatorial Algorithms, and Symmetric Functions.
204 ESCOFIER. Galois Theory.
205 FÉLIX/HALPERIN/THOMAS. Rational Homotopy Theory. 2nd ed.
206 MURTY. Problems in Analytic Number Theory.
Readings in Mathematics
207 GODSIL/ROYLE. Algebraic Graph Theory.
208 CHENEY. Analysis for Applied Mathematics.
209 ARVESON. A Short Course on Spectral Theory.
210 ROSEN. Number Theory in Function Fields.
211 LANG. Algebra. Revised 3rd ed.
212 MATOUŠEK. Lectures on Discrete Geometry.
213 FRITZSCHE/GRAUERT. From Holomorphic Functions to Complex Manifolds.
214 JOST. Partial Differential Equations. 2nd ed.
215 GOLDSCHMIDT. Algebraic Functions and Projective Curves.
216 D. SERRE. Matrices: Theory and Applications.
217 MARKER. Model Theory: An Introduction.
218 LEE. Introduction to Smooth Manifolds.
219 MACLACHLAN/REID. The Arithmetic of Hyperbolic 3-Manifolds.
220 NESTRUEV. Smooth Manifolds and Observables.
221 GRÜNBAUM. Convex Polytopes. 2nd ed.
222 HALL. Lie Groups, Lie Algebras, and Representations: An Elementary Introduction.
223 VRETBLAD. Fourier Analysis and Its Applications.
224 WALSCHAP. Metric Structures in Differential Geometry.
225 BUMP: Lie Groups.
226 ZHU. Spaces of Holomorphic Functions in the Unit Ball.
227 MILLER/STURMFELS. Combinatorial Commutative Algebra.
228 DIAMOND/SHURMAN. A First Course in Modular Forms.
229 EISENBUD. The Geometry of Syzygies.
230 STROOCK. An Introduction to Markov Processes.
231 BJÖRNER/BRENTI. Combinatorics of Coxeter Groups.
232 EVEREST/WARD. An Introduction to Number Theory.
233 ALBIAC/KALTON. Topics in Banach Space Theory.
234 JORGENSEN. Analysis and Probability.
235 SEPANSKI. Compact Lie Groups.
236 GARNETT. Bounded Analytic Functions.
237 MARTÍNEZ-AVENDAÑO/ROSENTHAL. An Introduction to Operators on the Hardy-Hilbert Space.
238 AIGNER. A Course in Enumeration.
239 COHEN. Number Theory – Volume I: Tools and Diophantine Equations.
240 COHEN. Number Theory – Volume II: Analytic and Modern Tools.
241 SILVERMAN. The Arithmetic of Dynamical Systems.
242 GRILLET. Abstract Algebra. 2nd ed.
243 GEOGHEGAN. Topological Methods in Group Theory.
244 BONDY/MURTY. Graph Theory.
245 GILMAN/KRA/RODRIGUEZ. Complex Analysis.
246 KANIUTH. A Course in Commutative Banach Algebras.
247 KASSEL/TURAEV. Braid Groups.
248 ABRAMENKO/BROWN. Buildings: Theory and Applications.
249 GRAFAKOS. Classical Fourier Analysis.
250 GRAFAKOS. Modern Fourier Analysis.
251 WILSON. The Final Simple Groups.
252 GRUBB. Distributions and Operators.
253 MACCLUER. Elementary Functional Analysis.
254 STICHTENOTH. Algebraic Function Fields and Codes, 2nd ed.

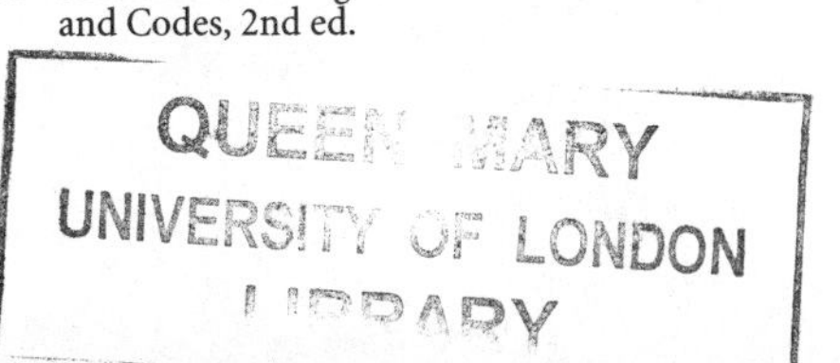

Printing: Krips bv, Meppel, The Netherlands
Binding: Stürtz, Würzburg, Germany

Printing: Krips bv, Meppel, The Netherlands
Binding: Stürtz, Würzburg, Germany